Vegetable Crop Production

Vegetable Crop Production

Editor: Alabaster Jenkins

R CALLISTO REFERENCE

www.callistoreference.com

Callisto Reference,
118-35 Queens Blvd., Suite 400,
Forest Hills, NY 11375, USA

Visit us on the World Wide Web at:
www.callistoreference.com

ISBN: 978-1-63239-785-0 (Hardback)

The publisher's policy is to use permanent paper from mills that operate a sustainable forestry policy. Furthermore, the publisher ensures that the text paper and cover boards used have met acceptable environmental accreditation standards.

Trademark Notice: Registered trademark of products or corporate names are used only for explanation and identification without intent to infringe.

Printed in the United States of America.

Cataloging-in-publication Data

Vegetable crop production / edited by Alabaster Jenkins.
    p. cm.
Includes bibliographical references and index.
ISBN 978-1-63239-785-0
1. Vegetables. 2. Horticulture. 3. Vegetables--Diseases and pests. I. Jenkins, Alabaster.
SB320.9 .V44 2017
635--dc23

Table of Contents

Permissions

List of Contributors

Index

Preface

Vegetable crop production is the cultivation of vegetables for consumption. This book aims to equip students and experts with the advanced topics and upcoming concepts in this area such as horticultural practices and methods for crop improvement. The aim of this book is to present researches that have transformed this discipline and aided its advancement. It is a compilation of chapters that discuss the most vital concepts and emerging trends in this field of crop production. It studies, analyses and upholds the pillars of vegetable production and its utmost significance in modern times. For someone with an interest and eye for detail, this book covers the most significant topics in the field of vegetable crop production.

Significant researches are present in this book. Intensive efforts have been employed by authors to make this book an outstanding discourse. This book contains the enlightening chapters which have been written on the basis of significant researches done by the experts.

Finally, I would also like to thank all the members involved in this book for being a team and meeting all the deadlines for the submission of their respective works. I would also like to thank my friends and family for being supportive in my efforts.

Editor

Variation in Broccoli Cultivar Phytochemical Content under Organic and Conventional Management Systems: Implications in Breeding for Nutrition

Erica N. C. Renaud[1]*, Edith T. Lammerts van Bueren[1], James R. Myers[2], Maria João Paulo[3], Fred A. van Eeuwijk[3], Ning Zhu[4], John A. Juvik[4]

1 Wageningen UR Plant Breeding, Plant Sciences Group, Wageningen University, Wageningen, The Netherlands, 2 Department of Horticulture, Oregon State University, Corvallis, Oregon, United States of America, 3 Biometris, Plant Sciences Group, Wageningen University, Wageningen, The Netherlands, 4 Department of Crop Sciences, University of Illinois, Urbana, Illinois, United States of America

Abstract

Organic agriculture requires cultivars that can adapt to organic crop management systems without the use of synthetic pesticides as well as genotypes with improved nutritional value. The aim of this study encompassing 16 experiments was to compare 23 broccoli cultivars for the content of phytochemicals associated with health promotion grown under organic and conventional management in spring and fall plantings in two broccoli growing regions in the US (Oregon and Maine). The phytochemicals quantified included: glucosinolates (glucoraphanin, glucobrassicin, neoglucobrassin), tocopherols (δ-, γ-, α-tocopherol) and carotenoids (lutein, zeaxanthin, β-carotene). For glucoraphanin (17.5%) and lutein (13%), genotype was the major source of total variation; for glucobrassicin, region (36%) and the interaction of location and season (27.5%); and for neoglucobrassicin, both genotype (36.8%) and its interactions (34.4%) with season were important. For δ- and γ-tocopherols, season played the largest role in the total variation followed by location and genotype; for total carotenoids, genotype (8.41–13.03%) was the largest source of variation and its interactions with location and season. Overall, phytochemicals were not significantly influenced by management system. We observed that the cultivars with the highest concentrations of glucoraphanin had the lowest for glucobrassicin and neoglucobrassicin. The genotypes with high concentrations of glucobrassicin and neoglucobrassicin were the same cultivars and were early maturing F_1 hybrids. Cultivars highest in tocopherols and carotenoids were open pollinated or early maturing F_1 hybrids. We identified distinct locations and seasons where phytochemical performance was higher for each compound. Correlations among horticulture traits and phytochemicals demonstrated that glucoraphanin was negatively correlated with the carotenoids and the carotenoids were correlated with one another. Little or no association between phytochemical concentration and date of cultivar release was observed, suggesting that modern breeding has not negatively influenced the level of tested compounds. We found no significant differences among cultivars from different seed companies.

Editor: Hany A. El-Shemy, Cairo University, Egypt

Funding: Funding was provided by Seeds of Change (www.seedsofchange.com) and monetary funding from Wageningen University, University of Illinois and Oregon State University. The funder had no role in the study design, data collection and analysis, decision to publish, or preparation of the manuscript.

Competing Interests: The principle researcher, Erica Renaud was employed by Seeds of Change and later Vitalis Organic Seeds, the organic division of Enza Zaden, a vegetable seed company.

* Email: E.Renaud@enzazaden.com

Introduction

Organic food consumption is in part driven by consumer perception that organic foods are more nutritious and simultaneously less potentially harmful to human health [1–2]. Studies, such as Smith-Sprangler et al. [3], have concluded that there is little evidence for differences in health benefits between organic and conventional products, but other studies have indicated that organic vegetables and fruits contain higher concentrations of certain plant phytochemicals associated with health promotion than those produced conventionally [4–8]. A number of these compounds are produced by plants in response to environmental stress or pathogen infection, providing a potential explanation of why concentrations of these compounds might be higher in plants grown in organic systems without application of pesticides [9]. In addition, higher phytochemical levels may be due to the effects

that different fertilization practices have on plant metabolism. Synthetic fertilizers used in conventional agriculture are more readily available to plants than organic fertilizers [10]. Nutrients derived from organic fertilizers need to be mineralized, and the availability of these nutrients depends on soil moisture, temperature and level of activity of soil organisms [11]. Conventional systems seek to maximize yields, resulting in a relative decrease of plant phytochemicals and secondary metabolites [12–15]. Correspondingly, compounds such as phenolics, flavonoids, and indolyl glucosinolates may be induced by biotic or abiotic stress [16–17].

Broccoli is an abundant source of nutrients, including provitamin A (β-carotene), vitamin C (ascorbate), and vitamin E (tocopherol) [18]. It is also a source of phytochemicals associated with health benefits and these include glucosinolates, carotenoids, tocopherols, and flavonoids [19–21]. Verhoeven et al. [22], Keck and Finley [23] and Here and Büchler [24], reported that diets rich in broccoli reduce cancer incidence in humans. Strong associations between consumption level and disease risk reduction exists for glucosinolates (anti-cancer), tocopherols (cardiovascular), and the carotenoids (eye-health) [25].

Sulfur containing glucosinolates are found in the tissues of many species of the *Brassicaceae* family. When glucosinolates are consumed, they are hydrolyzed into isothiocyanates (ITC) and other products that up-regulate genes associated with carcinogen detoxification and elimination. Aliphatic glucoraphanin (up to 50% of total glucosinolates) and the indolylic glucosinolates, glucobrassicin and neoglucobrassicin are abundant in broccoli florets [20,19,26]. Glucoraphanin is hydrolyzed either by the endogenous plant enzyme myrosinase [27–28] or by gut microbes to produce sulforaphane, an ITC. The indole glucosinolates are tryptophan-derived in a similar but alternate biosynthetic pathway [29]. The health promoting effects of the indolyl glucosinolates are attributed to indole-3-carbinol, a hydrolysis product of gluco-brassicin, N-methoxyindole-3-carbinol and neoascorbigen, hydro-lysis products from neoglucobrassicin, and the catabolic products derived from alkyl glucosinolates. Clinical studies have shown that the glucosinolate hydrolysis products reduce the incidence of certain forms of cancer (e.g., prostate, intestinal, liver, lung, breast, bladder) [30–35]. The lipophilic phytonutrients found in broccoli include the carotenoids lutein, zeaxanthin, β-carotene, and tocopherols (forms of vitamin E) [36–37]. In addition to their role as vitamins, these compounds are powerful antioxidants [38–39]. Consumption of vegetables high in tocopherols and caroten-oids has decreased the incidence of certain forms of cancer [40]. Lutein and zeaxanthin protect against development of cataracts and age-related macular degeneration [41]. Tocopherols have also been associated with reduced risk of cardiovascular disease by preventing oxidative modification of low-density lipoproteins in blood vessels [42].

The genetic potential for high nutrient content has long been a concern of the organic industry in order to meet the expectations of organic consumers. This has often been manifested by questioning whether modern elite cultivars may have lower levels of nutritional content than older open pollinated cultivars. Indirect evidence supporting this argument comes from Davis et al. [43], who compared USDA nutrient content data for 43 garden crops released between 1950 and 1999. Statistically significant decreases were noted for six nutrients (protein, calcium, potassium, iron, riboflavin, and ascorbic acid), with declines ranging from 6% for protein to 38% for riboflavin. Crop varieties in 1950 had been bred to be adapted to specific regions and a relatively low input agriculture system, but contemporary cultivars are selected for yield, disease resistance, broad adaptation to high input agriculture systems, and for increased 'shipability' and shelf life. Traka et al. [44] recommend breeding with greater genetic diversity when the goal is enhanced phytochemical content by exploiting wild crop relatives. The genotype is important in determining the level of nutrients in a crop cultivar [45–47]. What is unclear, however, is whether the nutritional content of a cultivar is associated with certain genotypic categorization, e.g. old versus modern, open pollinated versus F_1 hybrid cultivars. In addition, there is no clear differentiation as to what extent nutritional content in a crop is determined by genotypic or by field management factors or by the interaction of both. Some studies comparing performance of genotypes in organic and conventional production systems have shown that for certain agronomic traits, cultivars perform differently between the two production systems (e.g. for winter wheat: Murphy et al. [48], Baresel et al. [49]; for lentils: Vlachostergios et al. [50]; for maize: Goldstein et al.[51], while others have shown no differences in ranking performance (for

maize: Lorenzana and Bernardo [52]; for onions: Osman et al. [53]; for cereals: Przystalski et al.[54]). The results of these studies have profound implications for organic cultivar selection and breeding strategies and raise questions as to the need for cultivars to be bred with broad adaptability or specific adaptation for the requirements of regional organic production and for designing breeding programs that optimize phytochemicals in an adapted management system.

Previous studies comparing organically versus conventionally grown broccoli for nutritional quality have been 'market basket' (off-the-shelf) studies [55–56]. Harker [57] explained that the limitation of market basket studies is that they either have purchased the products from the store shelf and cannot relate differences to specific growing conditions or that the number of cultivars is too small to generalize the results. While other studies have compared cultivars from one production season time period to another, knowledge of the actual cultivar and production system (soil quality, temperature, rainfall) was not available [58,43]. The concentrations and form of health-promoting nutrients in *Brassica* vegetables have been reported to vary significantly due to (1) genotype (cultivar and genotypic class) [59,20,26,60,21,37,61,44], (2) environmental conditions such as season [62–67], light [19], max/min temperature, irrigation [68–69], (3) genotype by environment interactions [19,70–71]; (4) management system including soil fertility [72–73], organic versus conventional [13,74–75], days to harvest [63–64], and (5) post-harvest management [76–77]. Identifying specific growing conditions and genotypes that produce cultivars with varying phytochemical content and putative disease-prevention activity could offer value-added commercial opportunities to the seed and food industry.

In addition to research conducted on how broccoli genotypes, management system and environment interact for horticultural traits [78], we address in this paper the question of how do genotypes, management system and environment interact to determine the nutritional contributions of broccoli to the human diet. We studied the relative importance and interaction among genotypes (cultivars, genotypic classes) and environment {man-agement system [M: organic (O) or conventional (C)], season (S, a combination of year and season within year, i.e., fall 2006, spring 2007, fall 2007, spring 2008), location (E)} in a set of 23 broccoli cultivars for floret glucosinolate, tocopherol and carotenoid concentrations grown under organic and conventional production systems in two contrasting broccoli production regions of the US: Oregon and Maine. Specifically we addressed the following questions: (1) what is the impact of organic management system compared to the environmental factors including climatic region, season and their interactions [Genotype (G) x Environment (E) x Management System (M)]? (2) is there a significant difference in phytochemical content between different genotypes and genotypic classes (old and modern cultivars; open pollinated and F_1 hybrid cultivars; early and late maturing cultivars; and between different commercial seed sources)? (3) what is the best selection environ-ment for a broccoli breeding program for enhanced phytochem-ical content?

Materials & Methods

Plant Material and Field Trial Locations

Twenty-three broccoli cultivars including open pollinated (OP) cultivars, inbred lines, and F_1 hybrids were included in field trials (**Table 1**). Cultivars were grown in a randomized complete block design with three replicates in Maine (ME)-Monmouth (Latitude 44.2386°N, Longitude 70.0356°W); and Oregon (OR)-Corvallis (Latitude 44.5647°N, Longitude123.2608°W)] with each location including organically (O) and conventionally (C) managed treatments. Plots contained 36 plants, planted in three rows of 12 plants at 46 cm equidistant spacing within and between rows. The 2006 trials had only 18 of the 23 entries, and the Oregon 2006 trial had only two replicates at the organic location. Field trials were conducted for three consecutive years with one production cycle in Fall 2006, two production cycles in Spring and Fall 2007 and one production cycle in Spring 2008. The primary management differences between the organic and conventional field trial sites are outlined in **Table S1 in File S1**, which describes the production system, soils, fertility applications, the applied supplemental irrigation, and weather conditions for the area of study. Further details of the field design are reported in Renaud et al. [78].

Field Data Collection

As plots approached maturity they were evaluated three times a week for field quality and broccoli heads that had reached commercial market maturity (approximately 10 to 12 cm in diameter for most of the cultivars while retaining firmness). Field quality traits evaluated on a 1 to 9 ordinal scale included head color, bead size, and bead uniformity. Average head weight was determined by taking the mean of the five individual heads per plot. Head diameter averaged for five heads at harvest maturity from each plot. Maturity was based on days to harvest from transplanting date. Detailed procedures and horticulture trait performance data are reported in Renaud et al. [78].

Broccoli Floret Samples and glucosinolate, tocopherol, and carotenoid analysis

In order to analyse nutritional compounds of the broccoli heads, the following procedure was followed: As plots approached maturity, five broccoli head tissue samples were harvested fresh from each subplot at each trial location and were composited into a single sample per replication. The samples were frozen at $-20°C$ and shipped in a frozen state to the University of Illinois, Urbana-Champaign where they were freeze-dried and assessed for nutritional phytochemicals. Each sample was analyzed for the glucosinolates (glucoraphanin, glucobrassicin and neoglucobrassicin), carotenoids (β-carotene, lutein, and zeaxanthin), and tocopherols (δ-, γ-, α- tocopherol) by high-performance liquid chromatography (HPLC) analysis using analytical protocols described in Brown et al. [19] for glucosinolates, and Ibrahim and Juvik [37] for tocopherols and carotenoids. Glucosinolates in lyophilized floret tissue samples were extracted and analysed by HPLC using a reverse phase C18 column. Three hundred mg samples of broccoli floret tissue were weighed out for extraction and the HPLC quantification of the tocopherols and carotenoids.

Statistical Analysis

Various linear mixed models were used for the analysis of trait variation. We followed the same methodology as described in Renaud et al. [78], which was comparable to the approach followed by Lorenzana and Bernardo [52]. For fitting the linear

Table 1. Overview of commercially available broccoli cultivars, showing origin, main characteristics, included in paired organic - conventional field trials 2006–2008.

Cultivar	Abbreviation	Origin	Cultivar Type[a]	Date of Market Entry	Maturity Classification[b]
Arcadia	ARC	Sakata	F_1	1985	L
B1 10	B11	Rogers	F_1	1988	M
Batavia	BAT	Bejo	F_1	2001	M
Beaumont	BEA	Bejo	F_1	2003	L
Belstar	BEL	Bejo	F_1	1997	L
Diplomat	DIP	Sakata	F_1	2004	L
Early Green	EGR	Seeds of Change	OP	1985	E
Everest	EVE	Rogers	F_1	1988	E
Fiesta	FIE	Bejo	F_1	1992	L
Green Goliath	GRG	Burpee	F_1	1981	M
Green Magic	GRM	Sakata	F_1	2003	M
Gypsy	GYP	Sakata	F_1	2004	M
Imperial	IMP	Sakata	F_1	2005	L
Marathon	MAR	Sakata	F_1	1985	L
Maximo	MAX	Sakata	F_1	2004	L
Nutribud	NUT	Seeds of Change	OP	1990	E
OSU OP	OSU	Jim Myers, OSU	OP	2005	E
Packman	PAC	Petoseed	F_1	1983	E
Patriot	PAT	Sakata	F_1	1991	M
Patron	PAN	Sakata	F_1	2000	M
Premium Crop	PRC	Takii	F_1	1975	E
USVL 048	U48	Mark Farnham, USVL	Inbred	not released	L
USVL 093	U93	Mark Farnham, USVL	Inbred	not released	M

[a]Cultivar Type: F_1: hybrid; OP: Open Pollinated; Inbred.
[b]Maturity Classification: E: Early; M: Mid; L: Late.

mixed models, GenStat 15 (VSNi, 2012) was used. The models followed the set-up:

$$y = E + R(E) + G + G \times E + e.$$

Here y is the phytochemical response. Term E represents the environment in a very general sense, it includes all main effects and interactions of Season (S), Location (L) and Management (M). For analyses per location, the terms involving L were dropped. Similarly, for analyses regarding a specific management regime, the terms involving M were dropped. Term $R(E)$ is the effect of replicate within environment, and there were two or three replicates in individual trials. G and $G \times E$ are genotype and genotype by environment interaction effects, respectively. Finally e is a residual.

Variance components were reported as coefficients of variation, i.e., $CV = 100\sqrt{V}/x$ with V the variance corresponding to specific effects and x the trait mean. Repeatability was calculated from the variance components in its most general form as $H^2 = V_G/(V_G$ $V_{GL}/nL + + V_{GS}/nS + V_{GM}/nM + V_{GLS}/(nL.nS) + V_{GLM}/(nL.nM)$ $+ V_{GSM}/(nS.nM) + V_{GLSM}/(nL.nS.nM) + V_e/(nL.nS.nM.nR))$, where the variance components correspond to the terms in the mixed model above. The terms nL, nS, nM and nR stand for the number of locations (2: Maine and Oregon), number of 'seasons' (4: Fall 2006, Spring 2007, Fall 2007, Spring 2008), management (2; organic and conventional), and replicates (2 or 3).

Genotypic means were calculated by taking genotypic main effects fixed instead of random in the mixed models above. Pairwise comparisons between genotypic means were performed using GenStat procedure *VMCOMPARISON*. Correlations on the basis of genotypic means were referred to as genetic correlations. Genotypic stabilities under organic and conventional conditions were calculated as the variance for individual genotypes across all trials in the system.

To assess the feasibility of selection for organic conditions (the target environment) under conventional conditions, we calculated the ratio of correlated response (for organic conditions using conventional conditions), CR, to direct response (for organic conditions in organic conditions), DR, as the product of the genetic correlation between organic and conventional systems (r_G) and the ratio of the roots of conventional and organic repeatabilities (H_C and H_O respectively): $CR/DR = r_G H_C/H_O$. A ratio smaller than 1 indicates that selection is better done directly under organic conditions when the aim is indeed to improve the performance in organic conditions.

Results

Comparison of phytochemicals means over the environments

Glucosinolates. Across all trials, glucoraphanin levels were comparable between locations and seasons but were more variable at the individual location and season trial analysis level (**Table 2**). Glucoraphanin, glucobrassicin and neoglucobrassicin levels were comparable between organic and conventional treatments. Comparisons of organic versus conventional by location and season for the glucosinolate phytochemicals are presented in **Figure S1**. Comparable levels of glucosinolates were observed in the organic - conventional comparisons within locations and seasons.

Tocopherols. Across trials compared regionally, Oregon had higher levels of all three tocopherols compared to Maine (**Table 2, Figure S2**). The tocopherols δ- and γ- were higher in Fall compared to Spring, but not so for α-tocopherol (**Figure S2**).

Organic and conventional levels for all tocopherol concentrations were in the same range and not significantly different. When the three tocopherols were analysed by organic versus conventional within location and season, there were no clear significant differences in management system across the season and location combinations (**Table 2, Figure S2**).

Carotenoids. Overall, Oregon had higher levels of lutein and β-carotene compared to Maine (**Table 2, Figure S3**) and comparative levels of zeaxanthin (**Table 2, Figure S3**). Spring produced higher levels of all carotenoids compared to Fall levels in contrast to the glucosinolates and the δ- & γ- tocopherol concentrations. There were no significant differences between organic and conventional for any carotenoid measured. When carotenoids were analysed by management system within location and season, β-carotene showed significantly lower levels in Maine in the Fall compared to other location and season combinations (**Figure S3**).

Partitioning of variance components

Glucosinolates. For glucoraphanin across all trials in both regions, Genotype (G) main effect accounted for the largest proportion of variance, followed by G×L×S interaction (**Table 3**). There was no Management (M) main effect, but M contributed to the three (L×S×M and G×S×M) and four-way interactions (G×L×S×M). In contrast to glucoraphanin, Location (L) had the largest effect for glucobrassicin and neoglucobrassicin across all trials in both regions, followed by the L×S interactions. For neoglucobrassicin the S and G main effect was more important than for glucobrassicin. When trials were further partitioned by location, a G and S main effect was apparent for neoglucobrassicin in both locations; for glucobrassicin the S main effects was only apparent in Oregon and not in Maine (**Table S2 in File S1**). There was M main effect for glucobrassicin and neoglucobrassicin, but not for glucoraphanin, and no G×M interaction for all glucosinolates.

Tocopherols. For δ- and γ-tocopherol across all trials in both regions, the Season (S) main effect accounted for the largest proportion of variance (**Table 3**). In contrast the proportion of the variation associated with S for α-tocopherol across all trials was minor. For all three tocopherols there was minor to no M effect, but a large L main effect, being the greatest for γ-tocopherol. The G main effect showed a similar pattern to L.

Carotenoids. For all three carotenoids across all trials in both regions, the G main effect described a significant component of total variance and was of largest influence for lutein (**Table 3**). The S main effect played an important role for zeaxanthin, and to a lesser extent for lutein but not for β-carotene. For all three carotenoids the L effect was minor, but the L×S interaction for β-carotene was relatively large and mostly associated with Maine (**Table S2 in File S1**). There was no M main effect; only for β-carotene was there a small effect of the G×M interaction (mainly driven by Maine).

Repeatability, genetic correlation and ratio of correlated response to direct response

Organic versus conventional. In the present study, we were able to estimate the proportion of the genotypic variance relative to phenotypic variance, but because we did not have a genetically structured breeding population, we apply the term repeatability rather than broad sense heritability. Of the phytochemicals studied, repeatabilities for concentrations of seven of the nine were comparable or higher in organic compared to conventional systems (**Table 4**). Only for glucobrassicin and δ-tocopherol was repeatability under organic conditions lower than under conven-

Table 2. Trait means[1] of phytochemicals of 23 broccoli cultivars grown across four pair combinations of location (Maine/Oregon), season (Fall/Spring) two-years combined and management system (Conventional/Organic), 2006–2008.

	Maine					Oregon				
	Fall		Spring			Fall		Spring		
	2006–2007 Combined		2007–2008 Combined		Mean	2006–2007 Combined		2007–2008 Combined		Mean
	C	O	C	O		C	O	C	O	
Glucoraphanin	5.31 e	3.77 bc	3.56 b	4.06 c	4.18	3.46 b	3.03 a	4.64 d	4.51 d	3.91
Glucobrassicin	1.06 b	0.90 a	1.45 c	1.33 c	1.19	5.14 f	5.51 g	2.24 d	2.70 e	3.90
Neoglucobrassicin	0.46 a	0.40 a	2.16 c	1.85 b	1.22	2.34 c	3.20 d	4.32 e	5.10 f	3.74
δ-Tocopherol	2.34 c	2.77 d	1.91 b	1.70 a	2.18	3.53 e	3.66 e	1.91 b	2.24 c	2.83
γ-Tocopherol	4.67 c	4.40 c	2.63 a	2.98 b	3.67	8.48 d	8.73 d	3.31 b	3.22 b	5.94
α-Tocopherol	25.83 a	27.33 a	38.61 b	40.51 bc	33.07	43.04 c	43.20 c	40.52 bc	42.25 c	42.25
Lutein	11.49 a	12.47 a	15.53 b	15.93 b	13.85	15.91 b	16.04 b	16.48 b	17.81 c	16.56
Zeaxanthin	0.81 a	0.83 ab	0.87 ab	0.88 b	0.85	0.83 ab	0.84 ab	1.02 c	1.02 c	0.93
β-Carotene	12.98 a	13.25 a	28.73 c	29.71 c	21.16	29.10 c	30.10 c	25.16 b	25.80 b	27.54

[1]Values in the table are means. Means of the same letter in the same row are not significantly different at the $P<0.05$ level.

Table 3. Partitioning of variance components (%) presented as coefficients of variation for phytochemicals of 23 broccoli cultivars grown across eight pair combinations of location (Maine/Oregon), season (Fall/Spring) and management system (Conventional/Organic), 2006–2008.

	Location (L)	Season (S)	Management (M)	L×S	L×M	S×M	L×S×M	L×S×M×R Rep (R)	Genotype (G)	G×L	G×S	G×M	G×L×S	G×L×M	G×S×M	G×L×S×M	Residual
Glucoraphanin	0.01	5.45	0.00	7.20	0.01	0.00	11.86	1.56	17.45	0.01	0.01	0.00	0.01	15.97	7.49	12.62	11.94
Glucobrassicin	36.00	0.00	0.00	27.51	5.58	4.18	1.34	1.86	9.42	7.77	0.01	0.00	0.00	13.84	0.00	10.63	10.91
Neoglucobrassicin	36.81	13.51	0.00	34.36	8.47	6.84	4.50	4.76	15.16	6.24	0.01	0.00	0.00	16.40	0.01	13.40	15.80
δ-Tocopherol	6.83	35.22	0.43	7.87	0.01	0.01	3.39	0.01	5.57	5.65	6.01	0.00	0.01	13.65	0.00	12.21	12.74
γ-Tocopherol	12.02	19.09	0.01	12.11	0.01	3.64	4.47	2.12	13.79	4.85	15.82	0.00	0.00	12.95	0.01	11.08	12.03
α-Tocopherol	6.73	0.01	0.28	10.20	0.01	0.01	0.01	1.29	2.79	3.42	0.01	0.97	0.00	10.07	0.00	8.18	8.92
Lutein	3.71	4.91	0.00	7.52	0.01	2.70	1.85	1.40	13.03	4.14	0.01	0.01	0.73	10.76	0.01	9.21	9.95
Zeaxanthin	1.97	11.55	0.01	3.99	0.01	0.00	0.01	0.00	8.44	3.18	0.01	1.05	0.01	6.91	0.83	8.52	11.36
β-Carotene	4.61	0.00	0.00	17.84	0.01	0.70	0.01	0.71	8.41	4.45	0.00	2.31	4.63	11.32	0.65	12.83	10.99

tional. In the analyses δ- and α-tocopherol had relatively low repeatabilities. The highest repeatabilities were for glucoraphanin (0.82–0.84), neoglucobrassicin (0.75–0.76), γ-tocopherol (0.72–0.75), lutein (0.83–0.85) and zeaxanthin (0.76–0.77). Genetic correlations were high between organic and conventional for the glucosinolates, γ-tocopherol and lutein (0.84–0.95), while δ-tocopherol, α-tocopherol, zeaxanthin and β-carotene were lower (0.63–0.77). The ratio of the correlated response to direct response for selection in the organic system was less than 1.0 for all traits.

By location and season. For the glucosinolates, glucoraphanin and glucobrassicin repeatability at each location, season and treatment trial were comparable and generally high (0.83–0.97) between organic and conventional trials, while no clear trend for neoglucobrassicin repeatabilities was observed between organic and conventional aside from being much lower than glucoraphanin and glucobrassicin (**Table S3 in File S1**). For γ- and α-tocopherol, repeatabilities were comparable between organic and conventional, while for δ-tocopherol repeatabilities were comparable between systems or higher in conventional except for one paired trial. For the carotenoids, repeatabilities were comparable or higher in organic for all paired trials, while for lutein in seven of the eight paired trials organic was comparable or greater than conventional. Repeatabilities for zeaxanthin concentrations were comparable for six of the eight paired trials.

Comparison of cultivar ranking for phytochemical concentration and stability across trials

To determine trends in cultivars with both the highest concentration of phytochemical groups most stable across locations, seasons and production systems, phytochemical concentrations were plotted against stability per genotype across trials. A group of cultivars were identified as both highest in concentration and most stable and are indicated in the highlighted 'red circle' per phytochemical (**Figure 1A–I**). For glucoraphanin, the same group of cultivars had both the highest concentrations and were the most stable across production systems (**Figure 1A; Table S4 in File S1**). While for glucobrassicin, a different set of cultivars had the highest concentrations across production systems (**Figure 1B; Table S5 in File S1**). Overall stability of all cultivars across production system was less related to cultivar mean concentrations for glucobrassicin than for glucoraphanin. None of the cultivars with the highest concentration for neoglucobrassicin were in the top quartile for stability across trials; all cultivars with the highest neoglucobrassicin content were in the bottom half for stability (**Figure 1C; Table S6 in File S1**). Some but not all cultivars that had the highest concentrations of α-tocopherol were among the top group for δ- and/or γ-tocopherol. There was no relationship between δ-tocopherol concentrations and stability, but both γ- and α- tocopherols had higher concentrations associated with greater stability (**Figure 1D–F; Tables S7–9 in File S1**). Open pollinated and early maturing cultivars had the highest and most stable concentrations for all carotenoids. (**Figure 1G–I; Tables S10–12 in File S1**).

Comparison of phytochemical concentration by genotype classification

The open pollinated and F_1 hybrid cultivars were compared across trials for each phytochemical analysed (**Figure 2A**). The levels of glucoraphanin in F_1 hybrids tended to be higher than the open pollinated cultivars. But the inverse trend was observed for glucobrassicin, which was supported by the ranking and stability analysis where the F_1 hybrids showed higher levels and more stability across trials than the open pollinated cultivars for

Table 4. Repeatabilities, genetic correlation and ratio of correlated response to direct response for broccoli phytochemicals comparing organic versus conventional management systems over all trial season/location combinations, 2006–2008.

	Repeatability (H)		r_A [a]	CR_{org}/R_{org} [b]
	C	O		
Glucoraphanin	0.84	0.82	0.84	0.83
Glucobrassicin	0.70	0.64	0.88	0.84
Neoglucobrassicin	0.75	0.76	0.94	0.94
δ-Tocopherol	0.50	0.42	0.73	0.66
γ-Tocopherol	0.75	0.72	0.95	0.93
α-Tocopherol	0.23	0.35	0.61	0.76
Lutein	0.83	0.85	0.93	0.94
Zeaxanthin	0.76	0.77	0.77	0.78
β-Carotene	0.62	0.72	0.63	0.68

[a]Average genetic correlation between conventional and organic production systems across locations.
[b]Ratio of correlated response to direct response.

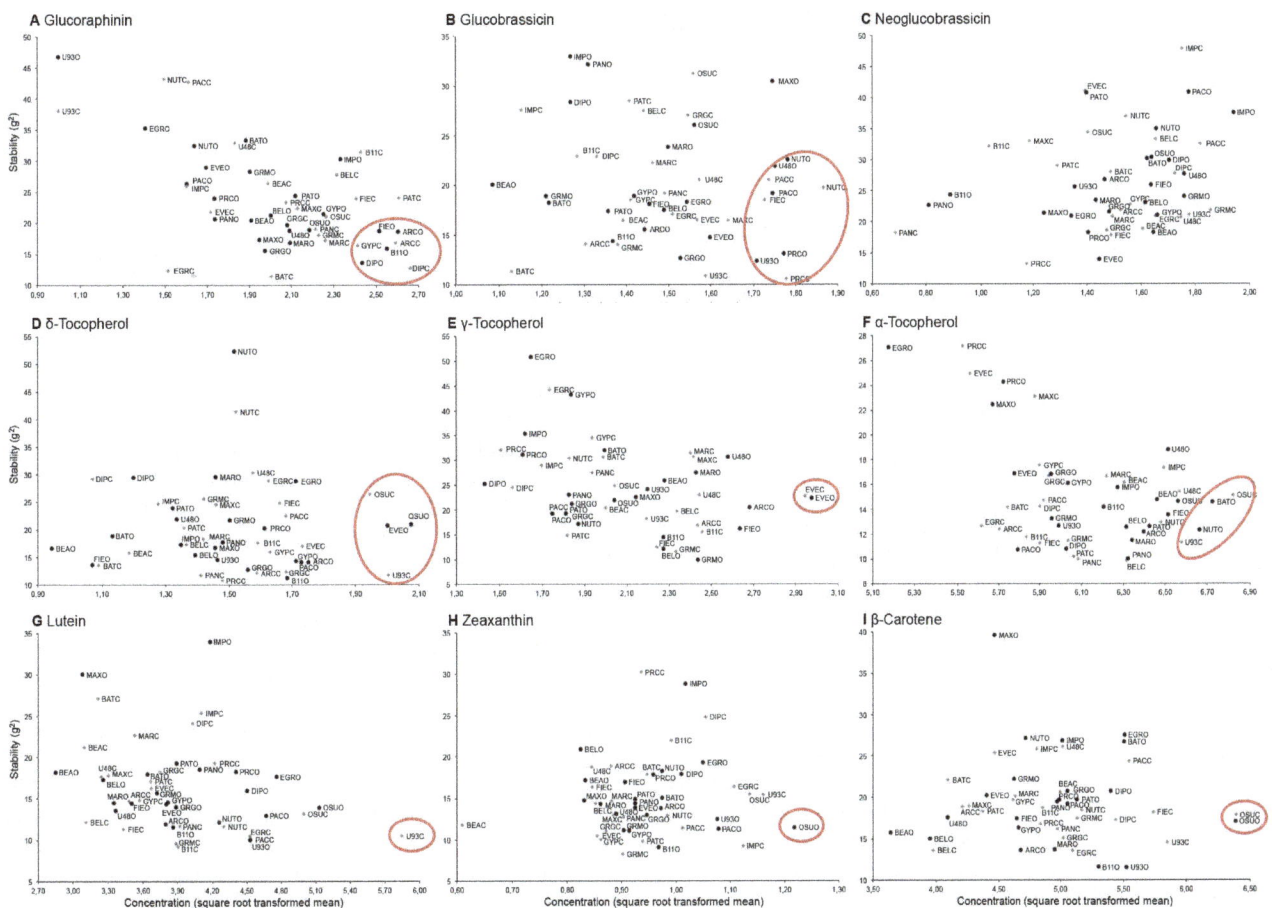

Figure 1. Broccoli cultivar stability from trials conducted in two locations over four seasons with two management systems plotted against phytochemical content. A. Glucoraphanin, B. Glucobrassicin, C. Neoglucobrassicin, D. δ-tocopherol, E. γ-tocopherol, F. α-tocopherol, G. Lutein, H. Zeaxanthin, I. β-carotene. See Table 1 for cultivar name abbreviations. The C or O at the end of the cultivar abbreviation indicates conventional or organic management system, respectively.

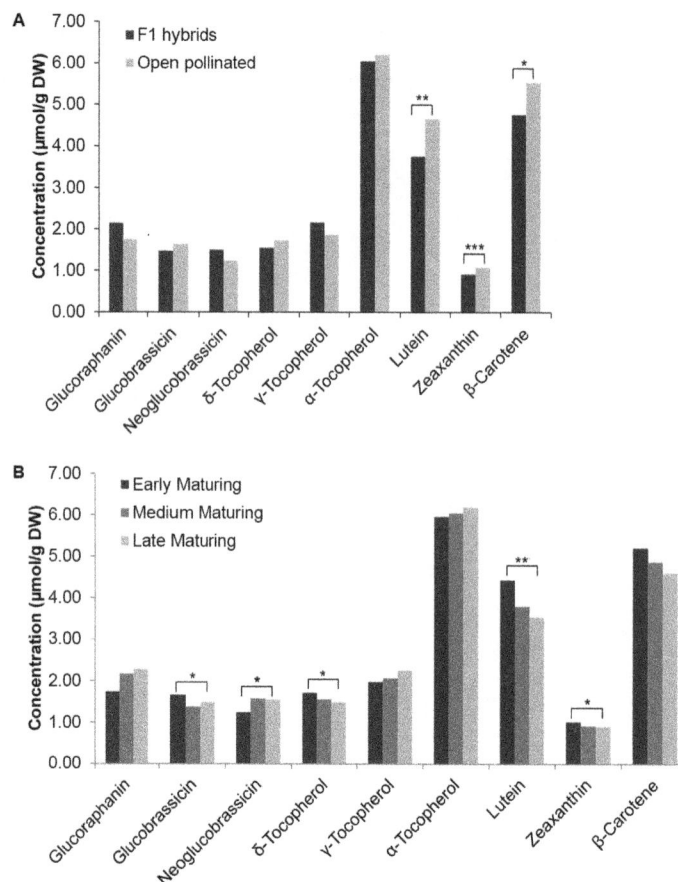

Figure 2. Mean phytochemical content of broccoli genotypic classes. A. Mean phytochemical content of broccoli F_1 hybrids versus open pollinated cultivars, and B. Mean phytochemical content of early, mid- and late-maturing cultivars grown across all trials at two locations (Maine and Oregon), in two seasons (Fall and Spring) and in two management systems (Conventional and Organic) and conventional management systems. See Table 1 for key to cultivar F1 hybrid versus open pollinated classification and maturity classification. Significance (* = P<0.05, ** = P<0.01, *** = P< 0.001).

glucoraphanin. The reverse was observed for glucobrassicin. For the carotenoids, the open pollinated cultivars had a significantly higher mean value of lutein and zeaxanthin and tended to be higher for β-carotene compared to the F_1 hybrids.

Based on the results of our field trials, the 23 cultivars of broccoli were grouped into three distinct maturity classes: Early (55–63 days); Mid (64–71 days); and Late (72–80 days) and analysed for the effect of the maturity class on phytochemical content (**Figure 2B**). For glucoraphanin, late maturing cultivars had significantly higher content levels, while for the carotenoids, early maturing cultivars tended to have higher concentrations and were significantly higher for lutein.

When cultivar performance between genetic material originating from two primary broccoli breeding companies was compared for phytochemical content there were no significant differences with the exception of lutein, where company 1's cultivars had significantly higher concentrations than those of company 2 (**data not shown**).

A negative correlation between the date of release and levels of glucobrassicin ($R^2 = 0.21$; $p = 0.03$) (**Figure 3**) was observed, but no significant correlations for any other phytochemical were seen when 21 cultivars (the total set minus the two inbred lines) were analysed by their date of commercial release (1975–2005).

Correlation analysis among phytochemicals and horticulture traits

Phytochemical correlation across trials. Correlation among phytochemicals indicated that glucoraphanin was significantly negatively correlated to glucobrassicin (**Table 5**). Correlations between the glucosinolates and the tocopherols were not significant. Glucoraphanin and neoglucobrassicin were negatively correlated to all carotenoids but only lutein and glucoraphanin were statistically significant. Glucobrassicin demonstrated a positive trend with all carotenoids. No statistically significant correlations were observed within tocopherols. Δ-tocopherol was positively correlated, while γ-tocopherol was negatively correlated to all carotenoids. There were no significant correlations for α-tocopherol with carotenoids. All carotenoids were highly positively correlated with one another.

Phytochemical correlation to horticulture traits across trials. A correlation analysis was conducted for six horticulture traits, derived from the field study component of this research, Renaud et al. [78], and the nine phytochemicals across trials. The results indicated that greater head weight and head diameter were significantly positively correlated with glucoraphanin and negatively correlated with glucobrassicin, δ-tocopherol and the carotenoids. Increasing days to maturity was positively correlated with glucoraphanin, and negatively correlated to carotenoids.

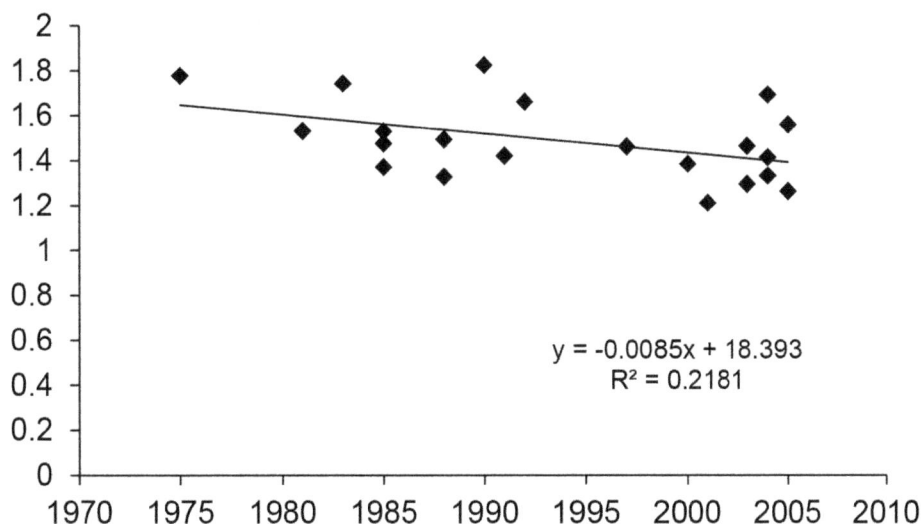

$$y = -0.0085x + 18.393$$
$$R^2 = 0.2181$$

Figure 3. Regression of broccoli floret glucobrassicin concentrations on date of cultivar release for 23 cultivars grown across all trials in two locations (Maine and Oregon), in two seasons (Fall and Spring), in two management systems (Conventional and Organic), 2006–2008.

Head color was significantly correlated with δ-tocopherol and the carotenoids, but not with glucosinolates or γ- and α-tocopherol. Bead size and bead uniformity were positively correlated with glucoraphanin, neoglucobrassicin and γ-tocopherol and negatively correlated with glucobrassicin and the carotenoids.

Principal component biplot analysis: correlation between phytochemicals and cultivars by production system

In the principal component analysis the first PC axis accounted for similar amounts of the total variation in both conventional and organic production systems (43.5% vs. 39.6%). The second PC axis showed a similar trend with 17.02% for conventional and 16.93% for organic (**Figures 4A and 4B**). The first two PC axes together accounted for 60.53% and 56.57% of total variation for conventional and organic, respectively. The PCA biplot analysis supported our findings that carotenoids were highly associated across systems, while tocopherols were highly associated in conventional, but not in organic (tocopherols demonstrated the largest shift between production systems). Glucoraphanin and neoglucobrassicin were associated with one another, but not with glucobrassicin across production systems. Glucoraphanin was associated with α-tocopherol in organic, but not in conventional treatments. Glucobrassicin was associated with δ- and α-tocopherol in conventional, but not in organic treatments. δ-tocopherol had a higher association with the carotenoids in organic than conventional. The biplots show response of both cultivars and phytochemical traits to environment. Those cultivars close to the origin reveal little about the relationship of cultivars and trait vectors, whereas those located near the extremes of trait vectors are those with the highest (or lowest) values for those traits.

Discussion

Impact of organic management system compared to environmental factors on phytochemical content

Few studies have specifically compared the levels of health promoting compounds in *Brassica* vegetable species grown under organic and conventional production systems [13,74–75]. To our knowledge, this investigation is the most comprehensive study with

the broadest range of phytochemical compounds (9) and a diverse set of broccoli cultivars (23) over regions (2), and management systems (2), with Fall and Spring season trials (2 each). In this study organic versus conventional management systems contributed the smallest source of variation compared to genotype, region and season. Within the phytochemicals studied individual compound concentrations responded differently. All compounds showed genetic variation, but also a substantial proportion of variance components were accounted for by high level interactions (**Table 3; Table S2 in File S1**). While M main effect was generally small, it had a substantial contribution in three- and four-way interactions. In particular, many G×L×S×M interactions were large relative to other variance components. This indicates that for the phytochemicals, M did have an influence on G, but that there were no consistent patterns across locations and seasons that would have shown up as significant G×M. Rather in each season and location, the paired organic and conventional environments differed significantly from one another but each situation was unique. In contrast to many comparisons between organic and conventional production systems [79], it should be noted that in our trials, yields averaged over the years did not differ significantly between the organic and conventional management systems [78].

Among the nine compounds, glucoraphanin was the most strongly influenced by genotype followed by lutein: supporting the findings of several other broccoli studies where variation in concentrations for glucoraphanin [19,70,65–66] and lutein [21,37] was primarily due to genotype. For γ-tocopherol, genotype was a large source of variation, but this compound was equally influenced by location and season (also found by Ibrahim and Juvik [37]). For glucobrassicin and neoglucobrassicin the location was the largest source of variation, but also L×S interaction was very influential, particularly for neoglucobrassicin, which is supported by Kushad et al. [20] and Schonhof et al. [26]. Jasmonic acid, a signal transduction compound in plants, is up-regulated under conditions of plant stress, wounding, and herbivory. Increased endogenous levels or exogenous application of this compound (or methyl jasmonate) increases biosynthesis and transport of neoglucobrassicin to broccoli florets. This up-

Table 5. Correlations coefficients (r) for six horticultural traits and nine phytochemicals, calculated using data standardized across trials.

	Head Weight	Head Diameter	Maturity	Head Color	Bead Size	Bead Uniformity	Glucoraphanin	Glucobrassicin	Neoglucobrassicin	δ-tocopherol	γ-tocopherol	α-tocopherol	Lutein	Zeaxanthin	β-Carotene
Head Weight															
Head Diameter	0.81														
Maturity															
Head Color															
Bead Size	0.63	0.48	0.69												
Bead Uniformity	0.49														
Glucoraphanin	0.47	0.44	0.43		0.63	0.51									
Glucobrassicin	−0.54	−0.50			−0.56	−0.64	−0.51								
Neoglucobrassicin			0.58		0.48										
δ-Tocopherol	−0.55			0.49	0.43										
γ-Tocopherol															
α-Tocopherol															
Lutein	−0.65		−0.70	0.56	−0.69		−0.41			0.55	−0.54				
Zeaxanthin	−0.68	−0.43	−0.62	0.49	−0.64					0.60	−0.42		0.95		
β-Carotene	−0.53		−0.54	0.59	−0.48					0.50	−0.43		0.90	0.90	

Correlation results include means from 23 cultivars, across eight pair combinations of location (Maine/Oregon), season (Fall/Spring) and management system (Conventional/Organic), 2006–2008[a].
[a]For empty cells, r is not significantly different from zero (P<0.05).

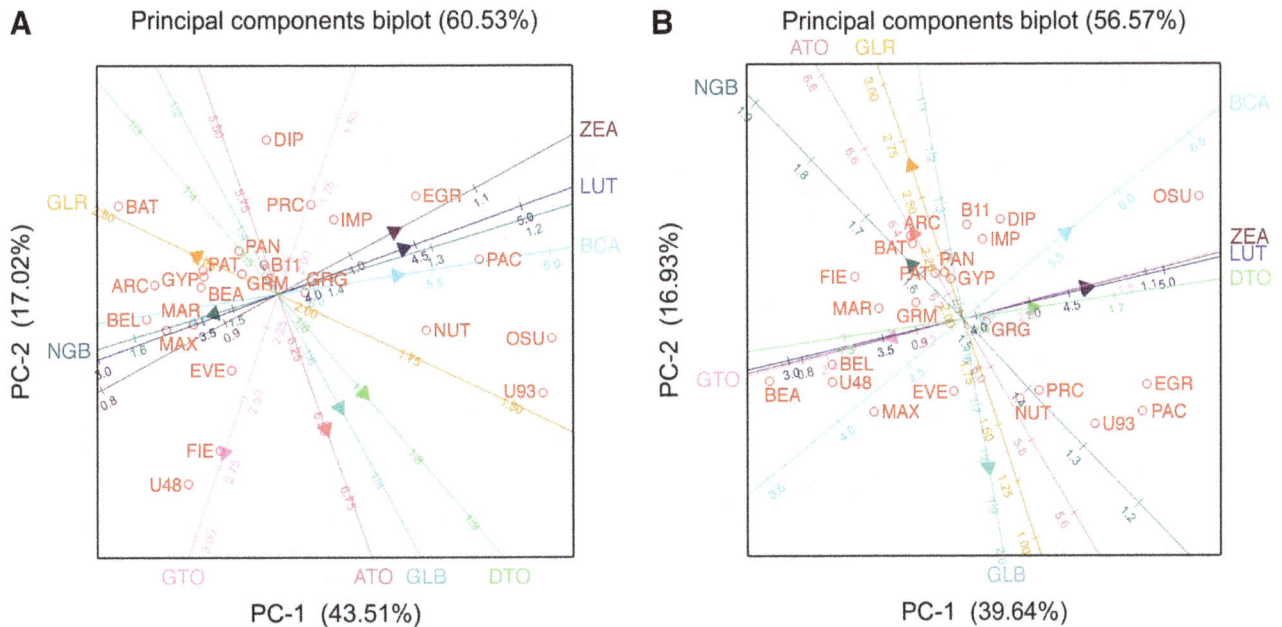

Figure 4. Principal components biplot of phytochemicals (vectors) and 23 cultivars (circles) grown in four seasons in Oregon and Maine. A. Biplot for conventional production, B. Biplot for organic production. See Table 1 for cultivar name abbreviations. Trait abbreviations: GLR: Glucoraphanin; GLB: Glucobrassicin; NGB: Neoglucobrassicin; DTO: δ-tocopherol; GTO: γ-tocopherol; ATO: α-tocopherol; LUT: Lutein; ZEA: Zeaxanthin; BCA: β-Carotene.

regulation was not observed for glucobrassicin biosynthesis [17]. This could explain why neoglucobrassicin was primarily under the control of Location and L×S interaction in our study. Season was the largest variance component for δ-tocopherol and zeaxanthin, which contrasts with the work of Ibrahim and Juvik [37] who found genotype had the largest influence on these compounds, followed by genotype by environment interaction although this study was constrained by the fact that the experiment was conducted in only one location over two growing seasons. For the other compounds such as α-tocopherol and β-carotene, L×S and the G×L×S interactions were most important.

Overall we found high genetic correlations between glucosinolates in organic and conventional trials. When trial locations were analysed separately, M main effect was present for glucobrassicin and neoglucobrassicin. The mean concentrations of glucobrassicin and neoglucobrassicin in broccoli from Oregon organic trials had higher concentrations compared to Oregon conventional trials, while Maine trials were comparable between management systems (**Table 2, Figure S1**). These results can be explained by the larger environment effect on glucobrassicin and genotype by environment effect on neoglucobrassicin found in the variance component analysis indicating sensitivity of these compounds to abiotic and/or biotic stresses. Our location specific findings are supported by those of Meyer and Adam [13] who performed a comparative study of the glucosinolate content of store bought organic and conventional broccoli and determined that the indolyl glucosinolates, glucobrassicin and neoglucobrassicin were significantly higher in the organically grown versus the conventionally grown. Evaluation of 10 broccoli genotypes over two years by Brown et al. [19] further supports our findings and those of Rosa and Rodrigues [62], Vallejo et al. [64], and Farnham et al. [70], that variation in concentration for glucoraphanin was primarily due to genetic variation, while differences in glucobrassicin was due to environmental variation (e.g. season, temperature) and genotype by environment interaction. The significantly higher

levels of glucobrassicin in Oregon in the Fall harvested trials compared to Maine could be attributed to the higher maximum temperatures and GDD in Oregon compared to Maine (**Table S1 in File S1**).

Compared to glucosinolates, there is substantially less research on the genotype by environment interaction of tocopherol and carotenoid phytochemical groups in broccoli, and no specific studies exploring the influence of organic production system. In our study, minor management system effect at the overall trial analysis level was observed for the tocopherols and for carotenoids, there was management system effect only for lutein in Oregon Spring trials. Picchi et al. [75] also did not find differences in levels of carotenoids in cauliflower in organic versus conventional systems. In the tocopherols, there were no significant differences in location, but for δ- and γ- tocopherol concentration levels were higher in the fall compared to the spring, while for α-tocopherol, concentration levels were higher in the spring compared to the fall. For the carotenoids, there were no significant location differences, however there was a seasonal trend that all carotenoids were higher in spring compared to fall. Ibrahim and Juvik [39] found significant environmental variation among 24 broccoli cultivars for carotenoids and tocopherols which they attributed to the stressful production environments. Factors explaining the genotype and genotype by environment interaction components of variation in the carotenoids and tocopherols could be clarified by the fact that environmental stimuli are both up- and down-regulating genes associated with carotenoid and tocopherol biosynthesis. There is evidence in the literature that there are coordinated responses of the carotenoid and tocopherol antioxidants *in vivo*. There was a reduction in rape seed (*Brassica napus*) tocopherol content in response to increased carotenoid levels due to over expression of the enzyme phytoene synthase [80]. This response could explain the negative correlation between γ-tocopherol concentration and the carotenoids observed in our trials.

Differences in phytochemical content between different genotypes and genotypic classes

The partitioning of variance indicated that genotype was an important source of variation for all glucosinolates. The cultivar ranking and rank correlation analysis demonstrated that there was a pattern in genotype content of glucosinolates where cultivars with the highest concentrations of glucoraphanin had the lowest levels for glucobrassicin (**Figures S1**). In our trials, the range in glucoraphanin concentrations across cultivars was (1.15–7.02 µmol/g DW, **Table S4 in File S1**), while glucobrassicin was 1.46–3.89 µmol/g DW, **Table S5 in File S1**). Several of the cultivars with the highest concentrations of neoglucobrassicin were those that had the highest concentrations of glucobrassicin. Range in neoglucobrassicin concentrations across cultivars was 0.68–4.54 µmol/g DW, **Table S6 in File S1**). In earlier studies, glucosinolate concentrations in broccoli have shown dramatic variation among different genotypes. Rosa and Rodriguez [62] studied total glucosinolate levels in eleven cultivars of broccoli and found ranges from 15.2–59.3 µmol/g DW. Among 50 accessions of broccoli Kushad et al. [20] found glucoraphanin content ranges from 0.8–22 µmol/g DW with a mean concentration of 7.1 µmol/g DW, while Wang et al. [61] found glucoraphanin content of five commercial hybrids and 143 parent materials ranging from 1.57–5.95 µmol/g for the hybrids and 0.06–24.17 µmol/g in inbred lines and Charron et al. [65] found ranges from 6.4–14.9 µmol/g DW. While the means in our study are somewhat lower, they are within the range of other studies.

A genotype effect was observed for tocopherols, but predominantly for γ-tocopherol. The PCA biplots (**Figure 4AB**) and the correlation analysis (**Table 5**) demonstrated the high positive correlations between δ-tocopherol, α-tocopherol and the carotenoids (α-tocopherol and β-carotene were also highly correlated in the Kushad et al. [20] study). The cultivar relationship to different phytochemicals was represented in the biplots as well as in the cultivar content and stability analysis (**Figure 1**). Many cultivars with the highest concentrations in the tocopherols and carotenoids were open pollinated cultivars, inbreds and early maturing, older F_1 hybrids. Many of this same group were also relatively high in glucobrassicin concentrations. Kurilich et al. [38] found that carotenoid and tocopherol concentrations among 50 broccoli lines were highly variable and primarily genotype dependent. Specifically, levels of β-carotene ranged from 0.4–2.4 mg/100 g FW. Ibrahim and Juvik [37] also found broad ranges for total carotenoid and tocopherol concentrations among 24 genotypes ranging from 55–154 µg/g DW and 35–99 µg/g DW, respectively. Farnham and Kopsell [21] studied the carotenoid levels of nine double haploid lines of broccoli. Similar to our findings, lutein was the most abundant carotenoid in broccoli ranging from 65.3–139.6 µg/g DM. The sources of variation for lutein were predominantly genotype, followed by environment and G×E interaction, which also supports our findings. No genotypic differences were found for β-carotene in Farnham and Kopsell [21], which is in contrast to our findings. Overall, they found that most of the carotenoids measured were positively and highly correlated to one another as was observed in our study (**Table 5**). Kopsell et al. [81] found lutein levels in kale of 4.8–13.4 mg/100 g FW where the primary variance components for both lutein and β-carotene were also genotype and season.

Our research aimed also to address the question whether the phytochemical content of broccoli cultivars is associated with certain genotypic classes, e.g. open pollinated vs. F_1 hybrids; older vs. newer cultivar releases; and between commercial sources. Broccoli is typically a cross-pollinated, self-incompatible crop species and cultivars are either open pollinated and composed of heterogeneous genetically segregating individuals, or F_1 hybrids produced by crossing of two homozygous inbred lines, resulting in homogeneous populations of heterozygous individuals. In the 1960's virtually all broccoli grown was derived from OPs. By the 1990's almost all commercial cultivars were hybrids [82].

In our trials with 18 F_1 hybrids (released between 1975–2005) and 3 open pollinated cultivars (released from 1985–2005), we found several interesting trends related to genotype and genotypic class performance as it related to the three groups of phytochemicals. When analysing F_1 hybrid and open pollinated cultivars, they also demonstrated different performance patterns depending upon the individual phytochemical or group of compounds analysed. When cultivars were ranked for content and stability per phytochemical, there were distinct trends for certain compounds such as late maturing, F_1 hybrids outperforming early maturing F_1 hybrids and open pollinated cultivars for glucoraphanin, while the inverse was found for glucobrassicin and all carotenoids studied. This analysis was further supported by the PCA biplots that showed a strong relationship for select cultivars to certain phytochemicals or groups of phytochemicals such as 'OSU OP' to the carotenoids. When the full set of cultivars was divided into F_1 hybrid and open pollinated groups and the means compared by phytochemical, the results further supported the individual cultivar analysis where F_1 hybrids had higher mean values for glucoraphanin than the open pollinated cultivars (**Figure 2A**). Clear cultivar performance differences were identified where early maturing versus late maturing cultivars performed differently depending upon the phytochemical (**Figure 2B**). We also found that late maturing cultivars had higher concentrations for glucoraphanin than early maturing lines (and the inverse for glucobrassicin and the carotenoids). Picchi et al. [75] studied the quantity of glucosinolates of an early and late maturing cultivar of cauliflower grown in one conventional and three organic production systems, and found a significantly higher level of glucoraphanin in the later maturing cultivar compared to the early maturing cultivar in the organic production system. Another interesting trend was that cultivars with higher concentration levels for those phytochemicals whose expression is heavily influenced by environmental factors were not necessarily the most stable across trial environments; as was the case with neoglucobrassicin, δ- and γ-tocopherol in our study. For traits where genotype played a more significant role in contributing to variation, cultivars with a higher concentration level tended to also be those that were most stable across environments as was seen for lutein and glucoraphanin concentrations.

No significant differences were found for cultivar performance in phytochemical concentrations between genetic materials originating from two distinct commercial sources, with the exception of lutein (data not shown). When the full set of broccoli cultivars were analyzed for a correlation between date of release and mean level of phytochemical content across trials, no significant correlation was found with the exception of a negative trend for glucobrassicin (**Figure 3**). Our data does not support the idea that modern breeding for high yield performance and disease resistance necessarily leads to a trade-off in level of phytochemicals. Previous reports examining the relationship between year of release and performance had focussed on wheat vitamin and mineral content [83–85], and mineral content in broccoli [86–87]. However these authors did not study phytochemical content and their results were equivocal on the question on an innate biological trade-off between increased yield and nutritional content.

Not many studies have included two or more groups of phytochemicals. In our study with three phytochemical groups we found that phytochemicals demonstrating a negative correlation with one another (e.g. glucoraphanin with the carotenoids), showed an inverse cultivar response: e.g. cultivars with highest concentrations of glucoraphanin were the lowest in the carotenoids and vice versa. When both horticultural traits and phytochemicals were analysed for their phenotypic correlation, head weight was significantly and positively correlated with glucoraphanin and negatively correlated with δ- and α-tocopherol and the carotenoids. Farnham and Kopsell [21] explained that negative correlations may occur as a result of increased biomass accumulation in a certain genotype that is not accompanied by increased carotenoid production, effectively lowering the carotenoid concentration in the immature broccoli florets when pigments are expressed. Comparatively, head color was highly correlated to the carotenoids and negatively correlated to the glucosinolates overall. The cultivar 'OSU OP' was explicitly bred for a dark green stem and head color, not only for a darker green dome surface but also for a dark green interior color between the florets of the dome and in the stem (personal communication, Jim Myers 2013). 'OSU OP' was the highest in overall carotenoid concentrations across trials as it is known that carotenoids are correlated with chlorophyll concentrations and the intensity of green pigmentation [88].

Perspectives on breeding broccoli for enhanced phytochemical content specifically for organic agriculture

Our study included predominantly broccoli cultivars selected for broad adaptability in conventional production systems and not purposely bred for high phytochemical content nor for adaptation to organic agriculture. What we can conclude from our data is that there has been little change in levels of several phytochemicals over three decades of breeding. This may indicate genetic variation for phytochemicals is limited in elite germplasm, or it may be the result of the lack of selection tools for these traits. This may be changing with recent efforts to introgress high glucoraphanin from *B. villosa* to produce the high-glucoraphanin F_1 cultivar 'Beneforté' [89–90,44]. The seed industry needs to exploit known sources of variation in the genus *Brassica* to enhance levels of other health-promoting phytochemicals and to broaden the genetic diversity of commercial broccoli germplasm. Our finding of a strong correlation between dark green color and high carotenoid levels provides breeders with a simple and efficient means of increasing carotenoids. The three groups of phytochemicals studied contribute to health promotion in different ways. As these groups are related to different metabolic pathways selecting for one compound does not necessarily inadvertently improve the other compounds, and may even result in negative correlation as we have seen in our data between glucoraphanin and the carotenoids. Although these compounds belong to different metabolic pathways, their production may be coordinated through regulatory feedback loops, or the structural and/or regulatory genes controlling these pathways may be genetically linked.

Designing a breeding program for broccoli high in glucosinolates would require the following considerations generated from our research: (1) Glucoraphanin is a highly genetically determined compound with minor location and season main effects but with substantial G×L×S interaction. (2) Comparatively, glucobrassicin and neoglucobrassicin are more impacted by location and season and L×S interaction with highest glucobrassicin concentrations and largest range in our Oregon Fall trials and neoglucobrassicin highest in Oregon Spring trials. (3) Cultivar performance for glucoraphanin and glucobrassicin and neoglucobrassicin was negatively correlated indicating that there may be a trade-off between glucoraphanin on the one hand, and glucobrassicin and neoglucobrassicin on the other hand. (4) Selection for glucoraphanin without consideration of horticultural traits would probably result in larger headed and later maturing cultivars. Conversely, selection for smaller headed, early maturing cultivars would favor glucobrassicin and neoglucobrassicin at the expense of glucoraphanin.

A breeding program for broccoli for high tocopherol content would require: (1) Overall the tocopherols were more season, location and L×S dependent and had lower overall repeatabilities compared to the glucosinolates. In a structured genetic population where additive genetic variance could be partitioned, narrow sense heritability would likely be low, and increasing tocopherol content would best be conducted with breeding methods suited to low heritability traits; (2) δ- and γ-tocopherols were both season dependent and fall grown broccoli had higher concentrations of these compounds across trials and a wider range of content levels, whereas levels of α-tocopherol were higher in spring but the range was comparable under both seasons. Thus, fall would be the preferred environment for breeding for these compounds; (3) there were no significant differences for location for δ- or γ-tocopherol, but the average levels of α-tocopherol levels were significantly higher in Oregon than Maine, suggesting greater potential for genetic gain in the Oregon environment.

If the goal is to design a breeding program for broccoli enhancing the levels of carotenoids it would require the following considerations: (1) For all three carotenoids studied, genotypic variation, particularly for lutein, was relatively more important than location and season; (2) however, zeaxanthin exhibited a large S (spring) and L×S interaction. For both β-carotene and lutein, spring grown broccoli had significantly higher levels than fall produced. Thus, selection for carotenoids would probably be more effective in spring than in fall; (3) early maturing and small headed cultivars had higher levels of carotenoids. Since most of the carotenoids are associated with the outer surfaces of the inflorescence, smaller broccoli heads with a greater surface area to volume ratio should show higher concentrations of these compounds; (4) because carotenoids have high G main effect good germplasm sources as indicated in **Figure 1** have high concentrations of carotenoids and demonstrated stability across environments. As all three carotenoids are highly correlated with one another, selecting for one should effectively select for all; (5) selection for darker green colour more widely distributed throughout the tissues of the head should allow the breeder to relatively efficiently increase carotenoid content in broccoli.

In closing, we want to address the question of selecting in an organic or a conventional environment. The argument commonly used to support selecting in productive environments is that heritabilities are higher compared to resource poor environments [91–92]. Organic is often considered a low-external input environment, resulting on average in 20% less yield compared to conventional production [79]. Nevertheless, in our trials repeatabilities for some phytochemicals were higher or comparable to conventional (**Table 3**). Narrow sense heritabilities would be expected to be significantly lower. For those traits where repeatabilities were higher or comparable, direct selection under organic systems could enhance selection gain. In all cases, the ratio of correlated response to direct response was less than one suggesting that direct selection would allow more rapid progress than correlated selection. Our data on phytochemicals did not show a wider range of levels under organic conditions as we found for horticultural traits in the same trials [78], however, in several cases, repeatabilities in organic production were higher than in conventional.

To maximize efficiency in a breeding program, commercial breeders may seek to combine breeding for both conventional and organic markets, and a combination of strategies can be proposed. Some studies that utilized highly heritable (agronomic) traits, where cultivar yield performance ranked similarly between organic and conventional management systems and which had high genetic correlations, suggested that early breeding be conducted under conventional conditions, with the caveat that advanced breeding lines be tested under organic conditions for less heritable traits (e.g. Löschenberger et al. [93]; Lorenzano and Bernardo, 2008) [52]. In studies where cultivar yield performance differed between management systems and there were significant differences in cultivar ranking, and in some cases low genetic correlations for lower heritability traits (e.g. Kirk et al. [94]; Murphy et al. [48]), these studies recommended that cultivars intended for organic agriculture be selected only under organic conditions. In our study of phytochemicals, we would recommend for organic purposes selection under organic conditions for the compounds where genetic correlations between organic and conventional were moderate.

Supporting Information

Figure S1 **Comparison of broccoli cultivars for glucosinolates (μmol/g DW) grown across all trials in two locations (Maine and Oregon), in two seasons (Fall and Spring), in two management systems (Conventional and Organic), and at the individual trial level, 2006–2008.** A. Glucoraphanin, B. Glucobrassicin, C. Neoglucobrassicin.

Figure S2 **Comparison of broccoli cultivars for tocopherols (μmol/g DW) grown across all trials in two locations (Maine and Oregon), in two seasons (Fall and Spring), in two management systems (Conventional and Organic), and at the individual trial level, 2006–2008.** A. δ-tocopherol, B. γ-tocopherol, C. α-tocopherol.

Figure S3 **Comparison of broccoli cultivars for carotenoids (μmol/g DW) grown across all trials in two locations (Maine and Oregon), in two seasons (Fall and Spring), in two management systems (Conventional and Organic), and at the individual trial level, 2006–2008.** A. Lutein, B. Zeaxanthin, and C. β-carotene.

File S1 **Supporting tables. Table S1.** Description of agronomic and environmental factors of the trial locations with paired organically and conventionally managed fields, 2006–2008. **Table S2.** Partitioning (%) of variance components for various traits of 23 broccoli cultivars grown across four pair combinations in Maine, season (Fall/Spring) and management system (Conventional/Organic), 2006–2008. Variance components reported as coefficients of variation. **Table S3.** Repeatability for broccoli for phytochemicals and per trial of 23 broccoli cultivars grown across eight pair combinations of location (Maine/Oregon), season (Fall/Spring) and management system (Conventional/Organic), 2006–2008. **Table S4.** Glucoraphanin level (μmol/g DW) of 23 cultivars grown under conventional (C) and organic (O) conditions in two locations (Maine and Oregon) in two seasons (Fall and Spring) from 2006–2008. **Table S5.** Glucobrassicin level (μmol/g DW) of 23 cultivars grown under conventional (C) and organic (O) conditions in two locations (Maine and Oregon) in two seasons (Fall and Spring) from 2006–2008. **Table S6.** Neoglucobrassicin level (μmol/g DW) of 23 cultivars grown under conventional (C) and organic (O) conditions in two locations (Maine and Oregon) in two seasons (Fall and Spring) from 2006–2008. **Table S7.** δ-tocopherol level (μmol/g DW) of 23 cultivars grown under organic (O) and conventional (C) conditions in two locations (Maine and Oregon) in two seasons (Fall and Spring) from 2006–2008. **Table S8.** γ-tocopherol level (μmol/g DW) of 23 cultivars grown under conventional (C) and organic (O) conditions in two locations (Maine and Oregon) in two seasons (Fall and Spring) from 2006–2008. **Table S9.** α-tocopherol level (μmol/g DW) of 23 cultivars grown under conventional (C) and organic (O) conditions in two locations (Maine and Oregon) in two seasons (Fall and Spring) from 2006–2008. **Table S10.** Lutein level (μmol/g DW) of 23 cultivars grown under conventional (C) and organic (O) conditions in two locations (Maine and Oregon) in two seasons (Fall and Spring) from 2006–2008. **Table S11.** Zeaxanthin level (μmol/g DW) of 23 cultivars grown under conventional (C) and organic (O) and conditions in two locations (Maine and Oregon) in two seasons (Fall and Spring) from 2006–2008. **Table S12.** β-carotene level (μmol/g DW) of 23 cultivars grown under conventional (C) and organic (O) conditions in two locations (Maine and Oregon) in two seasons (Fall and Spring) from 2006–2008.

Acknowledgments

For support in understanding the genotype selection and providing select elite inbred lines for the study, the authors thank Dr. Mark Farnham from the USDA, Charleston, South Carolina. For the Oregon trials, the authors wish to thank the organic growers Jolene Jebbia and John Eveland at Gathering Together Farm for providing the location and support for the organic broccoli trials. Deborah Kean, Faculty Research Assistant at the Oregon State University Research Station, and the students Hank Keogh, Shawna Zimmerman, Miles Barrett, and Jennifer Fielder provided support in data collection of the field trials. For the Maine trials, the authors wish to thank the students Heather Bryant, Chris Hillard and Greg Koller for support in data collection of the field trials. We also thank the University of Maine Highmoor Farm Superintendent, Dr. David Handley University of Maine Cooperative Extension. For the soil analysis, we thank Dr. Michelle Wander from the University of Illinois, Urbana and her students. At Wageningen University, we thank Paul Keizers and Dr. Chris Maliepaard for support with the statistical analysis. We thank Carl Jones for his valuable input on iterations of this research paper and Ric Gaudet for support in data organization.

Author Contributions

Conceived and designed the experiments: ER ELvB JM MP FvE NZ JJ. Performed the experiments: ER JM NZ JJ. Analyzed the data: ER ELvB JM MP FE JJ. Contributed reagents/materials/analysis tools: ER JM MP FvE NZ JJ. Wrote the paper: ER ELvB JM JJ FvE.

References

1. Saba A, Messina F (2003) Attitudes towards organic foods and risk/benefit perception associated with pesticides. Food Quality and Preferences 14: 637–645.

2. Stolz H, Stolze M, Hamm U, Janssen M, Ruto E (2011) Consumer attitudes towards organic versus conventional food with specific quality attributes. Netherlands Journal of Agricultural Science 58: 67–72.

3. Smith-Spangler C, Brandeau ML, Hunter GE, Bavinger JC, Pearson M, et al (2012) Are organic foods safer or healthier than conventional alternatives? A systematic review. Annals of Internal Medicine 157: 348–366.

4. Asami DK, Hong YJ, Barrett DM, Mitchell AE (2003) Comparison of the total phenolic and ascorbic acid content of freeze-dried and air-dried marionberry, strawberry, and corn grown using conventional, organic and sustainable agricultural practices. Journal of Agriculture & Food Chemistry 51(5): 1237–1271.

5. Chassy AW, Bui L, Renaud ENC, Mitchell AE (2006) Three-year comparison of the content of antioxidant microconstituents and several quality characteristics in organic and conventionally managed tomatoes and bell peppers. Journal of Agriculture and Food Chemistry 54: 8244–8252.

6. Brandt K, Leifert C, Sanderson R, Seal CJ (2011) Agroecosystem management and nutritional quality of plant foods: The case of organic fruits and vegetables. Critical Reviews in Plant Sciences 30: 177–197.

7. Hunter D, Foster M, McArthur JO, Ojha R, Petocz P, et al (2011) Evaluation of the micronutrient composition of plant foods produced by organic and conventional methods. Critical Reviews in Food Science and Nutrition 51: 571–582.

8. Koh E, Charoenpraset S, Mitchell AE (2012) Effect of organic and conventional cropping systems on ascorbic acid, vitamin C, flavonoids, nitrate, and oxalate in 27 varieties of spinach (Spinacia oleracea L.) Journal of Agriculture and Food Chemistry. 60(12): 3144–3150.

9. Crozier A, Clifford MN, Ashihara H (eds) (2006) Plant secondary metabolites: occurrence, structure and role in the human diet. Blackwell Publishing Ltd.

10. Bourn D, Prescott J (2002) A comparison of the nutritional value, sensory qualities and food safety of organically and conventionally produced foods. Critical Review Food Science Nutrition 42(1): 1–34.

11. Mäder P, Fliessbach A, Dubois D, Gunst L, Fried P, et al (2002) Soil fertility and biodiversity in organic farming. Science 296: 1694–1697.

12. Martinez-Bellesta MC, Lopez-Perez L, Hernandez M, Lopez-Berenguer C, Fernandez-Garcia N, et al (2008) Agricultural practices for enhanced human health. Phytochemical Review 7: 251–260.

13. Meyer M, Adam ST (2008) Comparison of glucosinolate levels in commercial broccoli and red cabbage from conventional and ecological farming. European Food Research Technology 226: 1429–1437.

14. Mozafar A (1993) Nitrogen fertilizers and the amount of vitamins in plants: a review. Journal of Plant Nutrition 16: 2479–2506.

15. Zhao X, Carey EE, Wang W, Rajashekar CB (2006) Does organic production enhance phytochemical content of fruit and vegetables? Current knowledge and prospects for research. HortTechnology 16(3): 449–456.

16. Dixon RA, Paiva NL (1995) Stress-induced phenylpropanoid metabolism. The Plant Cell 7: 1085–1097.

17. Kim HS, Juvik JA (2011) Effect of selenium fertilization and methyl jasmonate treatment on glucosinolate accumulation in broccoli floret. Journal of the American Society for Horticulture Science 136(4): 239–246.

18. USDA National Nutrient Database, 2011. http://ndb.nal.usda.gov/(last visited 14 April 2014).

19. Brown AF, Yousef GG, Jeffrey EH, Klein BP, Wallig MA, et al (2002) Glucosinolate profiles in broccoli: variation in levels and implications in breeding for cancer chemoprotection. Journal of American Society Horticulture Science 127(5): 807–813.

20. Kushad MM, Brown AF, Kurilich AC, Juvik JA, Klein BP, et al (1999) Variations of glucosinolates in vegetable crops of Brassica oleracea. Journal of Agriculture and Food Chemistry 47: 1541–1548.

21. Farnham MW, Kopsell DA (2009) Importance of genotype on carotenoid and chlorophyll levels in broccoli heads. HortScience 44(5): 1248–1253.

22. Verhoeven DT, Goldbohm PA, van Poppel GA, Verhagen H, van den Brandt PA (1996) Epidemiological studies on Brassica vegetables and cancer risk. Biomarkers Prevention 5: 733–748.

23. Keck A-S, Finley JW (2004) Cruciferous vegetables: cancer protective mechanisms of glucosinolate hydrolysis products and selenium. Integrative Cancer Therapies 3(1): 5–12.

24. Here I, Büchler MW (2010) Dietary constituents of broccoli and other cruciferous vegetables: implications for prevention and therapy of cancer. Cancer Treatment Review 36: 377–383.

25. Higdon JV, Delage B, Williams DE, Dashwood RH (2007) Cruciferous vegetables and human cancer: epidemiologic evidence and mechanistic basis. Pharmacology Research 55(3): 224–236.

26. Schonhof I, Krumbein A, Bruchner B (2004) Genotypic effects on glucosinolates and sensory properties of broccoli and cauliflower. Nahrung/Food 48(1): 25–33.

27. Fenwick GR, Heaney RK, Mulllin WJ (1983) Glucosinolates and their breakdown products in food and food plants. Critical Reviews in Food Science and Nutrition 18(2): 123–201.

28. Juge N, Mithen RF, Traka M (2007) Molecular basis for chemoprevention by sulforaphane: a comprehensive review. Cellular and Molecular Life Sciences 64: 1105–1127.

29. Mithen RF, Dekker M, Verkerk R, Rabot S, Johnson IT (2000) The nutritional significance, biosynthesis and bioavailability of glucosinolates in human foods. Journal of the Science of Food and Agriculture 80: 967–984.

30. Wang LI, Giovannucci EL, Hunter D, Neuberg D, Su L, et al (2004) Dietary intake of cruciferous vegetables, glutathione S-transferase (GST) polymorphisms and lung cancer risk in a Caucasian population. Cancer Causes Control 15: 977–985.

31. Hsu CC, Chow WH, Boffetta P, Moore L, Zaridze D, et al (2007) Dietary risk factors for kidney cancer in eastern and central Europe. American Journal of Epidemiology 166: 62–70.

32. Kirsh VA, Peters U, Mayne ST, Subar AF, Chatterjee N, et al (2007) Prospective study of fruit and vegetable intake and risk of prostate cancer. Journal of the National Cancer Institute 99: 1200–1209.

33. Lam TK, Ruczinski I, Helzlsouer KJ, Shugart YY, Caulfield LE, et al (2010) Cruciferous vegetable intake and lung cancer risk: a nested case-control study matched on cigarette smoking. Cancer Epidemiology, Biomarkers and Prevention 19: 2534–2540.

34. Bosetti C, Filomeno M, Riso P, Polesel J, Levi F, et al (2012) Cruciferous vegetables and cancer risk in a network of case-control studies. Annals of Oncology 23: 2198–2203.

35. Wu QJ, Yang Y, Vogtmann E, Wang J, Han LH, et al (2012) Cruciferous vegetables intake and the risk of colorectal cancer: a meta-analysis of observational studies. Annals of Oncology 00: 1–9. doi:10.1093/annonc/mds601.

36. Kopsell DA, Kopsell DE (2006) Accumulation and bioavailability of dietary carotenoids in vegetable crops. Trends in Plant Science 11(10): 499–507.

37. Ibrahim KE, Juvik JA (2009) Feasibility for improving phytonutrient content in vegetable crops using conventional breeding strategies: case study with carotenoids and tocopherols in sweet corn and broccoli. Journal of Agriculture and Food Chemistry 57: 4636–4644.

38. Kurilich AC, Tsau GJ, Brown A, Howard L, Klein BP, et al (1999) Carotene, tocopherol, and ascorbate contents in subspecies of Brassica oleracea. Journal of Agriculture and Food Chemistry 47: 1576–1581.

39. Kurilich AC, Juvik JA (1999) Quantification of carotenoid and tocopherol antioxidants in Zea mays. Journal of Agriculture and Food Chemistry 47: 1948–1955.

40. Mayne ST (1996) βeta carotene, carotenoids, and disease prevention in humans. The FASEB Journal. 10(7): 690–701.

41. Krinsky NI, Landdrum JT, Bone RA (2003) Biologic mechanisms of the protective role of lutein and zeazanthin in the eye. Annual Review Nutrition. 23: 171–201.

42. Kritchevsky SB (1999) β-Carotene, carotenoids and the prevention of coronary heart disease. Journal of Nutrition 129: 5–8.

43. Davis D, Epp MD, Riordan HD (2004) Changes in USDA food composition data for 43 garden crops, 1950 to 1999. Journal of the American College of Nutrition 23(6): 1–14.

44. Traka MH, Saha S, Huseby S, Kopriva S, Walley PG, et al (2013). Genetic regulation of glucoraphanin accumulation in Beneforte broccoli. New Phytologist, doi: 10.1111/nph.12232.

45. Munger HM (1979) The potential of breeding fruits and vegetables for human nutrition. HortScience 14(3): 247–250.

46. Welch RM, Graham RD (2004) Breeding for micronutrients in staple food crops from a human nutrition perspective. Journal of Experimental Botany 55(396): 353–364.

47. Troxell Alrich H, Salandanan K, Kendall P, Bunning M, Stonaker F, et al (2010) Cultivar choice provides options for local production of organic and conventionally produced tomatoes with higher quality and antioxidant content. Journal of Science Food and Agriculture 90: 2548–2555.

48. Murphy KM, Campbell KG, Lyon SR, Jones SS (2007) Evidence of varietal adaptation to organic farming systems. Field Crops Research 102: 172–177.

49. Baresel JP, Zimmermann G, Reents HJ (2008) Effects of genotype and environment on N uptake and N partition in organically grown winter wheat (Triticum aestivum L.) in Germany. Euphytica 163: 347–354.

50. Vlachostergios DN, Roupakias DG (2008) Response to conventional and organic environment of thirty-six lentil (Lens culinaris Medik.) varieties. Euphytica 163: 449–457.

51. Goldstein W, Schmidt W, Burger H, Messmer M, Pollak LM, et al (2012) Maize: Breeding and field testing for organic farmers. In: Lammerts van Bueren ET, Myers, JR editors. Organic crop breeding. West Sussex, UK, Wiley-Blackwell, John Wiley & Sons, Inc.

52. Lorenzana RE, Bernardo R (2008) Genetic correlation between corn performance in organic and conventional production systems. Crop Science 48(3): 903–910.

53. Osman AM, Almekinders CJM, Struik PC, Lammerts van Bueren, ET (2008) Can conventional breeding programmes provide onion varieties that are suitable for organic farming in the Netherlands? Euphytica 163: 511–522.

54. Przystalski M, Osman AM, Thiemt EM, Rolland B, Ericson L, et al (2008) Do cereal varieties rank differently in organic and non-organic cropping systems? Euphytica 163: 417–435.

55. Wunderlich SM, Feldmand C, Kane S, Hazhin T (2008) Nutritional quality of organic, conventional, and seasonally grown broccoli using vitamin C as a marker. International Journal of Food Sciences and Nutrition 59(1): 34–45.

56. Koh E, Wimalasin KMS, Chassy AW, Mitchell AE (2009) Content of ascorbic acid, quercetin, kaempferol and total phenolics in commercial broccoli. Journal of Food Composition and Analysis. 22(7–8): 637–643.

57. Harker FR (2004) Organic food claims cannot be substantiated through testing of samples intercepted in the marketplace: a horticulturist's opinion. Food Quality and Preference 15: 91–99.

58. Benbrook C (2009) The impact of yield on nutritional quality: lessons from organic farming. HortScience 44(1): 12–14.

59. Carlson DG, Daxenbichler ME, VanEtten CH, Kwolek WF, Williams PH (1987) Glucosinolates in crucifer vegetables: broccoli, brussels sprouts, cauliflower, collards, kales, mustard greens, and kohlrabi. Journal of the American Society for Horticulture Science 112(1): 173–178.

60. Farnham MW, Stephenson KK, Fahey JW (2005) Glucoraphanin level in broccoli seed is largely determined by genotype. HortScience 40(1): 50–53.

61. Wang J, Gu H, Yu H, Zhao Z, Sheng X, Zhang X (2012) Genotypic variation of glucosinolates in broccoli (Brassica oleracea var. italica) florets from China. Food Chemistry 133: 735–741.

62. Rosa EAS, Rodrigues AS (2001) Total and individual glucosinolate content in 11 broccoli cultivars grown in early and late seasons. HortScience 36(1): 56–59.

63. Vallejo F, Tomas-Barberán FA, García-Viguera C (2003a) Effect of climatic and sulphur fertilisation conditions, on phenolic compounds and vitamin C, in the inflorescences of eight broccoli cultivars. European Food Research Technology 216: 395–401.

64. Vallejo F, Tomas-Barberán FA, Gonzalez Benavente-Garcia A, García-Viguera C (2003b) Total and individual glucosinolate contents in inflorescences of eight broccoli cultivars grown under various climatic and fertilisation conditions. Journal of the Science of Food and Agriculture 83: 307–313.

65. Charron CS, Saxton A, Sams CE (2005a) Relationship of climate and genotype to seasonal variation in the glucosinolate – myrosinase system. I. Glucosinolate content in ten cultivars of *Brassica oleracea* grown in fall and spring seasons. Journal of the Science of Food and Agriculture 85: 671–681.

66. Charron CS, Saxton A, Sams CE (2005b) Relationship of climate and genotype to seasonal variation in the glucosinolate – myrosinase system. II. Myrosinase activity in ten cultivars of *Brassica oleracea* grown in fall and spring seasons. Journal of the Science of Food and Agriculture 85: 682–690.

67. Aires A, Fernandes C, Carvalho R, Bennett RN, Saavedra MJ, et al (2011) Seasonal effects on bioactive compounds and antioxidant capacity of six economically important brassica vegetables. Molecules 16: 6816–6832.

68. Pek Z, Daood H, Nagne MG, Berki M, Tothne MM, et al (2012) Yield and phytochemical compounds of broccoli as affected by temperature, irrigation, and foliar sulphur supplementation. HortScience 47(11): 1646–1652.

69. Schonhof I, Blandenburg D, Müller S, Krumbein A (2007) Sulfur and nitrogen supply influence growth, product appearance, and glucosinolate concentration of broccoli. Journal of Plant Nutrition and Soil Science 170: 65–72.

70. Farnham MW, Wilson PE, Stephenson KK, Fahey JW (2004) Genetic and environmental effects on glucosinolate content and chemoprotective potency of broccoli. Plant Breeding 123: 60–65.

71. Björkman M, Klingen I, Birch ANE, Bones AM, Bruce TJA, et al (2011) Phytochemicals of Brassicaceae in plant protection and human health – influences of climate, environment and agronomic practice. Phytochemistry 72: 538–556.

72. Robbins RJ, Keck AS, Banuelos G, Finley JW (2005) Cultivation conditions and selenium fertilization alter the phenolic profile, glucosinolate, and sulforaphane content of broccoli. Journal of Medicinal Food 8(2): 204–214.

73. Xu C, Guo R, Yan H, Yuan J, Sun B, et al (2010) Effect of nitrogen fertililization on ascorbic acid, glucoraphanin content and quinone reductase activity in broccoli floret and stem. Journal of Food, Agriculture and Environment 8(1): 179–184.

74. Naguib AM, El-Baz FK, Salama ZA, Hanna HAEB, Ali HF, et al (2012) Enhancement of phenolics, flavonoids and glucosinolates of broccoli (*Brassica oleracea*, var. Italica) as antioxidants in response to organic and bio-organic fertilizers. Journal of the Saudi Society of Agricultural Sciences 11: 135–142

75. Picchi V, Migliori C, Scalzo RL, Campanelli G, Ferrari V, et al (2012) Phytochemical content in organic and conventionally grown Italian cauliflower. Food Chemistry 130: 501–509.

76. Hansen M, Moller P, Sorensen H, Cantwell de Trejo M (1995) Glucosinolates in broccoli stored under controlled atmosphere. Journal of the American Society for Horticulture Science 120(6): 1069–1074.

77. Tiwari U, Cummins E (2013) Factors influencing levels of phytochemicals in selected fruit and vegetables during pre- and post-harvest food processing operations. Food Research International 50: 497–506.

78. Renaud ENC, Lammerts van Bueren ET, Paulo MJ, van Eeuwijk FA, Juvik JA, et al (2014) Broccoli cultivar performance under organic and conventional management systems and implications for crop improvement. Crop Science DOI: 10.2135/cropsci201.

79. De Ponti T, Rijk B, van Ittersum MK, (2012) The crop yield gap between organic and conventional agriculture. Agricultural Systems 108: 1–9.

80. Shewmaker CK, Sheehy JA, Daley M, Colburn S, Ke DY (1999) Seed-specific overexpression of phytoene synthase: increase in carotenoids and other metabolic effects. Plant Journal 20: 401–412.

81. Kopsell DA, Kopsell DE, Lefsrud MG, Curran-Celentano J, Dukach LE (2004) Variation in lutein, β-carotene, and chlorophyll concentrations among *Brassica oleracea* cultigens and seasons. HortScience 39(2): 361–364.

82. Hale AL, Farnham MW, Ndambe Nzaramba M, Kimberg CA (2007) Heterosis for horticultural traits in broccoli. Theoretical Applied Genetics 115(3): 351–60.

83. Murphy KM, Reeves PG, Jones SS (2008) Relationship between yield and mineral nutrient concentrations in historical and modern spring wheat cultivars. Euphytica 163: 381–390.

84. Hussain A, Larsson H, Kuktaite R, Johansson E (2010) Mineral composition of organically grown wheat genotypes: contribution to daily minerals intake. International Journal of Environmental Research and Public Health 7: 3443–3456.

85. Jones H, Clarke S, Haigh Z, Pearce H, Wolfe M (2010) The effect of the year of wheat variety release on productivity and stability of performance on two organic and two non-organic farms. Journal of Agricultural Science 148: 303–317.

86. Farnham MW, Keinath AP, Grusak MA (2011) Mineral concentration of broccoli florets in relation to year of cultivar release. Crop Science 51: 2721–2727.

87. Troxell-Alrich H, Kendall P, Bunning M, Stonaker F, Kulen O, et al (2011) Environmental temperatures influence antioxidant properties and mineral content in broccoli cultivars grown organically and conventionally. Journal of AgroCrop Science 2(2): 1–10.

88. Khoo H-E, Prasad KN, Kong K-W, Jiang Y, Ismail A (2011) Carotenoids and their isomers: color pigments in fruits and vegetables. Molecules 16(2): 1710–1738.

89. Faulkner K, Mithen R, Williamson G (1998) Selective increase of the potential anticarcinogen 4-methylsulphinylbutyl glucosinolate in broccoli. Carcinogenis 19(4): 605–609.

90. Mithen R, Faulkner K, Magrath R, Rose P, Williamson G, Marquez J (2003) Development of isothiocyanate-enriched broccoli, and its enhanced ability to induce phase 2 detoxification enzymes in mammalian cells. Theoretical Applied Genetics 106: 727–734.

91. Ceccarelli S (1996) Adaptation to low/high input cultivation. Euphytica 92: 203–214.

92. Ceccarelli S (1994) Specific adaptation and breeding for marginal conditions. Euphytica 77: 205–219.

93. Löschenberger F, Fleck A, Grausgruber H, Hetzendorfer H, Hof G, et al (2008) Breeding for organic agriculture: the example of winter wheat in Austria. Euphytica 163: 469–480.

94. Kirk AP, Fox SL, Entz MH (2012) Comparison of organic and conventional selection environments for spring wheat. Plant Breeding 131: 687–694.

2

Do We Produce Enough Fruits and Vegetables to Meet Global Health Need?

Karen R. Siegel[1,2]*, Mohammed K. Ali[2], Adithi Srinivasiah[3], Rachel A. Nugent[4], K. M. Venkat Narayan[1,2]

1 Nutrition and Health Sciences, Laney Graduate School, Emory University, Atlanta, Georgia, United States of America, **2** Hubert Department of Global Health, Emory University, Atlanta, Georgia, United States of America, **3** Emory College, Emory University, Atlanta, Georgia, United States of America, **4** Department of Global Health, University of Washington, Seattle, Washington, United States of America

Abstract

Background: Low fruit and vegetable (FV) intake is a leading risk factor for chronic disease globally, but much of the world's population does not consume the recommended servings of FV daily. It remains unknown whether global supply of FV is sufficient to meet current and growing population needs. We sought to determine whether supply of FV is sufficient to meet current and growing population needs, globally and in individual countries.

Methods and Findings: We used global data on agricultural production and population size to compare supply of FV in 2009 with population need, globally and in individual countries. We found that the global supply of FV falls, on average, 22% short of population need according to nutrition recommendations (supply:need ratio: 0.78 [Range: 0.05–2.01]). This ratio varies widely by country income level, with a median supply:need ratio of 0.42 and 1.02 in low-income and high-income countries, respectively. A sensitivity analysis accounting for need-side food wastage showed similar insufficiency, to a slightly greater extent (global supply:need ratio: 0.66, varying from 0.37 [low-income countries] to 0.77 [high-income countries]). Using agricultural production and population projections, we also estimated supply and need for FV for 2025 and 2050. Assuming medium fertility and projected growth in agricultural production, the global supply:need ratio for FV increases slightly to 0.81 by 2025 and to 0.88 by 2050, with similar patterns seen across country income levels. In a sensitivity analysis assuming no change from current levels of FV production, the global supply:need ratio for FV decreases to 0.66 by 2025 and to 0.57 by 2050.

Conclusion: The global nutrition and agricultural communities need to find innovative ways to increase FV production and consumption to meet population health needs, particularly in low-income countries.

Editor: Wagner L. Araujo, Universidade Federal de Vicosa, Brazil

Funding: There are no current funding sources for this study.

Competing Interests: The authors have declared that no competing interests exist.

* Email: krsiege@emory.edu

Introduction

Low fruit and vegetable (FV) intake is a leading risk factor for death and disability globally, estimated to contribute to approximately 16.0 million disability-adjusted life years and 1.7 million deaths worldwide annually [1]. According to a World Health Organization report, current global dietary guidelines recommend that individuals consume at least 5 servings of FV daily [2]. Recent cross-country evidence supports this recommendation, showing a strong dose-response relationship between higher FV consumption and lower all-cause mortality [3] as well as lower risk of major chronic diseases such as cardiovascular disease, diabetes, and certain cancers, which impact every region of the world [4–6].

Much of the world's population, however, does not consume the recommended five servings of FV daily. Data from 52 mainly low- and middle-income countries participating in the 2002–2003 World Health Survey reported that, overall, 77.6% of men and 78.4% of women surveyed consumed less than the recommended five daily servings of FV. The survey also showed that FV consumption patterns vary around the world, but lower-than-recommended reported consumption is common in high, middle, and low-income countries. For example, in a recent report, poor dietary habits, which includes low FV consumption, was *the* leading risk factor in the United States (U.S.), accounting for 26% of all deaths and 14% of all disability [7], and increasing individual FV consumption to up to 600 grams per day (slightly more than 5 servings per day) could reduce the total worldwide burden of disease by 1.8%, and reduce the burden of ischemic heart disease and ischemic stroke by 31% and 19% respectively [2].

Despite a wealth of research on behavioral determinants of FV, it remains unknown whether global production and supply of FV is actually sufficient to meet population needs. We used global population and agriculture databases to compare the global supply of ("supply") with recommended dietary intake (implied "demand", hereafter referred to as "need") globally and in individual countries. Using agricultural production and population projections data, we also project supply and need for FV for 2025 and 2050.

Methods

Data Sources

We used three main data sources for our analysis: (1) Food and Agricultural Organization (FAO) 2009 Food Balance Sheets [8], (2) age-specific FV intake recommendations for individuals [2], and (3) the United Nations (UN) World Population Prospects: The 2012 Revision [9].

The FAO 2009 Food Balance Sheets (the most recent year for which these data were available) report FV (excluding wine) supply by individual country for over 175 countries. These data are calculated by taking into account production, imports and exports, and food losses (through storage, transport, and processing; feed to livestock; or use as seeds and non-dietary purposes). The data reflect "formal" food production, and do not capture FV production from subsistence farming and production, which may not enter formal economies. For the FAO Food Balance Sheets, this estimated national food supply is divided by population size estimates to derive the reported per capita supply of FV (in kg/person/year).

For FV recommendations, we used a World Health Organization (WHO) report on the quantitative comparison of different health risks worldwide [2]. The report cited previously calculated and validated estimates for the average annual weight of the 5 recommended servings of FV per day: 330 grams per day for individuals aged 0–4 years, 480 grams per day for individuals aged 5–14 years, and 600 grams per day for all individuals aged 15 years and older. We converted these data into kilograms.

The UN World Population Prospects: 2012 Revision (the most recent version) provides country-level population estimates, in terms of the total population size as well as the proportion of each country's population by age. Calculations are done yearly using data classified by broad age groups (0–14 years, 15+ years) and for five-year periods (the latest years being 2005 and 2010) using data classified by more specific age groups, including 0–4 years, 5–14 years, and 15 years and older. To align our population estimates with age-specific FV recommendations, we used population estimates from 2010. This data source also provides population projections based on different scenarios for changing fertility levels for the period 2010–2100 for individual countries and globally.

Data Analysis

To calculate "supply" (in kg/year), we multiplied the FAO per-capita estimates by total population estimates for each country from the UN. The equation for supply is:

$$Supply = \frac{FV(kg)}{person} * population$$

To calculate "need" (assuming all individuals are able to meet their daily recommended intake of FV – "perfect need"), we multiplied the UN's age-specific population estimates by recommendations for FV servings per day for the same age-specific groups. Total country-specific population need (in kg/year) was then calculated by summing the recommended FV weights for all three age categories. The equation for "need" is:

$$Need = \left[popn(0-4 \; yrs) * \frac{0.33 \; kg}{persons(0-4 \; yrs)} \right]$$
$$+ \left[popn(5-14 \; yrs) * \frac{0.48 \; kg}{persons(5-14 \; yrs)} \right]$$
$$+ \left[popn(15+ \; yrs) * \frac{0.60 \; kg}{persons(15+ \; yrs)} \right]$$

Finally, we calculated a supply:need ratio by dividing supply by need, both expressed in kg/year, where a value greater than 1.0 signifies surplus, a value of 1.0 implies balance, and less than 1.0 signifies deficit. Supply, need, and supply:need ratios were calculated for each individual country and globally. We also calculated averages of these supply, need, and supply:need ratio indicators across varying country income levels, defined according to World Bank categories: low-income economies (per capita Gross Domestic Product [GDP] of $1,025 or less), lower-middle-income economies (per capita GDP of $1,026 to $4,035), upper-middle-income economies (per capita GDP of $4,036 to $12,475), and high-income economies (per capita GDP of $12,476 or more).

For the projections for 2025 and 2050, we calculated changes in production ("supply") using agricultural production growth rates to 2030 (1.6% for developing and 0.7% for developed countries) and 2050 (0.9% for developing and 0.3% for developed countries) as estimated by the FAO [10]. Similar to our calculations for current need, we calculated projected need by multiplying age-specific population projections for 2025 and 2050 by recommendations for FV servings per day for the same age-specific groups and summing across all three groups. For this projections analysis, we assumed a medium variant fertility scenario (2–3 children per woman).

All calculations were performed in Excel and data analysis was performed using Statistical Analysis Software (SAS) version 9.3. We used ArcMAP to illustrate the data geographically.

Sensitivity Analyses

To account for need-side food wastage at the household/individual level, we performed a sensitivity analysis to adjust these estimates to account for wastage of 33% in high-income regions/countries and 15% in low- to middle-income regions/countries [11]. For the projections, we also performed a sensitivity analysis in order to account for "best-case" (low fertility, or <2.1 children per woman) and "worst-case" (high fertility, or >5 children per woman) scenarios [9]. In addition to the main projections analysis, we also performed a sensitivity analysis assuming current levels of agricultural production.

Results

Table 1 shows descriptive statistics, overall and by country income level, for all countries for which all data were available (n = 170). Overall, the global supply (not including subsistence production that may not enter formal economies) of available FV falls 22% short of population's need according to nutritional recommendations, and as much as 95% short in some countries (overall supply:need ratio: 0.78 [range: 0.05–2.01]). This ratio varies widely by country income level, with a median supply:need ratio of 0.42 in low-income countries and a median supply:need ratio of 1.02 in high-income countries (Table 1). In a sensitivity analysis in which we accounted for need-side food wastage, similarly insufficient FV supplies were noted, to a slightly greater extent. The global supply:need ratio was 0.66 when need-side

Table 1. Descriptive Statistics of Fruit and Vegetable Supply, Need, and Supply: Need Ratio, Overall and by Country Income Level.

	n	Supply	Need	Supply:Need Ratio
Full Sample, all countries	170	1.15 (0.01–524.25)	1.90 (0.02–282.50)	0.78 (0.05–2.01)
Low Income	34	0.97 (0.05–7.50)	2.36 (0.13–30.18)	0.42 (0.05–0.99)
Lower-middle Income	43	1.01 (0.01–142.51)	1.49 (0.02–241.62)	0.63 (0.19–1.72)
Upper-middle Income	50	1.52 (0.01–524.25)	1.71 (0.02–282.50)	0.87 (0.24–2.01)
High Income	43	1.60 (0.04–71.63)	1.64 (0.05–64.59)	1.02 (0.55–1.86)

Notes: All numbers provided as median (range). Supply and Need are reported in billions of kilograms of fruits and vegetables. Country Income Level defined according to World Bank categories: Low-income economies ($1,025 or less), Lower-middle-income economies ($1,026 to $4,035), Upper-middle-income economies ($4,036 to $12,475), High-income economies ($12,476 or more).

wastage was accounted for, and this varied from 0.37 (low-income countries) to 0.77 (high-income countries) (see Table S1 for results by country and Table S2 for results across country income level).

The supply:need ratio also varies widely by geographical region. The highest ratios of greater than 1.0 (indicating more than sufficient supply to meet the population's needs) are seen in the Mediterranean/North African countries of Montenegro (supply:need ratio 2.01), Greece (1.86), Turkey (1.78), Egypt (1.72), Libya (1.67), Tunisia (1.52), Italy (1.50), and Portugal (1.48); Middle Eastern countries of Iran (1.78), Israel (1.56), and Lebanon (1.46); Caribbean countries of Bahamas (1.61) and Belize (1.50); Albania (1.59); and China (1.86). The countries with the greatest shortage, where need is far greater than supply, are primarily African countries such as Eritrea (0.05), Chad (0.09), Burkina Faso (0.10), Mozambique (0.12), Ethiopia (0.12).

Table 2 shows projected supply, need and supply:need ratios overall and by country income level, for all countries for which all data was available (n = 169). Assuming medium fertility (2–3 children per woman) and projected agricultural production growth rates, the global supply:need ratio for FV increases slightly to 0.81 by 2025 and to 0.88 by 2050. As with current, the projected supply:need ratio in 2025 and 2050 varies by country income level. The lowest ratio is seen in low-income countries, where it dips to 0.30 in 2050, assuming medium fertility. The projected supply:need ratio is higher in high income countries, where it ranges from an estimated 0.98 to 1.21. In a sensitivity analysis using current levels of FV production (ie, assuming no increase in production), the global supply:need ratio for FV decreases to 0.66 by 2025 and to 0.57 by 2050 assuming medium fertility (2–3 children per woman). As with current, the projected supply:need ratio in 2025 and 2050 varies by country income level, with the lowest ratio of 0.18 in low-income countries by 2050 and the highest ratio of 0.99 in high income countries (see Table S3).

Figure 1 illustrates current and projected supply:need ratios, highlighting the growing gap between supply and need in low-income countries over time.

Discussion

Within the formal agricultural sector, there is an estimated 22% supply gap in meeting current need for FV (34% when considering food wastage at the household/individual level), and this varies from 58% to 13% across low- and upper-middle income countries. High income countries appear to have sufficient supply (supply:need ratio is 1.02). Furthermore, these gaps between high/middle-income and low-income countries will worsen with time. Assuming medium fertility and projected increases in production of FV, the global supply:need ratio for FV increases slightly to 0.81 by 2025 and to 0.88 by 2050, but divergence occurs whereby we estimated

a supply gap of 70% and 65% in low-income countries by 2025 and 2050, respectively, while middle- and high-income countries approach a supply:need of 1.0, implying balance of supply and need. Without the projected increase in FV production, however, the global supply:need ratio could decrease to 0.66 by 2025 and to 0.57 by 2050, dipping as low as 0.18 in low-income countries.

There may be several reasons for these findings. Supply-side factors include subsidies and distribution systems for supply, and international trade for addressing imbalances in supply:need ratios across countries and country-income levels [12]. Many countries provide producer-end subsidies for grain crops and meat/dairy, incentivizing farmers to grow these items while dis-incentivizing FV production. In the U.S., the commodity crops receiving the largest amount of agricultural subsidies are grains, livestock, and dairy and under current agricultural policy, farmers are penalized for growing "specialty crops" (FV) if they have received federal farm payments to grow other crops [13,14]. As a result, grains, meat, and dairy are abundant [15], the supply of FV, at least in the US, is insufficient to meet population needs [16]. In low-income countries, where we found FV need to be greatest, the lack of adequate distribution systems may lead to supply-side wastage and disincentives for their production. This is an issue particularly in warm climates like India and Africa, where FV are prone to spoiling before reaching their market destinations [17].

In particular, international trade (and climates ideal for growing FV) could help explain the differences in findings across country-income groups and geographical regions. International trade in FV, which since the 1980s has expanded more rapidly than other agricultural commodities and was 17% of total agricultural trade in 2001, is also an important consideration for increasing supply of FV, particularly in countries where production may be high but supply low due to exports [18]. Climates ideal for growing FV is also a very important supply-side factor when considering FV production. As noted in the results section, there appear to be varying levels of agronomical potentials of countries located in different geographical regions, as highlighted by the large geographical variations in the supply:need ratio, with high ratios seen in many Mediterranean countries. For example, it is known that Mediterranean countries are great producers of fruits for the fresh market due to climatic conditions – drip irrigation combined with dry summers is a perfect scenario for producing high quality crops (although a substantial proportion of this production is exported to other countries).

On the need side of the equation, population size – and relatively large projected increases, particularly in certain low-income countries – helps to explain the large and growing gaps between supply and need in these countries. The projections data show that, assuming an estimated increase in FV production, the

Table 2. Projected Need and Supply:Need Ratios, Overall and by Country Income Level.

	n	2025			2050		
		Supply	Need	Supply:Need Ratio	Supply	Need	Supply:Need Ratio
Full Sample, all countries	169	1.45 (0.02–675.83)			1.79 (0.02–875.25)		
High fertility			2.21 (0.02–310.96)	0.79 (0.04–2.52)		2.74 (0.02–380.34)	0.78 (0.03–3.25)
Medium fertility			2.16 (0.02–302.40)	0.81 (0.04–2.59)		2.48 (0.02–335.52)	0.88 (0.03–3.69)
Low fertility			2.10 (0.02–293.83)	0.84 (0.04–2.67)		2.23 (0.02–293.93)	1.00 (0.03–4.21)
Low Income	34	1.20 (0.07–9.67)			1.55 (0.09–12.52)		
High fertility			3.65 (0.19–37.53)	0.34 (0.04–1.15)		5.89 (0.33–48.38)	0.26 (0.03–1.30)
Medium fertility			3.55 (0.19–36.28)	0.35 (0.04–1.18)		5.28 (0.30–42.11)	0.30 (0.03–1.47)
Low fertility			3.45 (0.18–35.03)	0.36 (0.04–1.22)		4.70 (0.27–36.43)	0.33 (0.03–1.68)
Lower-middle Income	42	1.32 (0.04–183.72)			1.71 (0.06–237.93)		
High fertility			2.35 (0.04–297.58)	0.62 (0.16–1.90)		3.49 (0.05–380.34)	0.58 (0.09–2.34)
Medium fertility			2.28 (0.04–288.77)	0.64 (0.16–1.95)		3.08 (0.05–335.52)	0.66 (0.10–2.65)
Low fertility			2.21 (0.04–279.95)	0.66 (0.17–2.02)		2.70 (0.04–293.93)	0.75 (0.11–3.01)
Upper-middle Income	50	1.96 (0.02–675.83)			2.54 (0.02–875.25)		
High fertility			1.85 (0.02–310.96)	0.94 (0.19–2.52)		1.86 (0.02–327.36)	1.03 (0.12–3.25)
Medium fertility			1.79 (0.02–302.40)	0.97 (0.19–2.59)		1.64 (0.02–290.93)	1.16 (0.14–3.69)
Low fertility			1.74 (0.02–293.83)	1.00 (0.20–2.67)		1.44 (0.02–257.35)	1.33 (0.15–4.21)
High Income	43	1.79 (0.04–80.09)			1.97 (0.04–88.05)		
High fertility			1.91 (0.06–74.60)	1.04 (0.59–2.04)		2.17 (0.07–92.40)	0.98 (0.52–2.15)
Medium fertility			1.86 (0.06–72.69)	1.06 (0.61–2.09)		1.96 (0.07–83.32)	1.08 (0.58–2.38)
Low fertility			1.81 (0.06–70.79)	1.09 (0.63–2.14)		1.76 (0.06–74.67)	1.21 (0.65–2.65)

Notes: All numbers provided as median (range). Need is reported in billions of kilograms of fruits and vegetables. Country Income Level defined according to World Bank categories: Low-income economies ($1,025 or less), Lower-middle-income economies ($1,026 to $4,035), Upper-middle-income economies ($4,036 to $12,475), High-income economies ($12,476 or more). Fertility is defined according to the United Nations World Population Prospects, 2012 Revision: high fertility (more than 5 children per woman), medium fertility (2–3 children per woman), and low fertility (less than 2.1 children per woman).

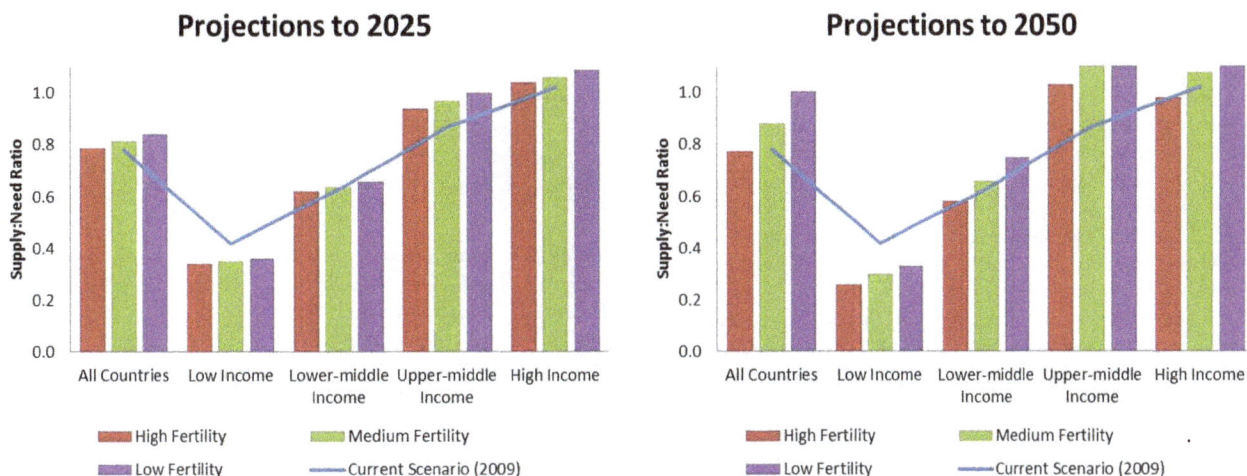

Figure 1. Projected Supply: Need Ratio, 2025 and 2050. Notes: Country Income Level defined according to World Bank categories: Low-income economies ($1,025 or less), Lower-middle-income economies ($1,026 to $4,035), Upper-middle-income economies ($4,036 to $12,475), High-income economies ($12,476 or more). Fertility is defined according to the United Nations World Population Prospects, 2012 Revision: high fertility (5 or more children per woman), medium fertility (2–3 children per woman), and low fertility (less than 2.1 children per woman).

supply:need ratio narrows on a global scale, but that it widens to a considerable extent in low-income countries, primarily as a reflection of higher fertility in these countries and agricultural production growth rates that cannot keep up with population growth. The ability to produce enough FV to meet the needs of large and growing populations, coupled with the supply-side limitations mentioned above, are of particular concern for these countries. In the 18th century, Malthus projected that human population growth would outpace expansion in food production. Since then, with the help of technological advances spurred by the Green Revolution, production and subsequent supply of carbo-hydrates and grains has increased to meet global population needs. Our projections analysis suggests that high-income countries may be making strides towards increasing production and subsequent supply of FV to meet their population's needs, but that the same cannot be said for the low-income countries, at least within the formal agricultural economy, where the gap in supply not taking subsistence farming into account could widen to 65% by 2050 if not addressed. Of greater concern, if projected increases in agricultural production of FV do not manifest, by 2025 and 2050 high- and low-income countries alike may not able to meet their population's needs for FV.

While ecological data suggests that food availability can influence food consumption patterns and in turn, cardiometabolic health outcomes like diabetes [19,20], to date there has been a relatively limited focus on production and supply of FV. Researchers at America's Farmland Trust investigated supply of FV in the United States (U.S.) alone; they concluded that an estimated 13 million more acres of farmland would be needed to produce a sufficient supply for the U.S. population [21]. Our analysis builds upon these results. The first study to incorporate empirical country-level data and age-specific recommendations for FV consumption to examine global and country-specific FV supply (in the formal sector) as it compares to need, our study highlights inadequate supply of FV as it compares to the population's nutritional needs, from the perspective of preventing chronic diseases, which currently place enormous burdens on countries around the world and are largely preventable through healthy diet and higher FV consumption [2,7].

These findings must be contextualized by limitations to our analysis. First, the data used were macro-level indicators collected at the country-level and may be prone to either over- or under-estimation. The data do not account for how much people actually access FV in various countries nor the quality and diversity of FV consumption, including how these FV are consumed (raw, cooked, or processed FV have different nutrient bioavailability), nor how much *individuals* actually consume. For example, many Medi-terranean and Caribbean countries, which were found to have high supply:need ratios, are great citrus producers, but in the latter fruits are processed (for juice) and not sold on the fresh market. Additionally, every fruit and vegetable does not have the same macro- and micro-nutrient content, and even the same fruit or vegetable grown in a different climate or soil may have differing amounts of macro- and micro-nutrients. Additionally, there may also be differences in the quality and validity of the data in high-versus low-income countries. However, the FAO Food Balance Sheets are the most commonly used source of food availability information at the national level, providing standardized estimates of the average amount of food available per person on a daily basis and a useful tool for international comparisons [22]. Second, our analysis is at the country level, and therefore does not take into account urban/rural differences in supply that may result from challenges in distribution (for example, transporting FV from the farm to urban areas. This may be a particular issue in resource-poor settings, where distributional infrastructure may be lacking. Further analyses could investigate these issues, analyzing potential heterogeneity of supply and need within countries and in urban versus rural settings.

A third limitation is that our analysis does not capture local food economies (ie, subsistence farming and food production) in individual countries. That is, it does not take into account the production of FV that may exist outside of the formal agricultural sector (i.e., home gardens), which may vary widely across countries. This may be an additional area of future research. For example, researchers could utilize the powerful technologies of Google Earth to look within countries, at the regional, city, district, or even household level, at the presence or absence of informal community or household gardens. Lastly, our analysis does not incorporate additional economic indicators such as the costs of production or the resulting prices of FV. Our results suggest that insufficient supply exists relative to population needs under current production conditions. We have not taken into account the potential for supply to increase due to technological improvements and supportive government policies. Both those factors could lower FV prices and increase consumption.

Our study adds unique value by underlining the importance of increasing supply of FV and sets the stage for further analyses to delve further into the policy levers for increasing production and supply. In particular, investigating the supply of FV resulting from subsistence farming could augment our analysis. At the same time, continuing efforts to improve demand for FV – for example, through public health education and health promotion programs, proposing taxes on foods of low nutritional value (e.g., soda, high-fat foods) or subsidies on foods of high nutrition value (e.g., FV), improved food labeling, and stricter controls on the marketing of foods [23–27] – is equally important. Without an accompanying increase in supply, however, these efforts may have limited reach. It is hoped that our straightforward analysis, highlighting inadequate formal supply of FV in the context of perfect need (assuming all individuals are able to meet their daily recommended intake of FV), may provide value by offering an understanding of the current and future global disconnect between nutritional recommendations and supply of FV, and guide conversations and future investigations to consider appropriate policy responses. The triumph of grains production over the doom and gloom forecast of Malthus is a major testament to the technological and organizational success of food production and distribution worldwide that has accompanied industrialization and modern development. The current state of affairs presents a challenge to the global nutrition and agricultural communities to increase FV production in the same way, especially in low-income countries. Change is possible.

Supporting Information

Table S1 List of Countries and Their Respective Supply, Need, and Supply:Need Ratios. Notes: All numbers provided as median (range). Need is reported in billions of kilograms of fruits and vegetables. Country Income Level defined according to World Bank categories: Low-income economies

References

1. Lim SS, Vos T, Flaxman AD, Danaei G, Shibuya K, et al. (2012) A comparative risk assessment of burden of disease and injury attributable to 67 risk factors and risk factor clusters in 21 regions, 1990?2010: a systematic analysis for the Global Burden of Disease Study 2010. The Lancet 380: 2224–2260.
2. Lock K, Pomerleau J, Causer L, McKee M (2004) Low Fruit and Vegetable Consumption. In: Ezzati M, Lopez AD, Rodgers A, Murray CJL, editors. Comparative Quantification of Health Risks: Global and Regional Burden of Diseases Attributable to Selected Major Risk Factors. Geneva: World Health Organization.
3. Bellavia A, Larsson SC, Bottai M, Wolk A, Orsini N (2013) Fruit and vegetable consumption and all-cause mortality: a dose-response analysis. Am J Clin Nutr 98: 454–459.
4. FAO/WHO (2003) Diet, nutrition and the prevention of Chronic Diseases. Geneva: Food and Agricultural Organization, World Health Organization.
5. Hung HC, Joshipura KJ, Jiang R, Hu FB, Hunter D, et al. (2004) Fruit and vegetable intake and risk of major chronic disease. J Natl Cancer Inst 96: 1577–1584.
6. Dauchet L, Amouyel P, Hercberg S, Dallongeville J (2006) Fruit and vegetable consumption and risk of coronary heart disease: a meta-analysis of cohort studies. J Nutr 136: 2588–2593.
7. (2013) The state of US health, 1990–2010: burden of diseases, injuries, and risk factors. JAMA 310: 591–608.
8. FAO (2009) Food and Agricultural Organization Food Balance Sheets. FAO.
9. UNDESA (2012) World Population Prospects: The 2012 Revision. United Nations, Department of Economic and Social Affairs: Population Division, Population Estimates and Projections Section.

($1,025 or less), Lower-middle-income economies ($1,026 to $4,035), Upper-middle-income economies ($4,036 to $12,475), High-income economies ($12,476 or more). Fertility is defined according to the United Nations World Population Prospects, 2012 Revision: high fertility (more than 5 children per woman), medium fertility (2–3 children per woman), and low fertility (less than 2.1 children per woman).

Table S2 Sensitivity Analysis of Fruit and Vegetable Supply, Need, and Supply:Need Ratio, Overall and by Country Income Level. Notes: All numbers provided as median (range). Supply and Need are reported in billions of kilograms of fruits and vegetables. Country Income Level defined according to World Bank categories: Low-income economies ($1,025 or less), Lower-middle-income economies ($1,026 to $4,035), Upper-middle-income economies ($4,036 to $12,475), High-income economies ($12,476 or more).

Table S3 Sensitivity Analysis of Projected Need and Supply:Need Ratios (Assuming Current Levels of Agricultural Production), Overall and by Country Income Level. Notes: All numbers provided as median (range). Need is reported in billions of kilograms of fruits and vegetables. Country Income Level defined according to World Bank categories: Low-income economies ($1,025 or less), Lower-middle-income economies ($1,026 to $4,035), Upper-middle-income economies ($4,036 to $12,475), High-income economies ($12,476 or more). Fertility is defined according to the United Nations World Population Prospects, 2012 Revision: high fertility (more than 5 children per woman), medium fertility (2–3 children per woman), and low fertility (less than 2.1 children per woman).

Author Contributions

Conceived and designed the experiments: KRS KMN. Performed the experiments: KRS AS. Analyzed the data: KRS MKA KMN RAN. Contributed reagents/materials/analysis tools: KRS AS. Wrote the paper: KRS.

10. Alexandratos N, Bruinsma J (2012) World Agriculture Towards 2030/2050: The 2012 Revision. Rome: Global Perspective Studies Team, FAO Agricultural Development Economics Division.
11. Joffe M, Robertson A (2001) The potential contribution of increased vegetable and fruit consumption to health gain in the European Union. Public Health Nutr 4: 893–901.
12. Nugent R (2011) Bringing Agriculture to the Table: How Agriculture and Food Can Play a Role in Preventing Chronic Disease. Chicago, IL: The Chicago Council on Global Affairs.
13. Jackson RJ, Minjares R, Naumoff KS, Shrimali BP, Martin LK (2009) Agriculture Policy Is Health Polciy. Journal of Hunger & Environmental Nutrition 4: 393–408.
14. Franck C, Grandi SM, Eisenberg MJ (2013) Agricultural subsidies and the american obesity epidemic. Am J Prev Med 45: 327–333.
15. Pollan M (2003) The (Agri)Cultural Contradictions of Obesity. The New York Times Magazine. New York City: The New York Times.
16. AFT (2010) 13 Milion More Acres.
17. Industry PMsCoTa (2010) Report on Food & Agro Industries Management Policy.
18. Wu Huang S (2004) Global Trade Patterns in Fruits and Vegetables. Washington, DC: Economic Research Service, United States Department of Agriculture.
19. Siegel KR, Echouffo-Tcheugui JB, Ali MK, Mehta NK, Narayan KM, et al. (2012) Societal correlates of diabetes prevalence: An analysis across 94 countries. Diabetes Research and Clinical Practice 96: 76–83.
20. Basu S, Yoffe P, Hills N, Lustig RH (2013) The relationship of sugar to population-level diabetes prevalence: an econometric analysis of repeated cross-sectional data. PLoS One 8: e57873.

21. AFT (2010) American Farmland Trust Says The United States Needs 13 Million More Acres of Fruits and Vegetables to Meet the RDA. Washington, DC: American Farmland Trust.

22. Sekula W, Becker W, Trichopoulou A, Zajkas G (1991) Comparison of dietary data from different sources: some examples. WHO Reg Publ Eur Ser 34: 91–117.

23. Pomerleau J, Lock K, Knai C, McKee M (2005) Interventions designed to increase adult fruit and vegetable intake can be effective: a systematic review of the literature. J Nutr 135: 2486–2495.

24. Wolfenden L, Wyse RJ, Britton BI, Campbell KJ, Hodder RK, et al. (2012) Interventions for increasing fruit and vegetable consumption in children aged 5 years and under. Cochrane Database Syst Rev 11: CD008552.

25. WHO/FAO (2003) Diet, nutrition and the prevention of chronic diseases. Geneva: World Health Organization and Food and Agricultural Organization.

26. Thow AM, Jan S, Leeder S, Swinburn B (2010) The effect of fiscal policy on diet, obesity and chronic disease: a systematic review. Bull World Health Organ 88: 609–614.

27. CDC (2011) Strategies to Prevent Obesity and Other Chronic Diseases: The CDC Guide to Strategies to Increase the Consumption of Fruits and Vegetables. Atlanta: U.S. Department of Health and Human Services.

3

Persistence and Dissipation of Chlorpyrifos in Brassica Chinensis, Lettuce, Celery, Asparagus Lettuce, Eggplant, and Pepper in a Greenhouse

Meng-Xiao Lu[1,2], Wayne W. Jiang[3], Jia-Lei Wang[1], Qiu Jian[4], Yan Shen[2], Xian-Jin Liu[1,2], Xiang-Yang Yu[1,2]*

1 Pesticide Biology and Ecology Research Center, Nanjing, Jiangsu, China, 2 Key Laboratory of Food Safety Monitoring and Management of Ministry of Agriculture, Nanjing, Jiangsu, China, 3 Department of Entomology, Michigan State University, East Lansing, Michigan, United States of America, 4 Institute for the Control of Agrochemicals, Ministry of Agriculture, Beijing, China

Abstract

The residue behavior of chlorpyrifos, which is one of the extensively used insecticides all around the world, in six vegetable crops was assessed under greenhouse conditions. Each of the vegetables was subjected to a foliar treatment with chlorpyrifos. Two analytical methods were developed using gas chromatography equipped with a micro-ECD detector (LOQ $= 0.05$ mg kg^{-1}) and liquid chromatography with a tandem mass spectrometry (LOQ $= 0.01$ mg kg^{-1}). The initial foliar deposited concentration of chlorpyrifos (mg kg^{-1}) on the six vegetables followed the increasing order of brassica chinensis<lettuce<celery<asparagus lettuce<eggplant <pepper. The initial deposition of chlorpyrifos showed differences among the six selected vegetable plants, ranging from 16.5 ± 0.9 mg kg^{-1} (brassica chinensis) to 74.0 ± 5.9 mg kg^{-1} (pepper plant). At pre-harvest interval 21 days, the chlorpyrifos residues in edible parts of the crops were <0.01 (eggplant fruit), < 0.01 (pepper fruit), 0.56 (lettuce), 0.97 (brassica chinensis), 1.47 (asparagus lettuce), and 3.50 mg kg^{-1} (celery), respectively. The half-lives of chlorpyrifos were found to be 7.79 (soil), 2.64 (pepper plants), 3.90 (asparagus lettuce), 3.92 (lettuce), 5.81 (brassica chinensis), 3.00 (eggplant plant), and 5.45 days (celery), respectively. The dissipation of chlorpyrifos in soil and the six selected plants was different, indicating that the persistence of chlorpyrifos residues strongly depends upon leaf characteristics of the selected vegetables.

Editor: Youjun Zhang, Institute of Vegetables and Flowers, Chinese Academy of Agricultural Science, China

Funding: This work was financially supported by the Independent Innovation Fund of Agricultural Sciences in Jiangsu Province (cx (12) 3090) and the Natural Science Foundation of China (31071719). The funders had no role in study design, data collection and analysis, decision to publish, or preparation of the manuscript.

Competing Interests: The authors have declared that no competing interests exist.

* Email: yu98190@gmail.com

Introduction

Chlorpyrifos [O,O-diethyl O-(3,5,6-trichloro-2-pyridyl) phosphorothioate] is an organophosphorous insecticide, acaricide, and nematicide used to control a broad spectrum of foliage and soil-born insect pests on a variety of food and feed crops [1–2]. It is ranked as one of the most extensively used insecticides all around the world. In China, since the use of several highly-toxic organophosphorous insecticides was banned in 2006, chlorpyrifos has been recommended as one of the alternative insecticides and broadly used in agriculture. Extensive use of chlorpyrifos has led to a potential risk of residues in various crops. Chlorpyrifos is of great environmental concerns due to its widespread use in the past several decades and its potential toxic effects on human health. Thus, the degradation study of chlorpyrifos has become increasing important in recent years [3–5]. In a market monitoring study conducted between 2007 and 2010, chlorpyrifos was detected in approximately 22.8% of 2082 samples of 17 vegetable commodities collected from Zhejiang Province, China with a highest residue of 3.47 mg kg-1. The residue levels in 1.4% of vegetable samples were found to be higher than the maximum residue limits (MRLs) of China [3].

Although most pesticides are effective to control pests in agricultural industry, inappropriate uses of pesticides may lead to public concerns on food safety and human health [5–10], environmental contamination [11–12], insect resistance and resurgence [12–14], etc. In China, the use of agrochemicals is critical to provide food supplies for its growing population with its limited arable land. Ideally, control the harmful organism efficiently, having no or minimum pesticide residues in the harvested crops, or at least lower than the statutory MRLs [3,15]. Therefore, the field dissipation studies of pesticide persistence in foods and pesticide residue behavior is of particular importance in order to find out which pesticide application strategies are efficient to control insect pests while leaving minimum residues [16,17]. There are many factors that influence the dissipation behavior of pesticides in plants, including the climate conditions (temperature, humidity, light intensity, etc) [18], the crop species [19–22] the nature of the chemicals, the formulations, and the application methods [23–24]. Due to the difference in the extension of the foliar, which resulted in the different initial pesticide deposition, and/or the difference in the metabolism system of different crops, dissipation of a pesticide on various crops may be markedly different [25–26]. A dissipation study of pesticides on leafy vegetables showed that among the tested vegetables, spinach and amaranth could incur higher pesticide deposition and that half-lives ($t_{1/2}$ = 1.37–5.17 days) of chlorpyrifos were from on different leafy vegetables [22]. For chiral pesticide malathion, the calculated $t_{1/2}$ values of the enantiomers were relatively short, ranging from 0.83 to 1.43 days in five plants. The degradation of the two enantiomers in Chinese cabbage (brassica chinesis), rape, and sugar beet was highly selective, while non-enantioselectivity was found in paddy rice and wheat [20–21]. For better understanding the possible residue risk of a pesticide, dissipation studies for different crop species in the specific growing conditions are necessary to test if the established application strategies are suitable.

It is concluded that agrochemicals enter into plants via two major pathways, which are either via foliar treatment - foliar deposition followed by entering into the inner parts of the crops, or via soil treatment - root uptake from the soil [27–28]. For the foliar pesticide applications, the agrochemicals are deposited directly on foliar surfaces of the crops and the excess agrochemicals will precipitate in the soil. The resulting deposition fractions are determined predominantly by crop species, growth stage of crop, pesticide formulation, and spraying technology [29]. Generally, optimization of chlorpyrifos uses is foliar application with minimizing losses of the applied pesticides from plants to soil. Spray loss of chlorpyrifos may lead to the soil environment pollution and extend its residue duration in plants. The chlorpyrifos residues were found to persist in soils, as the half-lives ranged generally between 50 and 120 days [30–32]. It was reported in the literature that the chlorpyrifos residues were found in soils for over one year after the applications. Pesticide persistence in soils may depend on the formulation, rate of application, soil type, climate, and other conditions [33–37]. Plant rhizosphere plays an important role in the degradation of pesticides in soils [21,36].

In the present study, six vegetables were selected, including two fruit vegetables (pepper and eggplant) and 4 whole plant edibles leafy vegetables (brassica chinensis, lettuce, celery, and asparagus lettuce). They were treated with chlorpyrifos by foliar application under the controlled conditions in a greenhouse. The initial deposition of chlorpyrifos on the crop foliar was analyzed by GC

and on the fruits by LC/MS/MS and the dynamics of pesticide residues in the plants and rhizosphere soils were to be monitored. The work was to evaluate the persistence and dissipation behavior of chlorpyrifos in different vegetables and soil. The results would help to provide understanding of the residue characteristics of chlorpyrifos in vegetables, and help to guide proper and safe use of pesticides on vegetables to ensure food safety.

Materials and Methods

Instruments and Reagents

Gas chromatography was an Agilent GC 7890 (Agilent Technologies, Santa Clara, CA, USA) equipped with a micro-ECD detector and an analytical column HP-5ms J&W Ultra Inert capillary column (30 m length×0.25 mm I.D.×0.25 μm film thickness, Agilent Technologies, USA). LC/MS/MS (Agilent Technologies, USA) contained a 1200 SL HPLC system coupled to an Agilent G6410A triple quadrupole mass spectrometer. The column was an Agilent ZORBAX SB-C18 (2.1×150 mm, 5 μm) analytical column. Chlorpyrifos-EC 40% (Hubei Xian Long Chemical Co., LTD) was purchased from a local pesticide store. The analytical reference substance, chlorpyrifos (certified analytical standard, 99.7%) was purchased from National Standard Company (Tianjin, China). Two stock solutions (1000 mg L^{-1}) were prepared by dissolving the chlorpyrifos standard (100 mg) in 100 mL of acetone (for GC) and acetonitrile (for LC/MS/MS), respectively. Working solutions were prepared by diluting the stock solution or a working solution using the organic solvents (acetone for GC and acetonitrile for LC/MS/MS). Acetone, n-hexane, acetonitrile, sodium chloride, and anhydrous sodium sulfate were of analytical grades, purchased from Kermel Chemical Reagent Co., Ltd (Tianjin, China). The 0.22 μm SCAA-104 membranes and 500 mg florisil SPE cartridges were purchased from Anpel Scientific Instrument Co., Ltd (Shanghai, China).

The six selected vegetables were brassica chinensis (*Brassicachinensis L.*), lettuce (*Lactuca sativa* spp), pepper (*Capsicum annuum* spp), eggplant (*Solanum melongena* L), celery (*Apium graveolens*), and asparagus lettuce (*Asparagus Lettuce* spp). The pepper and eggplant seeds were grown in a nursery tray before being transplanted to the field at the 2–3 leaf stage seedlings, and the other four vegetables were directly sowed in the field. The row spacing and inter-plant spacing were set to be 30 cm for the vegetables, except for celery which was sowed in rows with row spacing of 30 cm.

Experiment design

Field experiments were conducted in a controlled environment in a greenhouse at the Experiment Station of Jiangsu Academy of Agricultural Science (JAAS), Nanjing, China from September to November 2013. JAAS permitted the study in its greenhouse and this study did not use protected area of land or sea, neither with relevant protected wildlife. Since the work was completed in the greenhouse, no specific permission was required. This study did not involve endangered or protected species. The soil was of sandy loam texture dried and its contents contained 30% of sand, 53% silt, 15% clay, and 2% organic matter. There were four 30 m^2 trial plots to be selected for each vegetable, i.e., three replicates and one control. Between the plots there was a buffer strip of 0.5 m in width. Chlorpyrifos-EC 40% was applied by foliar spraying at a rate of 0.97 kg a.i./ha on October 21, 2013. At the time of spraying, brassica chinensis, lettuce, and asparagus lettuce were at the stage when the leaves overspread; celery was 15–16 cm of height and the leaves scattered; pepper and eggplant were at the stages of 30–40 cm of height and the leaves of the adjacent plants touched each other and overlapped. In this study, the plants were

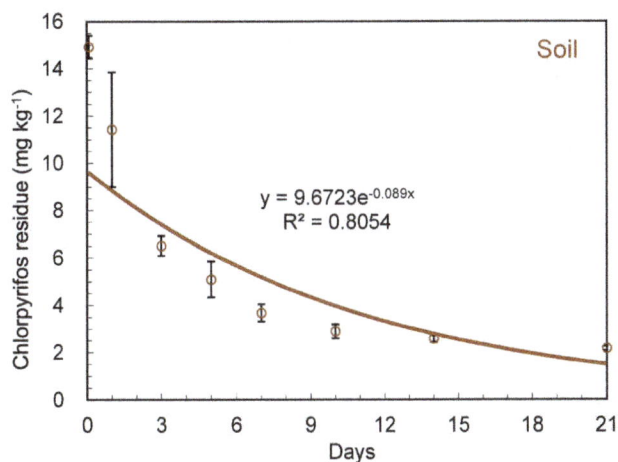

Figure 1. Dissipation dynamic of chlorpyrifos in soil.

Table 1. Dynamic equations, correlation coefficients and half-lives of chlorpyrifos in soil and six vegetable plants (foliar).

Matrix	Dynamic equation	Correlation coefficient (R^2)	Half-life (days)
Soil	$C_t = 9.672\ e^{-0.089t}$	0.8054	7.79
Brassica chinensis	$C_t = 11.74\ e^{-0.1192t}$	0.9618	5.81
Lettuce	$C_t = 17.30\ e^{-0.1769t}$	0.9359	3.92
Celery	$C_t = 40.62\ e^{-0.1271t}$	0.9648	5.45
Asparagus lettuce	$C_t = 31.97\ e^{-0.1775t}$	0.8850	3.90
Pepper	$C_t = 40.23\ e^{-0.2622t}$	0.9315	2.64
Eggplant	$C_t = 48.28\ e^{-0.2307t}$	0.9756	3.00

allowed to the leaves to overlap to cover most of the surface of soil and thus the loss of pesticides sprayed was minimum. The temperature inside the greenhouse was controlled between 16–26°C. Three representative samples of whole plant and rhizosphere soil were collected at 0 (2 h after application), 1, 3, 5, 7, 10, 14, and 21 days intervals after pesticides application. Pepper and eggplant fruit samples were collected on 0, 3, 7, 14, and 21 days.

Extraction and purification

All samples (plants and fruits) were homogenized using a Philips blender (Shanghai, China) and the ground samples were stored in a freezer (at −20°C) until analysis.

Plant samples. The extraction of chlorpyrifos residues from plants and fruits was carried out by the procedure as follows. Five (5.0) g of the homogenized sample was weighed into a 50 mL Teflon centrifuge tube. The extraction solvent (acetonitrile, 10 mL) was added. The samples were then mixed thoroughly for 1 min with a vortex mixer, followed by high-speed homogenizing for about 2 min. After addition of 2 g of sodium chloride, the samples were vortexed immediately for 1 min and centrifuged for 5 min at 5000 rpm. An aliquot of 1 mL of the supernatant was transferred into a 10 mL glass test tube, and then evaporated just

to dryness under a stream of nitrogen (40°C). The residue was dissolved in 1 mL of hexane and then subjected to Florisil SPE column clean-up. The SPE column was pre-conditioned by rinsing it with 5 mL of hexane. The extraction was added to the SPE column followed by eluting with 10 mL of a mixture containing acetone and n-hexane (9:1, v/v). The eluate was evaporated to dryness. The residues were redissolved with acetone to 1 mL. The final extract was filtered through a 0.2 μm SCAA-104 membrane followed by GC analysis.

Fruit samples. The method of sample extraction was a modification of a reference method [38]. Fruit sample (5.0 g) was weighted into a polypropylene centrifuge tube. An extraction solvent (10 mL of acetonitrile) was added. The sample was homogenized for 1 min using the homogenizer. The homogenizer probe was rinsed with a portion of 5 mL of the extraction solvent. The extracts were combined. The sample was centrifuged at 5000 rpm for 5 min. Transferred 2 mL of supernatant into a 15 mL centrifuge tube containing PSA (100 mg), ODS-C18 (100 mg) and florisil (100 mg). The sample was vortexed for 1 min and centrifuged at 5000 rpm for 2 min. After centrifugation, the supernatant was filtered using a 0.22 μm nylon filter into an autosampler vial for LC/MS/MS analysis.

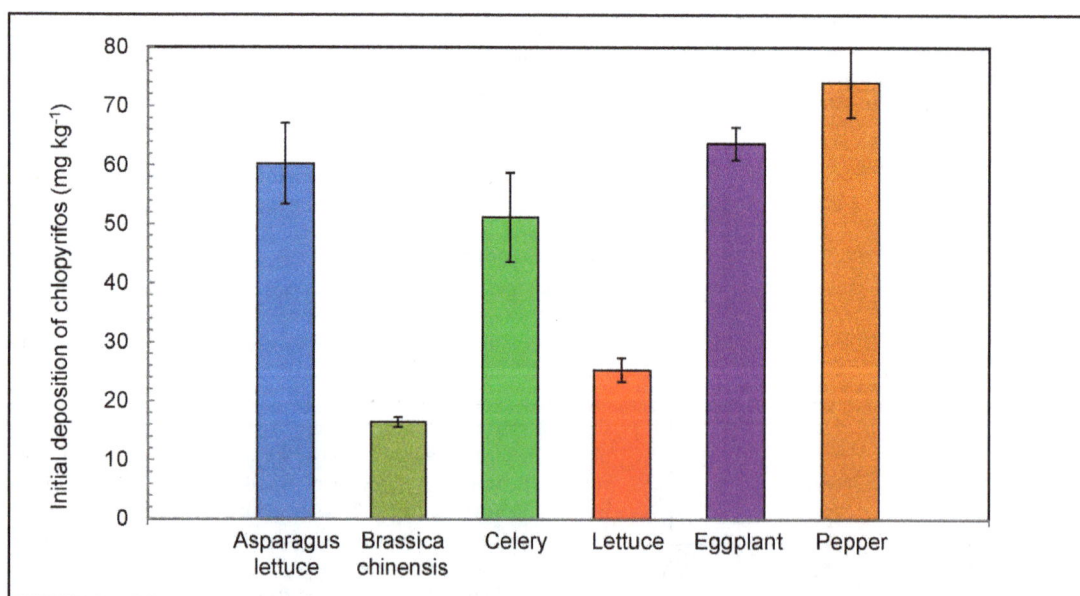

Figure 2. Initial depositions of chlorpyrifos on the six vegetable plants.

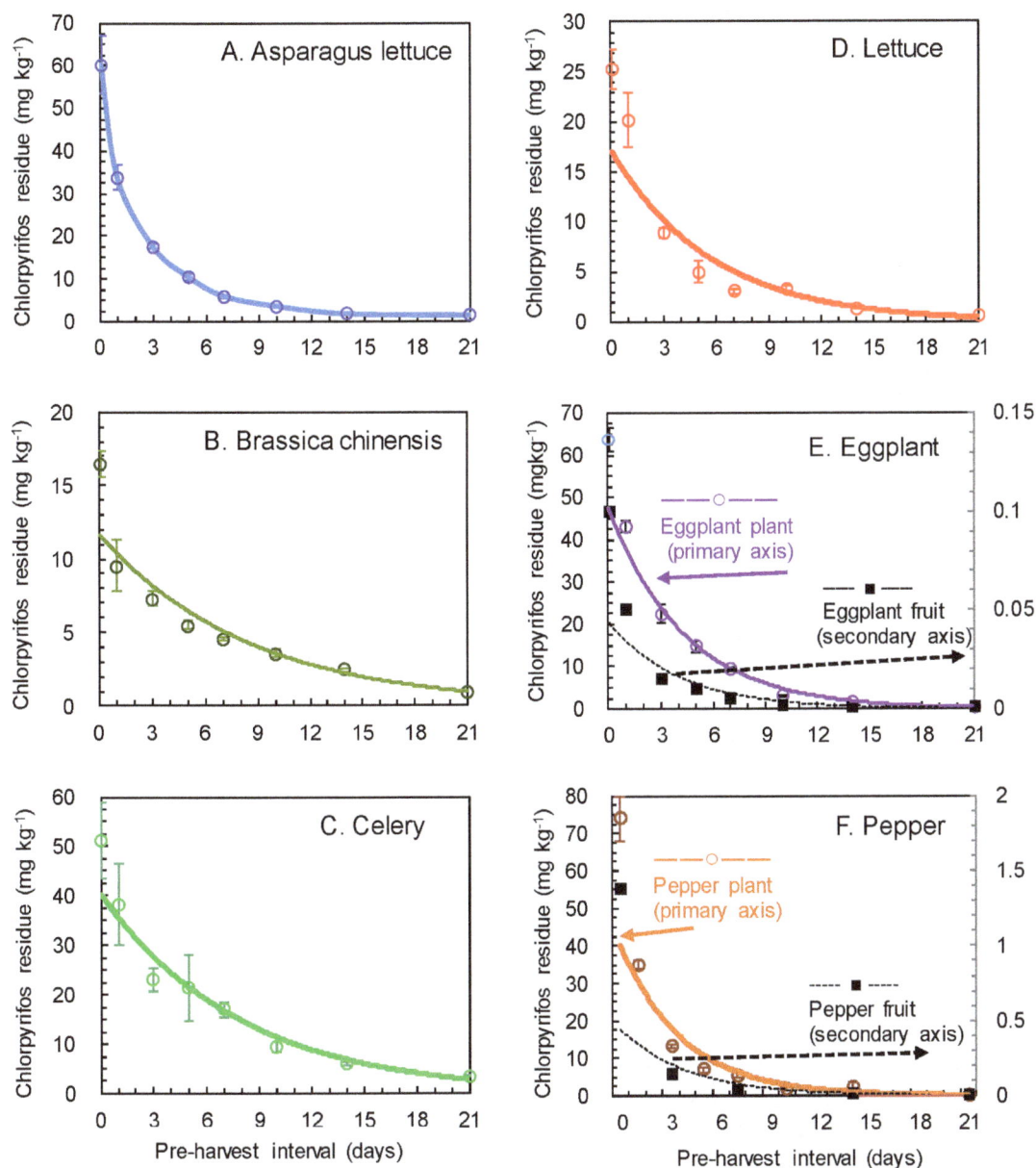

Figure 3. Dissipation dynamic of chlorpyrifos in vegetables and fruits. A. Asparagus lettuce, B. Brassica chinensis, C. Celery, D. Lettuce, E. Eggplant (eggplant plant - solid curve with marker ○ and primary axis; eggplant fruit – dashed curve with marker ■ and secondary axis), and F. Pepper (pepper plant - solid curve with marker ○ and primary axis; pepper fruit – dashed curve with marker ■ and secondary axis).

Soil. Five (5.0) g of soil was weighed into a 50 mL Teflon centrifuge tube and 5 g of sodium chloride was added. The contents were thoroughly mixed. Then, 20 mL of acetonitrile was added. The mixture was vortexed for 1 min, ultrasonically extracted at 40–45°C for 30 min, shaken on a rotary shaker for 2 h, and centrifuged at 5000 rpm for 5 min. Then 1 mL of the supernatant was transferred into a centrifuge tube and dried under a stream of nitrogen (40°C). The chlorpyrifos residues were redissolved in 1 mL of acetone followed by GC analysis.

GC and LC/MS/MS instrument analyses

GC. The conditions for the analysis were: detector temperature, 280°C; injector temperature 270°C; oven temperature program starting at 120°C, 3.67 min at 120–230°C (ramp 30°C min^{-1}), 5 min at 230°C, 2 min at 230–270°C (ramp 20°C min^{-1}), 2 min at 270°C; carrier gas, N_2 at 1 mL/min; injection volume 1.0 µL, in a splitless mode. A linear calibration curve was used and the calibration range was 0.01–5 mg kg^{-1}. Under these conditions chlorpyrifos retention times were approximately 7.52 min. The software was Agilent ChemStation Rev. B04.03 software for instrument control, data acquisition and processing.

LC/MS/MS. The instrument analysis method was an adaption of reference methods [38]. The HPLC conditions for the analysis were: mobile phase A: water containing 0.1% of formic acid (v/v); mobile phase B: acetonitrile containing 0.1% of formic acid (v/v); flow rate 0.4 mL min^{-1}; injection volume 10 µL; mobile phase gradients of binary pump: 10% B (0–1.50 min), 95% B (1.51–4.50 min), 95% B (4.51–6.00 min), and

Table 2. Comparison of chlorpyrifos residues in edible parts of brassica chinensis, lettuce, celery, asparagus lettuce, eggplant and pepper with Maximum Residue Limits (MRLs).*1.

Commodity	Crop group*2	Residue (mg kg⁻¹)		MRL (mg kg⁻¹)			
		PHI 7 days	PHI 21 days	CAC*3	China*4	USA*5	EU*6
Edible plants:							
Brassica chinensis	5B	4.55	0.97	1.0	0.1	1.0	0.5
Lettuce	4A	3.07	0.56	*7	0.1	*7	0.05
Celery	4B	16.9	3.50	*7	0.05	*7	0.05
Asparagus Lettuce	4B	5.96	1.47	*7	0.1	*7	*7
Edible fruits:							
Eggplant (fruit)	8–10B 8–10C	0.01	<0.01	*7	*7	*7	0.5
Pepper (fruit)	8–10B	0.03	<0.01	2.0 (sweet pepper)	*7	1.0	1.0

*1Please note that this work intended to study pesticide persistence instead of for MRL establishment. The comparison described in above the table is used to study the persistence and dissipation of chlorpyrifos in the selected crops. The high residues above were due to the high application rate and different formulations.
*2Code of Federal Regulations Title 40 Part 180.41 Crop group table (40 CFR 180.41). Group 4A/4B Leafy vegetables (except brassica vegetables), 5B Brassica leafy vegetable, 8–10B 8–10C Fruiting vegetable group.
*3CODEX Alimentarius: List of standards: http://www.codexalimentarius.org/standards/list-of-standards/en/?provide = standards&orderField = fullReference&sort = asc&num1 = CAC/MRL.
*4Chinese National Standards (GB27763-2012): Maximum Residue Limits for Pesticides in Foods.
*5The United States Tolerances and Exemptions for Pesticide Chemical Residues in Food: http://www.ecfr.gov/cgi-bin/text-idx?SID = e33dfa87fab3f5ddecad25dafa7028ea&node = 40:25.0.1.1.27.3.19.113&rgn = div8.
*6Pesticide EU-MRLs (Regulation (EC) No 396/2005, MRLs updated on 28/01/2014) http://ec.europa.eu/sanco_pesticides/public/index.cfm?event = substance.resultat&s = 1.
*7MRLs not currently established or registrations canceled.

10% B (6.01–7.00 min). The MS/MS conditions were: gas temperature 350°C; gas flow 10 L/min; nebulizer pressure 45 psi; and capillary voltage 4000 V. In order to achieve the highest sensitivity, the fragment, voltage and the collision energy were optimized. MRM (chlorpyrifos) transitions were 350.1>198 (quantitation, collision energy, CE, 20 V) and 350.1>97 (identification, CE 30 V). The retention time was 3.89 min. The software was Agilent MassHunter software for instrument control, data acquisition and processing.

Data analysis

The degradation rate constant and half-life were calculated using a first-order rate equation:

$$C_t = C_0 e^{-kt}$$

where C_t and C_0 represent the concentrations of the chlorpyrifos residues at the day t and day 0 (2 h), respectively, and k is the degradation rate constant. The half-life $(t_{1/2})$ is defined as the time required for the pesticide residue level to fall to the half of the initial residue level of day 0 (i.e., C_0) and was calculated using the following equation:

$$t_{1/2} = (ln\,2)/k$$

Results and Discussion

Method validation

GC method. Quantification was accomplished by using the standard curve constructed by plotting analyte concentrations against peak areas. Good linearity was achieved with the chlorpyrifos concentration and the correlation coefficient was 0.9995. Recoveries of chlorpyrifos at different fortification levels, i.e., 0.05, 1, and 10 mg kg^{-1}, were determined in three replicates for validation of the method. The recoveries of chlorpyrifos in the soil were 72.5%–89.6% with the relative standard deviation (RSD) 2.1%–7.2% and $R^2 = 0.9846$. For the six vegetable plants, the recoveries were from 79.3%–97.0% with RSD 3.0%–15% and $R^2 = 0.9910$. The limit of quantification (LOQ) was defined as the lowest fortification concentration whose signal-to-noise (S/N) ratio was equal to or greater than 10 and thus the LOQ of the GC analysis was 0.05 mg kg^{-1}. The sample extracts in which the chlorpyrifos residues were greater than 10 mg kg^{-1} were diluted 10 times prior to the evaporation/cleanup steps. And the diluted samples were re-analyzed to ensure the residues were in the acceptable ranges.

LC/MS/MS method. Three fortification levels, i.e., 0.01, 0.1, and 2 mg kg^{-1}, were analyzed. For pepper fruits, the average recoveries of chlorpyrifos residue in the soil were 87.1%–106% with the relative standard deviation (RSD) from 1.1% to 7.4% and $R^2 = 0.9991$. For eggplant fruits, the average recoveries were from 86.1% to 97.0%, with RSD 2.9%–5.9% and R2 = 0.9978. The LOQ of the LC/MS/MS analysis was set to be 0.01 mg kg^{-1}.

Soil and whole plant samples of the six selected crops were analyzed using GC (LOQ = 0.05 mg kg^{-1}). The fruit samples of pepper and eggplant were analyzed by LC/MS/MS (LOQ = 0.01 mg kg^{-1}). These methods were capable of conducting the analyses in this study.

Degradation of chlorpyrifos

Soil. The rhizosphere soil samples were collected at different intervals after chlorpyrifos was applied. The chlorpyrifos residues in soil were analyzed by GC. The concentration of chlorpyrifos in the soil decreased over time (Figure 1). The average initial deposition of chlopyrifos was 14.9±0.5 mg kg^{-1} (i.e., 2 h, day 0) and the final residue was 2.2±0.1 mg kg^{-1} on day 21. The first-order kinetic equation of chlorpyrifos dissipation is $C_t = 4.84e^{-0.089t}$ (Table 1) with correlation coefficient $R^2 = 0.8054$ and the half-life $t_{1/2} = 7.79$ days. The half-lives of chlorpyrifos in soil were in the range of 3–7 days reported by Singh et al. [30]. Singh et al. observed that chlorpyrifos persisted in a low pH soil, i.e., less than 3% of the pesticide had degraded after 10 days and more than 50% of chlorpyrifos was dissipated at a higher pH soil (pH 8.5). Chai et al. reported that the half-lives in humid tropical soils from Malaysia were typically 7–120 days [32]. However, Chai, et. al. also reported that some half-lives were 257 days in the soils containing less soil microbial populations [32]. In the literature, it was reported long environmental dissipation half-lives of chlorpyrifos, i.e., up to 4 years, depending on application rate, ecosystem, and pertinent environments [33]. Since chlorpyrifos presented low water solubility and a higher log K_{ow}, it had a strong tendency to sorb to organic matter and soil. Stability and effectiveness had made chlorpyrifos one of the most popular pesticides worldwide but on the other side its persistence had raised environmental concerns [34].

Initial Deposition. After foliar application, the whole plants of the six selected crops were collected. The chlorpyrifos residues in plants were analyzed by GC. The initial depositions of chlorpyrifos in the six selected plants are compared in Figure 2. As can be seen in Figure 2, the initial depositions (2 h, day 0) on the six plants were in an increasing order: 16.5±0.87 mg kg^{-1} (brassica chinensis), 25.3±2.0 mg kg^{-1} (lettuce), 51.2±7.6 mg kg^{-1} (celery), 60.3±6.8 mg kg^{-1} (asparagus lettuce) < 63.7±2.8 mg kg^{-1} (eggplant) <74.0±5.9 mg kg^{-1} (pepper), respectively.

It is presumed that the initially deposited chlorpyrifos amount mainly depended upon the surface area of the foliar which the pesticide was sprayed on in spite of the leaf characteristics of the plants, such as leaf roughness, content of cuticular waxes, etc. which were assumed to contribute little to the initial depositions. Therefore, the concentration of the initial deposition is directly proportional to the foliar area and inversely proportional to the biomass of the whole plant. The foliages of pepper plant, eggplant plant, celery, and asparagus lettuce were overlapping to maximize the effective foliar surface area. The shorter plants (lettuce and brassica chinensis) had an area of uncovered soils between rows. As a result, lettuce and brassica chinensis had the lowest initial depositions. Since it was lightest in weight, the pepper plant had over all the largest initial deposition of chlorpyrifos.

Dissipation. The dynamic equations of chlorpyrifos degradation are given in Table 1. The curves of dissipation in the six plants and two fruits are described in Figure 3. As can be seen in Figure 3, the dynamics curves demonstrated that the chlorpyrifos residues dissipated significantly in the first a few days and persisted in the crops for extended period of time. For example, at pre-harvest interval (PHI) 21 days, the chlorpyrifos residues in the six plants decreased to 1.47±0.22 mg kg^{-1} ((A) asparagus lettuce), 0.97±0.03 mg kg^{-1} ((B) brassica chinensis), 3.50±0.27 mg kg^{-1} ((C) celery), 0.56±0.06 mg kg^{-1} ((D) lettuce), 0.53±0.06 mg kg^{-1} ((E) eggplant), and 0.15±0.01 mg kg^{-1} ((F) pepper), respectively. As can be seen in Table 1, the half-lives (from low to high) were found to be: 0.91 days (pepper plants) <3.92 days (lettuce) <3.92

days (asparagus lettuce) <5.82 days (brassica chinensis) <3.00 days (eggplant plants) <5.46 days (celery).

The possible mechanisms are thought to be due to the difference in the activities of pesticide degradation enzymes and/ or pesticide degradation endophytes among these plants [25]. It is interesting to observe that the half-lives of the crop plants seemed to be a reverse order of the initial deposition. Pepper plants had the highest initial deposition but shortest half-live (0.92 day) while brassica chinensis had the lowest initial deposition but longest half-live (5.82 days). The difference in the calculated half-life values indicated that different vegetables had different degradation rates. One of the key factors could be photodegradation of chlorpyrifos [37]. The pepper plant had a greater effective foliar area which led to the highest pesticide position, and exposed to the ultraviolet lights. Also, after application, the leaf characteristics may affect how the pesticide would be retained on the surfaces of the leaves and then be penetrating into the plant tissues such as leaf surface roughness [39–40] and the content of water repellent cuticular waxes [41–44].

Edible parts of the vegetables. The edible parts are edible leaves and stems of brassica chinensis, lettuce, celery, and asparagus, and fruits of eggplant and pepper. The residue in pepper and eggplant fruits were analyzed by LC/MS/MS. The comparison of the residue data are given in Table 2. A number of existing Maximum Residue Limits (MRLs) and crop grouping are also included in Table 2. The higher residues of chlorpyrifos in the plants were due to high initial depositions. It was observed that the residues of foliages were significantly higher than those in the edible fruits (eggplant and pepper fruits), i.e., pepper had 0.03 mg kg^{-1} and <0.01 mg kg^{-1} of chlorpyrifos residues and eggplant had <0.10 mg kg^{-1} at PHI 7 days and PHI 21 days, respectively (Table 2). In Figure 3 (E and F), The half-lives were calculated to be 2.84 days (pepper fruits) and 3.15 days (eggplant fruits), respectively. It was observed that same residues exceeded the MRLs. However, it should be noted that the purpose of this work was to study the residues of chlorpyrifos change in the crops. Therefore, higher application rates were used in order to monitor such changes.

Because of chlorpyrifos' high hydrophobicity (high K_{ow} value), the pesticide would readily enter into the inter parts from the surfaces resulting in high residue levels. A portion of the residues may be transferred from leaves to the growing fruits. For eggplant and pepper fruits, the difference in chlorpyrifos residues mainly depended upon the composition of the surface waxes from pepper and eggplant [41,44]. Bauer et. al. [44] reported that the bell pepper contained 39% of fraction 1 (mainly C20–C35 of alkanes and aldehydes) and 61% of fraction 2 (15 various triterpenes) while the eggplant cultivars had 77% of fraction 1 and 23% of fraction 2. Fraction 2 consisted of 15 triterpenes, including α- and β-amyrin, lupeol, glutinol, 3β-friedelanol, friedelin, taraxerol, taraxasterol, δ-amyrin, germanicol, multi-florenol, ω-taraxasterol, isomultiflorenol, isobauerenol and bauerenol, as well as n-alkanoic acids 2-hydroxy-alkanoic acids [44]. These chemicals would enhance the dissipation of chlorpyrifos.

The comparisons of the chlorpyrifos residue data listed in Table 2 and presented in Figure 3 indicated that chlorpyrifos was relatively stable and persisted in the crops, especially leafy crops. The rate of degradation of pesticide residue is affected by environmental conditions, nature of the pesticide, application rate, formulation, and plant species, etc. [45]. The vegetables selected in this study are minor crops representing a range of various species and it was intended to promote the process. For the crop grouping, celery may be a good reprehensive species of leafy vegetable (group 4B, Table 2) with consideration of initial deposition and dynamic data.

Conclusions and Implications

The study investigated the residue behavior of chlorpyrifos in six vegetables in the greenhouse. The results of chlorpyrifos of initial depositions on different vegetables detected after pesticide application showed differences among the six selected crops. The half-lives of chlorpyrifos in the six vegetables were different indicating that different vegetables had different capacities for metabolizing chlorpyrifos.

Author Contributions

Conceived and designed the experiments: XYY MXL XJL. Performed the experiments: MXL. Analyzed the data: MXL WWJ QJ XYY. Contributed reagents/materials/analysis tools: MXL JLW YS. Contributed to the writing of the manuscript: WWJ MXL.

References

1. FAO (2000) Pesticide Residues in Food, 2000: Report of the Joint Meeting of the FAO Panel of Experts on Pesticide Residues in Food and the Environment and the WHO Core Assessment Group on Pesticide Residues, Geneva, Switzerland, 20–29 September 2000. 45–59 p.

2. Lemus R, Abdelghani A (2000) Chlorpyrifos: an unwelcome pesticide in our homes. Rev Env Health 15: 421–433.

3. Yuan Y, Chen C, Zheng C, Wang X, Yang G, et al. (2014) Residue of chlorpyrifos and cypermethrin in vegetables and probabilistic exposure assessment for consumers in Zhejiang Province, China. Food Control 36: 63–68.

4. Gao Y, Chen S, Hu M, Hu Q, Luo J, et al. (2012) Purification and Characterization of a Novel Chlorpyrifos Hydrolase from *Cladosporium cladosporioides* Hu-01. PLoS ONE 7(6): e38137.

5. Wentzell J, Cassar M, Kretzschmar D (2014) Organophosphate-Induced changes in the PKA regulatory function of Swiss cheese/NTE lead to behavioral deficits and neurodegeneration. PLoS ONE 9(2): e87526.

6. Rauh VA, Perera FP, Horton MK, Whyatt RM, Bansal R, et al. (2012) Brain anomalies in children exposed prenatally to a common organophosphate pesticide. PNAS 109: 7871–7876.

7. Janssens L, Stoks R (2013) Fitness Effects of Chlorpyrifos in the Damselfly *Enallagma cyathigerum* Strongly Depend upon Temperature and Food Level and Can Bridge Bridge Metamorphosis. PLoS ONE 8(6): e68107.

8. Canesi L, Negri A, Barmo C, Banni M, Gallo G, et al. (2011) The Organophosphate Chlorpyrifos Interferes with the Responses to 17β-Estradiol in the Digestive Gland of the Marine Mussel *Mytilus galloprovincialis*. PLoS ONE 6(5): e19803.

9. Sasikala C, Jiwal S, Rout P, Ramya M (2012) Biodegradation of chlorpyrifos by bacterial consortium isolated from agriculture soil. World J Microbio Biotech, 28(3), p1301.

10. Trunnelle KJ, Bennett DH, Tulve NS, Clifton MS, Davis MD, et al. (2014) Urinary pyrethroid and chlorpyrifos metabolite concentrations in northern California families and their relationship to indoor residential insecticide levels, Part of the study of use of products and exposure related behavior (SUPERB). Environ. Sci. Technol., 48 (3): 1931–1939.

11. Watts M (2012) Chlorpyrifos as a Possible Global Persistent Organic Pollutant. Pesticide Network North America, Oakland, CA, USA. Available: http://www.ipen.org/cop6/wp-content/uploads/2013/04/Chlorpyrifos_as_POP_final.pdf. Accessed on March 17, 2014.

12. Popp J, Peto K, Nagy J (2013) Pesticide productivity and food security. A review. Agron Sustain Dev 33: 243–255.

13. Zhang NN, Liu CF, Yang F, Dong SL, Han ZJ (2012) Resistance mechanisms to chlorpyrifos and F392W mutation frequencies in the acetylcholine esterase ace1 allele of field populations of the tobacco whitefly, Bemisia tabaci in China. J Insect Sci, Vol 12 Article 41.

14. Ouyang Y, Chueca P, Scott SJ, Montez GH, Grafton-Cardwell EE (2010) Chlorpyrifos Bioassay and Resistance Monitoring of San Joaquin Valley California Citricola Scale Populations. J Econ Ent 103(4): 1400–1404.

15. Mouron P, Heijne B, Naef A, Strassemeyer J, Hayer F, et al. (2012) Sustainability assessment of crop protection systems: SustainOS methodology and its application for apple orchards. Agri Syst 113: 1–15.

16. MacLachlan J, Hamilton D (2010) Estimation methods for maximum residue limits for pesticides. Reg Toxicol Pharmacol, 58: 208–218.

17. Malhat F, Kamel E, Saber A, Hassan E, Youssef A, et al. (2013) Residues and dissipation of kresoxim methyl in apple under field condition. Food Chem 140: 371–374.

18. Garau VL, Angioni A, Aguilera Del Real A, Russo MT, Cabras P (2002) Disappearance of azoxystrobin, cyprodinil, and fludioxonil on tomato in a greenhouse. J Agri Food Chem 50: 1929–1932.

19. Cabras P, Meloni M, Manca MR, Pirisi FM, Cabitza F, et al. (1988) Pesticide residues in lettuce. 1. Influence of the cultivar. J Agri Food Chem: 36: 92–95.

20. Wang M, Zhang Q, Cong L, Yin W, Wang M (2014) Enantioselective degradation of metalaxyl in cucumber, cabbage, spinach and pakchoi. Chemosphere 95: 241–256.

21. Sun H, Xu J, Yang S, Liu G, Dai S (2004) Plant uptake of aldicarb from contaminated soil and its enhanced degradation in the rhizosphere. Chemosphere. 54: 569–574.

22. Fan S, Zhang F, Deng K, Yu C, Liu S, et al. (2013) Spinach or amaranth contains highest residue of metalaxyl, fluazifop-p-butyl, chlorpyrifos, and lambda-cyhalothrin on six leaf vegetables upon open field application. J Agri Food Chem 61: 2039–2044.

23. Montemurro N, Grieco F, Lacertosa G, Visconti A (2002) Chlorpyrifos decline curves and residue levels from different commercial formulations applied to oranges. J Agri Food Chem 50: 5975–5980.

24. Cabras P, Meloni M, Gennari M, Cabitza F, ubeddu M (1989) Pesticide residues in lettuce. 2. Influence of formulations. J Agri Food Chem 37: 1405–1407.

25. Xia XJ, Zhang Y, Wu JX, Wang JT, Zhou YH, et al. (2009) Brassinosteroids promote metabolism of pesticides in cucumber. J Agri Food Chem 57: 8406–8413.

26. Itoiz ES, Fantke P, Juraske R, Kounina A, Vallejo AA (2012) Deposition and residues of zaoxystrobin and imidacloprid on greenhouse lettuce with impliacations of human consumption. Chemosphere 89: 1034–1041.

27. Collins C, Fryer M, Grosso A (2006) Plant uptake of non-ionic organic chemicals. Environ Sci Tech 40: 45–52.

28. Juraske R, Castells F, Vijay A, Muñoz P, Antón A (2009) Uptake and persistence of pesticides in plants: Measurements and model estimates for imidacloprid after foliar and soil application. J Hazard Mater 165: 683–689.

29. Hauschild M (2000) Estimating pesticide emissions for LCA of agricultural products. In: Weidema BP, Meeusen MJG. (Eds.), Agricultural Data for Life Cycle Assessments. Agricultural Economics Research Institute, The Hague, 64–79 p.

30. Singh BK, Walker A, Wright DJ (2006) Bioremedial potential of fenamiphos and chlorpyrifos degrading isolates: influence of different environmental conditions. Soil Biol Biochem 38: 682–93.

31. Chen S, Liu C, Peng C, Liu H, Hu M, et al. (2012) Biodegradation of Chlorpyrifos and Its Hydrolysis Product 3,5,6-Trichloro-2-Pyridinol by a New Fungal Strain *Cladosporium cladosporioides* Hu-01. PLoS ONE 7(10): e47205.

32. Chai LK, Wong MH, Hansen HCB (2013) Degradation of chlorpyrifos in humid tropical soils. J Environ Manage 125: 28–32.

33. Gebremariam SY, Beutel MW, Yonge DR, Flury M, Harsh JB (2012) Adsorption and Desorption of Chlorpyrifos to Soils and Sediments. Rev Environ Contam Tox 215: 123–175.

34. Kamrin MA (1997) Pesticide profiles toxicity, environmental impact, and fate. Lewis publishers: Boca Raton, FL 147–152 p.

35. US EPA (1999) Reregistration eligibility science chapter for chlorpyrifos fate and environmental risk assessment chapter. US EPA, office of prevention, pesticides and toxic substances, office of pesticide programs, environmental fate and effects division, US government printing office: Washington, DC.

36. Fang C, Radosevich M, Fuhrmann JJ (2001) Atrazine and phenanthrene degradation in grass rhizosphere soil. Soil Biol Biochem 33: 671–678.

37. Nieto LM, Hodaifa G, Vives SR, Casares JAG, Casanov MS (2009) Photodegradation of phytosanitary molecules present in virgin olive oil. J Photoch Photobio A 203: 1–6.

38. Liang Y, Wang W, Shen Y, Liu Y, Liu XJ (2012) Dynamics and residues of chlorpyrifos and dichlorvos in cucumber grown in greenhouse. Food Control 26: 231–234.

39. Gaskin RE, Steele KD, Foster WA (2005) Characterizing plant surfaces for spray adhesion and retention. New Zealand Plant Protection 58: 1790–183.

40. Hunche M, Bringe K, Schmitz-Eiberger M, Noga G (2006) Leaf surface characteristics of apple seedlings, bean seedlings and Kohlrabi plans and their impact on the retention and rainfastness of mancozeb. Pest Manage Sci 62: 839–847.

41. Bargel H, Koch K, Cerman Z, Neinhuis C (2006) Structure-function relationships of the plant cuticle and cuticular waxes – a smart material? Funct Plant Biol. 33: 893–910.

42. Wagner P, Furstner R, Barthlott W, Neinhuis C (2003) Quantitative assessment to the structural basis of water repellency in natural and technical surfaces. J Exp Bot 54: 1295–1303.

43. Malhat F, Badawy HMA, Barakat DA, Saber AN (2014) Residues, dissipation and safety evaluation of chromafenozide in strawberry under open field conditions. Food Chem 152: 18–22.

44. Bauer S, Schulte E, Their H-P (2005) Composition of the surface waxes from bell pepper and eggplant. Eur Food Res Technol 220: 5–10.

45. Fantke P, Juraske R (2013) Variability of Pesticide Dissipation Half-Lives in Plants. Environ Sci Tech 47: 3548–3562.

4

Identification of the Aggregation Pheromone of the Melon Thrips, *Thrips palmi*

Sudhakar V. S. Akella[1], William D. J. Kirk[1], Yao-bin Lu[2], Tamotsu Murai[3], Keith F. A. Walters[4¤], James G. C. Hamilton[1]*

1 Centre for Applied Entomology and Parasitology, School of Life Sciences, Huxley Building, Keele University, Keele, Staffordshire, England, United Kingdom, **2** Institute of Plant Protection and Microbiology, Zhejiang Academy of Agricultural Sciences, Hangzhou, Zhejiang, China, **3** Laboratory of Applied Entomology, Faculty of Agriculture, Utsunomiya University, Utsunomiya, Tochigi, Japan, **4** Food and Environment Research Agency, Sand Hutton, York, North Yorkshire, England, United Kingdom

Abstract

The objective of this study was to identify the aggregation pheromone of the melon thrips *Thrips palmi*, a major pest of vegetable and ornamental plants around the world. The species causes damage both through feeding activities and as a vector of tospoviruses, and is a threat to world trade and European horticulture. Improved methods of detecting and controlling this species are needed and the identification of an aggregation pheromone will contribute to this requirement. Bioassays with a Y-tube olfactometer showed that virgin female *T. palmi* were attracted to the odour of live males, but not to that of live females, and that mixed-age adults of both sexes were attracted to the odour of live males, indicating the presence of a male-produced aggregation pheromone. Examination of the headspace volatiles of adult male *T. palmi* revealed only one compound that was not found in adult females. It was identified by comparison of its mass spectrum and chromatographic details with those of similar compounds. This compound had a structure like that of the previously identified male-produced aggregation pheromone of the western flower thrips *Frankliniella occidentalis*. The compound was synthesised and tested in eggplant crops infested with *T. palmi* in Japan. Significantly greater numbers of both males and females were attracted to traps baited with the putative aggregation pheromone compared to unbaited traps. The aggregation pheromone of *T. palmi* is thus identified as (*R*)-lavandulyl 3-methyl-3-butenoate by spectroscopic, chromatographic and behavioural analysis.

Editor: Michel Renou, INRA-UPMC, France

Funding: This research was funded by a European Union, Marie Curie, Incoming International Fellowship scheme to Dr S. Akella (Project 252258) and as a subcontract of a project, The integrated control of Thrips palmi Karny, made to Central Science Laboratory by the United Kingdom government Department of Food Environment and Rural Affairs. The funders had no role in study design, data collection and analysis or preparation of the manuscript.

Competing Interests: I have read the journal's policy and have the following conflicts. Kirk and Hamilton are co-inventors on a patent owned by Keele University, "Method of monitoring and/or controlling thrips." Priority date: 30 October 2013. International patent application published under the Patent Cooperation Treaty WO 2014/068303 <file://localhost/tel/2014%252F068303> AI. Geneva: World Intellectual Property Organization. The patent covers the aggregation pheromone studied in the paper.

* Email: j.g.c.hamilton@keele.ac.uk

¤ Current address: Centre for Integrated Pest Management, Harper Adams University, Newport, Shropshire, England, United Kingdom

Introduction

Thrips are small insects, typically only 1–2 mm long, belonging to the order Thysanoptera. Adults and larvae of many species cause serious commercial damage to crops grown in protected environments, such as glasshouses and polytunnels (tunnels covered with polythene), and also to open-field crops, through feeding and virus transmission. Most commercially important thrips pest species are in the genera *Thrips* and *Frankliniella*, which belong to the same sub-family (Thripidae: Thripinae).

The western flower thrips *Frankliniella occidentalis* (Pergande) [1,2] and *Frankliniella intonsa* (Trybom) [3] have male-produced aggregation pheromones that are attractive to both female and male conspecifics. *F. occidentalis* males form lek-like aggregations within which there are aggressive male–male interactions. Females arrive continually, mate, and leave immediately, so although both sexes arrive at the aggregations, they contain predominantly males [4]. The aggregation pheromone is probably used by males and

females to locate these mating aggregations [2]. In *F. occidentalis*, the pheromone has been tested in the field and identified as a single component, the monoterpene ester neryl (*S*)-2-methylbutanoate (N(*S*)2 MB) [2]. In *F. intonsa* the aggregation pheromone may be a two-component mix of N(*S*)2 MB with (*R*)-lavandulyl acetate ((*R*)LA) [3], but the effects of synthetic compounds have not yet been tested.

These pheromone components probably originate from glandular tissue underlying a series of structures on the underside of the abdomen of adult males, known as sternal pore plates [5–7]. Species in the genus *Thrips* also have male pore plates [6] and aggregations of males have been recorded in some species [8], leading to speculation that they may also produce an aggregation pheromone.

The melon thrips *Thrips palmi* Karny is a global pest of a wide range of plants, particularly in the Solanaceae and Cucurbitaceae, including important vegetable and ornamental crops such as eggplant (brinjal, aubergine), melon, cucumber, sweet pepper and chrysanthemum [9], causing significant damage both by feeding and as a vector of tospoviruses [10]. Since the late 1970s, it has spread around the world, probably originating from southeast Asia, and is now a pest across Asia and the Pacific and is also found in Florida, the Caribbean and parts of South America, Africa and Australia [9,11]. It has recently been recorded for the first time in Iran [12]. Although *T. palmi* is not currently a problem in Europe, its well documented global dispersal in association with the international trade in plants or plant products has added an extra dimension to its pest status, and it is considered to pose a considerable threat to the European horticulture industry [13,14]. It is commonly intercepted at points of entry on imports of cut flowers, fruit and vegetables throughout the world, and such dispersal pathways result in crop colonisation. Its behaviour results in it seeking out small enclosed spaces, which can make it difficult to detect. Short generation times result in rapid population increases and development of insecticide resistance can result in control failures. Some incursions into European crops have occurred, where outbreaks are subject to plant quarantine legislation, and have been successfully eradicated [9,15], but this has been more readily achieved if management actions commence soon after initial infestation when populations are small. Enhanced methods for early detection and control are thus of central importance to maintain biosecurity and enhance the effect of current control methods.

The objectives of this study were to identify any male-produced volatile compounds of *Thrips palmi* and test whether they act as an aggregation pheromone in the field, with a view to providing a tool for potential use in both commercial and quarantine pest detection and management.

Materials and Methods

Thrips

Field experiments and some thrips collections were carried out on private land and we confirm that the owner of the land gave permission to conduct the study on this site.

The thrips for olfactometer bioassays were reared from *T. palmi* collected from an eggplant crop (*Solanum melongena* L.) in Hangzhou, Zhejiang Province, China (N 30° 18.331′ E 120° 11.730′). They were reared on bean pods (*Phaseolus vulgaris* L.) in 4.5 L glass canning jars at 27±1°C, 65–75% r.h., 16:8 light:dark. Mixed-age adult thrips were collected arbitrarily from the colony. To obtain known-age virgin females, large numbers of second-instar larvae were collected from the colony and transferred individually into 0.5 ml microcentrifuge tubes containing a section of bean pod. They were examined daily and virgin females were used for the experiments 1–3 d after emergence.

All adult male and female *T. palmi* used for the collection of volatile chemicals were obtained from the leaves of commercial eggplant crops (*S. melongena* var. Senryo 2) grown in a polytunnel at Himuro near Utsunomiya, Tochigi Prefecture, Japan (N 36° 30.483′ E 139° 59.536′).

To identify any volatile compounds that might be produced exclusively by male *T. palmi*, the headspace volatiles of both adult males and females with appropriate controls were collected separately and analysed. *T. palmi* were transported from Utsunomiya University, Japan, in small plastic boxes (12 cm×8 cm×4 cm) lined with layers of moistened tissue on fresh sprouting broad bean seeds (*Vicia faba* L.) to the Central Science Laboratory (CSL) (now Fera), York, UK. The insects were held in secure quarantine facilities (license number: PHL 251B/

5328(02/2006) amended (04/2006)) at 23°C, 65% r.h., 16:8 light:dark until required.

Additional collections of adult male and female *T. palmi* were made in the field in Japan in August 2011 and again in October 2011 at Himuro for entrainment of headspace volatiles. These additional entrainments were undertaken on mixed male and female groups of *T. palmi* with the aim of allowing us to collect and store larger quantities of the target compound that was already identified from our solid phase micro extraction (SPME) entrainments at CSL. The additional material allowed us to carry out further comparisons with mass spectrometry (MS) data held in the coupled gas chromatography/mass spectrometry (GC/MS) library as well as further chiral and achiral chromatography. Collections were made between 10:00–16:00 h. Adults were aspirated from the eggplant leaves and held in clean glass containers (50 ml round bottomed (r.b.) flasks) and kept cool in an ice box. Approximately 1 g of eggplant leaves and petals were added to the r.b. flasks to provide a food source and maintain humidity levels until the thrips were used for experimentation. Four separate collections were made with the first containing approximately 35 males and 400 females, the second 20 males and 365 females, the third 26 males and 415 females and the fourth 18 males and 300 females. The imbalance of the sexes was because males are found much less frequently than females in the field.

To confirm that the collected thrips were *T. palmi*, representative samples were checked under a stereo microscope in the UK and Japan. The main characteristic features are: body colour yellow to white, antennae 7-segmented, macropterous with wing-vein setae interrupted, ocelli red, and ocellar setae III outside the ocellar triangle [16]. Females are distinguished from males by the pointed shape of the tip of the abdomen and the presence of an ovipositor. Males have no ovipositor and the tip of the abdomen is blunt.

Olfactometer Bioassays

The response of adult thrips to male- or female-produced volatiles was tested in a glass Y-tube olfactometer. This had a stem 60 mm long, two arms 60 mm long, separated from each other at an angle of 90°, and an internal diameter of 5 mm. Air, filtered through activated charcoal, humidified and split into two air streams, each of which was fed through a 50 ml glass flask and into one arm of the olfactometer was drawn through at a flow rate of 60 mm/s. The two flasks provided test and control odour (clean air) sources. The flasks were illuminated from above by four flourescent tubes and by one arm of a fibre-optic cold-light source at a distance of 40 mm from the Y-tube (total illumination was approximately 10,000 lux). Connections between the components of the olfactometer apparatus were made with Teflon tubes. Olfactometer experiments were carried out at 25±2°C. Forty mixed-age adult thrips were collected with a small aspirator, anaesthetized with carbon dioxide, the sex of individuals was checked under a microscope, and they were then transferred into the treatment flask as the odour source. Test thrips were transferred individually to the stem of the Y-tube with a fine brush. Each thrips was observed for a maximum of 3 min, and its choice for one of the two odour sources (treatment or control) was recorded when it crossed a line 20 mm down either arm. 'No choice' was recorded if the line was not crossed after 3 min. After five thrips were tested, odour sources entering the arms of the Y-tube were swapped to avoid any potential bias in the apparatus. Each odour comparison was repeated four or five times on different days, with a total of 15–20 thrips per day. The apparatus was cleaned before each test by rinsing with hexane and baking in an oven (200°C).

The data were analysed with IBM SPSS Statistics 19 (IBM Corp., USA). Responses were tested by a binomial test with exact two-tailed P values, with the null hypothesis that the two arms were chosen with equal probability. "No choices" were excluded from the analysis.

Headspace Volatile Collection

All glassware used in the collection of headspace volatiles was cleaned by first washing in a 5–10% detergent solution, then rinsing with distilled water, drying with acetone and finally heating at 200°C in a clean oven overnight to remove potential contaminants. Teflon tubing used in the portable entrainment apparatus was cleaned by first washing in a 5–10% detergent solution, rinsing with distilled water, drying with acetone and then leaving in a fume hood at room temperature overnight to allow solvent to evaporate fully.

Entrainments at CSL

T. palmi (males, females or larvae) were removed from the sprouting beans with a small aspirator, anaesthetised with carbon dioxide, and transferred into a clean glass container (volume 1.9 ml) that was then sealed with Teflon tape. The thrips were illuminated from above with a 60 W tungsten filament lamp to induce patrolling behaviour [1]. Headspace volatiles were collected on a divinylbenzene (DVB)/carboxen/polydimethylsiloxane (PDMS) SPME fibre assembly (57348-U, Supelco, Poole, UK) inserted into the glass container containing the thrips through the Teflon tape at 27°C for 4–18 h [1]. The numbers of males, females and larvae entrained in this way varied from 30 to 100 per replicate and in total four entrainments of each sex and stage were carried out. The following entrainments were carried out, males by themselves, females by themselves and larvae. As the males, females or larvae were removed from the bean sprouts and entrained away from this food source a separate SPME entrainment of bean sprouts was not done. After each entrainment the SPME fibre was sealed in a clean glass tube and transferred to Keele University for GC/MS analysis.

Entrainments in Utsunomiya

Entrainments of headspace volatiles were carried out in Utsunomiya to provide greater quantities of the male-specific compound identified by SPME entrainment of *T. palmi* at CSL described above. After field collection the 50 ml r.b. flasks containing the thrips were transferred to the laboratory and the headspace volatiles collected using a portable entrainment apparatus (Barry Pye, Kings Walden, Herts. UK). Air, pushed through the entrainment apparatus by a pump, was first cleaned by passing it through an activated charcoal filter and then into a r.b. flask containing the thrips and plant material (for the thrips to feed on) via a Drechsel head. A control entrainment of eggplant only was also carried out. The air exiting from the r.b. flask then passed into a glass column containing an adsorbent polymer (ORBO 402, Tenax-TA). All tubing and components within the entrainment apparatus were connected with Swagelok connectors or Teflon tubing joints and were sealed with Teflon tape (Sigma-Aldrich Company Ltd., Gillingham, UK) to eliminate leakage of air. Air flow at the outlet of the Tenax-TA tube was measured with a bubble flow meter and maintained at 5 ml/s by adjustment of a rotameter (GPE Ltd., Leighton Buzzard, UK) at the air inlet side of the apparatus.

The entrainment was run continuously for a period of 4 days. The Tenax-TA columns were replaced every 24 h with a fresh adsorbent column when fresh petals were also added to both r.b. flasks.

Volatiles were eluted from the Tenax-TA tubes using 2 ml of a 95:05 mixture of *n*-hexane (SupraSolv grade; Merck, Germany) and ethyl acetate (Chromatography/HPLC grade; Fisher Scientific, Loughborough, UK). The extracts from the four collections were concentrated under a gentle stream of air to 1 ml and returned to Keele University where they were combined and the volume reduced again to 100 µl for GC/MS analysis. The amount of monoterpene ester present in the *T. palmi* extracts from Japan was quantified by comparison of the peak area of the unknown ester with a known amount of neryl (S)-2-methylbutanoate by GC/MS analysis.

Coupled Gas Chromatography/Mass Spectrometry

GC/MS analyses were carried out on either a HP 5890 II+ GC coupled to a HP 5972A MS or an Agilent 7890 GC coupled to an Agilent 5973 MS (Agilent Technologies, Ipswich, UK). The 5972A was operated in electron impact (EI) (70 eV, 180°C) mode only. The 5973 instrument was operated in either EI (70 eV, 180°C) or chemical ionization (CI) mode. CI analyses were carried out using isobutane as the reagent gas.

For the 5890 GC the carrier gas was helium (1 ml/min) and the injector was a Merlin Microseal (Thames-Restek, High-Wycombe, UK) septum-less heated injector (180°C) fitted with a SPME glass injection sleeve (0.75 mm i.d.; Supelco). SPME samples were injected in the splitless mode and desorbed for 8 min before the fibre assembly was withdrawn. Non-SPME samples (≤ 1 µl) of *T. palmi* extracts were also injected via this injector set in the splitless mode using a standard 10 µl syringe (Sigma Aldrich, UK) to maintain sensitivity. An initial temperature of 40°C was held for 2 min, increased (10°C/min) to 120°C, then increased (6°C/min) to 180°C and then increased (10°C/min) to the final temperature of 250°C (held for 1 min). The MS transfer line was set at 280°C. Prior to each SPME thrips entrainment analysis, a blank fibre was analysed to check system performance for the presence of possible contamination.

For the 7890 GC the injector was a multimode inlet set in splitless mode at 180°C and the GC analytical conditions were as described above for the 5890 GC.

SPME-collected headspace volatiles, hexane:ethyl acetate extracts of Tenax-TA entrainment tubes and synthetic standards were analysed on both HP5MS (Supelco), DBWax (Supelco) and chiral CycloSil-B (Agilent J&W, Agilent, Wokingham, UK) fused silica analytical columns (30 m×0.25 mm i.d., 0.25 µm phase thickness) as appropriate.

The retention index (RI) of the *T. palmi* compound was calculated relative to the retention times of saturated hydrocarbons. The RI and mass spectrum of the *T. palmi* compound were then compared against the RIs and mass spectra of a library of 200 synthetic monoterpene C5 esters that included pentanoates (72 compounds), pentenoates (83 compounds), pentadienoates (12 compounds) and pentynoates (33 compounds). The library was prepared by synthesising 200 of the possible combinations of 19 C5 fatty acids and their isomers with 17 commercially available acyclic, monocyclic and bicyclic monoterpene alcohols and their isomers. The esters were then analysed individually according to the general methodology described below and RI and EI/MS data were collected for each compound and isomer on both DB5 and DBWax columns. RI and EI/MS data were also collected for molecules with chiral centres on a CycloSil-B column.

Chiral Chromatography of Monoterpene Esters

Analysis of the enantiomeric composition of synthetic standards and authentic thrips material was carried out on the Agilent 7890 coupled 5972 GC/MS with a CycloSil-B column. The carrier gas was helium (flow rate 1 ml/min). Samples were introduced via a heated multimode injector port (180°C) and the GC was temperature programmed with an initial 2 min at 55°C, an increase of 5°C/min to 115°C, held for 1 min, then an increase of 0.5°C/min to 165°C.

Chiral Chromatography of the Racemate and R and S enantiomers of Lavandulol (5-Methyl-2-(1-methylethenyl)hex-4-en-1-ol)

The GC was temperature programmed with an initial 2 min at 55°C, then an increase of 10°C/min to a temperature of 125°C, held for 1 min, then an increase of 5°C/min to a temperature of 200°C and then to the final temperature of 250°C (10°C/min). (S)-lavandulol eluted first at 19.86 min and (R)-lavandulol at 20.29 min.

Synthesis of Racemic Lavandulyl 3-Methyl-3-butenoate

Lavandulol (1 mmol; Sigma-Aldrich), 3-methyl-3-butenoic acid (1.2 mmol; Sigma-Aldrich), and 4-dimethylaminopyridine (DMAP) (0.05 mmol; Sigma-Aldrich) were dissolved in dry dichloromethane (2 ml), and the solution was stirred in an ice bath. N,N'-Dicyclohexylcarbodiimide (DCC) (1.2 mmol; Sigma-Aldrich) was added portionwise over 30 min, and stirring was continued for another 30 min with cooling and then for 3 h at room temperature. The N,N'-dicyclohexylurea reaction by-product was filtered off, and the precipitate was washed with petroleum ether. The filtrate was washed with saturated aqueous sodium bicarbonate solution, dilute hydrochloric acid, and water, dried over magnesium sulphate, and filtered. After concentration, the residue was purified by column chromatography on silica gel (40 g, 100–200 mesh) eluted with a mixture (98:2) of petroleum ether:ethyl acetate. Pure fractions, identified by thin-layer chromatography, were collected and concentrated giving the ester in 95.3% yield and 100% purity. The (R) and (S) enantiomers of lavandulyl 3-methyl-3-butenoate were partially separated by analysis of the reaction product on a CycloSil-B column (Fig. 1A). ^{1}H NMR (CDCl$_3$, 300 MHz): δ 5.05 (t, 1H, J = 6.9 Hz, H-4a (hydrogen number by position on the structure, see Fig. 2) CH$_3$C = CH), 4.90 (m, 1H, = CH-CH$_2$), 4.83 (m, 1H, H-4b CH$_2$C = CH), 4.83 (m, 1H, H-2''b CH$_3$C = CH), 4.70 (d, 1H, J = 0.8 Hz, H-2''a CHC = CH), 4.07 (dd, 2H, J = 7.5, 3.0 Hz, CH$_2$O_COR), 3.02 (s, 2H, CH$_2$C(= CH$_2$)(CH$_3$), 2.36–2.45 (m, 1H, CH-CH$_2$O), 2.00–2.20 (m, 2H, CH$_2$-CH-CH$_2$O), 1.80 (br s, 3H, CH$_3$-C =), 1.69 (br s, 3H, CH$_3$-C =), 1.68 (br s, 3H, CH$_3$-C =), 1.60 (br s, 3H, CH$_3$-C =). ^{13}C NMR (CDCl$_3$, 75 MHz) δ 171.45, 144.80, 138.59, 133.00, 121.57, 114.72, 112.52, 66.06, 46.10, 43.52, 26.55, 25.80, 22.50, 19.94, 17.84.

Synthesis of (R)- and (S)-Lavandulyl 3-Methyl-3-butenoate

The (R) enantiomer of lavandulol was obtained from racemic lavandulol (97% purity, Fluka) by a lipase-catalysed acylation using porcine pancreas lipase type ll [17]. Enantiomerically enriched (R)-lavandulol was obtained (0.68 g) and chiral chromatography on CycloSil-B capillary column showed an enantiomeric excess of 98%.

The (S) enantiomer was prepared by alkaline hydrolysis of (S)-lavandulyl acetate [18]. Enantiomerically enriched (S)-lavandulol was obtained (5 mg) and chiral chromatography on the CycloSil-B analytical capillary column showed an enantiomeric excess of 98%.

The (R) and (S) lavandulyl 3-methyl-3-butenoate esters were prepared separately by esterification of the alcohol with 3-methyl-

3-butenoic acid as described for the racemic lavandulol above. Both esters were then purified by column chromatography. The formation of both the (R) and (S) enantiomers was shown by analysis of the products by chiral chromatography on the CycloSil-B column (Fig. 1B and 1C).

Field Trials

The biological activity of the male-produced compound was tested in September 2012 in the same greenhouse (a polytunnel 50 m long×5 m wide×2.5 m high at the apex) near Utsunomiya, Japan that had been used in August and October 2011 to collect male and female T. palmi for headspace volatile analysis. The effect of the chemical was tested by comparing the number of thrips caught on traps with and without the synthetic aggregation pheromone in a crop of mature eggplant (Solanum melongena variety 'Senryo 2') with two rows of crop running along the length of the greenhouse.

Rectangular blue sticky traps, 10 cm×25 cm (Takitraps, Syngenta Bioline, UK), were suspended on wire hangers so that they were directly above the middle of the row and placed so that the base of the trap was about 10 cm above the canopy of the crop, which was at a height of about 1.2 m. Blue traps were used because they are widely reported as being highly attractive to T. palmi and thus the additional effect of the putative pheromone over that of an already very attractive trap would be tested [19]. Pre-sampling of the crop indicated that it was infested with two species of thrips: T. palmi and, to a lesser extent, F. intonsa. The paper protecting the north-facing side of each trap was removed to expose the sticky surface and a rubber septum (diam. 6.3 mm, length 10.8 mm, pre-cleaned, International Pheromone Systems Ltd., Deeside, UK) was stuck to the middle of the exposed side. The test septa were loaded with 30 µg of (R)-lavandulyl 3-methyl-3-butenoate in 30 µl hexane whereas the control septa were loaded with 30 µl hexane only. This dose was chosen because an equivalent dose of the aggregation pheromone of F. occidentalis had been shown to be biologically active in the field [2,20].

Pairs of test and control traps with the order randomised within each pair were set out along the length of the rows of eggplants. A series of four trials with six or seven pairs of traps per trial (25 pairs in total) was conducted over 8 days with the spacing between traps within each pair set at either 1.6 m or 4 m and the duration of the trial lasting either 1 day or 4 days. A similar range of trap spacings has been used successfully when testing the aggregation pheromone of F. occidentalis (unpublished data). New traps were set out and re-randomised for each of the four trials. The thrips on the traps were identified and sexed under a stereo microscope; the two species could be separated easily by colour, position of wing vein cilia and antennal segment number [16]. The results of the four trials were combined after confirming the absence of a treatment×trial interaction. The data were log$_{10}$(x+1) transformed to homogenise the variance and analysed by analysis of variance with trap pairs and trials considered as blocks, using Minitab version 16 (Minitab Inc., USA).

Results

Olfactometer Bioassays

Virgin females were attracted to the odour of 50 adult males, but not to the odour of 50 adult females (Table 1). Mixed-age adult females were also attracted to adult males and their preference for the odour side (67%) was similar to that of virgin females (65%). Mixed-age males were also attracted to adult male odours. The preference of mixed-age adult males for adult males (68%) was similar to that of females for males (67%). The

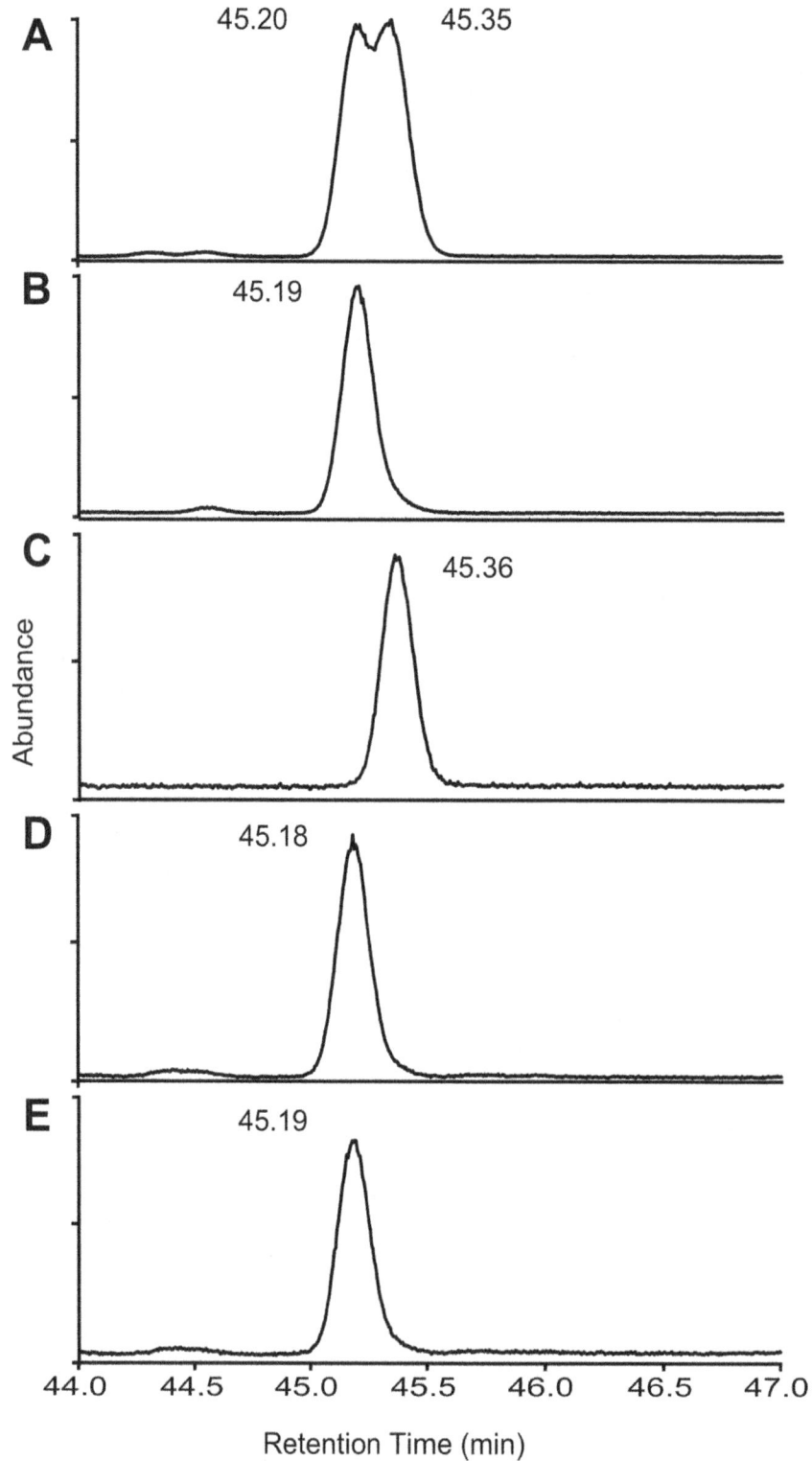

Figure 1. GC/MS analysis on a chiral column. Confirmation of *Thrips palmi* aggregation pheromone as the (*R*) enantiomer by GC/MS analysis on a CycloSil-B analytical column: (A) section of the TIC chromatogram from 44 to 47 min showing the two partly resolved peaks obtained from racemic lavandulyl 3-methyl-3-butenoate; (B) the peak obtained on injection of the (*R*)-lavandulyl 3-methyl-3-butenoate enantiomer; (C) the peak obtained on injection of the (*S*)-lavandulyl 3-methyl-3-butenoate enantiomer; (D) the peak obtained on injection of the *T. palmi* natural compound; (E) the enhanced peak obtained on co-injection of the *T. palmi* natural compound and (*R*)-lavandulyl 3-methyl-3-butenoate.

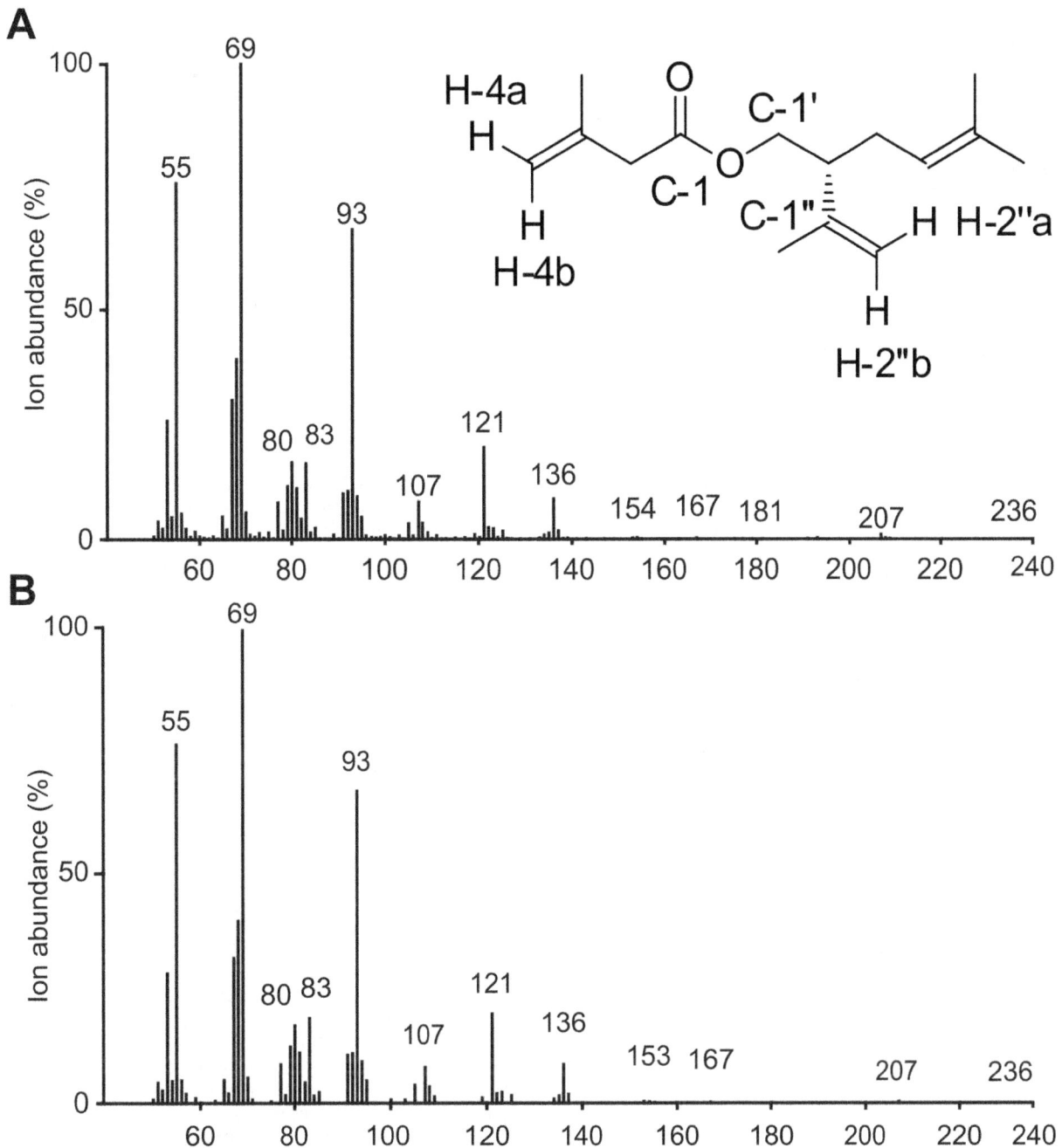

Figure 2. EI mass spectra of the major terpenoid component and the synthetic (R)-lavandulyl 3-methyl-3-butenoate. EI mass spectra (70 eV) of (A) the major terpenoid component (peak a) of the headspace volatiles of male *Thrips palmi* and (B) the mass spectrum of synthetic (R)-lavandulyl 3-methyl-3-butenoate. The inset shows the labelled structure of (R)-lavandulyl 3-methyl-3-butenoate.

proportion of thrips that made no choice within 3 min was low (5% overall, with a range of 3–8% across the four experiments).

GC/MS of SPME fibres and entrained extracts

Detailed comparison of TIC chromatograms obtained by GC/ MS analysis consistently showed that there was one compound present in SPME fibre entrainments of *T. palmi* adult males that was not present in the females or larvae. Examples of chromatograms of extracts from males, females and larvae are shown in Fig. 3 with the male-specific compound (peak a) present at 17.65 min. The mass spectrum of this compound resembled the

mass spectrum of the aggregation pheromone of *F. occidentalis*. Other peaks which were present in males and not in females in the example were not consistently present and are thus likely to be contaminants. Comparison of the area of peak a with a pentadecane standard suggested that it represents approximately 400 pg of material.

In total an estimated 2.2 μg of the male-specific *T. palmi* compound was collected by entrainments of headspace volatiles on Tenax-TA in Japan. A control entrainment in Japan of eggplant material without thrips confirmed that the compound was not produced by eggplants. The EI mass spectrum of the *T. palmi*

Table 1. Responses of adult *Thrips palmi* to volatiles produced by 50 adult males or females of the same species in a Y-tube olfactometer.

Test insects[a]	Odour source[b]	Number of choices		Preference for odour side (%)[c]	P[d]
		Odour side	Control side		
Virgin females	Females	29	43	40	0.12
Virgin females	Males	46	25	65	0.017
Females	Males	47	23	67	0.006
Males	Males	49	23	68	0.003

[a]All test and source insects were adults. Virgin females were 1–3 d post-emergence. Other adults were of mixed age.
[b]All odour sources consisted of 50 live adults of mixed age.
[c]The percentage of individual thrips that chose the odour side out of the total that made a choice.
[d]Exact probability based on null hypothesis of equal preference for the two sides.

compound is given in Fig. 2. It was similar to the mass spectrum of the *F. occidentalis* aggregation pheromone, neryl (*S*)-2-methylbutanoate (MW 238) [2] with characteristic ions at m/z 154 (0.4%), 136 (8%), 121 (18%), 93 (61%) and 69 (100%) suggesting a monoterpenoid substructure. An ion at m/z 236 (0.1%) suggested a molecular weight of 236 and ions at m/z 83 (16%) and 55 (75%) suggested the loss of $C_4H_7CO^+$ and $C_4H_7^+$ fragments respectively derived from a monounsaturated 5-carbon acid moiety. Isobutane CI analysis gave a strong ion at m/z 237 ($[M + H]^+$) confirming the molecular weight as 236. These data suggested that the compound was a monoterpene pentenoate.

The EI mass spectrum and retention index (RI) of the *T. palmi* compound were compared against those of a library of esters of monoterpene alcohols and pentenoic acids. Examples of RIs of three other monoterpene pentenoates are compared with the *T. palmi* compound and lavandulyl 3-methyl-3-butenoate in Table 2. The natural *T. palmi* compound had a retention time identical with that of lavandulyl 3-methyl-3-butenoate on both non-polar (HP5MS) and polar (DBWax) GC columns, and the mass spectra were superimposable. Co-injection of the *T. palmi* compound with lavandulyl 3-methyl-3-butenoate gave peak enhancement on both columns.

Chiral Chromatography

The *R* and *S* enantiomers of lavandulyl 3-methyl-3-butenoate gave two partially separated peaks with retention times (Rt) of 45.20 and 45.35 min respectively (Fig. 1A). (*R*)-lavandulyl 3-methyl-3-butenoate gave a single peak with a Rt of 45.19 min (Fig. 1B) with ions at m/z 154 (0.3%), 136 (8%), 121 (18%), 93 (65%), 83 (20%), 81 (10%), 69 (100%) and 55 (76%) and (*S*)-lavandulyl 3-methyl-3-butenoate also gave a single peak at 45.36 min (Fig. 1C). *T. palmi* pheromone eluted at 45.18 min (Fig. 1D), which suggested that the *T. palmi* compound was the *R* enantiomer. Co-injection of the *T. palmi* compound with (*R*)-lavandulyl 3-methyl-3-butenoate standard gave a single peak (Fig. 1E) confirming the *R* configuration for the *T. palmi* compound.

Field Trials

Test traps caught more *T. palmi* than control traps in all four trials and the effect was statistically significant across the four trials ($F_{1,24} = 13.05$, $P < 0.001$). The results were also significant for females ($F_{1,24} = 12.22$, $P = 0.002$) and males ($F_{1,24} = 6.71$, $P = 0.016$) when analysed separately (Table 3).

There was a trend towards higher numbers of *F. intonsa* on test traps than on control traps in all four trials, but this did not reach statistical significance ($F_{1,24} = 3.95$, $P = 0.059$), and the results were not significant for females ($F_{1,24} = 3.14$, $P = 0.089$) and males ($F_{1,24} = 1.97$, $P = 0.173$) when analysed separately (Table 3). Far fewer thrips of this species were present in the crop and on the traps.

Discussion

This paper presents the first identification of an aggregation pheromone in the genus *Thrips*. A previous study has shown that an aggregation pheromone was present in another thysanopteran genus *Frankliniella* [2]. Thus aggregation pheromones, which have now been confirmed in two different genera, may be more widespread in this commercially important group of insects than was previously recognised.

The olfactometer bioassays conducted in this study gave behavioural results that were of the same magnitude as those previously obtained with *F. occidentalis*, using similar apparatus [1]. Thus the percentage of *T. palmi* adult males and females that responded positively to the olfactometer arm with the male odour (65–68%) was similar to the percentage recorded in experiments with *F. occidentalis* (66–70%).

In this study we identified the aggregation pheromone of *T. palmi* as the monoterpene pentenoate ester (*R*)-lavandulyl 3-methyl-3-butenoate. The compound was found only in the headspace volatiles collected from males. Its structure was confirmed through comparison of its mass spectrum and three retention indices (RI) (collected on three different GC analytical columns, two achiral and one chiral) with a library of potential monoterpene pentenoate ester matches. Only one compound gave a positive match and this was confirmed by demonstration of peak enhancement. The absolute configuration was confirmed by RI matching with authentic *R* and *S* enantiomers on a chiral analytical GC column and by peak enhancement. The compound was previously obtained serendipitously during the synthesis of the sex pheromone of the vine mealybug *Planococcus ficus* [21].

Significant attraction, in the field, of both female and male *T. palmi* to sticky traps baited with (*R*)-lavandulyl 3-methyl-3-butenoate confirmed the identification of the aggregation pheromone. The pheromone increased trap catches on blue traps, which are already highly visually attractive. The percentage increases (62% females, 33% males) were similar to those found for *F. occidentalis* in experiments in which aggregation pheromone was added at the same dose rate to blue traps in pepper crops under

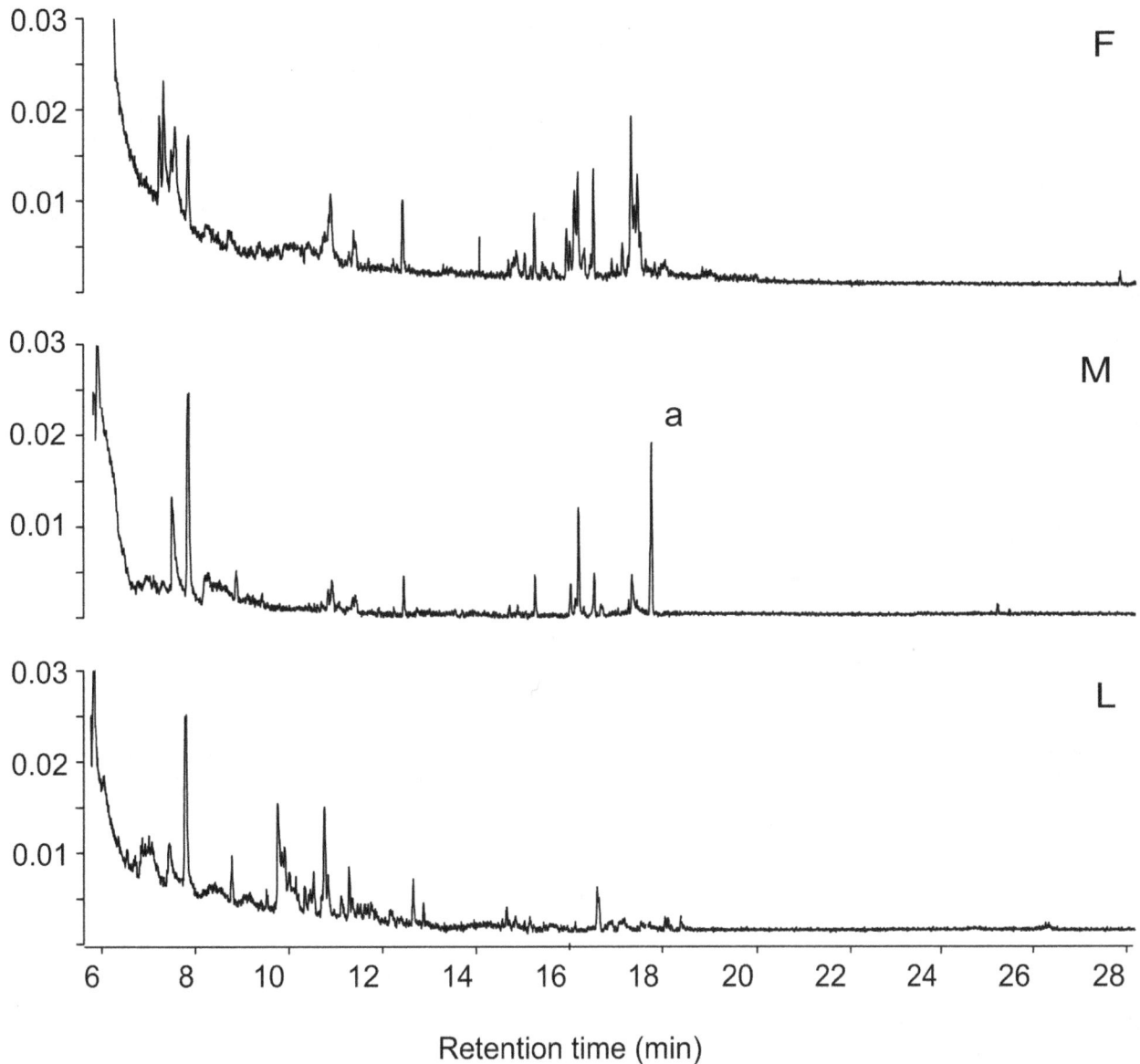

Figure 3. GC/MS traces of SPME-collected headspace volatiles from *Thrips palmi*. GC/MS total ion current (TIC) traces of SPME fibre collections of the headspace volatiles from mixed-age adult females (n = 40) (upper trace F), mixed-age adult males (n = 100) (middle trace M) and larvae (n = 40) (lower trace L) of *Thrips palmi* on a HP5MS column. The major male-specific compound at Rt = 17.65 min in the middle trace is indicated as peak a.

Table 2. GC retention indices (RI) of natural compound and examples for comparison of synthetic monoterpene pentenoates.

	RI	
	HP5MS	DBWax
natural *Thrips palmi* compound	1525	1854
lavandulyl 3-methyl-3-butenoate	1525	1854
neryl 3-methyl-3-butenoate	1599	1974
geranyl 3-methyl-3-butenoate	1624	2012
chrysanthemyl 3-methyl-3-butenoate (*cis* and *trans*)	1520 & 1530	1838 & 1844

Table 3. Catches of *Thrips palmi* and *Frankliniella intonsa* on blue sticky traps with and without the test compound.

Thrips species	No. on control traps[a]	No. on test traps[a]	Increase (%)	P (test vs control)
Trial 1. *Thrips palmi* females	24 (0.62±0.05)	44 (0.82±0.08)	83	-
Trial 2. *Thrips palmi* females	157 (1.43±0.03)	251 (1.60±0.08)	60	-
Trial 3. *Thrips palmi* females	57 (0.97±0.09)	68 (1.06±0.08)	19	-
Trial 4. *Thrips palmi* females	150 (1.37±0.09)	267 (1.64±0.06)	78	-
Thrips palmi females total	388 (1.08±0.07)	630 (1.26±0.08)	62	**
Trial 1. *Thrips palmi* males	9 (0.28±0.11)	33 (0.62±0.15)	367	-
Trial 2. *Thrips palmi* males	253 (1.60±0.08)	342 (1.71±0.10)	35	-
Trial 3. *Thrips palmi* males	80 (1.12±0.07)	112 (1.25±0.08)	40	-
Trial 4. *Thrips palmi* males	126 (1.29±0.10)	135 (1.29±0.13)	7	-
Thrips palmi males total	468 (1.04±0.11)	622 (1.20±0.10)	33	*
Thrips palmi total	856 (1.36±0.09)	1252 (1.54±0.09)	46	***
Frankliniella intonsa females	96 (0.58±0.07)	117 (0.69±0.05)	22	ns
Frankliniella intonsa males	15 (0.12±0.05)	27 (0.18±0.06)	80	ns
Frankliniella intonsa total	111 (0.60±0.07)	144 (0.75±0.06)	30	ns

[a]Total number of individuals caught followed, in brackets, by the mean catch per trap ±SE for the log-transformed data. Note that these standard errors include the variance between trap pairs and also between trials for the analysis of totals. They are therefore not appropriate for comparisons between test and control means. These extra variances are allowed for in the analysis of variance. Response of *Thrips palmi* and *Frankliniella intonsa* to blue sticky traps treated with lures loaded with 30 µg (R)-lavandulyl 3-methyl-3-butenoate in 30 µl hexane (test) or 30 µl hexane (control). Key to results of analysis of variance across trials:
*** = $P<0.001$;
** = $P<0.01$;
* = $P<0.05$;
ns = not significant ($P>0.05$); - = statistical comparison not carried out because of small number of replicates within the trial.

plastic (54% females, 38% males) [2]. These percentage increases in response to pheromone are low compared with those typically obtained in some other insects, such as moths, which can fly upwind for long distances, leading to speculation that the low percentages in thrips could be explained by a missing pheromone component. However, little is known about the role of thrips aggregation pheromones and it cannot be assumed that they act as long-range attractants in the same way as for moths. Higher percentage increases in thrips trap catches can be obtained when pheromones are used in conjunction with less visually attractive traps [22].

The *T. palmi* aggregation pheromone is structurally similar to the aggregation pheromone of *F. occidentalis*. In both cases the compounds are monoterpene esters; in *T. palmi* it is a pentenoate ester i.e. with a double bond in the fatty acid moiety and therefore with a reduced MW of 236, and in *F. occidentalis* it is a pentanoate ester (no extra double bond in the fatty acid moiety) with a MW of 238. In both cases the monoterpene is non-cyclic. Both molecules have chiral centres; in the *T. palmi* molecule the chiral centre is present in the monoterpene moiety and in the *F. occidentalis* molecule the chiral centre is present in the fatty acid moiety. The overall similarities are surprising because although *T. palmi* and *F. occidentalis* belong to the same sub-family Thripinae of the family Thripidae, they belong to two major and distinct groups: the *Thrips* genus-group and the *Frankliniella* genus-group. These are considered to be only distantly related [23],

probably having separated and diversified about 80–120 million years ago [24].

The source of the aggregation pheromone is unknown for *T. palmi* and other species. The sternal glands [6] are a possible source [5], but this remains to be confirmed. In experiments with *F. occidentalis*, the aggregation pheromone and the recently discovered contact pheromone, 7-methyltricosane ($C_{24}H_{50}$), were extracted from the surface over which male *F. occidentalis* moved [25]. The sternal glands, on the underside of the abdomen, would be conveniently positioned for depositing these chemicals.

The proportion of male *T. palmi* caught on the pheromone traps or control traps (50–55%) was markedly higher than the proportion of males found on the crop when collecting manually (5–8%). This phenomenon has been recorded before in *T. palmi* [26], *F. intonsa* [27] and *F. occidentalis* [28] and has been attributed to the greater activity of males [28].

The global trade in plants and plant products increases the risk of expansion of the range of *T. palmi* and the frequency of outbreaks within its current range. Where expansion does occur, control and eradication has been achieved so far, but it is recognised that successful management can be greatly facilitated by early intervention [9,15]. European Plant Health authorities are seeking improved methods for early detection as part of their contingency planning and deployment of pheromone-baited traps has been proposed as a potentially effective technique. Mass trapping experiments have confirmed this potential; field deployment of the aggregation pheromone of *F. occidentalis* doubled the catch on blue attractive traps [29], illustrating the possibility of using the technique to achieve earlier detection of the small populations typically present soon after the introduction of quarantine thrips.

In addition, field use of *F. occidentalis* aggregation pheromone in conjunction with the attractive traps, resulted in a combined reduction of 73% in thrips numbers and 68% in damage to strawberry crops, showing that in contained environments the approach can make a cost-effective contribution to population reduction in high-value crops, although further development work is required to achieve a more consistent outcome in commercial production systems [29]. The use of a thrips aggregation pheromone as part of an IPM programme has the advantages of removing both females and males, and may reduce both the rate of development of insecticide resistance and insecticide residues on crops.

Thus the enhanced trap catch of *T. palmi* provided by this aggregation pheromone may be a useful component of future pest management approaches, supporting improved quarantine detection, monitoring and control.

Acknowledgments

We thank Zhu Xiaoyun (Zhejiang Academy of Agricultural Sciences) for running the olfactometer bioassays, Prof. David Hall (NRI, University of Greenwich) for supplying (S)-lavandulyl acetate, Dr Masafumi Kobayashi (Utsunomiya University) for help in carrying out the field experiment and identification of male and female *T. palmi* on traps, and Dr Anthony Curtis (Keele University) for advising on NMR.

Author Contributions

Conceived and designed the experiments: SVSA WDJK TM JGCH YL. Performed the experiments: SVSA WDJK TM JGCH LU. Analyzed the data: SVSA WDJK TM JGCH LU. Contributed reagents/materials/analysis tools: TM KFAW. Wrote the paper: SVSA WDJK TM KFAW JGCH LU.

References

1. Kirk WDJ, Hamilton JGC (2004) Evidence for a male-produced sex pheromone in the western flower thrips *Frankliniella occidentalis*. J Chem Ecol 30: 167–174.
2. Hamilton JGC, Hall DR, Kirk WDJ (2005) Identification of a male-produced aggregation pheromone in the western flower thrips *Frankliniella occidentalis*. J Chem Ecol 31: 1369–1379.
3. Zhu X-Y, Zhang P-J, Lu Y-B (2012) Isolation and identification of the aggregation pheromone released by male adults of *Frankliniella intonsa* (Thysanoptera: Thripidae). Acta Entomol Sinica 55: 376–385.
4. Terry LI, Gardner D (1990) Male mating swarms in *Frankliniella occidentalis* (Pergande) (Thysanoptera: Thripidae). J Insect Behav 3: 133–141.
5. El-Ghariani IM, Kirk WDJ (2008) The structure of the male sternal glands of the western flower thrips, *Frankliniella occidentalis* (Pergande). Acta Phytopathol Entomol Hung 43: 257–266.
6. Mound LA (2009) Sternal pore plates (glandular areas) of male Thripidae (Thysanoptera). Zootaxa 2129: 29–46.
7. Sudo M, Tsutsumi T (2002) Ultrastructure of the sternal glands in two thripine thrips and one phlaeothripine thrips (Thysanoptera: Insecta). Proc Arthropod Embryol Soc Jpn 37: 35–41.
8. Kirk WDJ (1985) Aggregation and mating of thrips in flowers of *Calystegia sepium*. Ecol Entomol 10: 433–440.
9. Cannon RJC, Matthews L, Collins DW (2007) A review of the pest status and control options for *Thrips palmi*. Crop Prot 26: 1089–1098.
10. Whitfield AE, Ullman DE, German TL (2005) Tospovirus-thrips interactions. Annu Rev Phytopathol 43: 459–489.
11. Murai T (2002) The pest and vector from the East: Thrips palmi. In: Marullo R, Mound LA, editors. Thrips and Tospoviruses: Proceedings of the 7th International Symposium on Thysanoptera. Canberra: Australian National Insect Collection. 19–32.
12. Hamodi AA, Abdul-Rssoul MS (2012) New record of *Thrips palmi* Karny 1925 (Thysanoptera: Thripidae) in Iraq. Arab J Plant Prot 30: 142–144.
13. OEPP/EPPO (1989) Data sheets on quarantine organisms. No. 175. *Thrips palmi*. Bull OEPP 19: 717–720.
14. MacLeod A, Head J, Gaunt A (2004) An assessment of the potential economic impact of *Thrips palmi* on horticulture in England and the significance of a successful eradication campaign. Crop Prot 23: 601–610.
15. Cannon RJC, Matthews L, Collins DW, Agallou E, Bartlett PW, et al. (2007) Eradication of an invasive alien pest, *Thrips palmi*. Crop Prot 26: 1303–1314.
16. Palmer JM, Mound LA, du Heaulme GJ (1989) CIE Guides to insects of importance to Man. 2. Thysanoptera. Wallingford: CAB International.
17. Zada A, Dunkelblum E (2006) A convenient resolution of racemic lavandulol through lipase-catalyzed acylation with succinic anhydride: Simple preparation of enantiomerically pure (R)-lavandulol. Tetrahedron-Asymmetr 17: 230–233.
18. Theodorou V, Skobridis K, Tzakos AG, Ragoussis V (2007) A simple method for the alkaline hydrolysis of esters. Tetrahedron Lett 48: 8230–8233.
19. Song J-H, Han H-R, Kang S-H (1997) Color preference of *Thrips palmi* Karny in vinylhouse cucumber. RDA J Crop Prot 39: 53–56.
20. Gómez M, García F, GreatRex R, Lorca M, Serna A (2006) Preliminary field trials with the synthetic sexual aggregation pheromone of *Frankliniella occidentalis* on protected pepper and tomato crops in south-east Spain. IOBC/WPRS Bull 29: 153–158.
21. Hinkens DM, McElfresh JS, Millar JG (2001) Identification and synthesis of the sex pheromone of the vine mealybug, *Planococcus ficus*. Tetrahedron Lett 42: 1619–1621.
22. Sampson C, Hamilton JGC, Kirk WDJ (2012) The effect of trap colour and aggregation pheromone on trap catch of *Frankliniella occidentalis* and associated predators in protected pepper in Spain. IOBC/WPRS Bull 80: 313–318.
23. Buckman RS, Mound LA, Whiting MF (2013) Phylogeny of thrips (Insecta: Thysanoptera) based on five molecular loci. Syst Entomol 38: 123–133.
24. Austin AD, Yeates DK, Cassis G, Fletcher MJ, La Salle J, et al. (2004) Insects 'Down Under' - Diversity, endemism and evolution of the Australian insect fauna: examples from select orders. Aust J Entomol 43: 216–234.
25. Olaniran OA, Sudhakar AVS, Drijfhout FP, Dublon IAN, Hall DR, et al. (2013) A male-predominant cuticular hydrocarbon, 7-methyltricosane, is used as a contact pheromone in the western flower thrips *Frankliniella occidentalis*. J Chem Ecol 39: 559–568.
26. Kawai A (1986) Studies on population ecology and population management of *Thrips palmi* Karny [In Japanese with English summary]. Bull Veg & Ornam Crops Res Stn Japan C 9: 69–135.
27. Murai T (1988) Studies on the ecology and control of flower thrips, *Frankliniella intonsa* (Trybom). [In Japanese]. Bull Shimane Agric Exp Stn 23: 1–73.
28. Matteson NA, Terry LI (1992) Response to color by male and female *Frankliniella occidentalis* during swarming and non-swarming behavior. Entomol Exp Appl 63: 187–201.
29. Sampson C, Kirk WDJ (2013) Can mass trapping reduce thrips damage and is it economically viable? Management of the western flower thrips in strawberry. PLoS ONE 8: e80787.

5

Development of a Freeze-Dried Fungal Wettable Powder Preparation Able to Biodegrade Chlorpyrifos on Vegetables

Jie Liu[1][9], Yue He[2][9], Shaohua Chen[3], Ying Xiao[1], Meiying Hu[1], Guohua Zhong[1]*

1 Laboratory of Insect Toxicology, and Key Laboratory of Pesticide and Chemical Biology, Ministry of Education, South China Agricultural University, Guangzhou, P.R. China, 2 Guangdong Zhuhai Supervision Testing Institute of Quality and Metrology, Zhuhai, P.R. China, 3 Guangdong Province Key Laboratory of Microbial Signals and Disease Control, South China Agricultural University, Guangzhou, P.R. China

Abstract

Continuous use of the pesticide chlorpyrifos has resulted in harmful contaminations in environment and species. Based on a chlorpyrifos-degrading fungus *Cladosporium cladosporioides* strain Hu-01 (collection number: CCTCC M 20711), a fungal wettable powder preparation was developed aiming to efficiently remove chlorpyrifos residues from vegetables. The formula was determined to be 11.0% of carboxymethyl cellulose-Na, 9.0% of polyethylene glycol 6000, 5.0% of primary alcohol ethoxylate, 2.5% of glycine, 5.0% of fucose, 27.5% of kaolin and 40% of freeze dried fungi by response surface methodology (RSM). The results of quality inspection indicated that the fungal preparation could reach manufacturing standards. Finally, the degradation of chlorpyrifos by this fungal preparation was determined on pre-harvest cabbage. Compared to the controls without fungal preparation, the degradation of chlorpyrifos on cabbages, which was sprayed with the fungal preparation, was up to 91% after 7 d. These results suggested this freeze-dried fungal wettable powder may possess potential for biodegradation of chlorpyrifos residues on vegetables and provide a potential strategy for food and environment safety against pesticide residues.

Editor: Raul Narciso Carvalho Guedes, Federal University of Viçosa, Brazil

Funding: The authors are grateful to the National Natural Science Foundation (No.30871660 and No. 31371960) of China for the financial support. The funders had no role in study design, data collection and analysis, decision to publish, or preparation of the manuscript.

Competing Interests: The authors have declared that no competing interests exist.

* Email: guohuazhong@scau.edu.cn

[9] These authors contributed equally to this work.

Introduction

Chlorpyrifos is a broad–spectrum, moderately toxic organophosphorus pesticide [1]. Since been first introduced into market in 1960s, chlorpyrifos has been globally used in pest control in agriculture and home for its acute neurotoxic effects through acetylcholinesterase inhibition and consequent cholinergic hyper stimulation [2,3]. However, increasing evidences indicate that chronic exposure to chlorpyrifos can cause persisting neurobehavioural dysfunction [4], even with low doses, which can elicit chronic toxicity [5]. There is ample evidences that chlorpyrifos for controlling insects agriculturally or residentially can adversely affect non-target organisms like fish [6], bees [7], silkworms and rats [8]. These potential threats lead to a great concern of environment and food safety and human health for its potential toxic [9,10]. Although the strict limitation of chlorpyrifos residue launched in many countries, the use of the pesticide is still extensive. Thus, it is urgent to develop an efficient and convenient strategy able to eliminate the chlorpyrifos residues on vegetables and soils so that the food and environment safety can be guaranteed.

Biodegradation, using living microorganisms or active enzymes to detoxify and degrade pollutants, has received attention as a safe, efficient and cost-effective approach to clean up contaminated environment [11–15]. The degradation of chlorpyrifos by microorganisms in pure cultures has been well investigated [16,17] and all the published studies concluded that using active microorganisms could be a promising approach for chlorpyrifos degradation on farm products [16,18]. However, field conditions adversely influence microbial ingredients and impair the degradation capacity, so little is known on chlorpyrifos degradation by microorganisms in agricultural application. How to apply these pesticide-degrading microorganisms in field conditions has always been the greatest obstacle. In fact, with the reported methods [19], like using cell-free extracts [18], it was still difficult to reach the expected performance and hard to put into field application.

The practical application of microorganisms with pesticide degrading capacity could refer to the microbial pesticides like *Bacillus thuringiensis* (BT), which have already been developed many commercial formulations including oil solutions, wettable powder and effervescent tablets [20]. Development of microbial preparations capable of degrading pesticides mostly depends on the persistence of active compounds, which could be improved by the formulation design and selection of assistant agents. However, the extraction of active proteins increases the manufacture cost and the producing procedure may impact the enzyme ability [21], thus a cost-efficient, effective and safe method of microbial preparation should be established.

Table 1. Effects of protectants on chlorpyrifos degradation by freeze-dried fungi.

Protectants	Amount	Degradation rate
	(%)	(%)
Fucose	2.5	44.9 ± 5.3^{i}
	5.0	50.1 ± 6.6^{g}
	7.5	60.2 ± 5.2^{d}
Glycine	2.5	48.7 ± 1.6^{h}
	5.0	52.5 ± 1.8^{f}
	7.5	56.5 ± 6.3^{e}
Glycerol	2.5	68.1 ± 1.7^{c}
	5.0	72.0 ± 1.7^{b}
	7.5	78.8 ± 1.5^{a}
CK	N/A	34.6 ± 1.0^{j}

Note: The data presented are means of three replicates with standard deviation, which is within 5% of the mean. Different letters indicate significant differences ($p < 0.05$, LSD test). CK represents the control group without protectants.

In this study, using a chlorpyrifos degrading fungus *Cladosporium cladosporioides* strain Hu-01 (collection number: CCTCC M 20711), a fungal preparation able to eliminate chlorpyrifos residues was developed through a process of vacuum freeze dehydration with appropriate agents. In addition, the multiple factors could affect the preparation quality was evaluated and checked to ensure its being qualified for industrial manufacture and market promotion. The objectives of this study were to test and determine the capacity of the preparation to degrade chlorpyrifos residue on vegetables, and to explore a potential approach for chlorpyrifos degradation by microorganisms in field conditions.

Materials and Methods

Chemicals and fungus

Technical grade chlorpyrifos (96% purity) was obtained from Dow AgroSciences, USA. All other chemicals and solvents used were analytical grade.

Activated sludge samples were collected as inoculum from a pesticide-manufacturing wastewater treatment system and chlorpyrifos-degrading fungus *Cladosporium cladosporioides* Hu-01 was screened and isolated [16]. This strain was deposited in China Center for Type Culture Collection under the collection No. CCTCC M 20711. The constant-temperature culture method was referred to Gao et al. [22] and the conditions were inoculum amount 0.1% (wet fungi), 28°C and 150 rpm on a rotary shaker for 5 days. Previously, the degradation of chlorpyrifos by strain Hu-01 was determined. 92.7% of chlorpyrifos at the initial concentration of 25 mg·L^{-1} was degraded in half hour, which revealed that strain Hu-01 can significantly reduce chlorpyrifos residue.

Chemical analysis

The analysis method of chlorpyrifos was referred to Gao et al. [22]. Chlorpyrifos was analyzed on an Agilent 1100 High Performance Liquid Chromatography (HPLC) (Agilent, USA) equipped with a Hypersil ODS2 C$_{18}$ reversed phase column (4.6 nm × 250 mm, 5 mm). A mixture of methanol and water (90:10, v/v) was used as the mobile phase at a flow rate of 1.0 mL·min^{-1}. The injection volume was 10 μL.

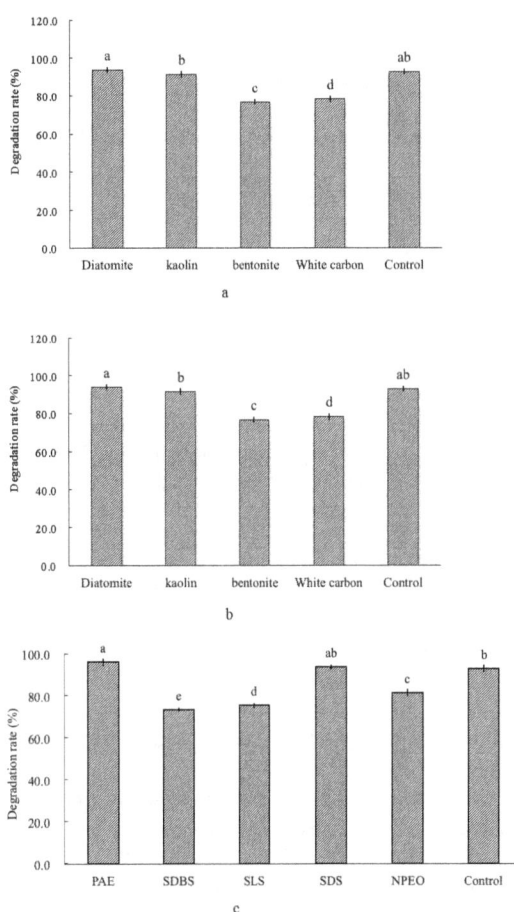

Figure 1. Effects of carriers, dispersants or surfactants on chlorpyrifos degradation added in freeze dried fungi. Note: a: effects of carriers; b: effects of dispersants; c: effects of surfactants.

Table 2. The dispersity of CMC-Na and PEG 6000 in different contents.

Dispersants	Content (%)	Dried fungi (%)	Dispersity(%)
CMC-Na	7	48	63.2±1.1[b]
	9	46	69.0±1.2[a]
	11	44	68.9±0.9[a]
PEG 6000	7	48	53.3±0.4[c]
	9	46	69.2±0.7[a]
	11	44	68.9±1.1[a]

Note: The data presented are means of three replicates with standard deviation, which is within 5% of the mean. Different letters indicate significant differences ($p <$ 0.05, LSD test).

Recoveries of 2.5, 5, 10, 25, and 50 mg·L^{-1} of chlorpyrifos were determined from 86.3% to 106.2% and relative standard deviation (RSD) ranged from 0.6% to 4.1%.

Procedure of freeze-dried fungi powder

After culturing for 7 d, hyphae were collected by filtering. Being pre-freezed at $-20°C$ for 24 h, the fungi samples were immediately vacuumed and freeze-dried for another 24 h. Then, those samples were kept in normal temperature and the preliminary preparation was done. The degradation of chlorpyrifos by dried powder was conducted by adding 0.1 g of fungi powder to 5 mL of 25 mg·L^{-1} chlorpyrifos in phosphate buffered solution and the residual concentration was determined in half hour by HPLC.

Determination of protectants and carriers

To maintain the microbial activity, protectants were added before dehydration and each fungi-protectant sample was moved to freeze dehydration machine for 24 h until the humidity fell to 2% or 3%. The degradation of 25 mg·L^{-1} chlorpyrifos was then investigated by HPLC. Control groups without protectants were set and the HPLC results leaded to the available protective agents and the suitable dosages.

The biocompatibility of carriers including diatomite, kaolin, bentonite and white carbon were investigated by mixing 50 mL of 10% fungi solution with 500 mg·L^{-1} carriers, adjusting chlorpyrifos at 25 mg·L^{-1} and degradation rate of chlorpyrifos was determined after 24 h. Carriers were individually mixed with freeze-dried fungi ($v: v = 2:3$) and then the status of blocking, fluidity and cost of each carrier-fungi powder sample was investigated by investigation and surveys. Control groups without carriers were set and the observation results leaded to the available carriers and the suitable dosages.

Determination and composition optimization of dispersants and surfactants

The biocompatibility of fungi powder with assistant agents was tested including dispersants: gelatin, polyethylene glycol 6000 (PEG 6000), polyvinylpyrrolidone (PVP), carboxymethyl cellulose-Na (CMC-Na), CMC and Polyvinyl alcohol (PVA); and surfactants: primary alcohol ethoxylate (PAE), sodium dodecyl benzene sulfonate (SDBS), sodium dodecyl sulfate (SDS) and nonylphenol ethoxylate (NPEO). 50 mL of 10% fungi solution was mixed with 50 mg·L^{-1} dispersant or surfactant, adjusting chlorpyrifos at 25 mg·L^{-1} and degradation rate of chlorpyrifos was determined.

For the dispersants, the selection was carried out by pouring 5.000 g of each dispersant into 250 mL of standard hard water at 30°C $±2°C$ and letting it sit for 30 min after oscillation. Then

move 9/10 of upper liquid out and hydrate the left 25 mL solution. Repeat 3 times. The weight of each sample was record and the dispersion rate was calculated by **Eq. (1)**.

$$W_1 = m_1 - m_2 \times 10 \times 100 \qquad (1)$$

where W_1 is dispersity (%), m_1 is the average mass of sample and m_2 is the average mass of hydrated sample of 25 mL left in bottom.

For the surfactants, wetting time of each one was determined. Pour 5.0 g of each surfactant-fungi powder sample into 100 mL of standard hard water. The duration it took to be fully wet was recorded and repeated 5 times. Then the average time was calculated and the available surfactants were determined.

Response surface methodology (RSM) based on the Central Composite Rotatable Design (CCRD) was applied to optimize the each key component as a variable [16]. Theoretically, the interactions among variables could significantly affect the dispersion of fungal powder preparation in solution and the degradation of chlorpyrifos can be influenced accordingly [23,24]. In this experiment, three assistant agents for optimizing were PAE (3.0% to 7.0%), CMC-Na (5.0% to 13.0%) and PEG 6000 (5.0% to 13.0%). The range and center point values of three independent variables were based on the results of preceding experiments. The dependent variable was the dispersion rate determined by adding each 5.000 g agent via **Eq. (1)**. The data were analyzed using RSM of the statistic analysis system (SAS) software (Version 9.0) to fit the following quadratic polynomial equation (**Eq. (2)**).

$$Y_i = b_0 + \sum b_i X_i + \sum b_{ii} X_i^2 + \sum b_{ij} X_i X_j \qquad (2)$$

where Y_i is the predicted response (dispersion (%)), X_i and X_j are variables, b_0 is the constant, b_i is the linear coefficient, b_{ij} is the interaction coefficient, and b_{ii} is the quadratic coefficient.

Preparation of wettable freeze-dried fungi powder

The procedure of fungi preparation was carried out as following. Applied protectants were individually mixed with hyphae filtrated from fermented solution before freeze dehydration, and then the freeze-dried powder sample was mingled with other assistant agents. After altering pH to 6~7 and smashing till 95% of it sieved to 63 μm, the final preparation was done and ready for degradation performance inspection.

Preparation quality inspection

Protein concentration was measured by the method of Coomassie blue staining method with bovine serum albumin as a standard using a spectrophotometer (Shimadzu, Japan).

Table 3. Central composite rotatable design (CCRD) matrix and the response of dependent variable for dispersity of preparation (Y).

Run	Contents of variables (%)			Dispersity (%)
	X_1	X_2	X_3	Y
1	11.0	11.0	6.0	61.42±0.84
2	11.0	7.0	4.0	61.13±0.66
3	7.0	11.0	4.0	64.09±1.25
4	7.0	7.0	6.0	65.09±0.58
5	11.0	11.0	4.0	60.57±1.84
6	11.0	7.0	6.0	62.49±0.48
7	7.0	11.0	6.0	65.62±1.10
8	7.0	7.0	4.0	64.37±0.55
9	11.0	11.0	4.0	56.87±1.62
10	11.0	7.0	6.0	63.11±1.14
11	7.0	11.0	6.0	65.04±1.04
12	7.0	7.0	4.0	65.82±1.68
13	11.0	11.0	6.0	61.87±1.22
14	11.0	7.0	4.0	61.23±1.73
15	7.0	11.0	4.0	64.20±0.98
16	7.0	7.0	6.0	68.67±1.05
17	9.0	9.0	5.0	65.92±2.97
18	9.0	9.0	5.0	69.83±0.91
19	9.0	9.0	5.0	68.58±0.79
20	9.0	9.0	5.0	71.04±0.85
21	5.0	9.0	5.0	59.98±0.69
22	13.0	9.0	5.0	72.07±2.02
23	9.0	5.0	5.0	59.05±1.93
24	9.0	13.0	5.0	70.41±0.54
25	9.0	9.0	3.0	60.68±17.99
26	9.0	9.0	7.0	72.66±2.09
27	9.0	9.0	5.0	72.04±0.27
25	9.0	9.0	3.0	60.68±17.99
26	9.0	9.0	7.0	72.66±2.09
27	9.0	9.0	5.0	72.04±0.27
28	9.0	9.0	5.0	71.83±0.40
29	9.0	9.0	5.0	72.60±0.49
30	9.0	9.0	5.0	72.51±0.63
31	9.0	9.0	5.0	72.46±0.67
32	9.0	9.0	5.0	71.60±0.85

Note: X_1: CMC-Na; X_2: PEG 6000; X_3: PAE; Y: dispersity. The data presented are means of three replicates with standard deviation, which is within 5% of the mean.

Wetting ability was determined by the time that 5.0 g of fungal powder took to be fully wet. Dispersion ability was inspected by the dispersion rate of fungal powder in solution. The size of powder granules was tested through the standard mesh of 325 ($\geq$ 95%). Humidity was detected by weighing 0.5 g of fungal preparation and culture vessel together. After 2 h in dryer at 100°C, calculate the average humidity content using the weight of cooled preparation and culture vessel and repeated 3 times.

Storage stability was checked by storing for 15 d and 1 to 5 months at 4°C and 25°C, respectively, then reacting with chlorpyrifos at 25 mg·L^{-1} for 30 min and the degradation performance was detected to illustrate the stability of the preparation. The stability in low temperature was tested at 0°C in 1 h and 7 d, respectively.

Degradation performance of freeze-dried fungal powder

Each test area of 4 m^2 was divided from a cabbage filed without any pesticides applied before and 45% chlorpyrifos emulsifiable concentrate (EC) was sprayed at 225, 350, 450, 700 and 900 g·L^{-1}, respectively, with the spray amount of 50 L·667 m^{-2}. After 3 d, the fungal preparation was sprayed at the amount of 200 g·667 m^{-2} against the controls sprayed only water. The samples investigated were taken randomly in 1, 3 and 7 d. 25.0 g of each vegetable sample soaking in 50.0 mL of

Table 4. Analysis of variance (ANOVA) for the fitted quadratic polynomial model for dispersity of preparation.

Source	DF	SS	MS	F value	Pr>F*
Model	9	718.08	35.90	6.805	0.001
Linear	3	334.91	66.98	12.695	0.000
Square	3	342.93	68.59	12.999	0.000
Interaction	3	40.24	4.02	0.763	0.661
Error	13	58.04	5.28		
Total	22	776.12			

Note: DF refers to degrees of freedom; SS refers to sum of sequences; MS refers to mean square.
*P Level less than 0.05 indicate the model terms are significant.

acetonitrile were decolored by 0.2 to 0.8 g of active carbon. After filtrating, partitioning with NaCl, clean-up, and concentration, residues dissolved in n-hexane was analyzed by Gas Chromatography (GC) HP-6890 with ECD and a capillary column of BD-1701, 30 m×0.32 mm×1.0μm. The temperatures of injection port, column and detector were 100°C, 220°C and 250°C, respectively. The flow of gas was as follows: nitrogen, 60 mL·min^{-1}; air, 60 m L·min^{-1}; hydrogen, 30 m L·min^{-1}.

Data analysis

All of the experiments were carried out in triplicate, and the results were the means of three replicates. Standard deviations were also determined using Statistic Analysis System (SAS). The significance ($p<0.05$) of differences was treated statistically by one-, two-, or three-way analysis of variance (ANOVA) and evaluated by post hoc comparison of means using lowest significant differences (LSD).

Ethics Statement

No specific permits were required for the described field studies. No specific permissions were required for these locations. We confirm that the location is not privately-owned or protected in any way. We confirm that the field studies did not involve endangered or protected species.

Results

Chlorpyrifos degradation by freeze-dried fungi

After freeze dehydration, the degradation of 10% freeze dried fungi powder was determined and simply $40.9\pm0.3\%$ of chlorpyrifos at the initial concentration of 25 mg·L^{-1} was degraded, which was significantly lower than the degradation rate of $92.5\pm0.1\%$ before freeze dehydration.

Protectants and carriers

To preserve the biodegradation ability of strain Hu-01, the available protective agents were selected. From the results in **Table 1**, it was shown that three protectants, fucose, glycin and glycerol, largely protected the freeze fungi from inactivation. With the increase of protectant content, the degradation of chlorpyrifos was gradually enhanced, which was significantly differed from the controls. 7.5% of glycerol revealed the best protective ability and nearly 80% of chlorpyrifos was degraded. However, the sample with glycerol was difficult to smash and dehydrate, so glycerol was not the available option. Fucose and glycine also presented as good protectants and more than half of chlorpyrifos was degraded when they reached the maximum testing amount. Concerning the factors above, 5.0% of fucose and 2.5% of glycine were selected to be the protective compounds in this fungal preparation for their protective performance, effective cost and a reasonable distribution of assisted agents.

The biocompatibility of tested carriers in **Fig. 1(a)** indicated that they were compatible with active fungi and no obvious

Table 5. Effect estimates for the fitted quadratic polynomial model for dispersity of preparation.

Term	Estimate	Standard Error	t value	Pr>|t|*
X_1	2.64	0.47	5.633	0.000
X_2	1.66	0.47	3.550	0.005
X_3	1.83	0.47	3.909	0.002
X_1*X_1	−1.94	0.42	−4.566	0.001
X_1*X_2	0.17	0.57	0.301	0.769
X_1*X_3	0.12	0.57	0.201	0.844
X_2*X_2	−2.26	0.42	−5.330	0.000
X_2*X_3	0.22	0.57	0.391	0.703
X_3*X_3	−1.78	0.42	−4.186	0.002

Note: X_1: CMC-Na; X_2: PEG 6000; X_3: PAE.
*P Level less than 0.05 indicate the model terms are significant.

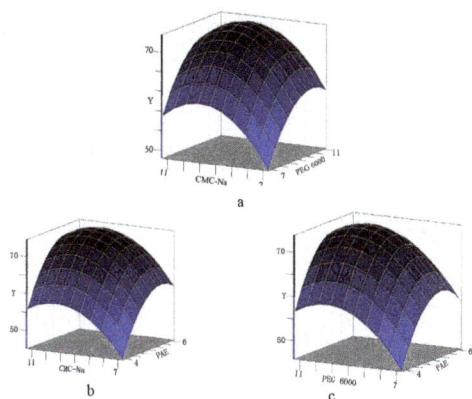

Figure 2. **Response surface plots showing the effects of three variables on dispersity of the fungal preparation.** Note: a: the effects of CMC-Na (X_1) and PEG 6000 (X_2) on dispersity (Y) of the preparation; b: the effects of CMC-Na (X_1) and PAE (X_3) on dispersity (Y) of the preparation; c: the effects of PEG 6000 (X_2) and PAE (X_3) on dispersity (Y) of the preparation.

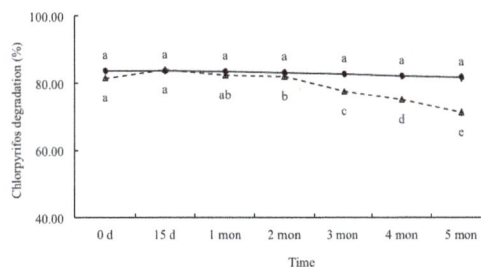

Figure 3. **Chlorpyrifos degradation trends reflecting the stability of fungal preparation in normal and low temperature for different periods.** Note: The data presented are means of three replicates with standard deviation, which is within 5% of the mean. Different letters indicate significant differences ($p < 0.01$, LSD test). ●, chlorpyrifos degradation by fungal preparation stored at 4°C; △, chlorpyrifos degradation by fungal preparation stored in normal temperature.

disadvantages observed. From the observed results, the samples pre-treated by 27.5% of kaolin were easy to smash into powder with good fluidity while other three carriers were lack of fluidity or easily, even seriously blocking. As a result, kaolin was chosen to be the carrier in this process.

Selection and composite optimization of dispersants and surfactants

The biocompatibility of dispersants and surfactants was investigated at first. The results in **Fig. 1(b and c)** illustrated that two categories of assistant agents were compatible with active fungi and no obvious disadvantages observed, which revealed that these agents in the present study contributed more or less to remain the degradation capacity of the strain and were suitable for further screening.

Then, four surfactants, SDS, SDBS, PAE and NPEO were evaluated. The time that each surfactant took to be fully wet was 63.7 ± 2.8 s, 49.3 ± 3.4 s, 10.6 ± 0.7 s and 14.1 ± 2.0 s, respectively, which indicated that PAE's wetting period was shorter than the other three surfactants, so PAE was selected to be the surfactant in the fungal preparation.

Dispersants in the microbial preparation remarkably improve the thermodynamic stability. The results indicated that the samples with 5% of CMC-Na or PEG 6000 were the best in dispersion and ones with CMC, PAE and NPEO followed with obvious sediments or floccules. Hence, CMC-Na and PEG 6000 were recommended to be the available dispersants. The results from **Table 2** showed that 9% or 11% of CMC-Na or PEG 6000 revealed more excellent performance in dispersion with no significant differences.

Based on CCRD, RSM was employed to investigate the interactive effects of significant variables including CMC-Na (X_1), PEG 6000 (X_2) and PAE (X_3) on dispersion performance. The design matrix and experimental responses for ingredient formulation are shown in **Table 3**. Subsequently, the data from **Table 3** were assessed by response surface regression procedure of SAS software package, and also the results of the quadratic polynomial model fitting in the term of ANOVA were shown in **Table 4**. By applying the multiple regression analysis on the experimental data, the following quadratic polynomial model equation (**Eq. (3)**) was delivered to explain the dispersion (Y) of

dried fungal preparation ($p > 0.05$):

$$Y = -64.04 + 8.50 \times X_1 + 9.20 \times X_2 + 16.16 \times X_3 - 0.43 \times X_1{}^2$$
$$+ 0.043 \times X_1 \times X_2 + 0.057 \times X_1 \times X_3 - 0.51 \times X_2{}^2 \quad (3)$$
$$+ 0.112187 \times X_2 \times X_3 - 1.58 \times X_3{}^2$$

where Y is the predicted dispersity (%) of the fungal preparation; X_1, X_2 and X_3 are the coded values for the CMC-Na, PEG 6000 and PAE, respectively.

The statistical significance of **Eq. (3)** was also evaluated by performing F-test and t-test (**Table 4 and 5**). The statistical analysis indicated that the model linear term coefficient of X_1, X_2 and X_3 and the quadric term coefficient of X_1, X_2 and X_3 showed significant effects ($p < 0.05$) on dispersion. Thus, the quadratic polynomial equation (**Eq. (4)**) was modified to be:

$$Y = -64.04 + 8.50X_1 + 9.20X_2 + 16.16X_3 - 0.43X_1{}^2$$
$$- 0.51X_2{}^2 - 1.58X_3{}^2 \quad (4)$$

The adequacy of two models was examined by the determination coefficients ($R^2 = 0.9252$), which suggested that the predicted values of the models were well correlated with the experimental values. The three-dimensional (3D) response surfaces in **Fig. 2 (a, b and c)** show the respective effects of X_1 and X_2, X_1 and X_3, X_2 and X_3 on dispersion of fungal preparation while keeping the third variable at 0. The model predicted a maximum dispersion of 70.3% at the stationary point. Accordingly, the optimized formulation was determined by the predict model and previous experiment results to be 11.0% of CMC-Na, 9.0% of PEG 6000, 5.0% of PAE, 2.5% of glycine, 5.0% of fucose, 27.5% of kaolin and 40% of freeze dried fungi.

Preparation quality inspection

To evaluate the quality of fungal wettable powder, the degradation of chlorpyrifos by optimized preparation was checked first and the results showed that it eliminated $76.1 \pm 0.4\%$ of chlorpyrifos at the initial concentration of 25 mg/L in 30 minutes, which can meet the practical requirements. As the amount and concentration of total protein indirectly reflected the content of active protein, the equation (**Eq. (5)**) of the standard curve was obtained ($R^2 = 0.9966$) by Coomassie blue staining method:

Table 6. The size, suspensibility and chlorpyrifos degradation of fungal preparation in normal and low temperature storage.

	Size(%)	Dispersity (%)	chlorpyrifos Degradation (%)
0 h	95.4±0.6[a]	77.8 n±1.6[b]	83.4±2.1[c]
0°C, 1 h	95.3±1.0[a]	74.7±1.4[b]	85.7±2.2[c]
0°C, 7 d	94.9±1.7[a]	74.6±1.3[b]	85.6±2.4[c]

Note: The data presented are means of three replicates with standard deviation, which is within 5% of the mean. Different letters indicate significant differences ($p<$ 0.05, LSD test). Size: the amount of preparation granules through standard sieve in 325 meshes.

$$y = 1964.6x - 26.689 \qquad (5)$$

where y is corresponding concentration of protein (mg·L^{-1}), and x is the value of OD$_{595}$. The value of R^2 suggested this model was stable and credible in the range of testing concentrations and the protein amount of freeze-dried fungi powder was 10.6 mg·g^{-1}.

The results in **Table 6** indicated that low temperature (0°C) storage rarely affected the dispersing and degrading performances, which was non-significantly different ($p>0.05$) from ones stored in normal temperature. Meanwhile, stored at 4°C in periods of 15 d, 1, 2, 3, 4 and 5 months, the degrading performance was neither degenerated nor significantly different from ones before low temperature storage. Although after 5 months, the degradation rate dropped a little, the value remained over 70%, which could be expected to meet the practical needs (**Fig. 3**). Furthermore, other quality features of the chlorpyrifos-degrading preparation were investigated individually including wettability (<1 min), dispersion (77.8±1.6%), humidity (≤1.8±0.8%), fineness (≥95% of granules sieved through 63 μm) and pH value (6 to 7). All the results indicated that the fungal preparation was physically stable and available for practical use.

Chlorpyrifos degradation on cabbage by freeze-dried fungal powder preparation

The recoveries of chlorpyrifos on cabbage were obtained by adding chlorpyrifos at concentration of 0.01, 0.05 and 0.10 mg·kg^{-1} by GC and the results were 77.0%, 88.9% and 84.7%, respectively. As RSD ranged from 2.0% to 7.9%, the pre-

treatment and detection methods of chlorpyrifos residue on cabbage were reliable and up the pesticide residue detection standards.

From **Fig. 4**, the residual concentrations started at 2.12, 2.31 and 2.45 mg·kg^{-1}, respectively in areas treated with 48% chlorpyrifos EC at the concentrations of 225, 350 and 450 g·ha^{-1}. After applied with the microbial preparation in 1 d, the residual concentrations remarkably dropped to 0.31, 0.53 and 0.81 mg·kg^{-1}, respectively while the residual concentrations of control groups were still over 2.0 mg·kg^{-1}. In 3 d, residual concentrations decreased to 0.28, 0.37 and 0.76 mg·kg^{-1} while the control groups were 1.75, 1.85 and 2.38 mg·kg^{-1}, respectively. In addition, the degradation rates grew up to nearly 87% within three days. In a period of 7 d, the chlorpyrifos residual concentrations constantly declined to 0.26, 0.21 and 0.29 mg·kg^{-1} while the control groups of 1.50, 1.41 and 1.59 mg·kg^{-1}, respectively. Moreover, the degradation rates of each area increased to 87.3%, 91.0% and 88.2%, respectively.

Discussion

Given the fact that biodegradation is an enzymatic reaction, the microbial activity of pesticide-degrading microorganisms is easily affected by external factors. Consequently, the technical difficulties of preparation design become the main obstacles of practical application and promotion. On the purpose of enhancing degradation activity and availability, the assistant agents, referred to the design of bio-pesticides, were introduced in this study. Assistant agents are all ingredients except the active compounds in microbial preparation and they help remain the extracted enzymes active and available [25,26].

In this study, four categories of assistant agents including protectants, carriers, dispersants and surfactants were compared within group and the best options were determined from the results of four independent tests. Firstly, protectants preserved the fungi cells from cell damage caused by the process of freeze dehydration. The selection of available protectants was conducted with fucose, glycine and glycerol. From the test result, glycine and fucose were chosen out among applied agents. Glycine is a well-known protective compound for preventing tissue injury and enhancing anti-oxidative ability [26,27] while fucose is safe in wound healing and protective against cytotoxicity [28,29]. Secondly, carriers are the basic composites which are required to be high absorptive, fluid and cost-efficient [30]. By the test of physical features, kaolin was the best option for the carrier of this microbial preparation. Then, concerning the actual function of a final preparation, the wettability is vital [31]. This preparation was supposed to be sprayed on the surface of vegetable leaves and it was hard to soak the leaf efficiently because most plant blades are covered with wax which is the main reason of low surface energy on leaf surface. As a result, the surfactants were introduced. Common surfactants are anionic surfactants like ABS and PAE

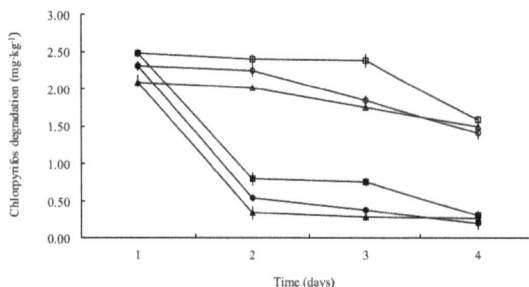

Figure 4. The trends of degradation of chlorpyrifos by fungal wettable powder on cabbage at different initial concentrations. The data presented are means of three replicates with standard deviation, which is within 5% of the mean: ■, degradation of chlorpyrifos at initial concentration of 225 g·ha^{-1}; ●, degradation of chlorpyrifos at initial concentration of 350 g·ha^{-1}; ▲, degradation of chlorpyrifos at initial concentration of 450 g·ha^{-1}. Solid legends are treatments with fungal wettable powder while hollow ones are control groups.

while some are nonionic surfactants like tween. Some natural products could be used as fine surfactants as well, for instance, lignosulfonates and saponin [32]. Through tests, PAE was selected for its shortest wetting time. Lastly, dispersants are regular assistant agents, which significantly decrease the gathering of granules in dispersion system. Generally, smashed preparations are not thermodynamically stable in water, which results in the automatical gathering of dispersed granules, called flocculation, to form a stable thermodynamic system [33]. However, the granules can be efficiently stopped from flocculating by dispersants which keep the active granules suspending in solution and remarkably improve the dispersibility and penetrability. Usually, dispersants are anionic surfactants with polycyclic structure, like lignosulfonates and alkylbenzene sulfonates (ABS). In the present study, CMC-Na and PEG 6000 were chosen out from tested dispersants because of their finest dispersity and compatibility in aqueous solution.

The strategy of response surface methodology (RSM) is commonly used in formula improvement and optimization [16,34–36]. Since the function of fungal preparation is mostly relied on its property of dispersity, this strategy was applied for a better design of chosen surfactant (PAE) and dispersants (CMC-Na and PEG 6000). The mathematical model **(Eq. (4))** was obtained from the statistics analysis, which could be effectively used to predict and optimize the dispersity of fungal wettable powder preparation within chosen factors. Consequently, the optimum preparation formula was determined to be 11.0% of CMC-Na, 9.0% of PEG 6000, 5.0% of PAE, 2.5% of glycine, 5.0% of fucose, 27.5% of kaolin and 40% of freeze dried fungi. Under this formula, the dispersion rate was above 70.0% and approximately 76.1% of chlorpyrifos was degraded within half an hour in lab condition.

Assessment of preparation quality is necessary. Apparently, the impact of its physical and biological properties on the degradation performance of fungal preparation cannot be underestimated or ignored. Therefore, several features of product quality were checked separately and the results showed that the fungal preparation would meet the practical demands. However, the results still revealed an inevitable fact that microbial preparations are more possibly affected by external factors than the chemosynthetic ones. Moreover, this shortage was also evident in the field tests. Because of the vulnerability of active ingredients, a less efficient degradation performance of this fungal preparation was detected after one day, which can be considered as an indirect consequence of environmental factors on the active enzymes. As a result, these limitations will definitely enlarge the difficulty of its practical application and market promotion.

Undeniably, the results from the field tests showed a degradation of chlorpyrifos on cabbage more than 85% within one day, which indicated that the removal was efficient by this fungal preparation. In addition, it revealed the potential of degradation within a broad range of initial chlorpyrifos concentrations, which can be necessary because of the inappropriate and excessive use of chlorpyrifos in some farming areas in China. During the field experiments, there were no adverse effects observed on cabbage growth, which suggested that the fungal preparation was safe to the plants and can be an option for large-scale use in crops treated with chlorpyrifos. Lastly, as chlorpyrifos residues were constantly reduced, it was suggested that even though the efficiency was weakened after one day, this fungal preparation showed an important feature of persistently degrading the target compound.

In conclusion, the results obtained in the present study indicated that freeze-dried fungal wettable powder preparation of *C. cladosporioides* Hu-01 efficiently reduced the residues of chlorpyrifos on cabbage, which can be used as an effective and promising approach for chlorpyrifos bioremediation on vegetables. Besides, this study also provided a safe and cost-effective procedure for microbial preparation, which can be regarded as a potential strategy for better utilization of active microorganisms in bioremediation.

Author Contributions

Conceived and designed the experiments: JL YH MH GZ. Performed the experiments: JL YH SC YX. Analyzed the data: JL YH. Contributed reagents/materials/analysis tools: JL YH MH GZ. Wrote the paper: JL SC GZ.

References

1. Racke KD (1994) Environmental fate of chlorpyrifos. Rev Environ Contam Toxicol 131: 1–150.

2. Amitai G, Moorad D, Adani R, Doctor BP (1998) Inhibition of acetylcholinesterase and butyrylcholinesterase by chlorpyrifos-oxon. Biochem Pharmacol 56: 293–299.

3. Sandahl JF, Baldwin DH, Jenkins JJ, Scholz NL (2005) Comparative thresholds for acetylcholinesterase inhibition and behavioral impairment in coho salmon exposed to chlorpyrifos. Environ Toxicol Chem 24: 136–145.

4. Eaton DL, Daroff RB, Autrup H, Bridges J, Buffler P, etal. (2008) Review of the toxicology of chlorpyrifos with an emphasis on human exposure and neurodevelopment. Crit Rev Toxicol 382: 1–125.

5. Haviland JA, Butz DE, Porter WP (2010) Long-term sex selective hormonal and behavior alterations in mice exposed to low doses of chlorpyrifos in utero. Reprod Toxicol 29: 74–79. dio: 10.1016/j.reprotox.2009.10.008.

6. Kavitha P, Rao JV (2008) Toxic effects of chlorpyrifos on antioxidant enzymes and target enzyme acetylcholinesterase interaction in mosquito fish, *Gambusia affinis*. Environ Toxicol Pharmacol 26: 192–198.

7. Shafiq-Ur-Rehman RS, Waliullah MIS (2012) Chlorpyrifos-induced neurooxidative damage in bee. Toxicol Environ Health Sci 4: 30–36.

8. Betancourt AM, Burgess SC, Carr RL (2006) Effect of developmental exposure to chlorpyrifos on the expression of neurotrophin growth factors and cell-specific markers in neonatal rat brain. Toxicol Sci 92: 500–506.

9. Bicker W, Lammerhofer M, Genser D, Kiss H, Lindner W (2005) A case study of acute human chlorpyrifos poisoning: novel aspects on metabolism and toxicokinetics derived from liquid chromatography-tandem mass spectrometry analysis of urine samples. Toxicol Lett 159: 235–251.

10. Sandhu MA, Saeed AA, Khilji MS, Ahmed A, Latif MSZ, etal. (2013) Genotoxicity evaluation of chlorpyrifos: a gender related approach in regular toxicity testing. J Toxicol Sci 38: 237–244.

11. Chen S, Hu M, Liu J, Zhong G, Yang L, etal. (2011) Biodegradation of betacypermethrin and 3-phenoxybenzoic acid by a novel *Ochrobactrum lupini* DG-S-01. J Hazard Mater 187: 433–440.

12. Chen S, Chang C, Deng Y, An S, Dong YH, etal. (2014) Fenpropathrin biodegradation pathway in *Bacillus* sp. DG-02 and its potentials for bioremediation of pyrethroid-contaminated soils. J Agric Food Chem 62: 2147–2157.

13. Chen S, Yang L, Hu M, Liu J (2011) Biodegradation of fenvalerate and 3-phenoxybenzoic acid by a novel *Stenotrophomonas* sp. strain ZS-S-01 and its use in bioremediation of contaminated soils. Appl Microbiol Biotechnol 90: 755–767.

14. Cycoń M, Wojcik M, Piotrowska-Seget Z (2011) Biodegradation kinetics of the benzimidazole fungicide thiophanate-methyl by bacteria isolated from loamy sand soil. Biodegradation 22: 573–583.

15. Arora PK, Sasikala C, Ramana CV (2012) Degradation of chlorinated nitroaromatic compounds. Appl Microbiol Biotechnol 93: 2265–2277.

16. Chen S, Liu C, Peng C, Liu H, Hu M, etal. (2012) Biodegradation of chlorpyrifos and its hydrolysis product 3,5,6-trichloro-2-pyridinol by a new fungal strain *Cladosporium cladosporioides* Hu-01. Plos One 7: e47205.

17. Lu P, Li Q, Liu H, Feng Z, Yan X, etal. (2013) Biodegradation of chlorpyrifos and 3,5,6-trichloro-2-pyridinol by *Cupriavidus* sp. DT-1. Bioresour Technol 127: 337–342.

18. Yu YL, Fang H, Wang X, Wu XM, Shan M, etal. (2006) Characterization of a fungal strain capable of degrading chlorpyrifos and its use in detoxification of the insecticide on vegetables. Biodegradation 17: 487–494.

19. Park J, Lee S, Lee J (2013) Sampling and selection factors that enhance the diversity of microbial collections: application to biopesticide development. Plant Pathol J 29: 144–153.

20. Carpio C, Dangles O, Dupas S, Lery X, Lopez-Ferber M, etal. (2013) Development of a viral biopesticide for the control of the Guatemala potato tuber moth *Tecia solanivora*. J Invertebr Pathol 112: 184–191.

21. Copping LG, Menn J (2000) Biopesticides: a review of their action, applications and efficacy. Pest Manag Sci 56: 651–676.

22. Gao Y, Chen S, Hu M, Hu Q, Luo J, etal. (2012) Purification and characterization of a novel chlorpyrifos hydrolase from *Cladosporium cladosporioides* Hu-01. Plos One 7: e381376.

23. Zhang C, Wang S, Yan Y (2011) Isomerization and biodegradation of beta-cypermethrin by *Pseudomonas aeruginosa* CH7 with biosurfactant production. Bioresour Technol 102: 7139–7146.

24. Chen S, Dong YH, Chang C, Deng Y, Zhang XF, etal. (2013) Characterization of a novel cyfluthrin-degrading bacterial strain *Brevibacterium aureum* and its biochemical degradation pathway. Bioresour Technol 132: 16–23.

25. Suciu NA, Ferrari T, Ferrari F, Trevisan M, Capre E (2012) Pesticide removal from waste spray-tank water by organoclay adsorption after field application: an approach for a formulation of cyprodinil containing antifoaming/defoaming agents. Environ Sci Pollut Res 19: 1229–1236.

26. Wang W, Wu Z, Dai Z, Yang Y, Wang J, etal. (2013) Glycine metabolism in animals and humans: implications for nutrition and health. Amino Acids 45: 463–477.

27. Hamburger T, Broecker-Preuss M, Hartmann M, Schade FU, de Groot H, etal. (2013) Effects of glycine, pyruvate, resveratrol, and nitrite on tissue injury and cytokine response in endotoxemic rats. J Surg Res 183: E7–E21.

28. Yamaki K, Goto M, Takano-Ishikawa Y (2009) Inhibitory effects of fucose-related sugar compounds on oxidised low-density lipoprotein uptake in macrophage cell line J774.1. Food Agric Immunol 20: 355–362.

29. Peterszeg G, Robert AM, Robert L (2003) Protection by L-fucose and fucose-rich polysaccharides against ROS-produced cell death in presence of ascorbate. Biomed Pharmacother 57: 130–133.

30. Shukla PG, Kalidhass B, Shah A, Palaskar DV (2002) Preparation and characterization of microcapsules of water-soluble pesticide monocrotophos using polyurethane as carrier material. J Microencapsul 19: 293–304.

31. Zhang G, Zhan J, Li H (2011) Selective binding of carbamate pesticides by self-assembled monolayers of calixarene lipoic acid: wettability and impedance dual-signal response. Org Lett 13: 3392–3395.

32. Matsushita Y, Yasuda S (2005) Preparation and evaluation of lignosulfonates as a dispersant for gypsum paste from acid hydrolysis lignin. Bioresour Technol 96: 465–470.

33. Dinner AR, Sali A, Smith IJ, Dobson CM, Karplus M (2000) Understanding protein folding via free-energy surfaces from theory and experiment. Trends Biochem Sci 25: 331–339.

34. Chen S, Hu W, Xiao Y, Deng Y, Jia J, etal. (2012) Degradation of 3-phenoxybenzoic acid by a *Bacillus* sp. Plos One 7: e50456.

35. Chen S, Luo J, Hu M, Lai K, Geng P, etal. (2012) Enhancement of cypermethrin degradation by a coculture of *Bacillus cereus* ZH-3 and *Streptomyces aureus* HP-S-01. Bioresour Technol 110: 97–104.

36. Chen S, Luo J, Hu M, Geng P, Zhang Y (2012) Microbial detoxification of bifenthrin by a novel yeast and its potential for contaminated soils treatment. Plos One 7: e30862.

6

Hierarchicality of Trade Flow Networks Reveals Complexity of Products

Peiteng Shi[1], Jiang Zhang[1]*, Bo Yang[2], Jingfei Luo[1]

1 School of Systems Science, Beijing Normal University, Beijing, China, **2** Ministry of Commerce of the People's Republic of China, Beijing, China

Abstract

With globalization, countries are more connected than before by trading flows, which amounts to at least 36 trillion dollars today. Interestingly, around $30-60$ percents of exports consist of intermediate products in global. Therefore, the trade flow network of particular product with high added values can be regarded as value chains. The problem is weather we can discriminate between these products from their unique flow network structure? This paper applies the flow analysis method developed in ecology to 638 trading flow networks of different products. We claim that the allometric scaling exponent η can be used to characterize the degree of hierarchicality of a flow network, i.e., whether the trading products flow on long hierarchical chains. Then, it is pointed out that the flow networks of products with higher added values and complexity like machinary, transport equipment etc. have larger exponents, meaning that their trade flow networks are more hierarchical. As a result, without the extra data like global input-output table, we can identify the product categories with higher complexity, and the relative importance of a country in the global value chain by the trading network solely.

Editor: Dante R. Chialvo, National Scientific and Technical Research Council (CONICET)., Argentina

Funding: This work received funding from the National Natural Science Foundation of China under Grant No.61004107, www.nsfc.gov.cn, and Beijing Higher Education Young Elite Teacher Project under Grant No.YETP0291, http://www.bjedu.gov.cn. The funders had no role in study design, data collection and analysis, decision to publish, or preparation of the manuscript.

Competing Interests: The authors have declared that no competing interests exist.

* E-mail: zhangjiang@bnu.edu.cn

Introduction

As the process of globalization accelerates, countries in the world are more connected and collaborative unprecedentedly under the background of an integrated global markets of capital, labor force and products. Consequently, some cross-border production chains, which comprise several countries or regions, emerged inevitable as the result of international labor force division and collaboration in the global level [1–3]. However, due to the heterogeneities of products, the production networks are very inhomogeneous. Some products in the electronics and automotive industries, say PCs or automobiles, can be broken down into several independent components, and easily transported and assembled in different countries [1]. Therefore, a large fraction of imports for these products are not for final consumption but re-production with higher value-added and exports [1,4,5]. On the other hand, the networks for agriculture or raw material products may have much shorter production chains. Thereafter the major imports of these products are for final consumption.

Differentiating these products according to their production chains and level of added-values is of importance for countries' long term development strategy. Conventional method [6–8] tries to build the value flow networks among different products directly by incorporating the international input-output tables [9–11]. Although the whole picture of production networks can be captured in detail, obtaining the accurate raw data on the global level is not easy [8,10]. On the other hand, the highly detailed international trade flow data for various products among countries are well documented for a long history [12,13]. Particularly, all the bilateral trade flows are classified by different products according to the SITC (Standard International Trade Coding) or other equivalent coding methods. Therefore, a unique flow structure of one product category can be extracted from the international trade data.

World wide trade network as a specific instance of complex network has been studied for several years [14–16]. These early works always focus on country positions or node centralities on the network. Except network structure, recent works focus more on dynamics, weights and different trade networks by products. The longitudinal studies of trade networks reveal how the network structure such as the centrality of entire network changes along time to reveal the potential influences of globalization [17,18]. Weights standing for trade flows between countries hide important information which cannot be uncovered by network structure solely [19,20]. If the kinds of products exported by countries are considered, the problem of a country's industrial structure and export strategy can be studied by a country-product bipartite network model [21–23]. In this paper, however, we take the consideration of network structure, flows and different products in the same time. We construct weighted multi-networks of different products [24] from the trade flow data. For each product, there is a unique flow network which can be used to reflect the characteristic of the product. Therefore, we can discriminate products on their level of complexity and value-added by identifying their unique trade flow structures. It can work because trade networks contain the information of global production networks - almost all the cross-border product flows in the global value chain are recorded in the international trade data. This analysis is done in several levels of products classification because the trade flow datasets provide the hierarchical classification information.

Our methodology is to compare the allometric scaling exponents among the flow networks of different products [25,26]. The allometric scaling pattern is found to be ubiquitous for trees spanned by binary networks [26,27], like food webs [27,28], trade webs [29] and biological networks [30]. Our previous work has incorporated the flow analysis methods developed in ecology to reveal the common nature of the flow networks in general [28,31]. It is natural to extend this method to trade flows, in which, the allometric scaling exponent is given a new explanation, the degree of hierarchicality. It characterizes whether the product flows along a long hierarchical chain or not. We calculate the allometric exponent of each flow network in different product classifications, and find that the manufacture products with higher added values have larger exponents. Furthermore, most exponents are larger than one, meaning that the networks are hierarchical. While, the networks of the primary products with relative low added values have smaller exponents and the networks are flat. Hierarchicality always means inequality and monopoly. We further calculate the relative importance of each country in a product trading network, and compare the heterogeneities of country impact distribution for different products by GINI coefficient of country's impact. Finally, the dynamics of allometric scaling exponents along time is shown, and the globalization process can be read.

Results

Trade Flow Networks

We use two data sets to study and compare for eliminating the potential discrepancy from the data. The fist one is from Feenstra, et al's "World Trade Flows: 1962–2000" dataset based on the United Nations COMTRADE database (abbreviated by UN data set) [12](see Section 1 and Table S1 in File S1). This data set covers the bilateral trade flows of about 800 kinds of products according to the SITC 4 (Standard International Trade Classification system, Rev.4) classification standard from 1963 to 2000. And the results of 2000 year are mainly shown and discussed in the main text. Another data set (OECD data set) is the bilateral trade data in 2009 which was complied by the Organization of Economic Co-operation and Development (OECD) [13](see Table S2 and S3 in File S1). The OECD data set contains only the OECD member countries so that the total number of countries is smaller than the UN data set. However, these countries dominate about 90% trade volume in the world. The products classification standard of the OECD data set is ISIC Rev.3 (International Standard Industrial Classification of All Economic Activities, Rev.3), which is slightly different from the SITC 4 classification. Please see detailed discussions of the data sets in File S1.

The SITC4 codes are hierarchical, meaning that the categories with longer codes are sub-categories of the ones with shorter codes if they share the same prefix. For example, the product category 7 in SITC4 stands for the category of machinery and transport equipment, so this is a very generalized classification. While, 71, 72 are two sub-categories of 7 representing the power machinery product and vehicle respectively.

Allometric Scaling of Trade Networks

For each product trade network, we can define an exponent η to characterize the hierarchicality of the flow network. At first, we need to calculate two vertex specific variables, namely, T_i and C_i.

T_i called the trading volume of country i, is defined as the maximum of i's total import or export. It reflects the capacity of

trade flows through i. Next, C_i is the impact of i on the entire network. It is defined as the total changes of trading volume of other nodes on the network after the hypothetical delete of i. The concrete calculation of these two variables are referred to the method section.

Usually, for various empirical trade networks, C_i and T_i have a strong correlation which can be described by a power law,

$$C_i \sim T_i^{\eta}, \qquad (1)$$

where, η is the allometric scaling exponent. This equation is extended from the empirical allometries from river basin, vascular networks and food webs [26,27]. The previous studies on spanning trees show that the exponent η can be used to reflect the hierarchicality or flatness of a tree. For example, two extreme cases of spanning trees can be shown in Figure 1. The star network which has the smallest exponent 1 is the flattest tree, while the chain network which has the largest exponent 2 is the most hierarchical tree.

This calculation can be extended to general flow networks [28,31], nevertheless the exponent is not bound in [1,2]. However, we can also define the exponent as the hiearchicality of a general flow network. Because it will contain long flow chains if its exponent is larger (see method section).

It turns out that the allometric scaling pattern (Equation 1) is very general for all the studied trade networks in all classification levels but their exponents are not similar. Figure 2 shows the allometric scaling patterns of two products in two digits level: power-generating equipment (SITC4 code: 71) and vegetables & fruits (SITC4 code: 05).

In Figure 2, each data point stands for a country participating the international trade of this product. The pairs of T_i and C_i form a straight line on the log-log coordinate which means a power law relationship between the two variables exist (i.e., Equation 1). The exponents for these two products are distinct indicating that the power generating trade network is more hierarchical than the network of fruit and vegetable. In another word, the production for power generating machines is along a longer value-added chain than the fruit and vegetable.

This point can be visualized by the network plots of these two products shown in Figure 3. Although only the backbone links are shown and other links are faded as backgrounds, it is clear that the upper network has many long chains which always root from some major exporters of power generating machine(e.g. U.S. and Japan). However, the lower network is more fragmental. Although several large countries (e.g. U.S.) still occupy a large fraction of fruit trade, most of them are importers. That implies the whole network is lack of center and more flat. Intuitively, that is the

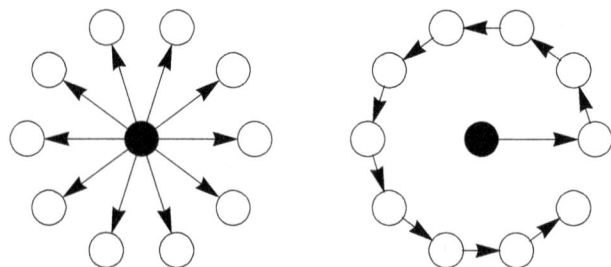

Figure 1. Two special spanning trees with minimum allometric exponent 1 (left, a star network) and maximum exponent 2 (right, a directed chain).

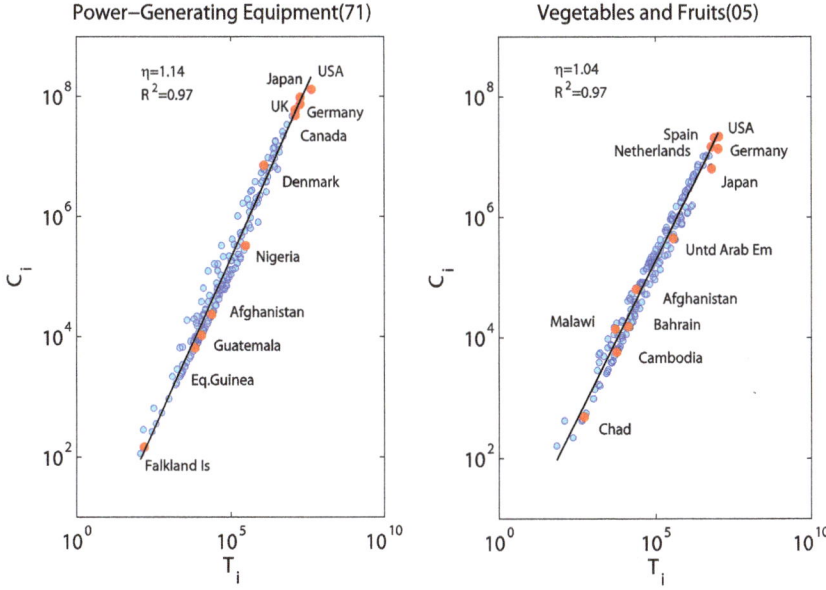

Figure 2. The allometric scaling law between T_i (in U.S. dollar) and C_i (in U.S. dollar) of two networks are shown. The left figure shows a super-linear scaling law (with exponent larger than 1) for power generating product, while the right one shows a sub-linear scaling law (with exponent smaller than 1) for fruit and vegetable.

reason why the exponent of the first network is larger than the latter one.

Exponents Comparison and Distributions

We further compare the exponents among different networks of products in the coarse classification level in a more systematic way. In Table 1 and 2, we list exponents for all 1-digit products in UN data set and OECD data set to compare.

Both tables show large gaps of exponents for different products ([1.001,1.136] for UN-Comtrade data set and [0.944,1.146] for OECD data set). Although some slight differences between SITC4 classification and ISIC Rev.3. classifications exist, the products of machinery, equipment, chemicals et al. are of higher exponents than the products of foods, mining and agriculture. This unique observation can be further confirmed and extended to finer classifications.

Figure 4 shows the exponents distribution of all products with 4-digits classification (the finest level in our dataset) in UN data set. The frequency curve has a bell-shape peaked at 1.09, which means most product networks are hierarchical. The stacked color bars show the distributions of all 1-digit classifications (Figure 4 left). Note that most blue bars locate in the right side of the bell-shaped curve, while, the green and yellow bars locate on the left side, indicating that the machinery and manufactured products have larger exponents than the food, beverage products. This phenomenon can be better illustrated by the right subplot of Figure 4, in which we simply classify the products as primary products (SITC4 codes prefix with 0,1,2,3,4) and manufacture products (SITC4 codes prefix with 5,6,7,8,9). The similar results can be derived for Leamer products classification standard (see SI section 3, Table S5 and Figure S2 in File S1).

Allometric Exponent and Product Complexity

According to the observations, we know that the allometric exponents of trade flow network can reflect the basic properties of products. The manufacture products with higher added-value and complex production process always have larger exponents. Therefore, we conjecture that a positive correlation between the exponents and the nature of products (complexity or value added) may exist.

To test our hypothesis we do two correlation analysis on both data sets. For the UN data set, we correlate the exponents with PRODY, one of the measurements of product complexity. It is calculated as the average income level of the exporters (measured by the GDP percapita) of this product weighted by the comparative advantage of this product in different exporters [32]. It is calculated as:

$$PRODY(p) = \sum_c Y_c RCA(c, p), \qquad (2)$$

where, Y_c is the GDP per capita of country c, and $RCA(c,p)$ is the comparative advantage of country c exporting p. The summation is taken for all the countries exporting p. $RCA(c,p)$ can be calculated as $RCA(c,p) = \dfrac{E(c,p)/\sum_p E(c,p)}{\sum_c (E(c,p)/\sum_p E(c,p))}$, where $E(c,p)$ is the total export value of c on p. The numerator of the weight, $E(c,p)/\sum_p E(c,p)$, is the value-share of the product p in the country c's overall export basket. The denominator of the weight, $\sum_c (E(c,p)/\sum_p E(c,p))$, aggregates all the value-shares across all countries. Therefore, the weight measures the relative comparative of product p in country c. And $PRODY(p)$ measures the average income level of p. It is a proxy of the product's complexity.

Figure 5 shows the relationship between exponent η and PRODY of each product in 2-digits classification of UN data set. The correlation coefficient of these two variables is 0.37 and it can be improved to 0.44 if the three outliers (triangles) in Figure 5 are omitted.

For the OECD data set, the domestic and foreign value-added for each product-country combinations are available (see the

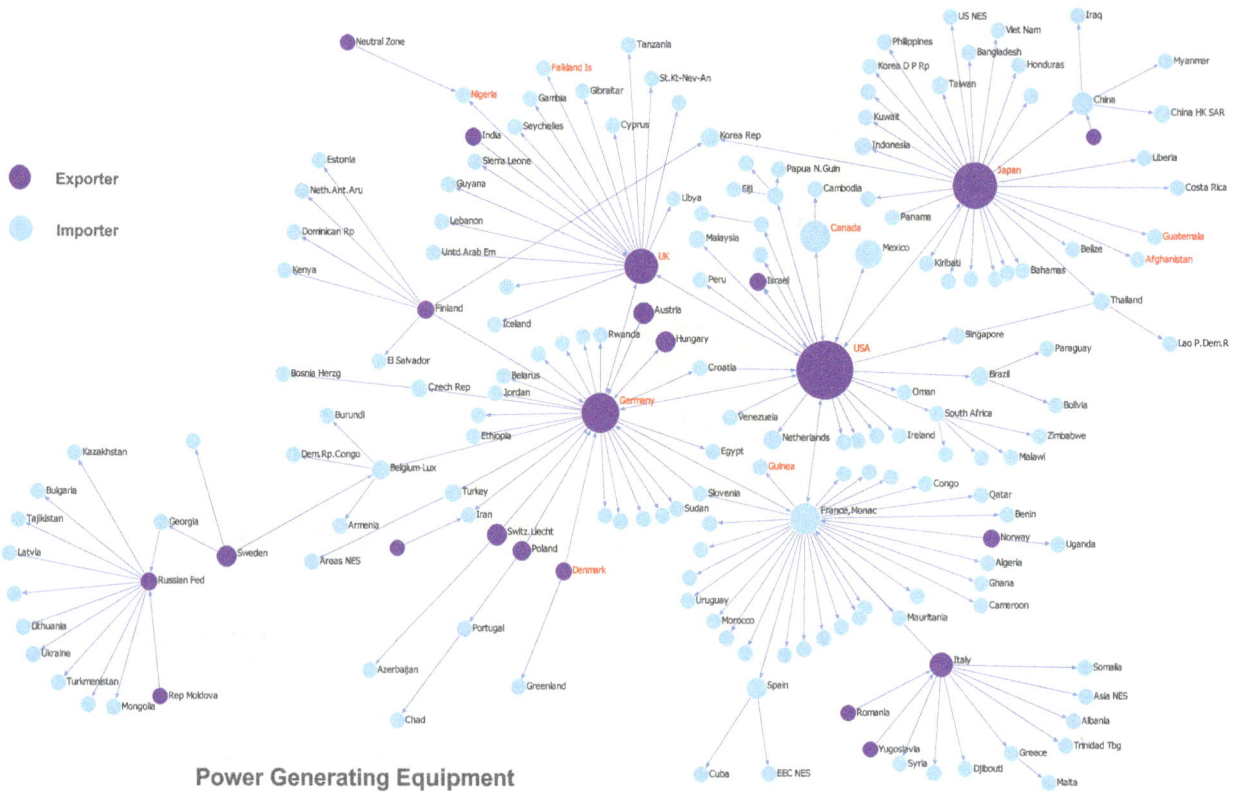

Power Generating Equipment

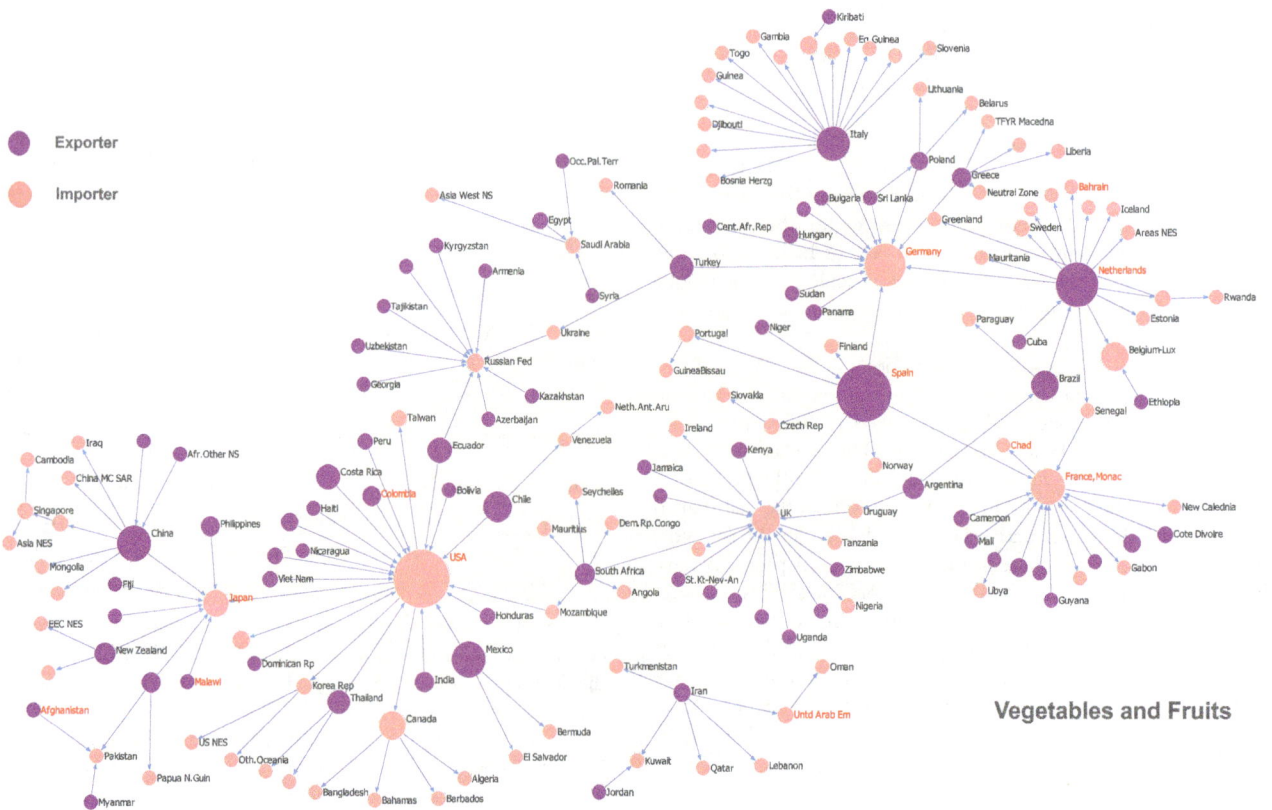

Vegetables and Fruits

Figure 3. Visualization of trade flow network for power generating equipment (upper) and fruit and vegetable (lower). We use different colors to distinguish nodes as importer (import is larger than its export) and exporter (export is larger than import). The size of node denotes the total volume of trade. In these two networks, only the backbones are shown as the main parts and all other un-important links are hidden as backgrounds. The backbone extracting method is according to [35].

discussion in Section 2 and Table S4 in File S1). This enables us to correlate exponents with average foreign value-added ratio of each product. Here, the proportion of foreign value-added is the ratio between the total value-added and gross export for all countries that exporting this product [1]. The relationship between η and foreign value-added proportion is shown in the right plot of Figure 5. There is a clear positive correlation between them, and the correlation coefficient is 0.692.

Consequently, we conclude that the allometric exponent η of each trade flow network can characterize the complexity and value-added proportion of given product. When a product needs more complex production processes, more countries must be involved to form a long value chain, so that more value is added on the product. All of these properties must be reflected in the flow structure of the product trade network. That is the reason why allometric exponent η can be distinct for different products.

Discussion

Country Impacts

Besides the structural properties of the entire network, node positions in the global value chain are also of importance and interests. In our study, C_i, the total impact of country i toward the entire network, can be viewed as a vertex centrality indicator because it measures the degree of the entire network is influenced if node i was removed. This understanding is in accordance with the standard HEM (Hypothetical Extraction Method) [33,34] in input-output analysis once the trade flow networks are understood as an input-output matrix.

Figure 6 shows the distributions of C_i for trade networks of all products and several selected products both in UN and OECD data sets. Also, top 10 countries are listed in Table S6 and S7 in File S1.

Centrality and Inequality

In our previous works of allometric scaling on ecological flow networks [31], the exponent η is explained as the degree of centrality, i.e., whether several big nodes dominate a disproportional impact on the entire network. This explanation can also be

extent to this study. The networks with higher ηs are more centralized. So, a few large countries can impact the entire network, in which the impact's degrees C_i are disproportional to their direct trade flow T_i.

For example, we have three flow networks with same $T_i = \{1,2,3,4,5\}$ but different $\eta = \{1,1/2,2\}$. Then, their C_i s are $C_i^{(1)} = \{1,2,3,4,5\}$ for $\eta = 1$, $C_i^{(2)} = \{1,1.4,1.7,2,2.2\}$ for $\eta = 1/2$ and $C_i^{(3)} = \{1,4,9,16,25\}$ for $\eta = 2$ respectively. As a result, the largest country (the node with largest T_i) dominates $5/(1+2+3+4+5) \approx 33\%$, $2.2/(1+1.4+1.7+2+2.2) \approx 27\%$ and $25/(1+4+9+16+25) \approx 45\%$ impacts of the entire networks respectively. Therefore, the third network is much more centralized than the second one.

However, the inequality of exporting products is mainly from the heterogeneity of the resource distribution but not the network effect which is characterized by η. For example, petroleum export is heterogenous due to the unevenness of fossil fuel resource distribution geographically. Therefore, new indicator is needed.

We use the GINI coefficient of C_i distribution to characterize the overall inequality of the flow network structure. C_i distribution can account for both inequality origins: natural resource distribution and network effect. First, it is obvious that the natural inequality of resource distribution can be reflected by T_i distribution. Suppose T_i follows a Zipf law, $T_i(r) \sim r^{-\alpha}$, where, α is the Zipf exponent, and r is the rank order of i. We know that there is a power law relationship between T_i and C_i according to Equation 1. Thus, C_i also follows the Zipf law: $C_i(r) \sim r^{-\beta} = r^{-\alpha\eta}$, where $\beta = \alpha\eta$ is its exponent. Therefore, the distribution of C_i (η) contains both information: natural heterogeneities (α) and network effect(η).

Although C_i does not follow the Zipf distribution in our empirical data (shown in Figure 6), the previous conclusion that the distribution of C_i contains both information, is still correct. Usually, GINI coefficient (bounded by [0,1]) can be used to characterize the inequality of a variable no matter what kind of distribution it follows.

In the last column of Table 1, we show the GINI coefficients of all 1-digit product categories. Most products have similar rank order by GINI as the order by η. But the order of manufactured

Table 1. Exponents of 1-digit SITC4 categories in UN data set.

Code	Classification	η	R^2	GINI
7	Machinery and transport equipment	1.136 ± 0.026	0.974	0.889
6	Manufactured goods classified chiefly by materials	1.120 ± 0.026	0.962	0.830
5	Chemicals and related products	1.117 ± 0.034	0.972	0.877
1	Beverages and tobacco	1.116 ± 0.033	0.958	0.868
4	Animal and vegetable oils, fats and waxes	1.077 ± 0.029	0.973	0.847
0	Food and live animals	1.043 ± 0.032	0.971	0.798
3	Mineral fuels, lubricants and related materials	1.042 ± 0.018	0.954	0.821
2	Crude materials, inedible, except fuels	1.001 ± 0.020	0.988	0.815
-	All Products	1.022 ± 0.030	0.965	0.817

The categories of 8 (Miscellaneous) and 9(Not classified) are ignored in this table, The last row shows the allometry of all products as an integrated network.

Table 2. Exponents for different products in OECD data set.

Code	Classification	η	R^2	GINI
29	Machinery and equipment, nec	1.146 ± 0.072	0.947	0.656
23T26	Chemicals and non-metallic mineral products	1.129 ± 0.079	0.937	0.563
34T35	Transport equipment	1.124 ± 0.075	0.941	0.669
30T33	Electrical and optical equipment	1.112 ± 0.070	0.948	0.667
27T28	Basic metals and fabricated metal products	1.092 ± 0.080	0.974	0.568
40T41	Electricity, gas and water supply	1.075 ± 0.054	0.931	0.649
36T37	Manufacturing nec; recycling	1.074 ± 0.078	0.967	0.684
15T16	Food products, beverages and tobacco	1.073 ± 0.081	0.931	0.553
20T22	Wood, paper, paper products, printing and publishing	1.051 ± 0.088	0.926	0.589
10T14	Mining and quarrying	1.019 ± 0.041	0.911	0.721
01T05	Agriculture, hunting, forestry and fishing	1.019 ± 0.070	0.978	0.576
17T19	Textiles, textile products, leather and footwear	0.998 ± 0.065	0.949	0.705
-	All industries	0.941 ± 0.072	0.924	0.474

The products in different industries coded by ISIC Rev.3 coding system for industries is shown. Industries of financial intermediation, business services, wholesale and retail trade, transport and storage, post and telecommunication, hotels and restaurants, and construction are ignored because their trades do not stand for goods flows. The last row shows the allometry of all industries as an integrated network.

goods (Code 6) falls down from No. 2 (by η) to No. 5 (by GINI coefficient), and the order of Food and live animals falls down from No.6 to the bottom. That indicates that these two kinds of products are not so unequal as predicted by the exponent η because the average trading volumes (T_i) distribute evenly among countries although their trading networks are more centralized. In the last column of Table 2, the GINI coefficients of all industries of OECD data set are shown. There is a large deviation of the orders by η from the GINI coefficients. Some industries like mining and textiles have high ranks of GINI coefficients but low ranks of η. That means these industries are resource monopolized. While

basic metals and chemicals have high ranks of η but low ranks of GINI coefficients which means the trade networks of these products are centralized.

Another interesting finding is the exponent of the integral trade network that consists of all trading products is 1.02 (It is 0.94 in OECD data set). This value is less than the mean exponent by averaging all individual products. It can be also observed for GINI coefficients. That implies international trade of all products in general becomes much more decentralized than each single product's trade. Therefore, trade on diverse kinds of products can

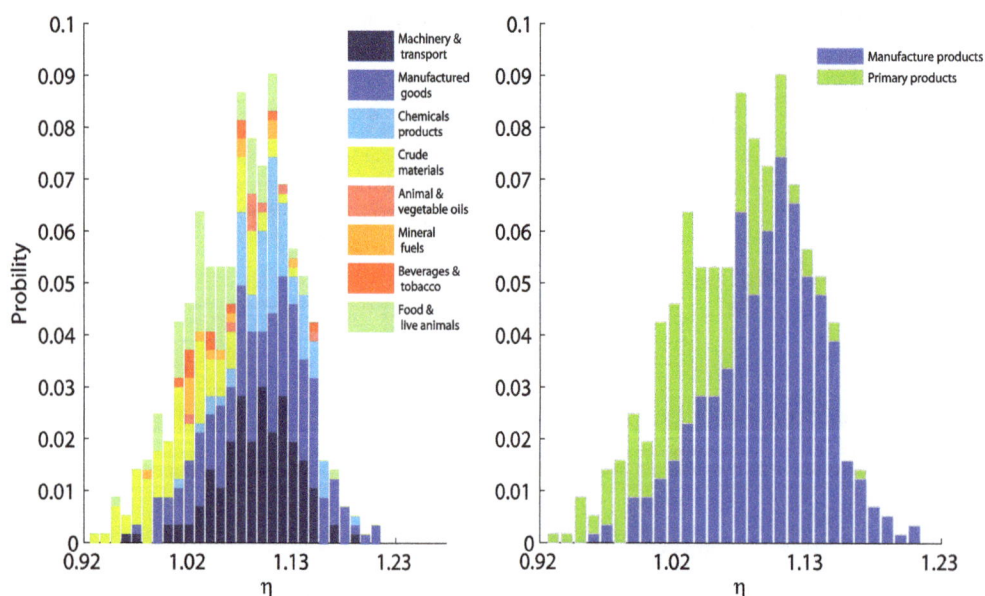

Figure 4. Exponents Distribution for All 4-digit SITC4 Product Categories. The stacked bar charts of different colors correspond to 1-digit SITC4 categories (left) and primary and manufacture classifications (right). For one specific 1-digit classification (say 0 for food and living animals), we can calculate the frequencies on each exponent intervals for all products with 0 prefix, then these frequencies as little bars are stacked on the tops of existing bars.

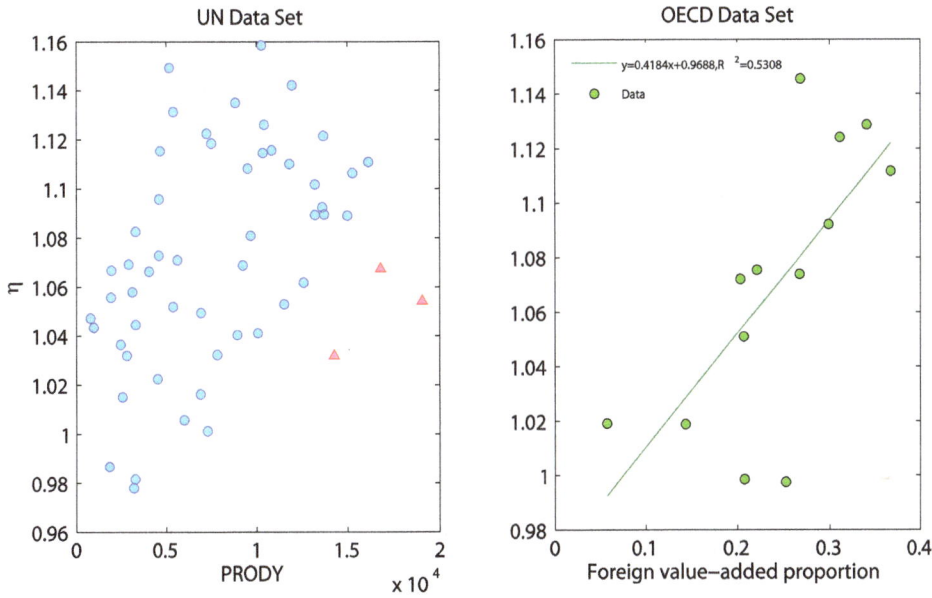

Figure 5. The relationship between η **and PRODY of each 2-digit classification(left) in UN data set and** η **versus mean proportion of foreign value added for products in OECD data set.**

make the world flatter. Though we still don't know in what degree this conclusion could be true. This will left for further investigation.

Exponents in Different Years

The UN data set records the international trade data historically from year 1962 to 2000. This enables us to study the dynamics of exponents. In Figure 7, we show how these exponents change along time.

Most exponents are almost stable. However, machinery, transport equipment and manufactured goods by materials have big changes. The latter has very large exponents before 1982, but the former climbs to the top 1 after around 1982. Note that some

cross-boarder companies emerged in around 1980s. Therefore, the product machinery and transport equipment which depends on vertical labor division but not material is of the largest exponent. While, the manufactured goods which is more independent on global cooperation change in an opposite direction. Hence, the dynamics of the exponents may reflect the globalization process.

Methods

Flow Network Model

A flow network model can be built for each product category. Nodes on the network are countries, directed edges are trading

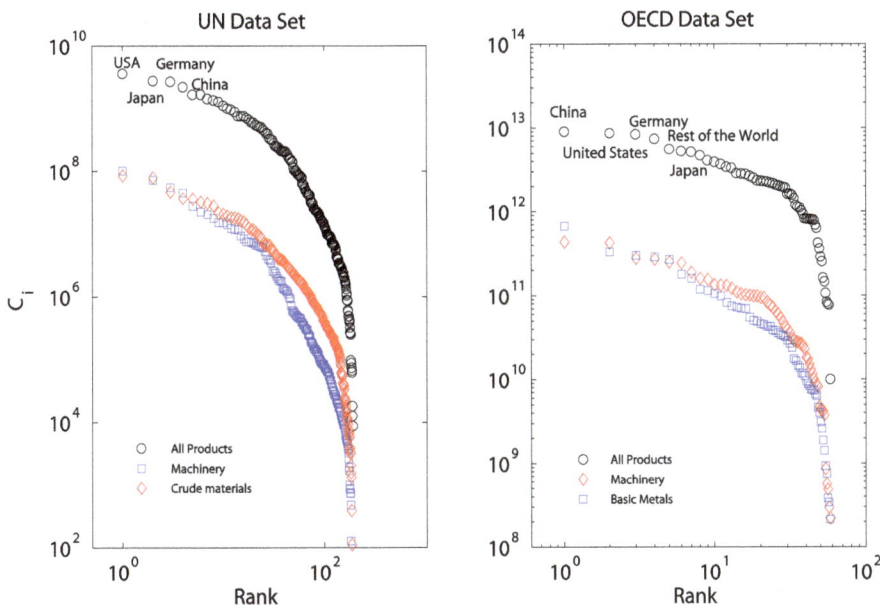

Figure 6. C_i **Distributions of Both Data Sets.** The unit of C_i is U.S.dollar.

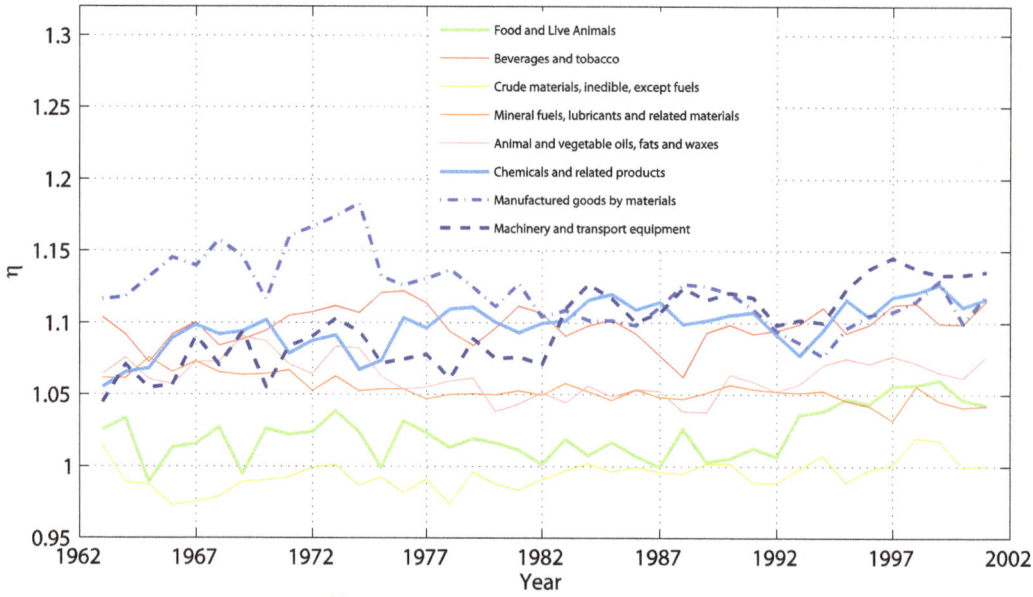

Figure 7. Allometric exponents η s of 1-digit classification products change with time.

relationships between countries and weights on edges are trading flows measured by the unified money units (It is U.S. dollar in our data sets).

If there are totally N countries participating trade of the focus product p, then a flow network can be represented by an $N \times N$ flux matrix F^p, in which the element f_{ij}^p stands for the trade flow of p from i to j. The superscript p will be omitted to facilitate our expression. And all the variables as well as the trade networks in the following sections are defined for one specific product.

From Trees to General Flow Networks

Previous studies on network allometry can only be applied to directed trees. In which T_i is the total number of nodes in the sub-tree rooted from i and C_i is the summation of all T_is in the sub-tree rooted from i [26,27] as shown in Figure 8 (a).

It is very difficult to generalize this definition for flow networks because the concept of sub-tree is vague due to the existence of

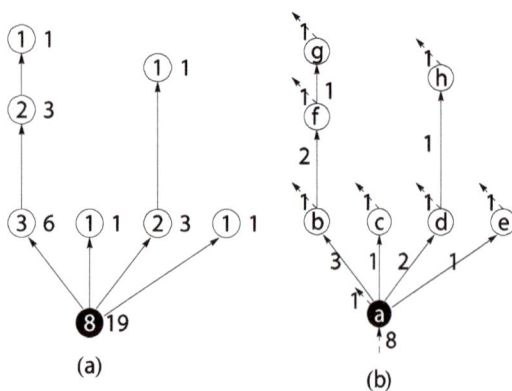

Figure 8. T_i and C_i for trees. In (a), the numbers inside the nodes are T_is and the numbers beside nodes are C_is. (b) is a flow network constructed according to (a), in which numbers represent flows. And dotted lines stand for dissipations.

loops. However, we can understand the directed tree as a flow network as shown in Figure 8(b) by assuming each node has one unit dissipation out of the network. Therefore, T_i is just the flux through node i. And C_i is the total flows reduced by the hypothetic removal of node i. For example, if we remove node b in Figure 8(b), then all the flows in the sub-tree rooted from b disappear. The total amounts of these flows are 6. Therefore $C_b = 6$. In this way, we can extend the definitions of T_i and C_i for general flow networks although the calculation of C_i is not easy. The detailed discussion of this method can be referred to [28,31].

Trading Volume and Impact

In this subsection, we will show the method on computing T_i and C_i in detail. Firstly, T_i defined as the trading volume of country i, is the maximum value of either import or export,

$$T_i = \max \left(\sum_{j=1}^{N} f_{ji}, \sum_{j=1}^{N} f_{ij} \right). \tag{3}$$

It measures the amount of product p flows through country i. T_i reflects the flow capacity that country i can import or export p.

C_i is defined as the total reduction of trade volume of all countries if i is deleted in the network. Although its definition is clear, the calculation is difficult. We will adopt the method of HEM (hyperthermic extraction method) [33,34] in input-output theory to compute.

Before C_i is defined, we should introduce another important matrix M in advance. It is the analogy of technical coefficient matrix in input-output theory,

$$m_{ij} = \frac{f_{ij}}{T_i}. \tag{4}$$

So, m_{ij} measures the ratio of the export from i to j to the total trade volume of i. Then, the following identity can be derived

$$T = MT + S, \qquad (5)$$

where, $T = (T_1, T_2, \cdots, T_N)^T$, $S = (S_1, S_2, \cdots, S_N)^T$. And

$$S_i = T_i - \sum_{j=1}^{N} f_{ji} \qquad (6)$$

can be viewed as the total domestic value-added from i (see the discussion in SI section 2, Figure S1, S3 in File S1). Then, we can obtain an important identity from Equation 5:

$$T = (I - M)^{-1} \cdot S, \qquad (7)$$

where, I is the identity matrix. Now, suppose node i is deleted in the network, then the ith column in M, and also S_i will be set to 0 according to the HEM method [33,34]. Suppose M turns into M' and S turns into S'. Then, the new total trade volume vector can be computed if we believe the identity Equation 7 is also hold for M', S' and T':

$$T' = (I - M')^{-1} \cdot S'. \qquad (8)$$

Then the total amount of trade volume reduction in the entire network is defined as C_i,

$$C_i = (1, 1, \cdots, 1) \cdot (T - T'). \qquad (9)$$

To ease our calculation, we always use the following equation

$$C_i = \sum_{k=1}^{N} \sum_{j=1}^{N} S_j \frac{u_{ji} u_{ik}}{u_{ii}}, \qquad (10)$$

where, $U = (I - M)^{-1}$. It can be proved that Equation 10 equals Equation 9 (see section 6 in File S1).

Network Allometry

Allometric scaling is a universal pattern of transportation networks including rivers, vascular networks, etc. The allometric exponents for trees are bounded in between 1 and 2. The minimum exponent can be obtained by a star-liked network, in which all links are from the root to other nodes, while, the maximum exponent is gotten by a chain as shown in Figure 1. These two special trees stand for two extremes for all directed trees. The star-liked tree is flat because every node except the root is equivalent. However, the chain-liked tree is hierarchical because

the nodes in the upper level dominate the other nodes in the lower level.

According to the discussion in the previous sections and our previous works [28,31], the network allometry is extended for general flow networks. Although the range of η is not bounded to [1,2], η can be still a good indicator for the level of hierarchicality of the flow structure because the relative speed of C_i can increase faster than T_i in a network with larger exponent. The network is more like a chain if its exponent is large. Therefore, some long flow chains can be revealed in these networks.

We distinguish networks as hierarchical ($\eta > 1$), neutral ($\eta \approx 1$) and flat ($\eta < 1$) by the exponent.

Conclusions

The most interesting finding of this paper is that the properties of a trade product can be reflected by the distinct flow structure of its trading network. Especially, the complexity or the level of value-added of a product can be characterized by the hierarchicality of the flow network which is measured by the allometric exponent. This conclusion is hold for different datasets in different coarse-grained levels of product classifications. Therefore, the information of production chain for different products and the relative positions of countries in the chain can be read from the international trade network.

Supporting Information

File S1 This file includes Table S1-Table S7 and Figure S1-Figure S3. Table S1, The dataset form in UN dataset. Table S2, The trade data in OECD dataset. Table S3, The value added data in OECD dataset. Table S4, The result of η computed according to (4) and (5). Table S5, Exponents of Leamer Classification Standard. Table S6, The top ten C_i of different products in UN dataset. Table S7, Top ten countries of different industries in the OECD Dataset. Figure S1, Balanced value flow of one country. Figure S2, Exponents Distribution for All 4-digit Leamer Classification Standard. Figure S3, The relationship between η and the mean proportion of foreign value added.

Acknowledgments

We thank for the useful advices from Prof. Y.G. Wang and Q.H. Chen. We also acknowledge the impressive discussions with the active members in Swarm Agents Club.

Author Contributions

Conceived and designed the experiments: JZ. Performed the experiments: PS JL. Analyzed the data: PS BY. Wrote the paper: JZ. Algorithm design: JL.

References

1. UNCTAD (2013) Global value chains and development. Available: http://unctad.org/en/PublicationsLibrary/diae2013d1_en.pdf.
2. Gibbon P, Bair J, Ponte S (2008) Governing global value chains: an introduction. Economy and Society 37: 315–338.
3. Gereffi G, Humphrey J, Sturgeon T (2005) The governance of global value chains. Review of International Political Economy 12: 78–104.
4. Kotha S, Srikanth K (2013) Managing a global partnership model: Lessons from the boeing 787 'Dreamliner' program. Global Strategy Journal 3: 41–66.
5. Rainnie A, Herod A, McGrath-Champ S (2013) Global production networks, labour and small firms. Capital & Class 37: 177–195.
6. Tukker A, Dietzenbacher E (2013) Global multiregional input-output frameworks: An introduction and outlook. Economic Systems Research 25: 1–19.
7. Lenzen M, Moran D, Kanemoto K, Geschke A (2013) Building eora: A global multi-region input-output database at high country and sector resolution. Economic Systems Research 25: 20–49.
8. Koopman R, Powers W, Wang Z, Wei SJ (2010) Give credit where credit is due: Tracing value added in global production chains. Working Paper 16426, National Bureau of Economic Research. Available: http://www.nber.org/papers/w16426.
9. Leontief W (1966) Input-Output Economics. Oxford University Press.
10. Miller RE, Blair PD (2009) Input-output analysis: foundations and extensions. Cambridge [England]; New York: Cambridge University Press.
11. Raa Tt (2005) The economics of input-output analysis. Cambridge: Cambridge University Press.

12. Feenstra RC, Lipsey RE, Deng H, Ma AC, Mo H (2005) World trade flows: 1962-2000. Working Paper 11040, National Bureau of Economic Research. Available: http://www.nber.org/papers/w11040.

13. Zhu S, Yamano N, Cimper A (2011) Compilation of bilateral trade database by industry and end-use category. OECD science, technology and industry working papers, Organisation for Economic Co-operation and Development, Paris. Available: http://www.oecd-ilibrary.org/content/workingpaper/5k9h6vx2z07f-en.

14. Snyder D, Kick EL (1979) Structural position in the world system and economic growth, 1955-1970: A multiple-network analysis of transnational interactions. American Journal of Sociology 84: 1096–1266.

15. Nemeth RJ, Smith DA (1985) International trade and world-system structure: A multiple network analysis. Review (Fernand Braudel Center) 8: 517–560.

16. Smith DA, White DR (1992) Structure and dynamics of the global economy: Network analysis of international trade 1965-1980. Social Forces 70: 857–893.

17. Fagiolo G, Reyes J, Schiavo S (2010) The evolution of the world trade web: a weighted-network analysis. J Evol Econ 20: 479–514.

18. Kali R, Reyes J (2007) The architecture of globalization: a network approach to international economic integration. J Int Bus Stud 38: 595–620.

19. Garlaschelli D, Loffredo MI (2004) Fitness-dependent topological properties of the world trade web. Physical Review Letters 93: 188701.

20. Serrano Mn, Bogu M, Vespignani A (2007) Patterns of dominant flows in the world trade web. J Econ Interac Coord 2: 111–124.

21. Hidalgo CA, Klinger B, Barabsi AL, Hausmann R (2007) The product space conditions the development of nations. Science 317: 482–487.

22. Hidalgo CA, Hausmann R (2009) The building blocks of economic complexity. PNAS 106: 10570–10575.

23. Caldarelli G, Cristelli M, Gabrielli A, Pietronero L, Scala A, et al. (2012) A network analysis of countries's export flows: Firm grounds for the building blocks of the economy. PLoS ONE 7: e47278.

24. Barigozzi M, Fagiolo G, Garlaschelli D (2010) Multinetwork of international trade: A commodity-specific analysis. Phys Rev E 81: 046104.

25. West G, Brown JH, Enquist BJ (1997) A general model for the origin of allometric scaling laws in biology. Science 276: 122–126.

26. Banavar J, Maritan A, Rinaldo A (1999) Size and form in efficient transportation networks. Nature 399: 130–132.

27. Garlaschelli D, Caldarelli G, Pietronero L (2003) Universal scaling relations in food webs. Nature 423: 165–168.

28. Zhang J, Guo L (2010) Scaling behaviors of weighted food webs as energy transportation networks. Journal of Theoretical Biology 264: 760–770.

29. Duan W (2007) Universal scaling behavior in weighted trade networks. Eur Phys J B 59: 271–276.

30. Herrada EA, Tessone CJ, Klemm K, Eguiluz VM, Hernandez-Garcia E, et al. (2008) Universal scaling in the branching of the tree of life. Plos one 3: e2757.

31. Zhang J, Wu L (2013) Allometry and dissipation of ecological flow networks. PLoS ONE 8: e72525.

32. Hausmann R, Hwang J, Rodrik D (2007) What you export matters. J Econ Growth 12: 1–25.

33. Cella G (1984) The input-output measurement of interindustry linkages. Oxford Bulletin of Economics and Statistics 46: 7384.

34. Song Y, Liu C, Langston C (2006) Linkage measures of the construction sector using the hypothetical extraction method. Construction Management and Economics 24: 579–589.

35. Foti NJ, Hughes JM, Rockmore DN (2011) Nonparametric sparsification of complex multiscale networks. PLoS ONE 6: e16431.

7

In Silico Insights into Protein-Protein Interactions and Folding Dynamics of the Saposin-Like Domain of *Solanum tuberosum* Aspartic Protease

Dref C. De Moura[1,2], Brian C. Bryksa[2], Rickey Y. Yada[1,2]*

1 Biophysics Interdepartmental Group, University of Guelph, Guelph, Ontario, Canada, **2** Department of Food Science, University of Guelph, Guelph, Ontario, Canada

Abstract

The plant-specific insert is an approximately 100-residue domain found exclusively within the C-terminal lobe of some plant aspartic proteases. Structurally, this domain is a member of the saposin-like protein family, and is involved in plant pathogen defense as well as vacuolar targeting of the parent protease molecule. Similar to other members of the saposin-like protein family, most notably saposins A and C, the recently resolved crystal structure of potato (*Solanum tuberosum*) plant-specific insert has been shown to exist in a substrate-bound open conformation in which the plant-specific insert oligomerizes to form homodimers. In addition to the open structure, a closed conformation also exists having the classic saposin fold of the saposin-like protein family as observed in the crystal structure of barley (*Hordeum vulgare* L.) plant-specific insert. In the present study, the mechanisms of tertiary and quaternary conformation changes of potato plant-specific insert were investigated *in silico* as a function of pH. Umbrella sampling and determination of the free energy change of dissociation of the plant-specific insert homodimer revealed that increasing the pH of the system to near physiological levels reduced the free energy barrier to dissociation. Furthermore, principal component analysis was used to characterize conformational changes at both acidic and neutral pH. The results indicated that the plant-specific insert may adopt a tertiary structure similar to the characteristic saposin fold and suggest a potential new structural motif among saposin-like proteins. To our knowledge, this acidified PSI structure presents the first example of an alternative saposin-fold motif for any member of the large and diverse SAPLIP family.

Editor: Eugene A. Permyakov, Russian Academy of Sciences, Institute for Biological Instrumentation, Russian Federation

Funding: This work was supported by the Natural Sciences and Engineering Research Council of Canada and the Canada Research Chairs Program. The funders had no role in study design, data collection and analysis, decision to publish, or preparation of the manuscript.

Competing Interests: The authors have declared that no competing interests exist.

* Email: ryada@uoguelph.ca

Introduction

Pepsin-like aspartic proteases (APs) constitute a family of endopeptidases found in all kingdoms of life [1]. APs of plant origin are generally homologous to members of the A1 family of APs (http://merops.sanger.ac.uk) [2] sharing similar primary and bilobal tertiary structures wherein two lobes are separated by a large active site cleft containing two catalytic aspartic acid residues, and low pH optima. However, plant APs are unique in that they frequently contain an extra 100-residue domain inserted in the C-terminal lobe, distinguishing them from their microbial and animal counterparts [3]. This extra domain, termed the plant-specific insert (PSI) or plant-specific sequence (PSS) [4–7], belongs to the saposin-like protein (SAPLIP) family and contains the Saposin B (Sap B) protein domain architecture [8]. Physiologically, SAPLIPs exhibit varied functionalities manifested primarily in their abilities to target, bind and/or perturb membranes, sometimes involving the ability to permeabilize and/or induce vesicle fusion [8–11]. Examples of SAPLIP function include sphingolipid degradation and antigen presentation [12], haemolytic activity (*Na*-SLP-1 and *Ac*-SLP-1) [13], antimicrobial and cytolytic activity (NK-lysin and granulysin) [14,15] and fusion of large unilamellar anionic vesicles *in vitro* (Sap C and recombinant PSI expressed without the parent AP) [9–11]. Recombinantly expressed free-form potato PSI has been shown to display potentially useful functionalities *in vitro* including antimicrobial activity against both plant and human pathogens [16], as well as anticancer activity against leukaemia cells without having lymphocyte toxicity [17].

Although SAPLIPs share low sequence identity and exhibit a multitude of functions in a variety of organisms, there have only been two discrete SAPLIP conformations observed to date. With the exception of granulysin [18], all known SAPLIPs have a characteristic pattern of 6 cysteines that form 3 disulfide bridges. The predominant tertiary structure is a substrate-free closed form first elucidated by NMR structure determination of porcine NK-lysin, and subsequently observed for all known SAPLIPs (see References [13,14,18–22]). This closed form, the classic saposin fold, is distinguished by a compact globular structure consisting of a 4 or 5 α-helix distorted bundle packed into an oblate spheroid.

By contrast, a second SAPLIP structural variant exists having an extended open conformation, first observed for Sap C bound to SDS micelles, that resembles two side-by-side boomerangs [19].

Unlike the compact structure seen in closed SAPLIPs, lipid-bound Sap C opens in a jackknife-like fashion thereby exposing the normally buried hydrophobic core to accommodate lipid interactions [23]. This V-shaped configuration has also been shown in Sap A bound to various amphiphiles [24]. The V-shaped SAPLIP configuration is also observed in Sap C homodimers in the absence of bound lipids [22]. As with Sap C bound to SDS micelles and Sap A lipoprotein discs, ligand-free Sap C jackknifes open at hinge points at the helix-helix junctions between the first two and last two helices. These hinge points allow Sap C monomers to open up and adopt an extended V-shape configuration forming domain-swapped homodimers. Hydrophobic regions are thus sequestered from their aqueous environment as the two interfaces come together yielding a hydrophobic core within the dimer. This jackknife opening mechanism serves to demonstrate the conformational flexibility of some SAPLIPs afforded by the helix-helix junctions. The ability to open and close allows for both membrane interactions and oligomerization [20,22,24].

Only two PSI structures (PDB IDs 1QDM and 3RFI), both resolved by X-ray crystallography, have been elucidated thus far: the inactive precursor structure of barley (*Hordeum vulgare* L.) phytepsin (HvAP) [25] and recombinant PSI of *Solanum tuberosum* (potato) AP [11]. Barley PSI was shown to have the archetypical compact saposin fold wherein the N- and C-termini remained attached to the C-terminal lobe of its parent phytepsin. By contrast, free form potato PSI adopts the less commonly observed open conformation as a homodimer analogous to that of ligand-free Sap C. Like their SAPLIP homologues, PSIs have been shown to induce vesicle leakage and fusion as well as having roles in plant vacuolar targeting [3,11,26,27].

To gain insight into the structural determinants of the PSI's pH-dependence for activity, the present study sought to elucidate and compare the protein dynamics and structural characteristics of the PSI in active and inactive pH conditions. Furthermore, the folding dynamics of the PSI open extended SAPLIP structure were investigated to clarify how the PSI tertiary structure relates to the typical closed SAPLIP fold.

Results

The PSI dimer forms a stable complex regardless of pH

Equilibrium molecular dynamics simulations of the dimer complex in acidic (pH 3.0) and neutral (pH 7.4) conditions revealed that the PSI dimer is stable regardless of pH, evidenced by low and relatively constant root-mean-square deviation (RMSD) values of the dimer trajectories when fitted to the initial coordinates of the crystal structure (Figure 1). As measured at the centre of mass (COM) of the individual monomers within the dimer complex, each monomer maintained steady contact at the dimer interface throughout the time course of each simulation (Figure 2). Further examination of the trajectories showed little fluctuation in the residues comprising helical regions. The C_α root-mean-square fluctuation (RMSF) for helices remained consistent through 100 ns and deviated little from the crystal structure (PDB ID 3RFI) providing further evidence of dimer stability (Figure 3).

Influence of pH on PSI dimer dissociation

Analogous to AFM pulling [28] and optical tweezers experiments [29], steered molecular dynamics (SMD) [30] simulations can be used to direct behaviour within a reduced number of degrees of freedom towards a particular state or phenomenon of interest. The efficiency of this technique can be exploited to study phenomena not normally accessible by conventional timescales

due to the computational expense of traditional MD simulations. SMD simulations typically use pulling velocities that are orders of magnitude higher than those used in AFM pulling or optical tweezers experiments, resulting in comparatively higher pulling forces. SMD is useful in exploring underlying processes involved in the dissociation of a dimer complex as evidenced in previous studies on unbinding pathways of proteins and substrates [31–35].

As a function of the distance between two molecules, the 1D potential of mean force (PMF) along a desired reaction coordinate (ξ) can be calculated [36,37]. In particular, it is the ability of the PMF, or free energy, to quantitatively describe $\Delta G_{dissociation}$ of a protein-ligand or dimer complex of interest. Although a number of ways of determining PMF exist [38,39], the umbrella sampling method was chosen for its efficient sampling along the reaction coordinate [40]. Using the umbrella sampling method in the context of dimer dissociation, an umbrella biasing potential was applied to restrain one monomer at increasing distances from the second reference monomer as measured between the respective centres of mass. As opposed to conventional MD simulations, the use of the restraining potential allowed for increased sampling of conformational space at defined positions along the reaction coordinate, resulting in a series of biased histograms. The weighted histogram analysis method (WHAM) was then used to combine the individual distributions and extract the unbiased PMF in a manner similar to [41].

To assess the potential influence of pH on the dissociation of the PSI dimer and gain insight into the unbinding mechanisms of the dimer complex, SMD simulations were performed in combination with umbrella sampling and WHAM. Equilibrium MD simulation structures were used as starting configurations for SMD, and pulling simulations were performed for acidic (active; pH 3.0) and neutral (inactive; pH 7.4) conditions in which one monomer (chain B) was pulled away from an immobile reference peptide (chain A) along the z-axis such that the distance between their centres of mass increased as the two peptides were pulled apart from one another. Although the PSI is optimally active at pH 4.5, pH 3.0 was used here since the pH 4.5 dimer is not sufficiently soluble to conduct ongoing monomer-dimer equilibrium experiments whose preliminary data indicate that PSI exists as a dimer under acidic conditions (unpublished data). The resultant trajectories were then used to generate the windows for umbrella sampling, and WHAM was used to extract PMF associated with dimer dissociation (Figure 4). The PMF profiles indicated that a significant amount of free energy was required to instigate dissociation with $\Delta G_{dissociation}$ values of 108.8 kJ mol^{-1} at acidic pH and 95.7 kJ mol^{-1} at neutral pH. The high free energy barrier to dissociation was suggestive of strong intermolecular protein-protein interactions. This was expected and reasonable as the probability of water contacting the hydrophobic undersides of the respective PSI monomers is minimized by maintaining strong contacts at the dimer interface [42], thereby sequestering hydrophobic residues from solvent and thus stabilizing dimer quaternary structure.

While the PMF profiles describing both the acidic and neutral pH dimer dissociations were similar, it should be noted that the $\Delta G_{dissociation}$ for the PSI at pH 3.0 was almost 13% larger than at neutral pH. Although the two monomers maintained similar contact distances throughout the time course of the simulation, the difference in $\Delta G_{dissociation}$ (Figure 2) may be explained by differences in ability to preserve contact at the hydrophobic interfaces. Compared to the pH 7.4 simulation in which all residues were in their standard state, histidine as well as all glutamic and aspartic acid residues were protonated in the acid simulation resulting in charge neutralization and mitigation of electrostatic repulsion among the expansive number of negatively

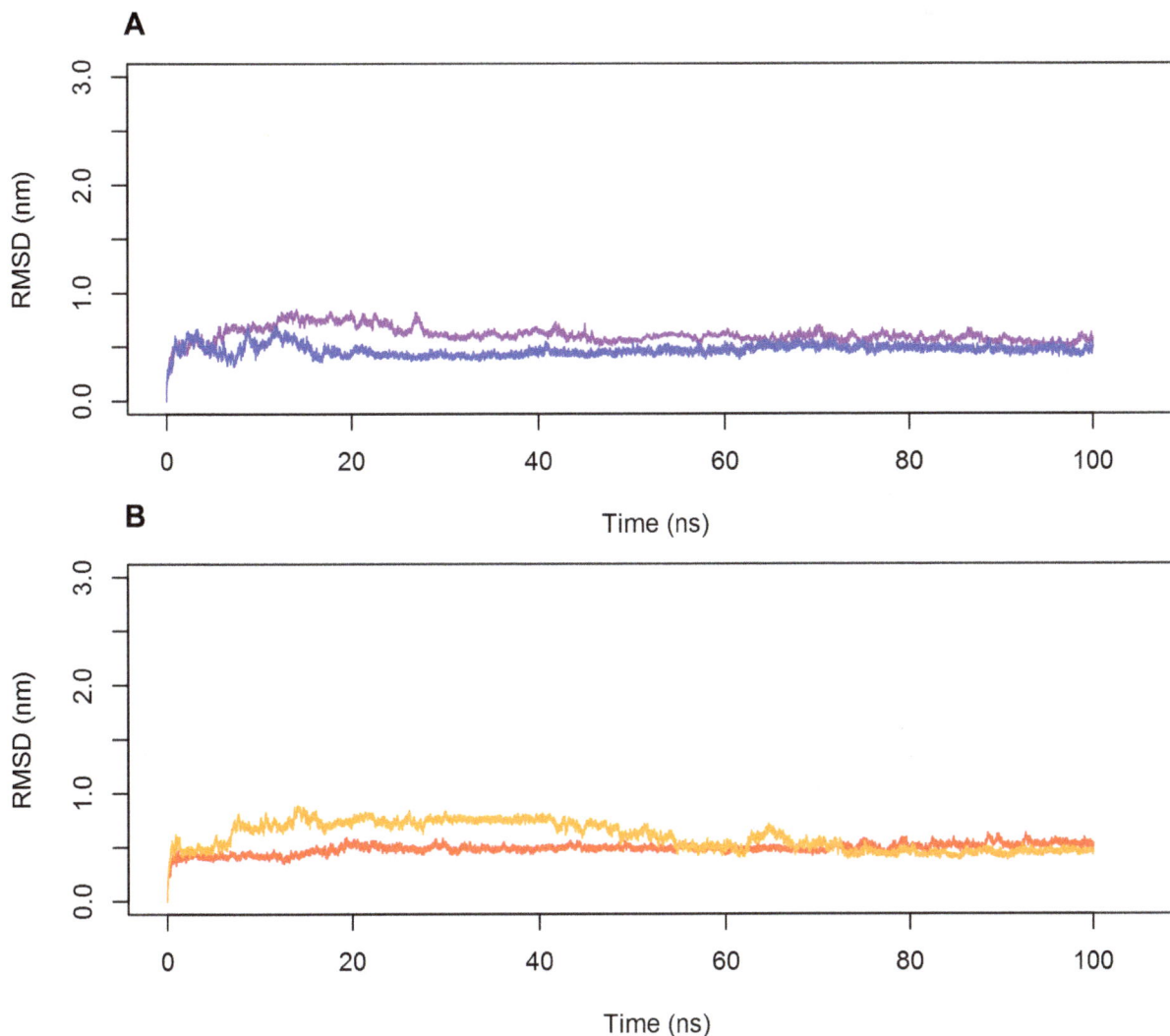

Figure 1. The backbone RMSD of the PSI dimer as a function of time. Backbone root-mean-square deviation (RMSD) of the PSI dimer at pH 3.0 (**A**) and pH 7.4 (**B**) indicated little deviation, evidenced by the low RMSD of the PSI backbone atoms for both peptides comprising the PSI homodimer. Colours identify the individual peptide chains within the dimer.

charged residues. This would result in the stabilization of the dimer as movement of the two monomers away from each other would be restricted by the dominant hydrophobic interactions at the dimer interface and the higher free energy requirement for dissociation.

The closed saposin-fold conformation is the dominant structure adopted by monomeric PSI

Principal component analysis (PCA) is a robust tool for identifying and separating the large-scale, and usually slowest, collective motions of atoms to reveal the largest contributors to atomic fluctuation of protein structures from the fast random internal motions [43,44]. To examine tertiary structure dynamics of monomeric PSI, and assess potential influence of pH on protein folding, unrestrained MD simulations were performed on the extended PSI monomer in solution at both active and inactive pH values. PCA was then applied to the unrestrained MD simulations and conformational changes were examined. Monomer conformational stability was evaluated by calculating backbone RMSD after least-square fitting by superposing MD trajectories onto the

PSI crystal structure. Simulations at both active and inactive pH produced similar trends in the evolution of RMSD, remaining stable with fluctuations in RMSD by approximately 0.2 nm – 0.8 nm until approximately 230 ns (pH 4.5) and 198 ns (pH 7.4), suggesting that the PSI deviated little from the crystal structure. At these times, a transition in tertiary structure occurred in which the RMSD brusquely increased 1.2 nm –1.4 nm after which the RMSD remained stable upon adopting a new conformation (Figure 5). The extended conformation closed in on itself and adopted the closed saposin fold characteristic of other SAPLIPs [13,14,18–25] irrespective of pH. Hence, the simulations essentially described a spontaneous tertiary structure transition from the open to closed state. As one might expect, the1D mode described for the first PC in either simulation corresponded to the closing motion of the PSI, accounting for approximately 78.8% and 74.2% of the overall motions for simulations at active and inactive pH, respectively (Figure 6). This closing motion corresponded to helices α1/α4 collapsing onto helices α2/α3, hinging at the flexible helix-helix junctions formed between α1/α2 and α3/α4 (Figure 7). At pH 7.4, the second PC was characterized by a slight twisting

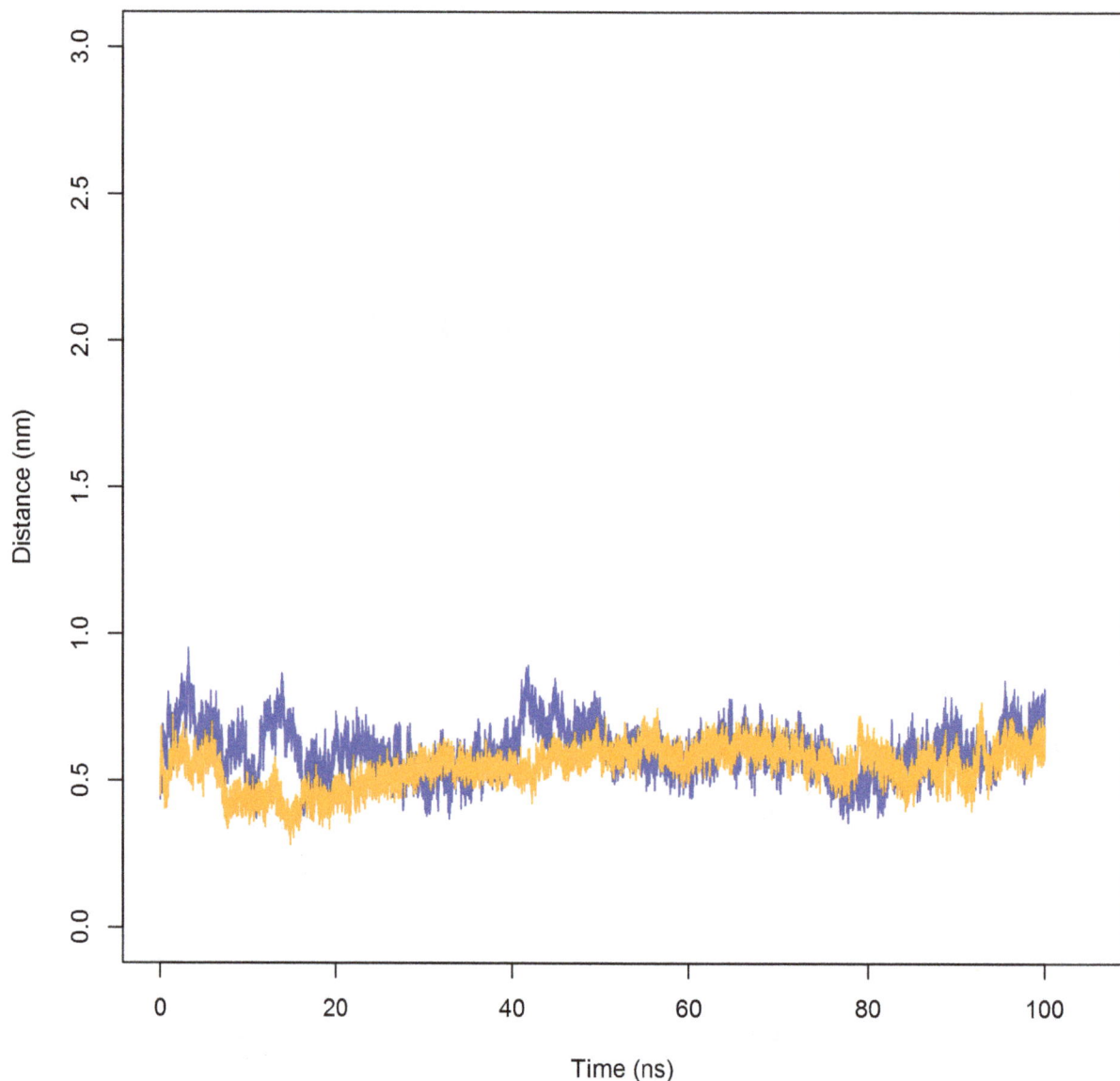

Figure 2. Distance between monomers of the PSI homodimer measured at the centres of mass (COMs) of each peptide. Distance between the two monomers of the PSI dimer revealed that the peptides maintained steady contact at the dimer interface regardless of pH. Both pH 3.0 (blue line) and pH 7.4 (orange line) simulations maintained an average distance of approximately 6 Å throughout the trajectories.

motion of the terminal helices (α1 and α4) relative to helices α2/α3 and was responsible for the characteristic distortion of the α-helix bundle typical for the saposin fold (Figure 8), a phenomenon that was not observed in the active pH simulation. Subsequent PCs showed diminished contributions of conformational changes to overall motions of the PSI.

Two-dimensional projections of the active and inactive trajectories onto their respective first and second PCs showed that the PSI explores a wide range of conformational space (Figure 9). The conformer plot for the inactive simulation (Figure 9B) revealed that the PSI transits through three distinct conformational states corresponding to three distinct minima, after which it becomes trapped in a third and final state. The first conformational state corresponds to the extended open crystal structure. After sampling the essential subspace near the starting conformation, the monomer then transitions to a second, discreet state as the protein begins to jackknife. This intermediate conformation corresponds to a quasi-folded tertiary structure in which helices α1 and α4 begin to collapse onto helices α2 and α3, thus forming the beginnings of the characteristic 4-helix bundle observed for all known SAPLIPs [8]. The third and final cluster is the most densely populated and closely packed cluster corresponding to a distorted helix bundle tertiary structure like that of the characteristic saposin-fold. Similarly, the active pH simulation showed a transition from the initial open extended structure to a quasi-folded, compact 4-helix bundle tertiary structure (Figure 9A). However, unique to the active pH simulation were several microstates sampled along the second PC before finally becoming trapped in the densely populated final cluster.

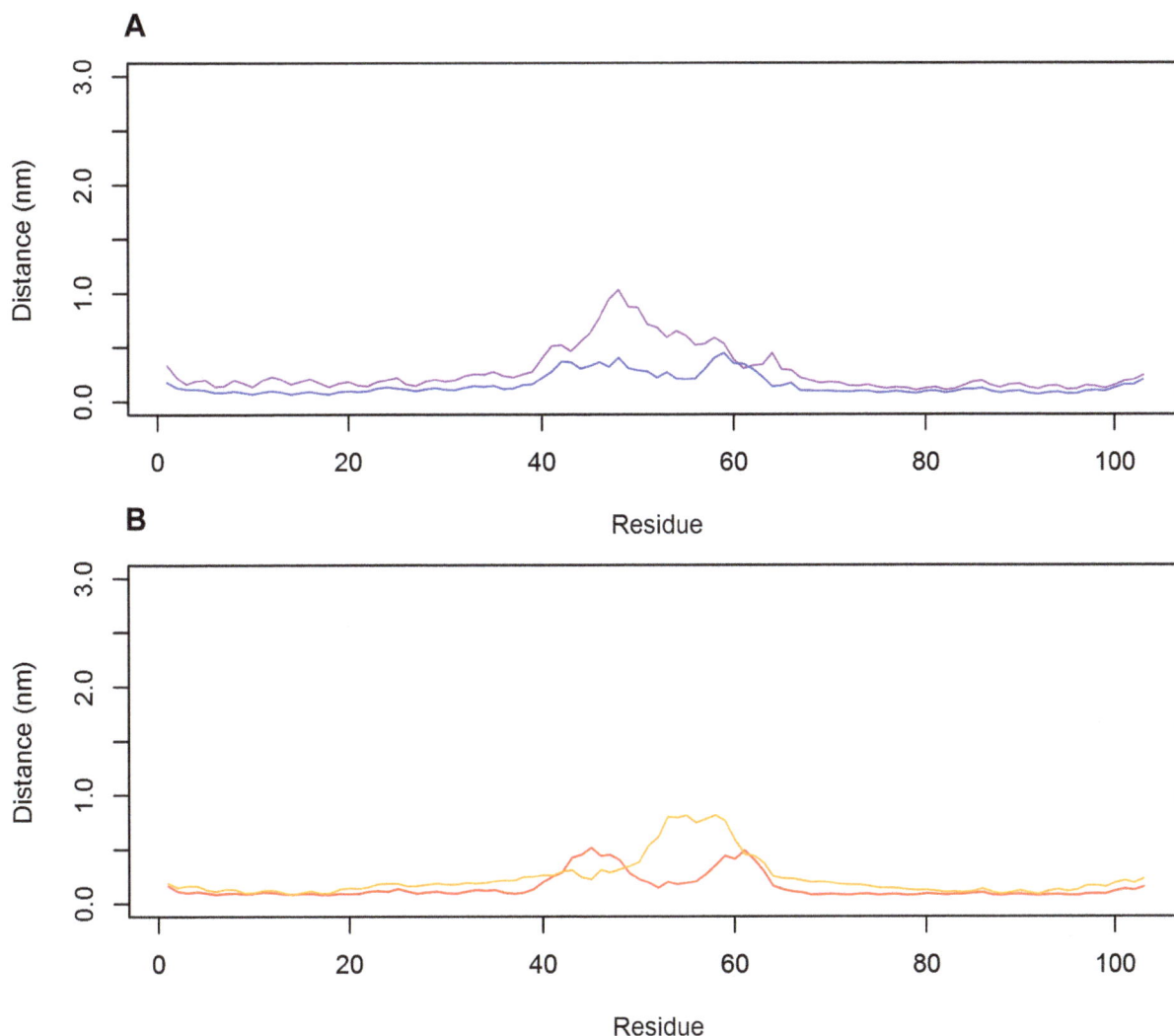

Figure 3. The C$_\alpha$ root-mean-square fluctuations (RMSFs) of the PSI dimer as a function of time. The C$_\alpha$ RMSFs for helices at pH 3.0 (**A**) and pH 7.4 (**B**) were consistent throughout the time course of the simulations, remaining below 5 Å for the helical regions. Fluctuations of up to 10 Å were noted for the flexible linker region, in agreement with the hypothesis that the linker region is intrinsically disordered, providing evidence of PSI dimer secondary structure stability regardless of pH. Colours identify the individual peptide chains within the dimer.

Radius of gyration measurements for investigating PSI compaction

To monitor compaction of the PSI monomer as indicated from the principal component analyses, the radius of gyration (R$_g$) was determined for hydrophobic residues located in helical regions (Figure 10). For both the active and inactive pH simulations, initial R$_g$ corresponded to fluctuations in the PSI open conformation. At approximately 200 ns, a sharp decrease in R$_g$ occurred corresponding to folding events related to the hydrophobic collapse of the concave face in which the stem formed between the N- and C-termini folded over onto helices $\alpha 2/\alpha 3$. This process corresponded to the large, abrupt changes in RMSD as well as the first PC observed at this time. Post-collapse, the lowered R$_g$ values and the scarcity in R$_g$ deviation throughout the remainder of the simulations were consistent with the adoption of a stable tertiary structure.

The adoption of the saposin fold-like tertiary structure for monomeric PSI is made possible by the hinges formed at the flexible helix-helix junctions between $\alpha 1/\alpha 2$ and $\alpha 3/\alpha 4$. Similar to

the orthorhombic saposin crystal structure (PDB ID 2QYP) [22], extended PSI had obtuse opening angles of 109° and 110° for the active and inactive pH simulations, respectively, as measured at the Cα atoms of Pro66, Glu85 and Lys101 in which the hinge between helices $\alpha 3$ and $\alpha 4$ open about Glu85 (Figure 7A). Upon closing, the opening angles closed to approximately 23° and 33° for the pH 4.5 and 7.4 simulations, respectively, in agreement with the 34° opening angle measured at the Cα atoms of Pro67, Asn86 and Arg102 of the resolved portion of the closed HvAP PSI crystal structure [25]. Some α-helical secondary structure was lost at the helix-helix junctions during folding, transitioning to random coil to accommodate the movement of side chains towards the hydrophobic concave face of the PSI. The adoption of a saposin-like fold is further reinforced by the low RMSD between the resolved 4-helical bundle of the HvAP PSI and closed StAP PSI structures generated through MD (1.060 Å and 0.6477 Å for pH 4.5 and pH 7.4, respectively). All data are available upon request.

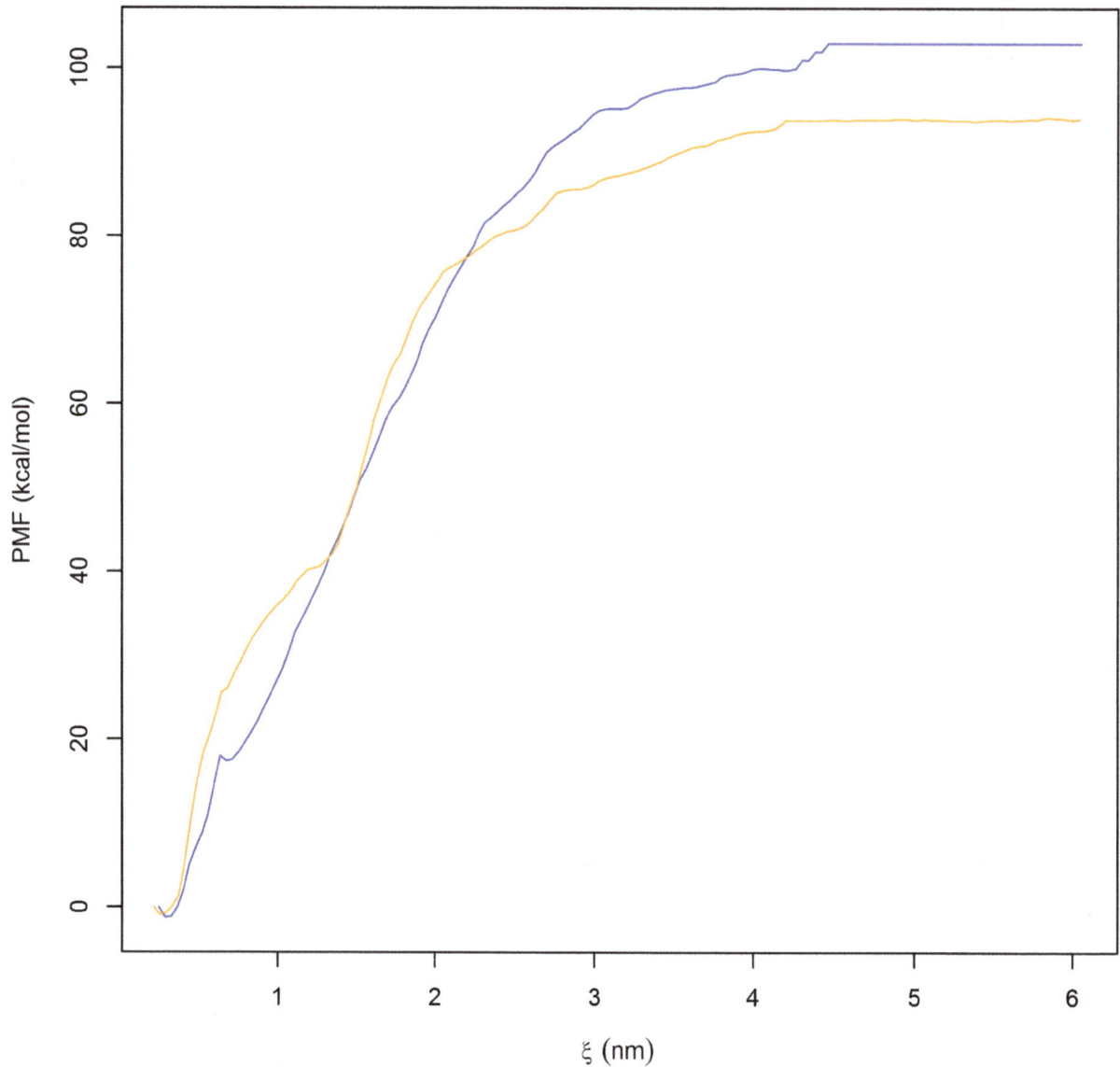

Figure 4. The potential of mean force (PMF) as a function of the distance between the COMs of PSI monomers. The PMF as a function of the intra-peptide distance between the PSI dimer monomers at pH 3.0 (blue line) and pH 7.4 (orange line) revealed that dissociation of the dimer requires increased energy as pH is lowered from pH 7.4 (95.7 kJ mol^{-1}) to pH 3.0 (108.8 kJ mol^{-1}), possibly the result of charge neutralization of carboxylate groups at acidic pH thereby minimizing charge-charge repulsion.

Discussion

Stability and dissociation of the PSI dimer

The calculated PMFs describing the dissociation of the PSI dimer at pH 3.0 and pH 7.4 gave $\Delta G_{\text{dissociation}}$ values of 108.8 kJ mol^{-1} and 95.7 kJ mol^{-1}, respectively. As expected, the large free energy requirement can be attributed to the need to sequester the hydrophobic concave face of the PSI from solution, thereby minimizing entropy associated with the exposure of hydrophobic residues. Though both dimers form stable conformations, it should be noted that greater binding between the monomers is achieved at acidic pH, and the major reason for this is likely charge neutralization of carboxylic acid groups on glutamic and aspartic acid residues. It would be expected that electrostatic repulsion between monomers would thus be lowered allowing hydrophobic interactions at the dimer interface to dominate. An analogous

phenomenon is seen with membrane-bound Sap C where neutralizing its negatively charged electrostatic surface removes membrane-protein charge-charge repulsion [19] thereby mitigating the unfavourable introduction of charges into the bilayer apolar hydrophobic environment. Furthermore, the calculated RMSD, RMSF and minimum distance maintained between the two monomers (see Figures 1–3) were consistent with the stability gained by folding the two monomers into a compact globular quaternary structure, limiting potential changes in tertiary structure.

Maintaining hydrophobic contact between the residues lining the concave face of the PSI is the driving force for preserving the quaternary structure and stability of the overall dimer, which may be related to the relative stability of the dimer at acidic pH relative to that at neutral pH. As the dimer experiences a lower free energy barrier to dissociation at neutral pH, the dimer quaternary

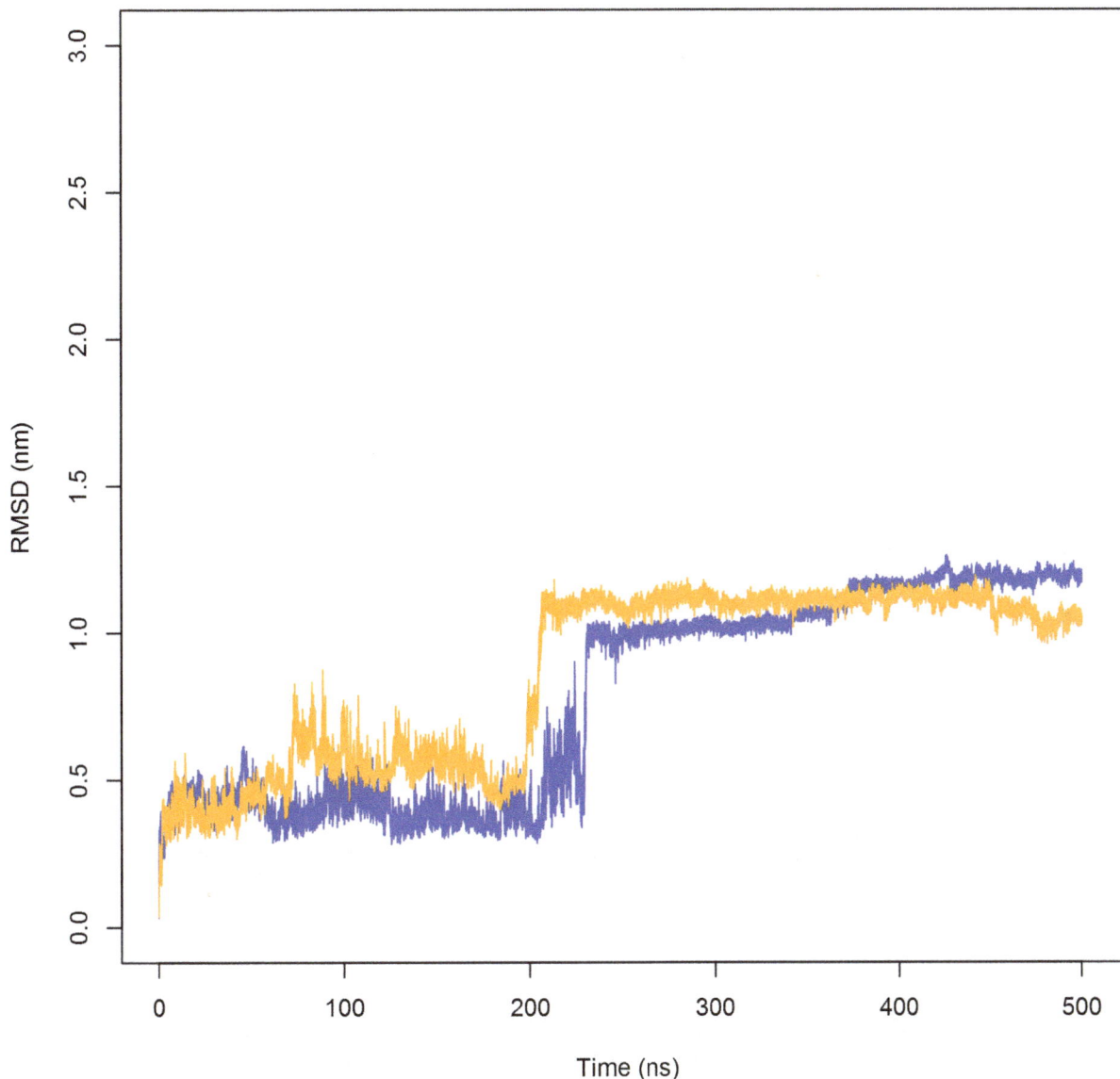

Figure 5. The backbone RMSD of PSI monomers as a function of time. Backbone RMSD of the PSI monomer at pH 4.5 (blue line) and pH 7.4 (orange line) are presented. The PSI monomer maintains its overall tertiary structure at both pH 4.5 and pH 7.4 similar to that of the native dimer structure until an abrupt change in RMSD at 230 ns and 198 ns for pH 4.5 and pH 7.4, respectively. At these times, the PSI jackknifes closed and adopts saposin-like fold characteristic of all known SAPLIP members.

structure can be interpreted as being less stable leading to eventual dissociation.

The larger free energy requirement for dissociation at acidic pH may be indicative of a physiological necessity. It has been established *a priori* that the PSI is active against bilayers at acidic pH [11,25,27,45]. We hypothesise that the dimer formation at acidic pH may represent a particular functional quaternary structure. Baoukina and Tieleman [46,47] previously concluded that covalently linked antiparallel lung surfactant protein B (SP-B) dimers, analogous to StAP PSI dimer, mediate faster kinetics of monolayer folding. SP-B dimers promoted bilayer folding and eventual formation of hemifusion-like stalk connections similar to those observed in vesicle fusion [46,47]. It is theorised in the present study that PSI dimers may also function in a similar manner. Such a pH-dependence for a quaternary structure-function relationship is further supported by the observed StAP

PSI capacity to induce bilayer fusion of large unilamellar vesicles (LUV) at acidic pH, causing both membrane disruption and fusion [11], and is supported by previous research examining the roles of the PSI in vesicle disruption and membrane targeting [26,27,45]. The idea that the dimer serves a functional role in bilayer disruption and fusion is also consistent with the "clip-on" model for Sap C-mediated vesicle fusion, proposed by Wang *et al.* [10] and further appended by Rossmann *et al.* [22]. This model hypothesises that Sap C dimers can bind to two vesicles, interacting with the membrane in a similar fashion to Sap C monomers through domain swapping, and thereby bring adjacent bilayers close enough to mediate fusion. Considering the similarities in structure and dimer stability pH-dependence, it stands to reason that our proposed model suggests a possible commonality between the Sap C "clip-on" model and the PSI mode of membrane interaction.

A

B

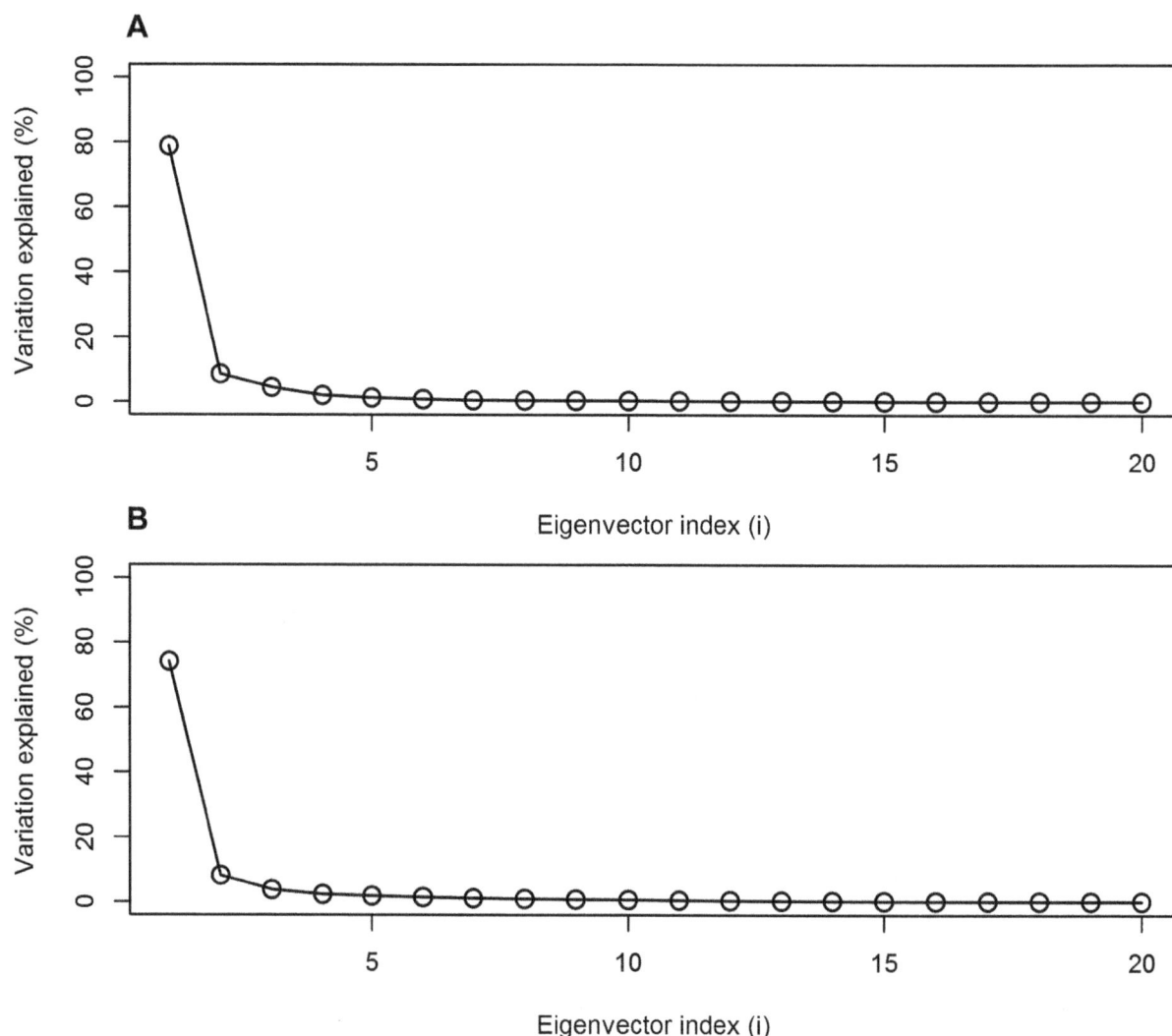

Figure 6. Contribution of the first twenty PCA eigenvectors to the overall closing motion of the PSI. The contribution of the first 20 PCA eigenvectors to the closing motion of the PSI at active (**A**) and inactive (**B**) pH are presented. The first eigenvectors contribute 78.8% and 74.2% of the overall motions for the pH 4.5 and pH 7.4 simulations, respectively, and correspond to the collapse of helices α1/α2 onto helices α3/α4.

Conformational flexibility and adoption of the saposin-fold

Insight into conformational changes can be gained by projecting the MD trajectory onto the subspace spanned by the two largest (typically the first and second) principal components [43,48]. In doing so, it is possible to characterize the transitions from the open, extended conformation of the PSI monomer to the closed saposin fold-like structure seen in the unrestrained simulations. As well, any possible intermediate structures that may be adopted during the opening-to-closing transition can be observed, providing a map of the overall structural variability of the PSI. The resultant conformer plots thus provide the means to interpret the conformational changes sampled by the unrestrained MD simulations and express the relationships between these conformers. Unrestrained MD simulations of the extended PSI monomer suggested that the PSI adopts a closed saposin-like conformation independent of pH.

Principal component analysis performed on the MD trajectories revealed that the first PC corresponds to the closing of the PSI, accounting for 78.8% and 74.2% of the overall motion of the

protein for the active and inactive pH simulations, respectively. Analogous to the PSI dimer, it is postulated that the closing motion observed in monomeric PSI arises from the need to reduce the entropy gained from exposure of these hydrophobic residues to water. Two-dimensional projections of the first two PCs revealed that the PSI transitions from an extended state to one or more intermediates before finally closing in on itself. Although the 2D projections sampled similar conformational space, the conformer plots of the active and inactive pH simulations differed in that the active pH simulation sampled several microstates before settling into a minimum and adopting a saposin-like closed motif. This differed from the inactive pH simulation in which the PSI sampled only three distinct states corresponding to energy minima for the initial structure, a molten globular structure, and finally a closed saposin-like tertiary structure. It is at this state that the concave face of the PSI has formed a hydrophobic core at its centre. These differences may be attributed to the differing electrostatic makeup of the two systems; negative charges on Glu and Asp are at least partially neutralized at active pH resulting in an overall positively charged protein, whereas both negative and positive residues exist

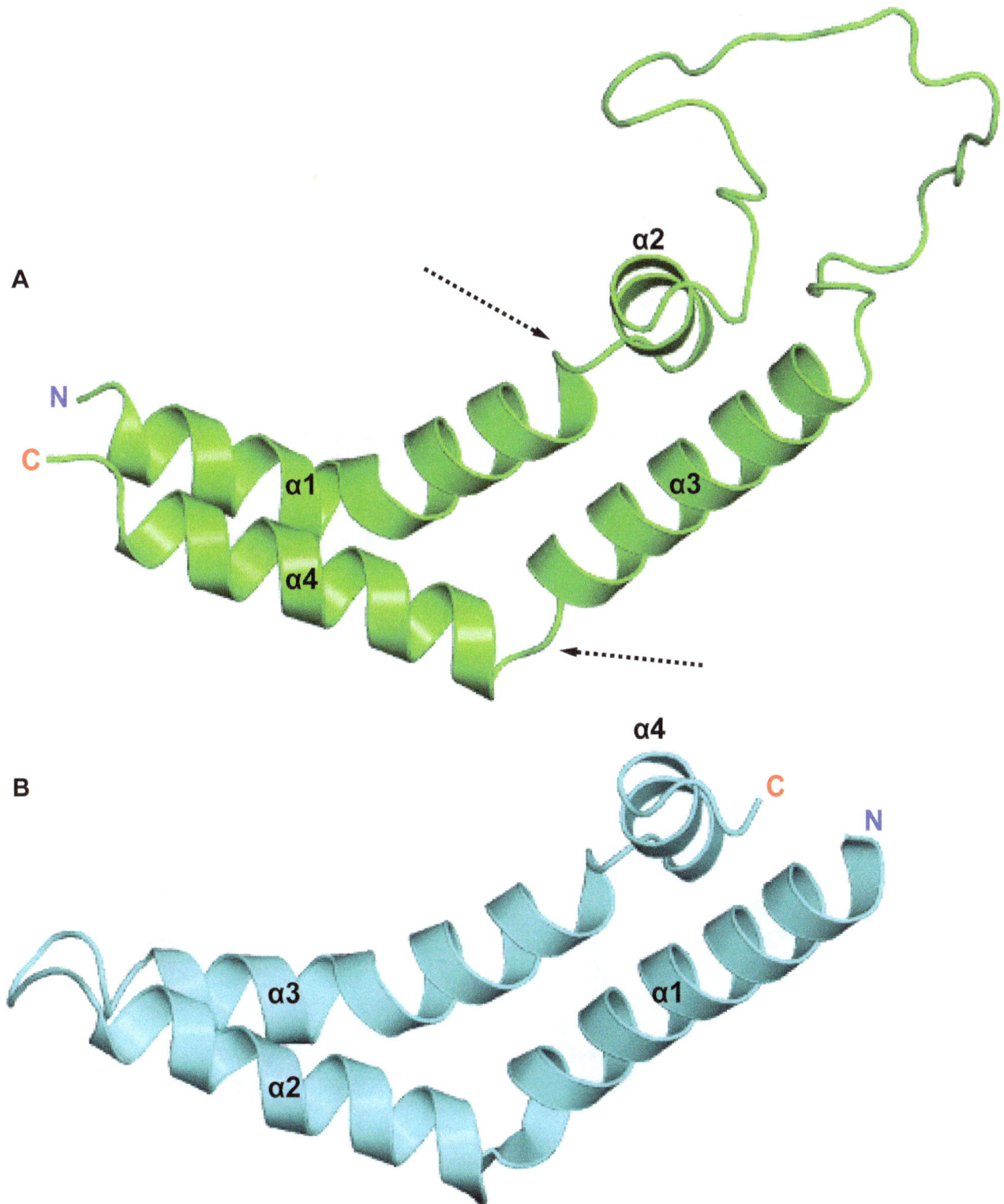

Figure 7. Structures of the PSI and orthorhombic Sap C. The crystal structure of potato (*Solanum tuberosum*) PSI (PDB ID 3RFI, **A**), with the missing linker region modelled and orthorhombic Sap C (PDB ID 2QYP, **B**) are presented. Like its Sap C homologue, potato PSI was crystalized as an extended dimer. The hinge-bending capability of the PSI is made possible by the flexible helix-helix junctions formed between α1/α2 and α3/α4, indicated by dashed arrows.

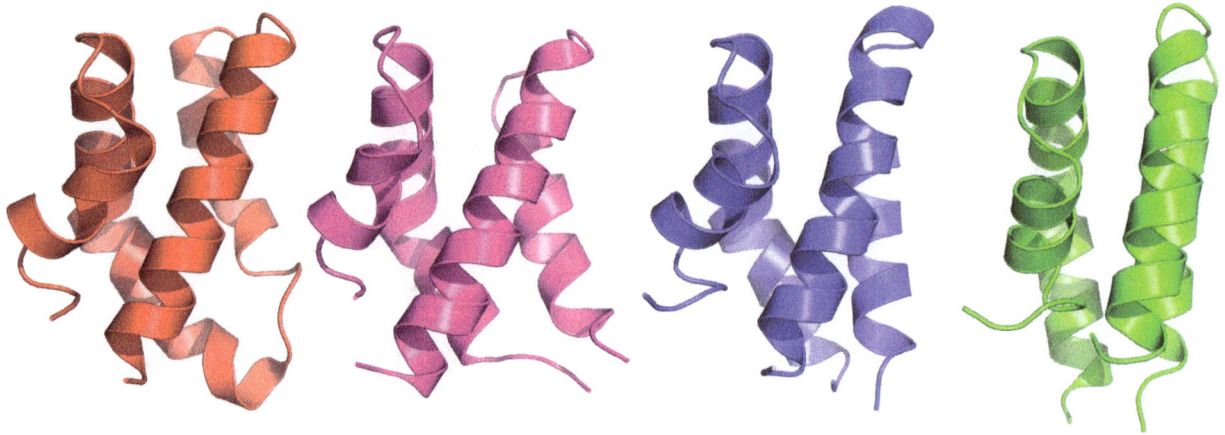

Figure 8. Comparison of folded potato PSI to other SAPLIPs. Structural comparison of the folded potato PSI at pH 4.5 (blue) and pH 7.4 (green), averaged over the last 200 ns of the simulation trajectories, to the crystal structure of barley PSI (PDB ID 1QDM, magenta) and the crystal structure Sap C (PDB ID 2GTG, red). Potato PSI simulated at pH 7.4, simulated with parameters closely resembling the experimental parameters used for both barley PSI and Sap C, exhibited a compact globular structure consisting of a distorted four-α-helix bundle characteristic of other SAPLIPs. Potato PSI simulated at pH 4.5 adopted a compact four-α-helix bundle structure not previously observed for any SAPLIP. The linker regions of potato PSI are omitted for clarity.

in the neutral pH simulations allowing for potential intra-peptide salt bridging or potentially different hydrogen bonding patterns.

The conformational changes adopted by the PSI as it transits from extended to closed conformation were attributed to the high degree of conformational flexibility at the hinge-bending regions of the helix-helix junctions, similar to what is observed in other SAPLIPs. The latter is a common characteristic for saposin members of the SAPLIP family which have been shown to have the capacity to exist both in substrate-free closed and in extended lipid- or peptide-bound conformations [22–24]. For the PSI, this flexibility is made possible in part by the local dynamics of side chains. Hydrophobic residues located in the helical regions orient

themselves such that their side chains are involved in the formation of the tight dimer interface (as observed in the unrestrained dimer simulations), induced by the presence of inter-protein hydrophobic interactions. The helix orientations in this packing motif are mimicked by monomeric PSI as it closes in a domain-swapped fashion in that helices α3 and α4 twist about their helical axes thereby maximizing intra-protein hydrophobic contacts. This folding process is marked by the hydrophobic regions of the four helices collapsing on themselves thereby minimizing contact with the polar environment and concomitantly maximizing aqueous contact with the polar outer surfaces.

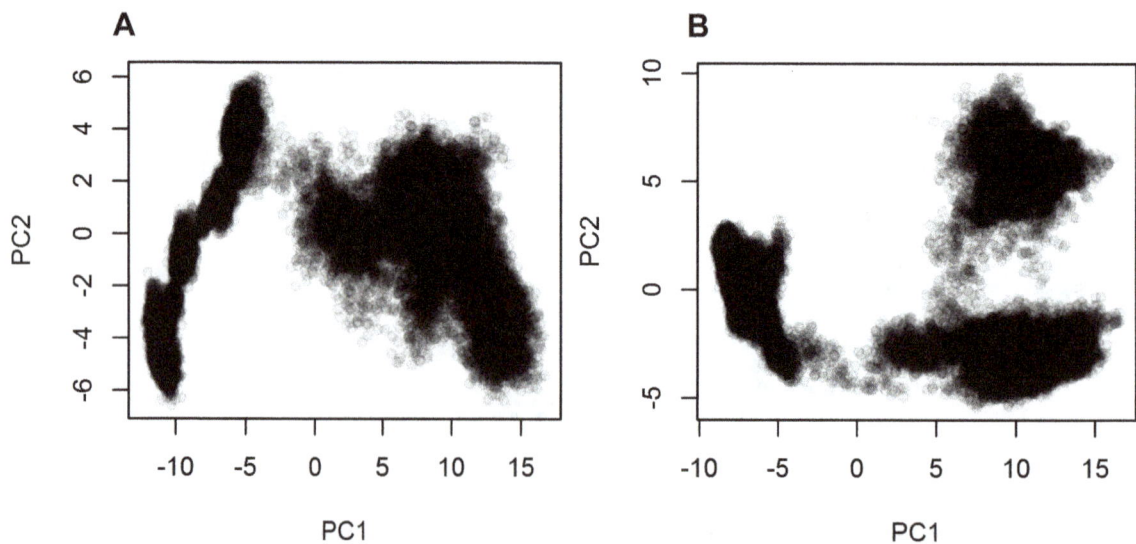

Figure 9. Two-dimensional projections of the first two eigenvectors of the PSI monomer. Projection of the first two eigenvectors of the unbiased PSI simulations at pH 4.5 (**A**) and pH 7.4 (**B**). Both simulations transited from the extended dimer-like structure to a saposin fold-like conformation over the course of the 500 ns trajectories. The inactive pH simulation transited through three distinct clusters whereas the active pH simulation transitioned through several microstates before becoming trapped in the last densely populated cluster. The differences in the essential subspace sampled by the two differing pH ranges may be due to unspecific (hydrophobic) interactions sampled in the pH 4.5 simulations where charge neutralization minimizes like-charge repulsions.

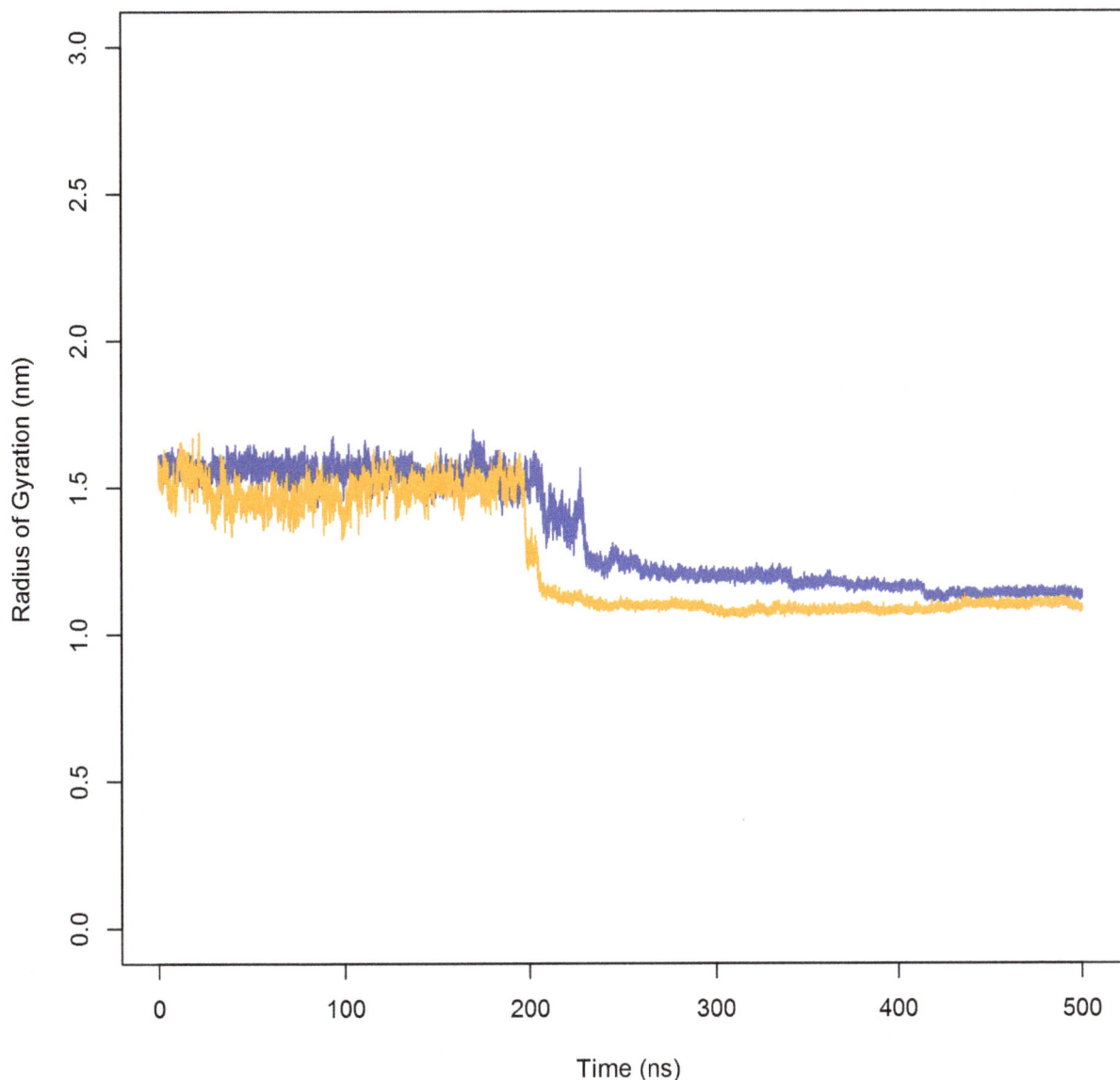

Figure 10. Radius of gyration (R$_g$) of the PSI over the time course of the simulations. R$_g$ of the PSI at pH 4.5 (blue line) and pH 7.4 (orange line) as a function of time. In either case, the PSI was free to move in the extended state. Upon adoption of a saposin-like fold, the collapse of the hydrophobic concave face of the PSI onto itself limits movement, thereby restricting water access to the hydrophobic core.

The dynamics of PSI closure suggest two possible structures for the PSI, and that pH influences these conformational differences. To date, the only SAPLIP pH-structure report has been for Sap C in which a reduction in pH from 6.8 to 5.4 did not result in observable conformational changes [23]. It should be noted, however, that Sap C acidification occurred with monomeric protein already folded to a local minimum having adopted the characteristic saposin fold. This is in contrast to the present study in which the open PSI structure was allowed to explore a large degree of conformational space as it closed to a local energy minimum. The similar structure for the neutral StAP PSI ensemble in the present study, and that for HvAP PSI [25], as well as the low RMSDs, may indicate that the classical saposin-fold is pH-dependant for at least some SAPLIP cases. The neutral pH StAP PSI simulation and HvAP crystallography [25] used similar experimental parameters (i.e., 100 mM NaCl and neutral pH) suggesting that the acidic pH saposin-like fold observed in the

present study likely presents a derivative of the classic saposin fold and it would be expected to be adopted by other SAPLIPs having similar structures and pH-function dependencies.

The present study undertook a comprehensive analysis of the PSI to identify conformational changes due to differences in pH and to assess the potential impact that these changes may have on protein function. Free energy changes for PSI dimer dissociation at acidic and neutral pH were predicted by steered MD simulations in combination with umbrella sampling. These identified key differences in binding affinities indicating that the PSI has a preference for maintaining the dimer quaternary structure at acidic (active) pH due to the higher free energy requirement for dissociation. In conclusion, we postulate that the preference for dimerization may be indicative of a functional structure that plays a role in membrane binding and vesicle fusion. PCA of unrestrained MD simulations of the PSI monomer after separation from the dimer complex was then used to assess conformational

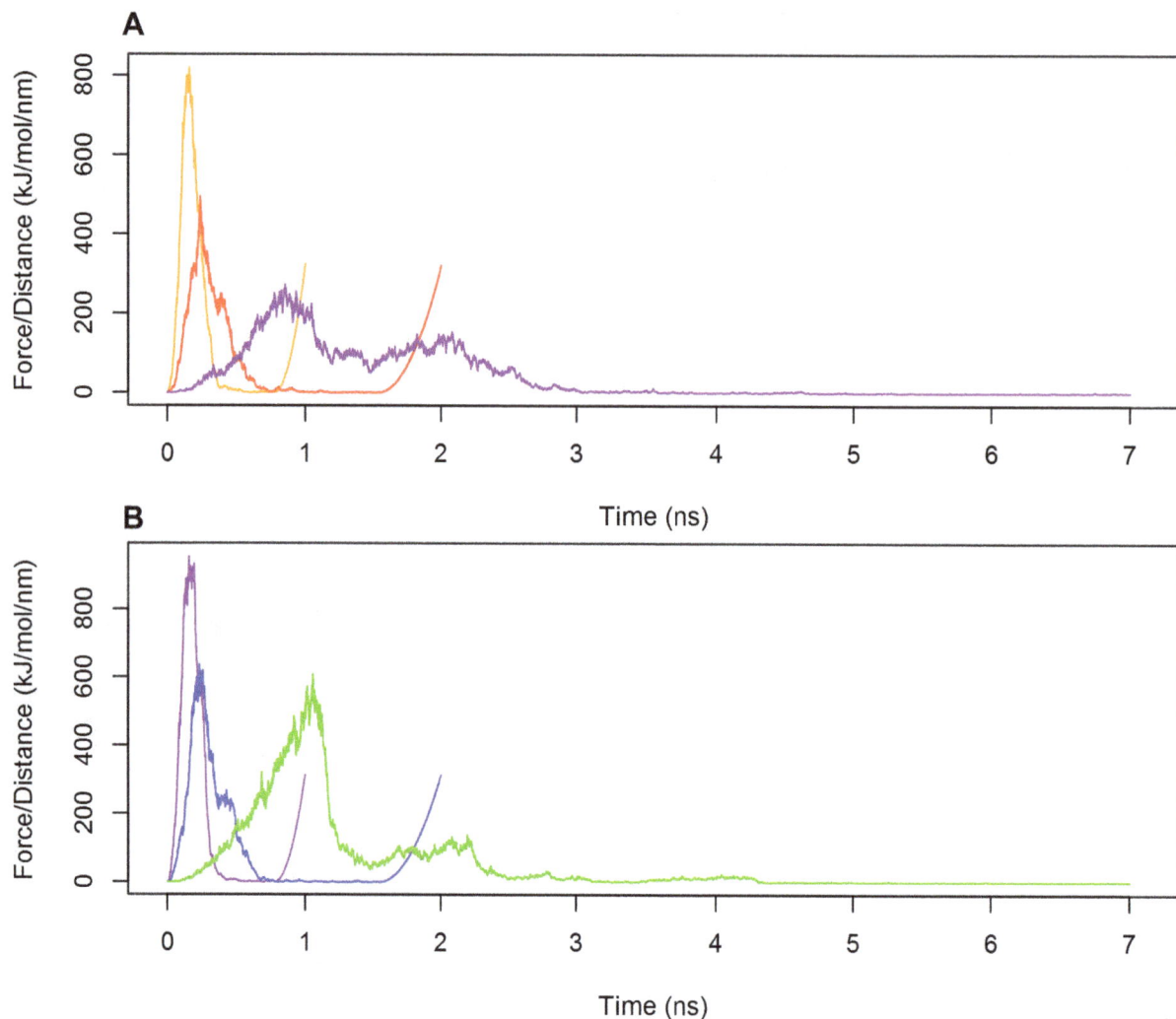

Figure 11. Force-Time curves of the dissociation of the PSI dimers using differing pull rates. Force-Ttime curves for pH 3.0 (**A**) and pH 7.4 (**B**) pulling simulations using 10 nm ns^{-1} (orange line in **A**, purple line in **B**), 5 nm ns^{-1} (red line in **A**, blue line in **B**) and 1 nm ns^{-1} (purple line in **A**, green line in **B**). The curves exhibit similar shapes overall suggesting that pull rate does not appreciably affect dimer dissociation.

changes adopted by the monomers. Although monomeric PSI folded to a closed conformation regardless of pH, the final closed structures differed in that the pH 7.4 PSI adopted a tertiary structure consistent with the characteristic saposin-fold whereas a distinct saposin-like fold was observed at pH 4.5. This acidified PSI structure presents the first example of an alternative saposin-fold motif for any member of the large and diverse SAPLIP family.

Methods

Initial models

In the present study, the high resolution (1.9 Å) X-ray crystal structure of extended potato (*Solanum tuberosum*) PSI (PDB ID 3RFI) [11] was used as the template structure for SMD simulations. Chain A was used for the unrestrained MD simulations of the PSI monomer. The linker region (residues 40–63) connecting helices α1/α2 to α3/α4 was not resolved in the original crystal structure. As such, MODELLER 9v8 was used to build the missing linker region *ab initio* and modeled as random coil [49,50]. Hydrogen atoms were added for all titratable residues in accordance to their calculated protonation states as determined

using the H++ web server [51–53] using an internal protein dielectric constant and solvent dielectric constant of 10 and 80, respectively, with sodium chloride added at 140 mM or 100 mM for the SMD dimer dissociation (pH 3.0 and 7.4) and unrestrained MD monomer (pH 4.5 and 7.4) simulations, respectively.

Unrestrained molecular dynamics system setup

All simulations and analyses were carried out using the GROMACS software suite, Version 4.5.5 [54–56] employing the Amber99sbnmr1-ILDN force field [57–59]. For each simulation, periodic boundary conditions were applied in all dimensions. The PSI was centred in a cubic box such that the protein was positioned at least 1.2 nm from the box edge and hydrated using the TIP3P explicit water model [60] to solvate the system. Sodium and chloride counterions were added at 140 mM or 100 mM concentrations for the SMD and unrestrained MD simulations, respectively, to produce electroneutral systems. Short-range electrostatic interactions were cut off at 8 Å whilst long-range electrostatic interactions were calculated using the particle-mesh Ewald (PME) summation method [61] with fourth order B-spine interpolation and a maximum grid spacing of 1.2 Å. A twin-range van der Waals cut-off was employed (0.8/1.0 nm) and an integration time step of 2 fs was used with neighbour searching performed every 5 steps with all bond lengths being constrained using the linear constraint solver (LINCS) algorithm [62].

Each simulation was prepared in 3 phases before production runs were performed. In the first phase, the protein was energy minimized using the steepest decent algorithm with position restraints placed on all heavy atoms ($k_{PR} = 1000$ kJ mol^{-1} nm^{-2}) until the maximum force converged to ≤ 500 kJ mol^{-1} nm^{-1}. In the next phase, the system was equilibrated for 1 ns with position restraints placed on all heavy atoms in the canonical ensemble using the Berendsen weak coupling method [63] with temperature maintained at 303.15 K ($\tau_T = 0.1$ ps). This equilibration was followed by another 1 ns position-restrained simulation in the isobaric-isochoric ensemble. Again, the Berendsen weak coupling method was used to maintain temperature at 303.15 K and pressure isotropically coupled at 1 bar ($\tau_P = 1.0$ ps). The isothermal compressibility of the system was set to 4.5×10^{-5} bar^{-1}. For the production unrestrained MD simulations, position restraints were removed. The velocity rescale (v-rescale) algorithm [64] was used to maintain the temperature of the system at 303.15 K ($\tau_T = 0.1$ ps) and the pressure was again maintained at 1 bar using the Parrinello-Rahman [65,66] barostat ($\tau_P = 2.0$ ps) in the isobaric-isochoric ensemble with long-range dispersion correction applied for both the energy and pressure terms. Production simulations were conducted for 500 ns.

Steered molecular dynamics

Equilibrated starting structures of the PSI dimer for the pH 3.0 and pH 7.4 SMD simulations were generated following the same procedure used for the unrestrained MD simulations with production MD conducted for 100 ns. The resultant structures were then used as the starting configuration for the corresponding SMD pulling simulations. The PSI dimer was placed in a rectangular box large enough to accommodate separation of the dimer along the z axis whilst satisfying the minimum image convention. The dimer was then subjected to energy minimization and equilibration in both the canonical and isobaric-isochoric ensemble again as described above. For the SMD pulling simulations, position restraints were removed from chain B of the PSI dimer while heavy atoms of chain A were harmonically restrained ($k_{PR} = 1000$ kJ mol^{-1} nm^{-2}) in a similar fashion to that used in the equilibration phases. Chain A was used as an immobile reference for chain B pulling. The 1D reaction coordinate was chosen to be the distance along the z axis between the COMs of

the two PSI monomers. Chain B was pulled away from chain A along the z axis for 1 ns with a constant velocity of 10 nm ns^{-1} using an elastic spring ($k = 1000$ kJ mol^{-1} nm^{-2}) positioned at the COM of the peptide. Trajectories at slower pulling rates (5 ns nm^{-1} and 1 ns nm^{-1}) were also tested to assess the influence of pulling forces on the structure as force is applied. These slower pulling rates resulted in similar force-time curves and similar overall trajectories (Figure 11) [67]. As such, the faster pulling rate was used for experimental SMD simulations to minimize usage of computational resources.

Umbrella sampling and determination of PMF

The pH 3.0 and pH 7.4 trajectories from SMD pulling simulations were used to generate sampling windows along the reaction coordinate. Windows were spaced between 0.5–2.0 Å for the first 2.5 nm followed by approximately 2.0 Å spacing until the overall distance between the COMs between chains A and B was approximately 6.0 nm. This resulted in 43 and 45 sampling windows being selected for the pH 3.0 and pH 7.4 simulations, respectively. MD was conducted for each window for 15 ns with a harmonic restraint ($k_{PR} = 1000$ kJ mol^{-1} nm^{-2}) applied to chain B to fix the peptide along the reaction coordinate and then the PMF was constructed. The unbiased PMF was calculated using WHAM and the $\Delta G_{dissociation}$ was evaluated as the difference in energy between the plateau and energy minimum along the PMF curve [68].

Principal component analysis

PCA was used to identify principal modes of motion sampled during the unrestrained MD simulations of the extended PSI monomer at pH 4.5 and pH 7.4. A covariance matrix of the backbone atoms in the monomer was constructed using the PSI trajectories. The matrices were then diagonalized yielding eigenvectors and their corresponding eigenvalues revealing both the directions and amplitudes of motion, respectively. Projection of the two largest eigenvectors onto 2D space was then used to quantitatively compare the ability of each ensemble to sample varying regions of conformational space.

Author Contributions

Conceived and designed the experiments: DCD BCB RYY. Performed the experiments: DCD. Analyzed the data: DCD. Wrote the paper: DCD BCB RYY.

References

1. Davies DR (1990) The structure and function of the aspartic proteinases. Annu Rev Biophys Biophys Chem 19: 189–215.

2. Rawlings ND, Barrett AJ, Bateman A (2010) MEROPS: The peptidase database. Nucleic Acids Res 38: D227–D233.

3. Runeberg-Roos P, Törmäkangas K, Östman A (1991) Primary structure of a barley-grain aspartic proteinase. Eur J Biochem 202: 1021–1027.

4. Cordeiro MC, Xue Z, Pietrzak M, Salomé Pais M, Brodelius PE (1994) Isolation and characterization of a cDNA from flowers of Cynara cardunculus encoding cyprosin (an aspartic proteinase) and its use to study the organ-specific expression of cyprosin. Plant Mol Biol 24: 733–741.

5. Simões I, Faro C (2004) Structure and function of plant aspartic proteinases. Eur J Biochem 271: 2067–2075.

6. Törmäkangas K, Kervinen J, Östman A, Teeri T (1994) Tissue-specific localization of aspartic proteinase in developing and germinating barley grains. Planta 195: 116–125.

7. Guruprasad K, Törmäkangas K, Kervinen J, Blundell TL (1994) Comparative modelling of barley-grain aspartic proteinase: A structural rationale for observed hydrolytic specificity. FEBS Lett 352: 131–136.

8. Bruhn H (2005) A short guided tour through functional and structural features of saposin-like proteins. Biochem J 389: 249–257.

9. Vaccaro AM, Tatti M, Ciaffoni F, Salvioli R, Serafino A, et al. (1994) Saposin C induces pH-dependent destabilization and fusion of phosphatidylserine-containing vesicles. FEBS Lett 349: 181–186.

10. Wang Y, Grabowski GA, Qi X (2003) Phospholipid vesicle fusion induced by saposin C. Arch Biochem Biophys 415: 43–53.

11. Bryksa BC, Bhaumik P, Magracheva E, De Moura DC, Kurylowicz M, et al. (2011) Structure and mechanism of the saposin-like domain of a plant aspartic protease. J Biol Chem 286: 28265–28275.

12. Matsuda J, Vanier MT, Saito Y, Tohyama J, Suzuki K, et al. (2001) A mutation in the saposin A domain of the sphingolipid activator protein (prosaposin) gene results in a late-onset, chronic form of globoid cell leukodystrophy in the mouse. Hum Mol Genet 10: 1191–1199.

13. Willis C, Wang CK, Osman A, Simon A, Pickering D, et al. (2011) Insights into the membrane interactions of the saposin-like proteins Na-SLP-1 and Ac-SLP-1 from human and dog hookworm. PLoS ONE 6: e25369.

14. Liepinsh E, Andersson M, Ruysschaert J, Otting G (1997) Saposin fold revealed by the NMR structure of NK-lysin. Nat Struct Mol Biol 4: 793–795.

15. Anderson DH, Sawaya MR, Cascio D, Ernst W, Modlin R, et al. (2003) Granulysin crystal structure and a structure-derived lytic mechanism. J Mol Biol 325: 355–365.

16. Guevara MG, Veríssimo P, Pires E, Faro C, Daleo DR (2004) Potato aspartic proteases: induction, antimicrobial activity and substrate specificty. J Plant Pathol 86: 233–238.

17. Mendieta JR, Fimognari C, Daleo GR, Hrelia P, Guevara MG (2010) Cytotoxic effect of potato aspartic proteases (StAPs) on Jurkat T cells. Fitoterapia 5: 329–335.

18. Jongstra J, Schall TJ, Dyer BJ, Clayberger C, Jorgensen J, et al. (1987) The isolation and sequence of a novel gene from a human functional T cell line. J Exp Med 165: 601–614.

19. de Alba E, Weiler S, Tjandra N (2003) Solution structure of human saposin C: pH-dependent interaction with phospholipid vesicles. Biochemistry 42: 14729–14740.

20. Ahn VE, Faull KF, Whitelegge JP, Fluharty AL, Privé GG (2003) Crystal structure of saposin B reveals a dimeric shell for lipid binding. Proc Natl Acad Sci U S A 100: 38–43.

21. Ahn VE, Leyko P, Alattia J, Chen L, Privé GG (2006) Crystal structures of saposins A and C. Protein Sci 15: 1849–1857.

22. Rossmann M, Schultz-Heienbrok R, Behlke J, Remmel N, Alings C, et al. (2008) Crystal structures of human saposins C and D: Implications for lipid recognition and membrane interactions. Structure 16: 809–817.

23. Hawkins CA, Alba Ed, Tjandra N (2005) Solution structure of human saposin C in a detergent environment. J Mol Biol 346: 1381–1392.

24. Popovic K, Holyoake J, Pomès R, Privé GG (2012) Structure of saposin A lipoprotein discs. Proc Natl Acad Sci U S A 109: 2908–2912.

25. Kervinen J, Tobin GJ, Costa J, Waugh DS, Wlodawer A, et al. (1999) Crystal structure of plant aspartic proteinase prophytepsin: Inactivation and vacuolar targeting. EMBO J 18: 3947–3955.

26. Törmäkangas K, Hadlington JL, Pimpl P, Hillmer S, Brandizzi F, et al. (2001) A vacuolar sorting domain may also influence the way in which proteins leave the endoplasmic reticulum. Plant Cell 13: 2021–2032.

27. Egas C, Lavoura N, Resende R, Brito RMM, Pires E, et al. (2000) The saposin-like domain of the plant aspartic proteinase precursor is a potent inducer of vesicle leakage. J Biol Chem 275: 38190–38196.

28. Binnig G, Quate CF, Gerber C (1986) Atomic force microscope. Phys Rev Lett 56: 930–933.

29. Ashkin A, Dziedzic JM, Bjorkholm JE, Chu S (1986) Observation of a single-beam gradient force optical trap for dielectric particles. Opt Lett 11: 288–290.

30. Izrailev S, Stepaniants S, Isralewitz B, Kosztin B, Lu H, et al. (1999) Steered molecular dynamics. In: Deuflhard P, Hermans J, Leimkuhler B, Mark A, Skeel RD, Reich S, editors. Computational Molecular Dynamics: Challenges, Methods, Ideas. Berlin: Springer-Verlag. 39–65.

31. Sotomayor M, Schulten K (2007) Single-molecule experiments *in vitro* and *in silico*. Science 316: 1144–1148.

32. West DK, Brockwell DJ, Olmsted PD, Radford SE, Paci E (2006) Mechanical resistance of proteins explained using simple molecular models. Biophys J 90: 287–297.

33. González A, Perez-Acle T, Pardo L, Deupi X (2011) Molecular basis of ligand dissociation in β-adrenergic receptors. PLoS ONE 6: e23815.

34. Kalikka J, Akola J (2011) Steered molecular dynamics simulations of ligand-receptor interaction in lipocalins. Eur Biophys J 40: 181–194.

35. Cuendet MA, Michielin O (2008) Protein-protein interaction investigated by steered molecular dynamics: The TCR-pMHC complex. Biophys J 95: 3575–3590.

36. Torrie GM, Valleau JP (1977) Nonphysical sampling distributions in monte carlo free-energy estimation: Umbrella sampling. J Comput Phys 23: 187–199.

37. Roux B (1995) The calculation of the potential of mean force using computer simulations. Comput Phys Commun 91: 275–282.

38. NategholEslam M, Holland BW, Gray CG, Tomberli B (2011) Drift-oscillatory steering with the forward-reverse method for calculating the potential of mean force. Phys Rev E Stat Nonlin Soft Matter Phys 83: 021114.

39. Kosztin I, Barz B, Janosi L (2006) Calculating potentials of mean force and diffusion coefficients from nonequilibrium processes without Jarzynski's equality. J Chem Phys 124: 064106.

40. Kästner J (2011) Umbrella sampling. Wiley Interdiscip Rev Comput Mol Sci 1: 932–942.

41. Kumar S, Bouzida D, Swendsen RH, Kollman PA, Rosenberg JM (1992) The weighted histogram analysis method for free-energy calculations on biomolecules. I. The method. J Comput Chem 13: 1011–1021.

42. Zhang BW, Brunetti L, Brooks CL (2011) Probing pH-dependent dissociation of HdeA dimers. J Am Chem Soc 133: 19393–19398.

43. Amadei A, Linssen ABM, Berendsen HJC (1993) Essential dynamics of proteins. Proteins: Struct Funct Genet 17: 412–425.

44. Berendsen HJ, Hayward S (2000) Collective protein dynamics in relation to function. Curr Opin Struct Biol 10: 165–169.

45. Frazão C, Bento I, Costa J, Soares CM, Veríssimo P, et al. (1999) Crystal structure of cardosin A, a glycosylated and arg-gly-asp-containing aspartic proteinase from the flowers of *Cynara cardunculus* L. J Biol Chem 274: 27694–27701.

46. Baoukina S, Tieleman DP (2010) Direct simulation of protein-mediated vesicle fusion: Lung surfactant protein B. Biophys J 99: 2134–2142.

47. Baoukina S, Tieleman D (2011) Lung surfactant protein SP-B promotes formation of bilayer reservoirs from monolayer and lipid transfer between the interface and subphase. Biophys J 100: 1678–1687.

48. Barrett CP, Hall BA, Noble MEM (2004) Dynamite: A simple way to gain insight into protein motions. Acta Crystallogr D Biol Crystallogr 60: 2280–2287.

49. Eswar N, Webb B, Marti-Renom MA, Madhusudhan MS, Eramian D, et al. (2006) Comparative protein structure modeling using Modeller. Curr Protoc Bioinformatics Chapter 5: Unit 5.6.

50. Fiser A, Do RK, Sali A (2000) Modeling of loops in protein structures. Protein Sci 9: 1753–1773.

51. Gordon JC, Myers JB, Folta T, Shoja V, Heath LS, et al. (2005) H++: A server for estimating pKas and adding missing hydrogens to macromolecules. Nucleic Acids Res 33: W368–W371.

52. Anandakrishnan R, Aguilar B, Onufriev AV (2012) H++3.0: Automating pK prediction and the preparation of biomolecular structures for atomistic molecular modeling and simulations. Nucleic Acids Res 40: W537–W541.

53. Myers J, Grothaus G, Narayanan S, Onufriev A (2006) A simple clustering algorithm can be accurate enough for use in calculations of pKs in macromolecules. Proteins 63: 928–938.

54. Van Der Spoel D, Lindahl E, Hess B, Groenhof G, Mark AE, et al. (2005) GROMACS: Fast, flexible, and free. J Comput Chem 26: 1701–1718.

55. Hess B, Kutzner C, van der Spoel D, Lindahl E (2008) GROMACS 4: Algorithms for highly efficient, load-balanced, and scalable molecular simulation. J Chem Theory Comput 4: 435–447.

56. Pronk S, Páll S, Schulz R, Larsson P, Bjelkmar P, et al. (2013) GROMACS 4.5: A high-throughput and highly parallel open source molecular simulation toolkit. Bioinformatics 29: 845–854.

57. Lindorff-Larsen K, Piana S, Palmo K, Maragakis P, Klepeis JL, et al. (2010) Improved side-chain torsion potentials for the amber ff99SB protein force field. Proteins 78: 1950–1958.

58. Long D, Li DW, Walter KF, Griesinger C, Brüschweiler R (2011) Toward a predictive understanding of slow methyl group dynamics in proteins. Biophys J 101: 910–915.

59. Li DW, Brüschweiler R (2010) NMR-based protein potentials. Angew Chem Int Ed Engl 49: 6778–6780.

60. Jorgensen WL, Chandrasekhar J, Madura JD, Impey RW, Klein ML (1983) Comparison of simple potential functions for simulating liquid water. J Chem Phys 79: 926–935.

61. Darden T, York D, Pedersen L (1993) Particle mesh Ewald: An $N·\log(N)$ method for Ewald sums in large systems. J Chem Phys 98: 10089–10092.

62. Hess B, Bekker H, Berendsen HJC, Fraaije JGEM (1997) LINCS: A linear constraint solver for molecular simulations. J Comput Chem 18: 1463–1472.

63. Berendsen HJC, Postma JPM, van Gunsteren WF, DiNola A, Haak JR (1984) Molecular dynamics with coupling to an external bath. J Chem Phys 81: 3684–3690.

64. Bussi G, Donadio D, Parrinello M (2007) Canonical sampling through velocity rescaling. J Chem Phys 126: 014101.

65. Parrinello M, Rahman A (1981) Polymorphic transitions in single crystals: A new molecular dynamics method. J Appl Phys 52: 7182–7190.

66. Nosé S, Klein ML (1983) Constant pressure molecular dynamics for molecular systems. Mol Phys 50: 1055–1076.

67. Lemkul JA, Bevan DR (2010) Assessing the stability of Alzheimer's amyloid protofibrils using molecular dynamics. J Phys Chem B 114: 1652–1660.

68. Chen P, Kuyucak S (2011) Accurate determination of the binding free energy for KcsA-charybdotoxin complex from the potential of mean force calculations with restraints. Biophys J 100: 2466–2474.

8

Exogenous Methyl Jasmonate Treatment Increases Glucosinolate Biosynthesis and Quinone Reductase Activity in Kale Leaf Tissue

Kang-Mo Ku[1], Elizabeth H. Jeffery[2], John A. Juvik[1]*

1 Department of Crop Sciences, University of Illinois at Urbana-Champaign, Urbana, Illinois, United States of America, 2 Department of Food Science and Human Nutrition, University of Illinois at Urbana-Champaign, Urbana, Illinois, United States of America

Abstract

Methyl jasmonate (MeJA) spray treatments were applied to the kale varieties 'Dwarf Blue Curled Vates' and 'Red Winter' in replicated field plantings in 2010 and 2011 to investigate alteration of glucosinolate (GS) composition in harvested leaf tissue. Aqueous solutions of 250 µM MeJA were sprayed to saturation on aerial plant tissues four days prior to harvest at commercial maturity. The MeJA treatment significantly increased gluconasturtiin (56%), glucobrassicin (98%), and neoglucobrassicin (150%) concentrations in the apical leaf tissue of these genotypes over two seasons. Induction of quinone reductase (QR) activity, a biomarker for anti-carcinogenesis, was significantly increased by the extracts from the leaf tissue of these two cultivars. Extracts of apical leaf tissues had greater MeJA mediated increases in phenolics, glucosinolate concentrations, GS hydrolysis products, and QR activity than extracts from basal leaf tissue samples. The concentration of the hydrolysis product of glucoraphanin, sulforphane was significantly increased in apical leaf tissue of the cultivar 'Red Winter' in both 2010 and 2011. There was interaction between exogenous MeJA treatment and environmental conditions to induce endogenous JA. Correlation analysis revealed that indole-3-carbanol (I3C) generated from the hydrolysis of glucobrassicin significantly correlated with QR activity ($r = 0.800$, $P < 0.001$). Concentrations required to double the specific QR activity (CD values) of I3C was calculated at 230 µM, which is considerably weaker at induction than other isothiocyanates like sulforphane. To confirm relationships between GS hydrolysis products and QR activity, a range of concentrations of MeJA sprays were applied to kale leaf tissues of both cultivars in 2011. Correlation analysis of these results indicated that sulforaphane, NI3C, neoascorbigen, I3C, and diindolylmethane were all significantly correlated with QR activity. Thus, increased QR activity may be due to combined increases in phenolics (quercetin and kaempferol) and GS hydrolysis product concentrations rather than by individual products alone.

Editor: Hitoshi Ashida, Kobe University, Japan

Funding: The Boerner Fellowship in Crop Sciences partially supported the first author. No additional external funding received for this study. The funders had no role in study design, data collection and analysis, decision to publish, or preparation of the manuscript.

Competing Interests: The authors have declared that no competing interests exist.

* Email: juvik@illinois.edu

Introduction

Epidemiological studies have reported that the intake of *Brassica* vegetables is inversely correlated with cancer risk, and this association is stronger than those between cancer and fruit and vegetable consumption in general [1]. Kale (*Brassica oleracea* L. *acephala*) is a frequently consumed leafy vegetable. Young tender leafs are harvested for human consumption and older plant tissues for animal feed [2]. Kale is a good source of vitamins (Vitamin A, C, and E) and of health promoting phytochemicals including glucosinolates (GS), carotenoids, phenolics, and tocopherols. In certain regions like on the Iberian Peninsula, kale (*Brassica oleracea acephala* group) leaves and flower buds are grown and harvested for consumption throughout the year [2].

There are several types of kales. Among them, it was previously reported that GS composition of Siberian kale (*B. napus*) was distinct from 'Vates' (*B. oleracea*) type kale [3]. Red Russian and Siberian kales (*Brassica napus ssp. pabularia*) are typically more tender and have a milder flavor than the European "oleracea" kales whose young leaves are superior for use in salads. Napus

kales have good cold tolerance so that they can be grown anywhere in the US over a broader range of growing seasons and are also used as animal forage. Forage and root vegetable cultivars of *B. napus* show high levels of progoitrin [4] which can promote goitrogenic effects in mammals [5]. Although cultivars of *Brassica napus* are thought to have originated from a chance hybridization between *Brassica rapa* and *Brassica oleracea*, the Red Russian type of kales were bred by artificial hybridization (http://seedambassadors.org/Mainpages/still/napuskale/napuskale.htm). The 'Red Winter' cultivar was derived from Red Russian kale types.

B. oleracea kale is a rich source of flavonoids, possessing up to 47 mg of kaempferol and 22 mg of quercetin per 100 g of fresh leaf tissue. Kale contains the highest flavonoid content among all of the *Brassica oleracea* vegetables [6]. Phenolics have putative antioxidant, anticancer, and anti-cardiovascular disease activity [7–9]. Previous research revealed that MeJA treatments enhance total polyphenolic compounds and flavonoids in kale leaf tissues [10]. The response to MeJA treatment was more dramatically observed in young tissue (apical leaves) compared to old tissue (basal leaves) [10].

Besides phenolic compounds, kale is also good source of GS. Glucosinolates are a class of secondary metabolites found in cruciferous crops. The breakdown products have been shown to affect human health, insect herbivory, and plant resistance to pathogens [11–13]. Some GS breakdown products have a chemoprotective effect against certain cancers in humans [14].

Up-regulation of phase II enzyme detoxification activity has been suggested as a good strategy for cancer prevention [15,16]. Phase II detoxifying enzymes including glutathione S-transferase (GST) and quinone reductase (QR) can enhance detoxification and elimination of carcinogens from the body [15,16]. Hydrolysis products of GS, isothiocyanates such sulforaphane and phenethyl isothiocyanate (PEITC) have been shown to enhance quinone reductase (QR) and provide other chemopreventive activities [17,18]. Previous studies have reported that the hydrolysis products of the indolyl GS including glucobrassicin and neogluco-brassicin also have cancer chemopreventive activity. Hydrolysis products of glucobrassicin including indole-3-carbinol (I3C), diindolylmethane, and ascorbigen induce QR [17,19]. N-methox-yindole-3-carbinol (NI3C) and neoascorbigen (NeoASG), the hydrolysis product of neoglucobrassicin has been reported to induce cell cycle arrest in human colon cancer cell lines [20] and to induce QR activity [21].

The GS are also associated with insect defense in *Brassica* species. Jasmonic acid (JA), an endogenous plant signal transduction compound whose biosynthesis is up-regulated when *Brassica* plant species are attacked by herbivores, causes enhanced indolyl GS biosynthesis [22]. The increased GS concentrations induced by exogenous MeJA spray treatment was found to be a species-specific response [23]. MeJA treatment significantly increased gluconasturtiin and neoglucobrasicin in broccoli [24] and glucor-aphanin, glucobrassicin, and neoglucobrassicin in cauliflower [23]. In addition, MeJA treatment significantly increased QR inducing activity and nitric oxide production inhibitory activity in broccoli and cauliflower [21,23,25,26]. To date, GS compositional changes

of kale leaf tissue induced by exogenous MeJA treatments have not been previously reported in the literature.

Compared to other *Brassica* vegetables including broccoli, watercress, and Brussels sprouts, anti-cancer bioactivity information about kale is limited [18,19,27]. The objective of this research is to determine the QR inducing health promoting effect derived from elevated phytochemical concentrations induced by MeJA in two different kale types.

Materials and Methods

Plant Cultivation

The cultivars 'Red Winter' (RW, *Brassica napus ssp. pabularia*) and 'Dwarf Blue Curled Vates' (DBCV, *Brassica oleracea* L. var. *acephala*) used for these experiments were purchased from Burpee Seed Co. (Warminster, PA). Seeds of each kale genotype were germinated in 32 cell plant plug trays filled with sunshine LC1 (Sun Gro Horticulture, Vancouver, British Columbia, Canada) professional soil mix. Seedlings were grown in a greenhouse at the University of Illinois at Urbana-Champaign under a 25°C/15°C and 14 h/10 h: day/night temperature regime with supplemental lighting. Thirty days after germination, seedling trays were placed in a ground bed to harden off for a week prior to transplanting into field plots at the University of Illinois South Farm (40° 04′38.89″ N, 88° 14′26.18″ W). Experimental design was a split-plot arrangement in a randomized complete block (RCB) design with three replicates. The experimental plot was surrounded by a guard row to avoid border effects. Transplanting of kale seedlings was conducted on June 11, 2010 and June 13, 2011. Harvesting kale occurred on July 25 in 2010 and July 27 in 2011. Irrigation was only applied during the first week of cultivation for the establishment of transplanted seedlings. Weather conditions during the 2010 and 2011 growing seasons collected from Illinois State Water Service (http://www.isws.illinois.edu/warm/data/cdfs/cmiday.txt) are presented in Table 1.

Table 1. Weather information during the growing seasons of 2010 and 2011 for Champaign, Illinois.

Total solar radiation (MJ·m^{-2})			
Year	Jun	Jul	Sum
2010	720	730	1450
2011	667	790	1457
% of (2011/2010)	93	108	100.5

Precipitation (mm)			
Year	Jun	Jul	Sum
2010	199	91	290
2011	107	40	147
% of (2011/2010)	54	44	50.7

Growing degree days (°C)			
Year	Jun	Jul	Sum
2010	373	408	781
2011	362	430	792
% of (2011/2010)	97	105	101.4

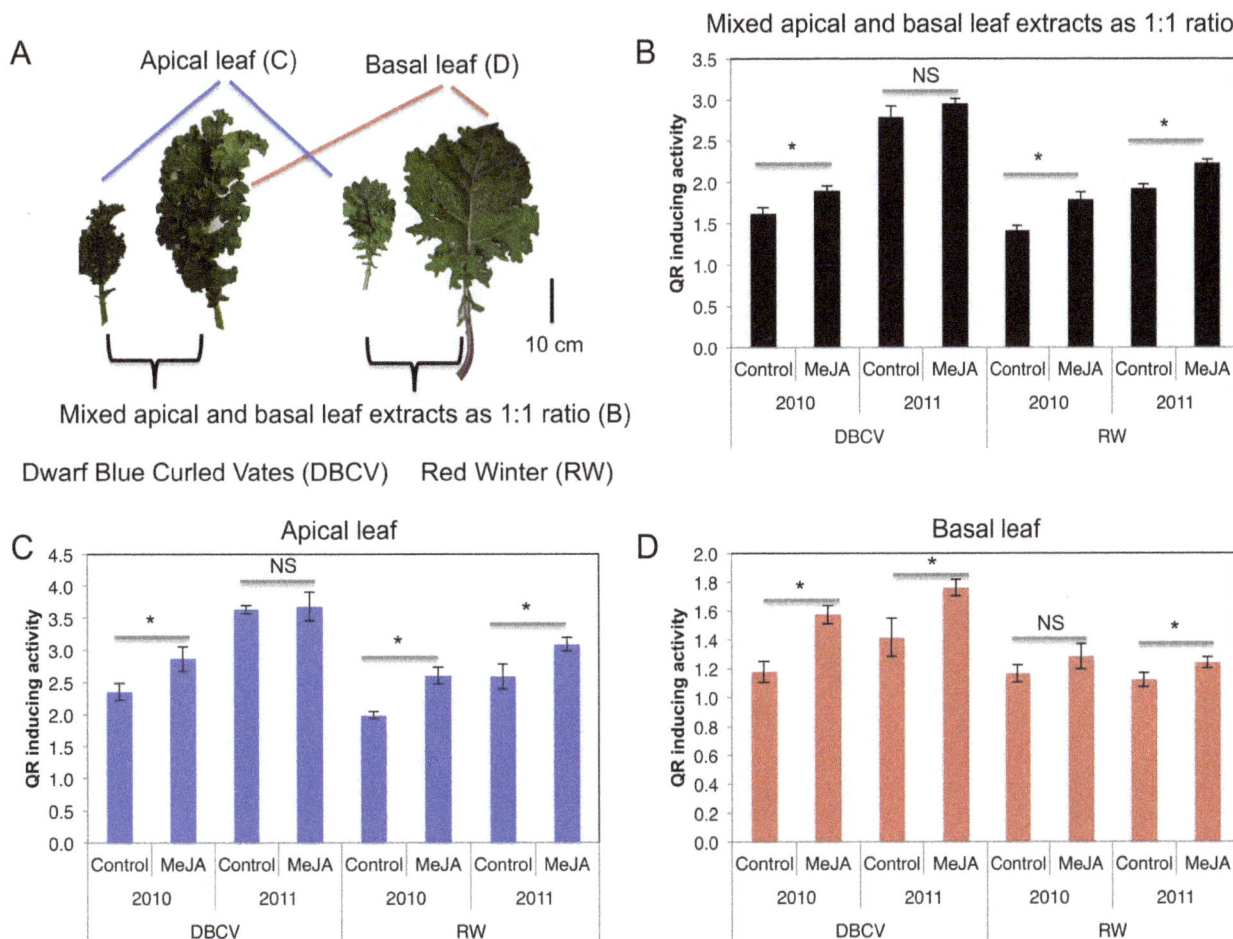

Figure 1. QR inducing activity of apical, basal and combined leaf tissue samples from two kale cultivars. A: Images of harvested apical and basal leaf samples. B: QR activity of mixed extract of 1:1 apical and basal leaf tissues. C: QR activity of apical leaf tissue. D: QR activity of basal leaf tissue. Student T-tests were conducted to determine significance at $P \leq 0.05$. NS and *indicate non-significance and significance at $P \leq 0.05$, respectively. Data are means $\pm$ SD (n = 3).

Kale Treatment with MeJA and Sample Preparation

An aqueous solution of 250 μM MeJA (Sigma-Aldrich, St. Louis, MO) and 0.1% Triton X-100 (Sigma-Aldrich) was sprayed on all aerial plant tissues to the point of runoff (approximately 100 mL) four days prior to harvest based on the result of experiments to determine when GS levels are optimized prior to harvest (Figure S1). Two different kale leaf samples (apical: three leaves from the below the meristematic growing point, at a minimum 8 cm in length; basal: three fully expanded leaves nearest the soil surface without discoloration or signs of senescence or damage) were harvested and bulked from five treated and control plants of each genotype for each replicate respectively (five leaves bulked for a replicate sample). Images of apical and basal samples of each kale cultivar are shown in Figure 1A. In order to confirm the relationship between increased hydrolysis products of GS and QR activity, 0, 50, 250, and 500 μM MeJA were sprayed on kale leaf tissue as described above in 2011. All kale leaf tissue samples were frozen in liquid nitrogen, and stored at −20°C prior to freeze-drying. Freeze-dried tissues were ground into a fine powder using a coffee grinder and stored at −20°C prior to chemical and bioactivity analyses.

Quinone Reductase (QR) Inducing Activity

Freeze-dried kale leaf powder (75 mg) was suspended in 1.5 mL of water in the absence of light for 4 h (time for the maximum concentration of indolyl GS hydrolysis products) at room temperature in a sealed 2 mL microcentrifuge tube (Fisher Scientific, Waltham, MA) to facilitate GSs hydrolysis by endogenous myrosinase. Slurries were then centrifuged at 12,000 ×g for 5 min and supernatants was used for QR assay. QR inducing activities were measured for individual apical and basal leaf tissue extracts and a pooled equal volume sample from both apical and basal leaf tissue extracts (Figure 1A). Hepa1c1c7 murine hepatoma cells (ATCC, Manassas, VA) were grown in alpha-minimum essential medium (α-MEM), enriched with 10% fetal bovine serum and maintained at 37°C in 95% ambient air and 5% CO_2. The cells were divided every three days with a split ratio of 7. Cells with 80–90% confluence were plated into 96-well plates (Costar 3595, Corning Inc, Corning, NY), 1×10^4 cells per well, and incubated for 24 h in antibiotic-enriched media (100 units/mL penicillin, 100 μg/mL streptomycin). The QR induction activities of different samples were determined by means of the protocol described by Prochaska & Santamaria [28]. After 24 h cells were exposed to the different sample extracts [0.25% final concentration (125 μg of freeze-dried kale/mL) in 200 μL of media] in new

Table 2. GS composition of different kale leaf tissue samples with or without MeJA treatment from two kale cultivars over two years.

Apical tissue

	DBCV				RW			
	2010		2011		2010		2011	
(μmol/g DW)	Control	MeJA	Control	MeJA	Control	MeJA	Control	MeJA
Glucoiberin	2.10	2.70^{ns}	9.62	9.20^{ns}	0.03	0.08^{ns}	0.00	0.04^{ns}
Progoitrin	0.12	0.04^{ns}	0.90	0.39^{ns}	22.02	22.97^{ns}	31.59	30.53^{ns}
Glucoraphanin	0.32	0.21^{ns}	0.37	0.32^{ns}	2.28	2.14^{ns}	5.48	3.52^{ns}
Gluconapin	0.61	0.85^{ns}	1.79	1.25^{ns}	0.98	0.81^{ns}	1.22	2.05^{ns}
Glucobrassicin	9.32	25.52*	22.44	29.56*	9.62	19.24*	5.93	19.50*
Gluconasturtiin	0.47	1.14*	1.22	1.44^{ns}	2.40	3.48*	2.05	3.57^{ns}
Neoglucobrassicin	2.30	9.66*	6.01	10.15*	7.88	20.72*	6.17	15.35*

Basal tissue

	DBCV				RW			
	2010		2011		2010		2011	
(μmol/g DW)	Control	MeJA	Control	MeJA	Control	MeJA	Control	MeJA
Glucoiberin	0.32	0.43^{ns}	0.71	0.81^{ns}	0.02	0.04^{ns}	0.00	0.00^{ns}
Progoitrin	0.30	0.10^{ns}	0.05	0.05^{ns}	1.19	2.04^{ns}	2.14	2.33^{ns}
Glucoraphanin	0.17	0.22^{ns}	0.11	0.12^{ns}	0.13	0.14^{ns}	0.05	1.02^{ns}
Gluconapin	1.05	0.43^{ns}	1.22	0.93^{ns}	0.19	0.31*	1.26	1.16^{ns}
Glucobrassicin	1.65	6.53*	2.53	4.54*	0.63	2.04*	0.61	1.36*
Gluconasturtiin	0.36	0.80^{ns}	0.47	0.53^{ns}	0.34	0.48*	0.35	0.36^{ns}
Neoglucobrassicin	0.69	1.67*	0.97	1.04*	1.04	2.56*	1.71	2.79*

Student T-tests were conducted to determine significance between control and MeJA treatment at $P \leq 0.05$. ^{ns} and *indicate non-significance and significance at $P \leq 0.05$, respectively. Data are mean value from triplicates.

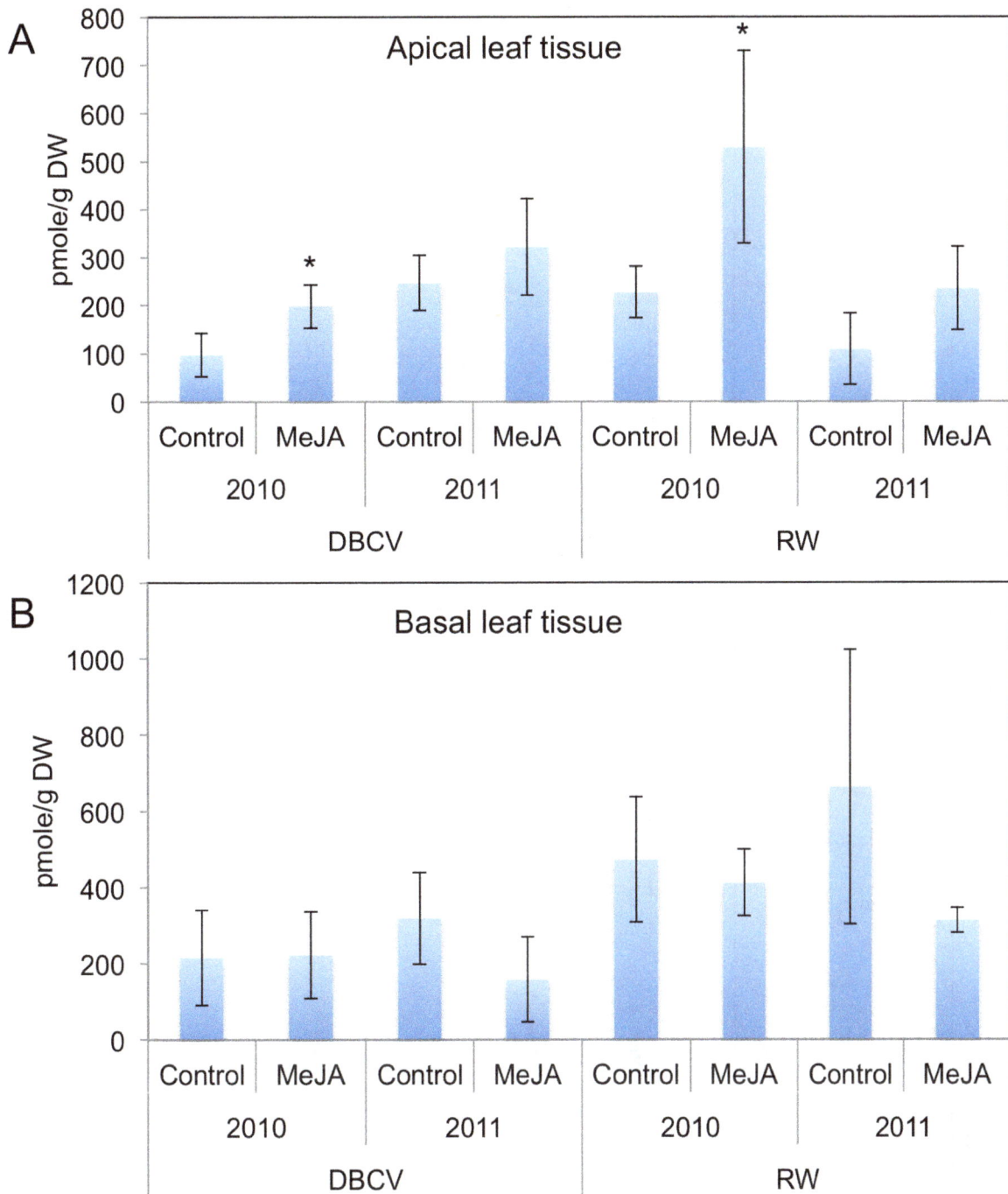

Figure 2. Effect of MeJA treatments on endogenous JA concentrations of apical and basal kale leaf tissue in two cultivars. Data are means ± SD (n = 3). *indicates significance at $P \leq 0.05$.

media for a further 24 h. Growth media alone was used as a negative control. Treated cells were rinsed with phosphate buffer at pH 7.4, lysed with 50 µL 0.8% digitonin in 2 mM EDTA, incubated and agitated for 10 min. A 200-µL aliquot of reaction mix [10 µM BSA, 82 µM Tween-20 solution, 927 µM glucose-6-phosphate, 1.85 µM NADP, 57 nM FAD, 2 units of glucose-6-phosphate dehydrogenase, 725 nM 3-(4,5-dimethylthiazo-2-yl)-2,5-diphenyltetrazolium bromide (MTT), and 50 µM menadione (dissolved in acetonitrile) in 25 mM Tris buffer] was added to the lysed cells. Readings were made at five time points, 50 s apart,

using a µQuant microplate reader (Bio-Tek Instruments, Winooski, VT) at 610 nm. Immediately after completion of the readings, 50 µL of 0.3 mM dicumarol in 25 mM Tris buffer was added into each well, and the plate was read again (five time points, 50 s apart) to determine non-specific MTT reduction. Total protein content was measured by the BioRad assay (Bio-Rad, Hercules, CA) using manufacture's instructions. Activity was expressed as QR specific activity (nmol MTT reduced/mg/min) ratio of treated to negative control cells. In order to measure QR inducing activity associated with phenolic compounds in extracts,

Table 3. Hydrolysis product composition of apical and basal leaf tissues with or without MeJA treatment from two kale cultivars over two years.

Apical tissue

	DBCV				RW			
	2010		2011		2010		2011	
(μmol/g DW)	Control	MeJA	Control	MeJA	Control	MeJA	Control	MeJA
I3C	2.27	1.68^{ns}	1.69	2.35*	0.92	1.05^{ns}	1.35	2.74*
DIM	0.32	0.35^{ns}	0.29	0.25^{ns}	0.20	0.16^{ns}	0.13	0.13^{ns}
NI3C	1.11	1.30^{ns}	0.73	1.37*	2.91	3.08^{ns}	2.79	2.78^{ns}
NeoASG	0.81	1.04^{ns}	0.55	0.89^{ns}	1.63	1.12^{ns}	0.88	1.45*
Sulforaphane	0.05	0.09^{ns}	0.07	0.06^{ns}	0.28	0.51*	0.58	0.84*

Basal tissue

	DBCV				RW			
	2010		2011		2010		2011	
(μmol/g DW)	Control	MeJA	Control	MeJA	Control	MeJA	Control	MeJA
I3C	0.70	0.93*	1.00	0.92^{ns}	0.80	0.97^{ns}	1.61*	1.11
DIM	0.14	0.26*	0.10	0.14^{ns}	0.38*	0.10	0.25	0.12^{ns}
NI3C	0.56	0.87*	0.30	0.38^{ns}	1.07	1.61*	0.74	0.96^{ns}
NeoASG	0.11	0.10^{ns}	0.16	0.16^{ns}	0.14	0.18^{ns}	0.70*	0.31
Sulforaphane	0.01	0.05^{ns}	0.01	0.03^{ns}	0.00	0.08*	0.00	0.02*

Student T-tests were conducted to determine significance between control and MeJA treatment at $P \leq 0.05$. ^{ns} and * indicate non-significance and significance at $P \leq 0.05$, respectively. Data are mean value from triplicates.

Table 4. Correlation analysis between intact GS, GS hydrolysis products and QR inducing activity from apical and basal leaf tissue extracts from 250 μM MeJA treated two different kale cultivars over two years.

Variable	1	2	3	4	5	6	7	8	9	10	11
1. Glucoraphanin											
2. Glucobrassicin	0.120										
3. Gluconasturtiin	**0.726**	**0.561**									
4. Neoglucobrassicin	**0.500**	**0.731**	**0.914**								
5. QR	0.305	**0.747**	0.415	0.431							
6. I3C	0.203	**0.627**	0.358	0.422	**0.800**						
7. DIM	−0.374	0.308	−0.209	−0.067	0.042	0.227					
8. NI3C	**0.810**	0.330	**0.880**	**0.767**	0.176	0.209	−0.199				
9. NeoASG	**0.584**	**0.590**	**0.802**	**0.754**	0.381	**0.548**	0.075	**0.788**			
10. Sulforaphane	**0.879**	0.325	**0.907**	**0.742**	0.361	0.382	−0.342	**0.853**	**0.682**		
11. Endogenous JA	−0.138	0.496	0.575	**0.849**	0.005	−0.502	−0.262	0.294	0.153	0.157	
	(−0.188)	(−0.672)	(−0.489)	(0.247)	(−0.371)	(−0.022)	(0.442)	(0.362)	(**0.775**)	(−0.320)	
12. Myrosinase	**0.606**	0.391	0.402	0.333	0.433	**0.724**	0.110	0.439	**0.546**	**0.541**	**−0.123**

Bold values indicate significant correlations among variables from apical and basal leaf tissue extracts based on the Pearson's correlation at $P \leq 0.05$ (n = 16). Upper and bottom values in endogenous JA row indicate correlation coefficients from apical and basal leaf tissue extracts, respectively (n = 8).

freeze-dried kale powder (0.2 g) and 4 mL of 70% methanol were added to 10 mL tubes (Nalgene, Rochester, NY) and heated on a heating block at 95°C for 10 min. After cooling on ice, the extract was centrifuged at 3,000×g for 10 min at 4°C. The supernatant (1 mL) was dried up using SpeedVac (Savant, Osterville, MA) and reconstituted with DMSO (1 mL). QR inducing activity of phenolic rich-extract was measured using the same procedures and concentrations described above.

Determination of Sample GS Concentrations

Extraction and quantification of GS using high-performance liquid chromatography was performed using a previously published protocol [29]. Freeze-dried kale powder (0.2 g) and 2 mL of 70% methanol were added to 10 mL tubes (Nalgene) and heated on a heating block at 95°C for 10 min. After cooling on ice, 0.5 mL benzylglucosinolate (1 mM) was added as internal standard (POS Pilot Plant Corp, Saskatoon, SK, Canada), mixed, and centrifuged at 3,000×g for 10 min at 4°C. The supernatant was saved and the pellet was re-extracted with 2 mL 70% methanol at 95°C for 10 min and the two extracts combined. A subsample (1 mL) from each pooled extract was transferred into a 2-mL microcentrifuge tube (Fisher Scientific, Waltham, MA). Protein was precipitated with 0.15 mL of a 1:1 mixture of 1 M lead acetate and 1 M barium acetate. After centrifuging at

12,000×g for 1 min, each sample was then loaded onto a column containing DEAE Sephadex A-25 resin (GE Healthcare, Piscataway, NJ) for desulfation with arylsulfatase (*Helix pomatia* Type-1, Sigma-Aldrich, St. Louis, MO) for 18 h and the desulfo-GS eluted. Samples (100 µL) were injected on to an Agilent 1100 HPLC system (Agilent, Santa Clara, CA), equipped with a G1311A bin pump, a G1322A vacuum degasser, a G1316A thermostatic column compartment, a G1315B diode array detector and an HP 1100 series G1313A autosampler. UV detector set at 229 nm wavelength. All-guard cartridge pre-column (Alltech, Lexington, Kentucky), and a LiChospher 100 RP-18 column (Merck, Darmstadt, Germany) were used for quantification. Desulfo-GS were eluted from the column over 45 min with a linear gradient of 0% to 20% acetonitrile at a flow rate of 1 mL/min. Benzylglucosinolate was used as an internal standard and UV response factors for different types of GS were used as determined by previous study [30]. The identification of desulfo-GS profiles were validated by LC-tandem MS using a Waters 32 Q-Tof Ultima spectrometer coupled to a Waters 1525 HPLC system and full scan LC-MS using a Finnigan LCQ Deca XP, respectively. The molecular ion and fragmentation patterns of individual desulfo-GS were matched with the literature for GS identification [31,32].

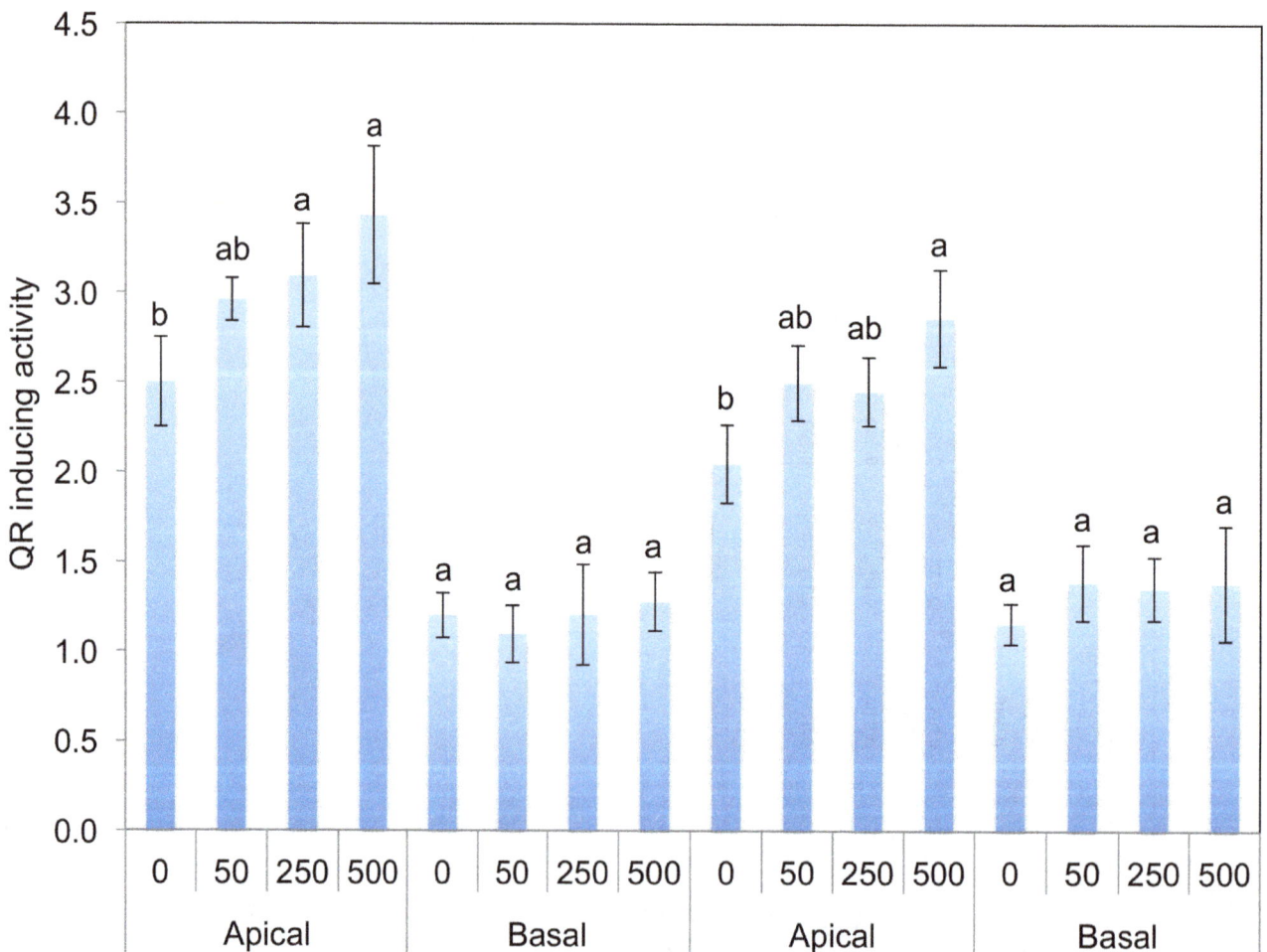

Figure 3. QR inducing activity from kale leaf tissue samples sprayed with varying concentrations of MeJA (0, 50, 250, and 500 µM). Different letters indicate significant differences among treatments based on Fisher's LSD test at $P \leq 0.05$. A: QR activity, B: GS profiles, and C: hydrolysis product profiles.

Table 5. GS concentrations from kale leaf tissue samples sprayed with varying concentrations of MeJA (0, 50, 250, and 500 μM).

DBCV

(μmol/g DW)	Apical				Basal			
	0	50	250	500	0	50	250	500
Glucoiberin	9.62 a	11.39 a	8.46 a	9.25 a	0.71 a	0.73 a	0.77 a	0.71 a
Progoitrin	1.03 a	0.39 a	0.29 a	0.22 a	0.05 a	0.03 a	0.06 a	0.02 a
Glucoraphanin	0.07 a	0.07 a	0.08 a	0.08 a	0.00 a	0.00 a	0.00 a	0.01 a
Gluconapin	1.28 a	1.35 a	1.20 a	1.19 a	1.22 a	1.10 a	0.99 a	0.82 a
Glucobrassicin	21.31 b	21.42 b	29.73 a	31.92 aa	2.53 a	2.96 a	3.24 a	3.47 a
Gluconasturtiin	0.97 a	1.50 a	1.29 a	0.86 a	0.47 a	0.68 a	0.54 a	0.47 a
Neoglucobrassicin	4.51 b	8.33 a	9.21 a	10.49 a	0.98 a	1.08 a	0.88 a	1.02 a

RW

(μmol/g DW)	Apical				Basal			
	0	50	250	500	0	50	250	500
Glucoiberin	0.00 a	0.03 a	0.03 a	0.03 a	0.00 a	0.00 a	0.00 a	0.00 a
Progoitrin	31.60 a	31.48 a	26.67 a	29.17 a	2.64 a	2.43 a	2.23 a	2.16 a
Glucoraphanin	5.48 a	4.24 a	3.39 a	3.76 a	0.08 a	0.04 a	0.37 a	0.07 a
Gluconapin	1.22 a	1.43 a	1.20 a	0.92 a	1.32 a	1.02 a	0.67 a	1.27 a
Glucobrassicin	5.93 b	14.83 a	15.86 a	15.81 a	0.85 b	0.93 b	1.24 ab	1.65 a
Gluconasturtiin	2.05 a	2.80 a	2.72 a	2.68 a	0.48 a	0.33 a	0.32 a	0.32 a
Neoglucobrassicin	6.17 b	12.30 a	12.61 a	13.58 a	2.03 b	1.83 b	2.57 ab	3.21 a

Different letters indicate significant differences among treatments based on Fisher's LSD test at $P \leq 0.05$. Data are mean value from triplicates.

Table 6. GS hydrolysis product concentrations from kale leaf tissue samples sprayed with varying concentrations of MeJA (0, 50, 250, and 500 µM).

DBCV	------Apical------				------Basal------			
(µmol/g DW)	0	50	250	500	0	50	250	500
I3C	1.32 a	1.20 a	0.65 b	0.60 b	0.17 a	0.21 a	0.13 a	0.12 a
DIM	0.18 a	0.13 a	0.14 a	0.14 a	0.09 a	0.10 a	0.11 a	0.10 a
NI3C	1.71 b	2.54 a	2.83 a	2.88 a	0.41 b	0.69 a	0.55 b	0.50 b
NeoASG	0.42 b	0.63 a	0.53 a	0.44 a	-	-	-	-
Sulforaphane	0.01 a	0.01 a	0.01 a	0.01 a	-	-	-	-

RW	------Apical------				------Basal------			
(µmol/g DW)	0	50	250	500	0	50	250	500
I3C	1.35 b	1.55 b	1.86 a	1.86 a	0.09 b	0.45 a	0.44 a	0.23 b
DIM	0.15 a	0.15 a	0.19 a	0.17 a	0.09 a	0.09 a	0.10 a	0.09 a
NI3C	2.83 b	4.00 a	4.15 a	4.50 a	0.88 b	0.98 b	1.49 a	1.78 a
NeoASG	0.44 b	0.44 b	0.65 a	0.96 a	-	-	-	-
Sulforaphane	0.54 b	0.60 a	0.71 a	0.62 a	0.02 a	0.02 a	0.03 a	0.03 a

Different letters indicate significant differences among treatments based on Fisher's.
LSD test at $P \leq 0.05$. Data are mean value from triplicates.

Table 7. Correlation analysis between intact GS, GS hydrolysis product and QR activity from kale leaf tissue across two kale cultivars sprayed with varying concentrations of MeJA (0, 50, 250, and 500 μM).

Variable	1	2	3	4	5	6	7	8	9
1. Glucoraphanin									
2. Glucobrassicin	0.081								
3. Gluconasturtiin	0.858	0.444							
4. Neoglucobrassicin	0.621	0.734	0.872						
5. QR	0.704	0.671	0.888	0.980					
6. I3C	0.768	0.504	0.920	0.841	0.856				
7. DIM	0.603	0.682	0.808	0.795	0.788	0.904			
8. NI3C	0.731	0.620	0.903	0.976	0.974	0.886	0.797		
9. NeoASG	0.581	0.742	0.845	0.914	0.918	0.890	0.867	0.897	
10. Sulforaphane	0.947	0.133	0.917	0.719	0.770	0.830	0.675	0.806	0.649

Bold values indicate significant correlations based on the Pearson's correlation at $P \leq 0.05$.

Analysis of endogenous JA in kale leaf tissues

Endogenous JA concentrations in kale leaf tissues were measured using a previously published method [33]. Samples (100 mg) of freeze-dried kale leaf powder were extracted with 1.5 mL methanol–water–acetic acid (90:9:1, v/v/v) and centrifuged for 1 min at 10,000 rpm. The supernatant was collected and the extraction repeated. Pooled supernatants were dried under N_2, resuspended in 200 μL of 0.05% acetic acid in water–acetonitrile (85:15, v/v), and filtered with a Millex-HV 0.45 μm filter from Millipore (Bedford, MA). Quantitation was estimated using external standards of a range of JA solutions (ranging from 1.25 to 10 pM). Analyses were carried out using a LC-tandem MS using a Waters 32 Q-Tof Ultima spectrometer coupled to a Waters 1525 HPLC system. All the analyses were performed using negative ion mode with a collision energy (CE) of −25 V. MRM acquisition was done by monitoring the 209/59 transitions for JA. An Eclipse XDB-C18 column (150×4 mm, particle size 5 μm, Agilent, Santa Clara, CA) was used at ambient temperature and the injected volume was 10 μl. The elution gradient was carried out with a binary solvent system consisting of 0.05% acetic acid in water (solvent A) and acetonitrile (solvent B) at a constant flow-rate of 0.6 mL/min. A linear gradient profile with the following proportions (v/v) of solvent B was applied (t (min), %B): (0, 15), (3, 15), (5, 100), (6, 100), (7, 15), (8, 15) with 5 min for re-equilibration.

Determination of Total Myrosinase Activity Using Glucose Release

Total myrosinase activity was measured with whole kale tissue using the ABTS-glucose assay [26] without protein extraction to avoid introducing any artifacts. Freeze-dried kale (50 mg) was weighed in duplicate into 2 mL tubes and one mL sinigrin (10 mM, Sigma) was added to each tube. After 10 s of vigorous vortexing, one of the paired samples was put directly into a heating block (95°C) for 10 min to inactivate the myrosinase enzyme (zero time blank). The second sample was incubated at 40°C for 30 min and then inactivated as outlined above. After inactivation, samples were cooled on ice for 5 min then centrifuged at 16,000 g for 2 min. The supernatants were diluted 96-fold and aliquots (30 μL) or glucose standards were added in a 96 well plate and followed by adding 200 μL of an ABTS-glucose solution [2.7 mM ABTS, 1,000 units peroxidase (Type VI-A, Sigma), and 1,000 units glucose oxidase in 100 mL], incubated for 20 min and absorbance measured at 630 nm in a μQuant plate reader (Bio-Tek instruments, Winooski, VT).

Analysis of Glucosinolate Hydrolysis Products

The extraction and analysis of isothiocyanates and other hydrolysis products was carried out according to previously published methods [21,26,34]. For the GS hydrolysis products, kale extracts were collected using the same protocol for the QR assay described above with sampling at 4 h and 24 h of incubation, which are hydrolysis duration periods that generate maximum concentrations for indolyl GS products and sulforaphane, respectively [21]. Freeze-dried kale leaf powder (75 mg) was suspended in 1.5 mL of water in the absence of light for 4 h and 24 h at room temperature in a sealed 2 mL microcentrifuge tube (Fisher Scientific, Waltham, MA) to facilitate GS hydrolysis by endogenous myrosinase. Slurries were then centrifuged at 12,000×g for 5 min and 0.5 mL of supernatants was transferred into a 2 mL microcentrifuge tube. Butyl isothiocyanate (0.5 mg/mL) and 4-methoxyindole (1 mg/mL) solutions were prepared and mixed in a 1:1 (v/v) ratio. An aliquot of this solution (40 μL) was added as the internal standards for sulforphane and the hydrolysis products of indolyl GS (I3C, DIM, NI3C, and NeoASG), to quantify respectively. After adding 0.5 mL of methylene chloride, tubes were shaken vigorously before being centrifuged for 2 min at 9,600 g. The methylene chloride layer (200 μL) was transferred to 350 μL flat bottom insert (Fisher Scientific, Pittsburgh, PA) in a 2 mL HPLC autosampler vial (Agilent, Santa Clara, CA) for mixing with 100 μL of a reagent containing 20 mM triethylamine and 200 mM mercaptoethanol in methylene chloride. The mixture was incubated at 30°C for 60 min under constant stirring, and then dried under a stream of nitrogen. The residue containing isothiocyanate derivatives (isothiocyanate-mercaptoethanol derivatives) and other hydrolysis compounds was dissolved in 200 μL of acetonitrile/water (1:1) (v/v), and 10 μL of this solution injected onto a Agilent 1100 HPLC system (Agilent, Santa Clara, CA), equipped with a G1311A bin pump, a G1322A vacuum degasser, a G1316A thermostatic column compartment, a G1315B diode array detector and an HP 1100 series G1313A autosampler. Extracts were separated on an Eclipse XDB-C18 column (150×4 mm, particle size 5 μm, Agilent, Santa Clara, CA) with an Adsorbosphere C18 all-guard cartridge pre-column (Grace, Deerfield, IL). Mobile phase A was water and B methanol. Mobile phase B was 0% at injection, increasing to 10% by 10 min, 100% at 35 min, and held 5 min, then decreased to 0% by 50 min. Flow rates were kept at 0.8 mL/min. The detector wavelength was set at 227 and 271 nm. Response factors of monomeric indolyl derivatives were used from a previous report [35]. Due to a lack of standards for NI3C and NeoASG the standard curve of I3C was applied for quantification of both NI3C and NeoASG. The quantity was expressed as I3C equivalent concentrations.

QR Inducing Activity Measurement of I3C

QR activity of hydrolysis product, I3C was measured to determine the concentrations required to double the specific activity of QR (CD value). Commercially purchased I3C (Sigma-Aldrich) was dissolved in DMSO, then seven concentrations (250, 125, 62.5, 31.3, 15.6, 7.8, and 3.9 μM) of I3C prepared by serial two fold dilutions and added to 96 well plates of cultured hepa1c1c7 cells. After 24 h incubation, QR activity was measured using the protocol described above.

Statistical Analysis

Analysis of variance (ANOVA) was conducted using JMP 10 statistical software program (SAS institute Inc., Cary, NC). Year, treatments, and genotype effects were considered as fixed factors. Block was considered as random. Analysis of variance was performed using the linear model: $Y_{ijklm} = m + G_i + Y_j + T_k + GY_{ij} + GT_{ik} + YT_{jk} + GYT_{ijk} + B_{l(j)} + \varepsilon_{ijklm}$, where Y_{ijklm} is the lth block of the phenotypic value of the kth treatment, i^{th} genotype in year j, m is the overall mean, G, Y, T, and B indicate the effects of genotype, year (weather), treatment and blocks nested in years, ε_{ijklm} is the experimental error associated with Y_{ijklm}, respectively. Fisher's Least Significant Difference (LSD) test, correlation analysis and Student's t-tests were also conducted using the JMP 10 software. All sample analyses were conducted in triplicate. The results are presented as means ± SD.

Results and Discussion

Effect of MeJA Treatment on QR Inducing Activity of Kale Leaf Tissues

MeJA treatment significantly increased QR activity in the combined apical and basal leaf extracts of the two different kale species extracts over two years except for the DBCV cultivar in 2011 (Figure 1B). There was significant year-to-year variation in QR activity with 2011 samples significantly greater than those in 2010. In 2010 apical leaf tissue extracts of MeJA treated kale increased 17% and 27% over QR activity for DBCV and RW controls, respectively, while in 2011, they increased only by 6% and 16%. QR activities of apical leaf tissue extract were up to 2-fold greater than extracts from basal leaves (Figure 1C and 1D), which is of relevance to vegetable growers where kale is harvested throughout the year.

Effect of MeJA Treatment on QR Inducing Activity Associated with Phenolic Rich-Extract of Kale Leaf Tissues

A previous study reported that MeJA treatment specifically increased phenolic and flavonoid concentrations in kale leaves primarily in the form of the flavonoids, quercetin and kaempferol [10]. In order to test if tissue phenolic concentrations induced by MeJA treatment contribute to QR inducing activity, we also measured QR inducing activity in phenolic rich-extracts after myrosinase inactivation by heating. Unlike aqueous extracts, there was no consistent QR activity increases associated with MeJA treatments in phenolic rich-extracts (Figure S2). The only significant increase was observed in apical leaf of RW in both years and basal leaf of DBCV in 2011. Aqueous kale extracts in this study have both GS hydrolysis products and water soluble phenolics. After subtracting QR inducing activity by phenolic rich-extracts from QR inducing activity by aqueous extract, we approximately calculated the contribution of phenolics to QR inducing activity. Averaged QR inducing activity derived by phenolic-rich, myrosinase-inactivated extracts of RW kale apical leaves accounted from 56% and 72% of the QR induction of aqueous extracts in 2010 and 2011, respectively. The phenolic rich-extracts of DBCV apical leaves accounted for 58% and 33% of QR inducing activity of aqueous extract in 2010 and 2011, respectively. Quercetin and kaempferol have been reported as QR inducers [36]. Also, glucoside forms of quercetin have been reported as QR inducers from onion [37]. Thus, it is possible that flavonoids in broccoli can contribute the QR induction. Since it is not feasible to completely inactivate myrosinase enzyme using water, we used 70% methanol to inactivate the enzyme. Using this different extraction solvent may lead to overestimating the contribution of phenolic compounds to QR activity because it can extract non-polar compounds as well. Nevertheless, this calculation suggests that phenolic compound concentrations induced by MeJA treatment partially contribute to QR inducing activity of kale leaf tissues. The magnitude of contribution is different based on the cultivar and year.

Effect of MeJA Treatment on GS Concentrations

Over both seasons, MeJA treatments significantly increased glucobrassicin and neoglucobrassicin concentrations in both apical and basal leaves. The treatment increased apical leaf concentrations of gluconasturtiin (56%), glucobrassicin (98%), and neoglucobrassin (150%) and basal leaf concentrations of gluconasturtiin (44%), glucobrassicin (166%) and neoglucobrassin (83%) averaged across cultivars and over years (Table 2). Total GS concentration in apical leaf tissues was up to seven fold greater than basal leaf tissues. This concentration difference can explain why apical leaf extracts induced higher QR activity.

From previous work, *B. napus* type kales (such as RW) have distinct GS compositional profiles compared with *B. oleracea* type kale [3]. As Figure S1 and Table 2 illustrate, the major GS in both DBCV and RW are glucobrassicin and neoglucobrassicin. However, DBCV contains a higher concentration of glucoiberin while RW is higher in progoitrin. Unlike DBCV, RW contains relatively high glucoraphanin concentrations.

MeJA mediated enhancement of GS concentrations in DBCV was greater in 2010 than in 2011 where glucobrassicin and total GS concentrations in apical leaf tissues were both 2.7 fold higher in 2010 compared to increases of 1.3 and 1.2 fold, respectively in 2011 (Table 2). MeJA treatments may be interacting with varying weather conditions in each season of application. This variation may be associated with reduced rainfall in 2011, which experienced only 51% of the precipitation recorded in the 2010 growing season (Table 1). The distribution of precipitation over the course of the growing season in 2010 and 2011 is presented in Figure S3. A recent study has shown that drought conditions were associated with increased concentrations of aliphatic GS in *Brassica juncea* without a reduction in leaf biomass yield [38]. Thus, drought conditions in 2011 may have increased GS concentrations in both kale cultivars, although the GS increased was species specific.

Effect of MeJA Treatment on Endogenous JA Concentrations

Endogenous JA concentrations in kale apical leaf tissue of two cultivars were significantly increased in 2010 by exogenous MeJA treatment but the treatment effect was not significant in 2011 (Figure 2A). Endogenous JA concentrations in apical leaves of control DBCV kale grown 2011 was significantly higher than control DBCV kale grown 2010. Endogenous JA has been observed to accumulate *in planta* under drought conditions [39]. The drought conditions in 2011 may have lead to the accumulation of endogenous JA which have could attenuate the effect of the exogenous MeJA treatment on DBCV. In contrast, endogenous JA concentrations in apical leaves of control RW kale harvested in 2011 was lower than control RW kale grown in 2010, implying endogenous JA concentration may be affected by other factors including insect activity and microenvironmental factors. Compared to DBCV, RW has relatively tender leaves and is more vulnerable to chewing insects. The RW cultivar displayed much more insect feeding activity by cabbage loopers [*Trichoplusia ni* (Hübner)] and flea beetles (*Phyllotreta cruciferae*) during the experiment (Figure S4). This insect activity and other environmental factors may compound drought effects on endogenous JA concentrations in RW and ultimately influence QR induction activity.

Unlike kale apical leaf tissue, MeJA treatment effects on basal leaf tissue did not have a significant influence on endogenous JA concentrations (Figure 2B). The concentration of endogenous JA in basal leaf tissues was higher than in apical leaf tissue, which may be related with insect feeding activity since basal leaves displayed greater insect damage. Reduction of endogenous JA in MeJA treatment groups in basal leaf tissue compared to control groups may be related with JA transport to apical leaf tissue in response to exogenous MeJA treatment. A recent study revealed that JA translocates from local damaged leaves systemically to other leaves in *Nicotiana tabacum* [40]. In another study, after radioactive JA application to one basal leaf, younger, apical leaves contained the most of total radioactivity in potato plants [41]. Higher JA accumulation in apical leaf tissue responding to exogenous MeJA may be related protection of younger tissues more important for plant reproduction and survival.

Effect of MeJA Treatment on Myrosinase Activity

Previously we reported that MeJA treatment enhanced myrosinase gene expression and enzyme activity using greenhouse grown broccoli [26]. Unlike this earlier study with broccoli, there

was no consistent response in myrosinase activity induced by MeJA treatments (Figure S5). This difference may be due to a tissue- (vegetative versus reproductive) or species-specific pattern of response. Myrosinase activity of field grown kale might also have been influenced by field biotic and abiotic factors, which also can change enzyme activity. There was significantly greater myrosinase activity in apical versus basal leaf tissue within each cultivar. In our previous research, we demonstrated a MeJA induced increase in myrosinase activity of greenhouse grown broccoli florets also contributed to QR inducing activity. However, in our two kale cultivars total myrosinase activity was observed to increase with MeJA treatment about 30% only in the apical leaf tissue of DBCV (2011 year). Increases were not seen in basal leaf tissue of either cultivar or in the apical leaf tissue of RW.

Effect of MeJA Treatment on GS Hydrolysis Products Concentrations

Only sulforaphane was significantly increased in both apical and basal leaf tissue of the RW cultivar by MeJA treatment over two years. Increased concentrations of other hydrolysis products were not consistently observed in all samples over two years (Table 3). Despite significant increases in glucobrassicin, I3C and DIM hydrolysis product concentrations in kale extracts were relatively low. I3C has been reported to be highly instable [42] and will react with other substrates generating by-products by condensation with ascorbic acid or through oligomerization [12]. Following hydrolysis of the parent GS, relatively higher levels of NI3C were observed than I3C. According to previous research ascorbigen is more unstable than neoascorbigen [43]. I3C may be less stable than NI3C.

Correlation Analysis between Intact GS or Hydrolysis Products, and Myrosinase activity

In order to elucidate the most active QR induction hydrolysis product in MeJA treated kale leaf tissue, correlation analysis was conducted between QR inducing activity and GS and GS hydrolysis product concentrations (Table 4). QR inducing activity significantly correlated with glucobrassicin ($r = 0.747$, $P = 0.001$) and I3C ($r = 0.800$, $P < 0.001$). However, there was no significant correlation between myrosinase activity and QR inducing activity, which suggests QR was influenced by hydrolysis products of MeJA induced GS. This suggests that myrosinase was not a limiting factor in QR inducing activity in field grown kales. Since there is a significant difference in endogenous JA concentration between apical and basal leaf tissue, correlation analyses were conduced separately. There was significant positive correlation ($r = 0.849$, $P = 0.008$) between endogenous JA and neoglucobrassicin within apical leaf tissue. This suggests that increased endogenous JA levels induced by exogenous MeJA treatment not only stimulates glucobrassicin biosynthesis but also promotes GS side chain modification from glucobrassicin to neoglucobrassicin. A previous study revealed a positive correlation between pupal mass and development time of *Pieris brassicae* and foliar GS composition, of which levels of neoglucobrassicin appeared to be the most important [44]. This suggests that the side chain modification from glucobrassicin to neoglucobrassicin with increased endogenous JA may be related with insect herbivore defense. However, intact GS only have bioactivity after hydrolysis by myrosinase. There were significant correlations between total myrosinase activity and GS hydrolysis products including I3C ($r = 0.724$, $P = 0.002$), NeoASG ($r = 0.546$, $P = 0.029$), and sulforaphane ($r = 0.541$, $P = 0.031$). Although MeJA treatment did not significantly increase total myrosinase activity in kale leaves, both GS concentration and total myrosinase activity in apical leaf tissue were observed to be higher than in basal leaf tissue. These correlations imply that these GS hydrolysis products may be closely related with insect defense in apical leaf tissue.

I3C as QR inducer in kale leaf tissue

Using different concentrations of commercial I3C, the CD value for I3C was observed to be 230 μM (Figure S6), which is a relatively weak QR induction agent compared to sulforaphane (0.2 μM), 7–Methylsulfinylheptyl isothiocyanate (0.2 μM), PEITC (5 μM), and brassinin (4 μM) [17,27]. Previously we reported that the QR CD value of NI3C was 35 μM and neoascorbigen was 38 μM from broccoli extracts [21]. Despite the significantly increased amount of NI3C and neoascorbigen, their contribution to enhanced QR inducing activity was relatively small. The CD value of I3C does not fully explain the increased QR activity from kale leaf tissue extracts (Figure 1).

MeJA Dose Dependent Induced GS and QR Activity in Kale Leaf Tissue

To further evaluate the association between induction of QR activity and GS concentrations in kale leaves tissues, a second experiment was conducted where different MeJA concentrations (0, 50, 250, and 500 μM) were applied to two kale cultivars as described above. As concentrations of MeJA treatment increased GS tissue concentrations (glucobrassicin and neoglucobrassicin), QR activity was increased in apical leaf extracts of both kale cultivars (Figure 3, Table 5). In addition, MeJA treatment significantly increased NI3C and NeoASG in apical leaf tissue of both kale cultivars (Table 6). Although there was dose dependent increase in I3C by MeJA treatment from apical leaf tissue of the RW cultivar, DBCV kale showed a reduction in I3C concentrations in response to MeJA treatment (Table 6). MeJA treatment not only changes GS biosynthesis but also hydrolysis related gene expression [26]. Although higher apical leaf tissue indolyl GS hydrolysis product concentrations were found in RW compared to DBCV (Table 6), QR induction activity by RW apical leaf tissue was relatively low (Figure 3). The low concentration of I3C in kale leaf tissue may be related with very low stability or its condensation/oligomerization [12,42]. Other hydrolysis products of glucobrassicin such as di(indol-3-yl)methane (DIM), brassinin, or 2,3-bis(indol-3-ylmethyl)-indole (TIR) which can induce QR activity at lower CD values [17,19] may also play an important role in QR induction in kale than I3C.

Correlation Analysis of GS, GS Hydrolysis Products and QR with Varying Treatment Concentrations of MeJA

Correlation of QR activity of the two kale cultivars over the two seasons were significant for gluconasturtiin ($r = 0.888$, $P < 0.001$), glucobrassicin ($r = 0.671$, $P = 0.001$), and neoglucobrassicin ($r = 0.980$, $P < 0.001$). The GS hydrolysis products I3C ($r = 0.856$, < 0.001), DIM ($r = 0.788$, $P < 0.001$), NI3C ($r = 0.974$, $P < 0.001$), NeoASG ($r = 0.918$, $P < 0.001$) and sulforaphane ($r = 0.770$, $P < 0.001$) also correlated with QR activity (Table 7). Sulforaphane is the predominant QR induction agent in MeJA treated broccoli extracts [21,26]. Similarly, sulforaphane may play an important role to induce QR activity in RW leaf extracts as in

broccoli extracts. In case of DBCV, this data suggests that the combination of I3C and its derivatives, NI3C and NeoASG induction contributed to enhanced QR activity of kale leaf tissue extracts. Since correlation analysis does not necessarily imply causation, further research is needed to address which compound or compounds are dominating QR induction in kale leaf tissue.

In a previous study we reported that MeJA treatment significantly increased QR inducing activity in cauliflower and broccoli, which was primarily associated with glucoraphanin and its hydrolysis product sulforaphane [21,23]. It is interesting that MeJA treatment can significantly increase QR inducing activity in the DBCV cultivar, which does not have high glucoraphanin concentrations in comparison to RW. This is at least partially due to increases in the flavonoid phenolics. In addition, apical leaf tissue of both RW and DBCV kale have significantly higher QR inducing activity and GS levels than basal leaves. These results will provide helpful information for kale production with enhanced consumer health promoting properties.

Acknowledgments

The authors thank Dr. Furong Sun and Dr. Kevin Tucker at the University of Illinois–School of Chemical Sciences Mass Spectrometry Laboratory for help with the optimization of the LC/MS method and analysis of the samples.

Author Contributions

Conceived and designed the experiments: KMK JAJ. Performed the experiments: KMK. Analyzed the data: KMK JAJ. Contributed reagents/materials/analysis tools: EHJ JAJ. Wrote the paper: KMK JAJ.

References

1. Michaud DS, Spiegelman D, Clinton SK, Rimm EB, Willett WC, et al. (1999) Fruit and vegetable intake and incidence of bladder cancer in a male prospective cohort. J Natl Cancer Inst 91: 605–613.

2. Velasco P, Cartea ME, Gonzalez C, Vilar M, Ordas A (2007) Factors affecting the glucosinolate content of kale (*Brassica oleracea* acephala group). J Agric Food Chem 55: 955–962.

3. Carson DG, Daxenbichler ME, VanEtten CH (1987) Glucosinolates in crucifer vegetables: broccoli, Brussels sprouts, cauliflower, collards, kale, mustard greens, and kohlrabi. J Amer Soc Hort Sci 112: 173–178.

4. Velasco P, Soengas P, Vilar M, Cartea ME, del Rio M (2008) Comparison of glucosinolate profiles in leaf and seed tissues of different *Brassica napus* crops. J Amer Soc Hort Sci 133: 551–558.

5. Mithen RF, Dekker M, Verkerk R, Rabot S, Johnson IT (2000) The nutritional significance, biosynthesis and bioavailability of glucosinolates in human foods. Journal of the Science of Food and Agriculture 80: 967–984.

6. U.S. Department of Agriculture ARS (2011) USDA database for the flavonoid content of selected foods, Release 3.0.

7. Dai J, Mumper RJ (2010) Plant phenolics: extraction, analysis and their antioxidant and anticancer properties. Molecules 15: 7313–7352.

8. Galati G, O'Brien PJ (2004) Potential toxicity of flavonoids and other dietary phenolics: significance for their chemopreventive and anticancer properties. Free Radic Biol Med 37: 287–303.

9. Morton LW, Caccetta RA-A, Puddey IB, Croft KD (2000) Chemistry and biological effects of dietary phenolic compounds: relevance to cardiovascular disease. Clin Exp Pharmacol Physiol 27: 152–159.

10. Ku KM, Juvik JA (2013) Environmental stress and methyl jasmonate-mediated changes in flavonoid concentrations and antioxidant activity in broccoli florets and kale leaf tissues. Hortscience 48: 996–1002.

11. Keum YS, Jeong WS, Kong AN (2005) Chemopreventive functions of isothiocyanates. Drug News Perspect 18: 445–451.

12. Agerbirk N, De Vos M, Kim JH, Jander G (2009) Indole glucosinolate breakdown and its biological effects. Phytochem Rev 8: 101–120.

13. Bednarek P, Piślewska-Bednarek M, Svatoš A, Schneider B, Doubský J, et al. (2009) A glucosinolate metabolism pathway in living plant cells mediates broad-spectrum antifungal defense. Science 323: 101–106.

14. Nestle M (1998) Broccoli sprouts in cancer prevention. Nutr Rev 56: 127–130.

15. Cuendet M, Oteham CP, Moon RC, Pezzuto JM (2006) Quinone reductase induction as a biomarker for cancer chemoprevention. J Nat Prod 69: 460–463.

16. Clapper ML, Szarka CE (1998) Glutathione S-transferases-biomarkers of cancer risk and chemopreventive response. Chem Biol Interact 111–112: 377–388.

17. Kang Y-H, Pezzuto JM (2004) Induction of quinone reductase as a primary screen for natural product anticarcinogens. In: Helmut S, Lester P, editors. Methods in Enzymology: Academic Press. 380–414.

18. Zhang Y, Talalay P, Cho CG, Posner GH (1992) A major inducer of anticarcinogenic protective enzymes from broccoli: isolation and elucidation of structure. Proc Natl Acad Sci U S A 89: 2399–2403.

19. Zhu CY, Loft S (2003) Effect of chemopreventive compounds from Brassica vegetables on NAD(P)H:quinone reductase and induction of DNA strand breaks in murine hepa1c1c7 cells. Food Chem Toxicol 41: 455–462.

20. Neave AS, Sarup SM, Seidelin M, Duus F, Vang O (2005) Characterization of the *N*-methoxyindole-3-carbinol (NI3C)' induced cell cycle arrest in human colon cancer cell lines. Toxicol Sci 83: 126–135.

21. Ku KM, Jeffery EH, Juvik JA (2013) Influence of seasonal variation and methyl jasmonate mediated induction of glucosinolate biosynthesis on quinone reductase activity in broccoli florets. J Agric Food Chem 61: 9623–9631.

22. Hopkins RJ, van Dam NM, van Loon JJ (2009) Role of glucosinolates in insect-plant relationships and multitrophic interactions. Annu Rev Entomol 54: 57–83.

23. Ku KM, Choi J-H, Kushad MM, Jeffery EH, Juvik JA (2013) Pre-harvest methyl jasmonate treatment enhances cauliflower chemoprotective attributes without a loss in postharvest quality. Plant Foods Hum Nutr 68: 113–117.

24. Kim HS, Juvik JA (2011) Effect of selenium fertilization and methyl jasmonate treatment on glucosinolate accumulation in broccoli florets. J Amer Soc Hort Sci 136: 239–246.

25. Ku KM, Jeffery EH, Juvik JA (2013) Optimization of methyl jasmonate application to broccoli florets to enhance health-promoting phytochemical content. J Sci Food Agric 94: 2090–2096. DOI:10.1002/jsfa.6529

26. Ku KM, Choi JH, Kim HS, Kushad MM, Jeffery EH, et al. (2013) Methyl jasmonate and 1-methylcyclopropene treatment effects on quinone reductase inducing activity and post-harvest quality of broccoli. PLoS One 8: e77127.

27. Rose P, Faulkner K, Williamson G, Mithen R (2000) 7-Methylsulfinylheptyl and 8-methylsulfinyloctyl isothiocyanates from watercress are potent inducers of phase II enzymes. Carcinogenesis 21: 1983–1988.

28. Prochaska HJ, Santamaria AB (1988) Direct measurement of NAD(P)H:quinone reductase from cells cultured in microtiter wells: a screening assay for anticarcinogenic enzyme inducers. Anal Biochem 169: 328–336.

29. Brown AF, Yousef GG, Jeffery EH, Klein BP, Wallig MA, et al. (2002) Glucosinolate profiles in broccoli: variation in levels and implications in breeding for cancer chemoprotection. J Amer Soc Hort Sci 127: 807–813.

30. Wathelet JP, Marlier M, Severin M, Boenke A, Wagstaffe PJ (1995) Measurement of glucosinolates in rapeseeds. Natural Toxins 3: 299–304.

31. Tian Q, Rosselot RA, Schwartz SJ (2005) Quantitative determination of intact glucosinolates in broccoli, broccoli sprouts, Brussels sprouts, and cauliflower by high-performance liquid chromatography-electrospray ionization-tandem mass spectrometry. Anal Biochem 343: 93–99.

32. Velasco P, Francisco M, Moreno DA, Ferreres F, García-Viguera C, et al. (2011) Phytochemical fingerprinting of vegetable Brassica oleracea and Brassica napus by simultaneous identification of glucosinolates and phenolics. Phytochem Anal 22: 144–152.

33. Segarra G, Jauregui O, Casanova E, Trillas I (2006) Simultaneous quantitative LC-ESI-MS/MS analyses of salicylic acid and jasmonic acid in crude extracts of *Cucumis sativus* under biotic stress. Phytochemistry 67: 395–401.

34. Wilson EA, Ennahar S, Zhao M, Bergaentzle M, Marchioni E, et al. (2011) Simultaneous determination of various isothiocyanates by RP-LC following precolumn derivatization with mercaptoethanol. Chromatographia 73: 137–142.

35. Agerbirk N, Olsen CE, Sørensen H (1998) Initial and final products, nitriles, and ascorbigens produced in myrosinase-catalyzed hydrolysis of indole glucosinolates. J Agric Food Chem 46: 1563–1571.

36. Uda Y, Price KR, Williamson G, Rhodes MJC (1997) Induction of the anticarcinogenic marker enzyme, quinone reductase, in murine hepatoma cells in vitro by flavonoids. Cancer Letters 120: 213–216.

37. Williamson G, Plumb GW, Uda Y, Price KR, Rhodes MJC (1996) Dietary quercetin glycosides: antioxidant activity and induction of the anticarcinogenic phase I marker enzyme quinone reductase in Hepa1c1c7 cells. Carcinogenesis 17: 2385–2387.

38. Tong Y, Gabriel-Neumann E, Ngwene B, Krumbein A, George E, et al. (2014) Topsoil drying combined with increased sulfur supply leads to enhanced aliphatic glucosinolates in *Brassica juncea* leaves and roots. Food Chem 152: 190–196.

39. Creelman RA, Mullet JE (1995) Jasmonic acid distribution and action in plants: regulation during development and response to biotic and abiotic stress. Proc Natl Acad Sci U S A 92: 4114–4119.

40. Sato C, Seto Y, Nabeta K, Matsuura H (2009) Kinetics of the accumulation of jasmonic acid and its derivatives in systemic leaves of tobacco (*Nicotiana tabacum* cv. Xanthi nc) and translocation of deuterium-labeled jasmonic acid from the wounding site to the systemic site. Biosci Biotechnol Biochem 73: 1962–1970.

41. Yoshihara T, Amanuma M, Tsutsumi T, Okumura Y, Matsuura H, et al. (1996) Metabolism and transport of [2-14C](±) jasmonic acid in the potato plant. Plant Cell Physiol 37: 586–590.

42. Bradlow HL (2008) Indole-3-carbinol as a chemoprotective agent in breast and prostate Cancer. In Vivo 22: 441–445.

43. Yudina LN, Korolev AM, Reznikova MI, Preobrazhenskaya MN (2000) Investigation of neoascorbigen. Chem Heterocycl Compd 36: 144–151.

44. Harvey J, Dam N, Raaijmakers C, Bullock J, Gols R (2011) Tri-trophic effects of inter- and intra-population variation in defence chemistry of wild cabbage (*Brassica oleracea*). Oecologia 166: 421–431.

9

Involvement of Potato (*Solanum tuberosum* L.) MKK6 in Response to *Potato virus Y*

Ana Lazar[1]*, Anna Coll[1], David Dobnik[1], Špela Baebler[1], Apolonija Bedina-Zavec[2], Jana Žel[1], Kristina Gruden[1]

1 Department of Biotechnology and Systems Biology, National Institute of Biology, Ljubljana, Slovenia, **2** Laboratory for Molecular Biology and Nanobiotechnology, National Institute of Chemistry, Ljubljana, Slovenia

Abstract

Mitogen-activated protein kinase (MAPK) cascades have crucial roles in the regulation of plant development and in plant responses to stress. Plant recognition of pathogen-associated molecular patterns or pathogen-derived effector proteins has been shown to trigger activation of several MAPKs. This then controls defence responses, including synthesis and/or signalling of defence hormones and activation of defence related genes. The MAPK cascade genes are highly complex and interconnected, and thus the precise signalling mechanisms in specific plant–pathogen interactions are still not known. Here we investigated the MAPK signalling network involved in immune responses of potato (*Solanum tuberosum* L.) to *Potato virus Y*, an important potato pathogen worldwide. Sequence analysis was performed to identify the complete MAPK kinase (MKK) family in potato, and to identify those regulated in the hypersensitive resistance response to *Potato virus Y* infection. *Arabidopsis* has 10 MKK family members, of which we identified five in potato and tomato (*Solanum lycopersicum* L.), and eight in *Nicotiana benthamiana*. Among these, St*MKK6* is the most strongly regulated gene in response to *Potato virus Y*. The salicylic acid treatment revealed that St*MKK6* is regulated by the hormone that is in agreement with the salicylic acid-regulated domains found in the St*MKK6* promoter. The involvement of St*MKK6* in potato defence response was confirmed by localisation studies, where StMKK6 accumulated strongly only in *Potato-virus-Y*-infected plants, and predominantly in the cell nucleus. Using a yeast two-hybrid method, we identified three StMKK6 targets downstream in the MAPK cascade: StMAPK4_2, StMAPK6 and StMAPK13. These data together provide further insight into the StMKK6 signalling module and its involvement in plant defence.

Editor: Miguel A. Blazquez, Instituto de Biología Molecular y Celular de Plantas, Spain

Funding: The work was supported by the Slovenian Research Agency (https://www.arrs.gov.si/en/dobrodoslica.asp, grant nos.: P4-0165, J1-4268), The European Cooperation in Science and Technology, COST action FA0806 (http://www.cost.eu/domains_actions/fa/Actions/FA0806) and the Slovenian-Polish bilateral project (2010-2011). The funders had no role in study design, data collection and analysis, decision to publish, or preparation of the manuscript.

Competing Interests: The authors have declared that no competing interests exist.

* Email: ana.lazar@nib.si

Introduction

Mitogen-activated protein kinase (MAPK) cascades are conserved signalling modules in eukaryotes that transduce extracellular stimuli downstream from the receptors, thus mediating the intracellular responses. The plant MAPK cascades have pivotal roles in the regulation of plant development and in responses to a variety of stress stimuli, including pathogen infection, wounding, temperature, drought, salinity, osmolarity, UV irradiation, ozone and reactive oxygen species [1].

In a general model of the MAPK signalling cascade, activation of plasma membrane receptors activates the MAPK kinase kinases (MKKKs). These are serine or threonine kinases that phosphorylate a conserved S/T-X3−5-S/T motif of the downstream MAPK kinases (MKKs), which, in turn, phosphorylate MAPKs on threonine and tyrosine residues in a conserved T-X-Y motif of their activation loop [2]. Following this MKK alteration of the phosphorylation-dependent properties of their target proteins, the activated MAPKs translate the information further, which

eventually leads to changes in, e.g., gene expression, cellular redox state, or cell integrity [3].

The genome of the model plant *Arabidopsis thaliana* encodes 60 MKKKs, 10 MKKs and 20 MAPKs [4]. This indicates that the MAPK cascade might not simply consist of a single MKKK, MKK and MAPK connected together, but has the potential to be organised into many thousands of distinct MKKK–MKK–MAPK combinations, with some level of redundancy. To minimise unwanted cross-talk, the spatial and temporal activities of the different components must be strictly regulated [5].

Despite the potential multiplicity of MAPK cascades, only a small number of MAPK modules have been experimentally defined [6]. As the MKK family consists of a relatively small number of genes, their activity in different MAPK modules is widely dispersed [5]. In *Arabidopsis*, MKKs can be divided into four different groups (A–D) based on their sequence similarities [4,7]. Group A includes *Arabidopsis thaliana* AtMKK1, AtMKK2 and AtMKK6. AtMKK1 and AtMKK2 act upstream of AtMAPK4 in response to cold, salinity and pathogens [8,9]. AtMKK6 is involved in cytokinesis control and cell-cycle regulation [10]. The group B MKKs includes AtMKK3, which functions upstream of AtMAPK6 in the regulation of jasmonic acid (JA) signal transduction [11] and is involved in pathogen defence [12]; overexpression of AtMKK3 leads to enhanced tolerance to salt and increased sensitivity to abscisic acid [13]. The group C MKKs include AtMKK4 and AtMKK5, which act upstream of AtMAPK3 and AtMAPK6 in the regulation of plant development and defence responses [1,14–16]. The group D MKKs include AtMKK7, AtMKK8, AtMKK9 and AtMKK10. AtMKK9 is involved in ethylene signalling [17] and in leaf senescence [18], while AtMKK7 is involved in plant basal and systemic resistance [19].

During pathogen attack, MAPK signalling is an indispensable component of the host defence response, in a way that it is involved in the crosstalk between secondary messengers and hormones [20]. The key hormones in plant biotic interactions include the salicylates, jasmonates and ethylene [21], whereby their specific roles depend on the particular host–pathogen interaction. Many studies have indicated that salicylic acid (SA) is a key regulatory compound in disease resistance against fungi, bacteria and viruses (reviewed in [22]). The importance of SA in viral multiplication and symptom development has also been confirmed in potato (Solanum tuberosum L.) - Potato virus Y (PVY) interaction [23,24]. Depending on the virus, SA can induce inhibition of viral replication and cell-to-cell or long-distance viral movement (reviewed in [25]) and in agreement with this, SA is also a key component in the directing of events during and following hypersensitive resistance (HR) [21,24]. Hypersensitive resistance is an efficient defence strategy in plants, as it restricts pathogen growth and can be activated during host, as well as non-host, interactions. It involves programmed cell death and manifests in necrotic lesions at the site of pathogen attack (reviewed in [26]). Potato virus Y is a member of the Potyviridae family and it is an important potato pathogen worldwide. In potato, HR is conferred by the Ny genes (reviewed in [27]). The potato cultivar (cv.) Rywal carries the Ny-1 gene and it develops HR that is manifested as necrotic lesions on leaves 3 days following their inoculation with various PVY strains [28].

To date, a large number of members of the MAPK cascades from different species have been investigated, although to the best of our knowledge, there has been no systematic investigation of the MKK family and its function in defence signalling in potato. Moreover, no MAPK immune response network module has been defined for potato – PVY interactions.

We thus first performed sequence analysis of the complete MKK gene family in potato, and of its close relatives, where their genomes have been sequenced. Based on the present transcriptome data, StMKK6 was identified as the most responsive member after viral attack. We further investigated the role of StMKK6 in the response to PVY infection at the gene expression level, and studied its intracellular localisation and identified its downstream targets in the MAPK signalling cascade.

Materials and Methods

Bioinformatics analysis

Arabidopsis thaliana MKK gene family (The Arabidopsis Information Resource; TAIR: http://www.arabidopsis.org/) was blasted (tBLASTx algorithm) [29] against the potato [30], tomato (*Solanum lycopersicum* L.) and *Nicotiana benthamiana* (The SOL Genomics Network: http://solgenomics.net/) genomes to identify the *MKK* gene family in all three *Solanaceae* species. The names of the potato *MKK* genes were assigned based on the names of their apparent *A. thaliana* orthologues.

To identify and gather all of the available sequence information on St*MKK6* (GenBank accession number KF837127), St*MAPK4_1* (GenBank accession number KJ027594), St*MAPK4_2* (GenBank accession number KJ027595), St*MAPK6* (GenBank accession number KJ027596) and St*MAPK13* (GenBank accession number KJ027597), several database searches were performed. Sequences of our isolated genes were blasted (tBLASTx; [29]) against NCBI (National Centre for Biotechnology Information: http://www.ncbi. nlm.nih.gov/); UniProt (Universal Protein Resource: http://www. uniprot.org/); TAIR; PlantGDB (http://www.plantgdb.org/); POCI (Potato Oligo Chip Initiative: http://pgrc-35.ipk-gatersleben.de/pls/ htmldb_pgrc/f?p = 194:1); The Gene Index Project (http:// compbio.dfci.harvard.edu/tgi/); and TIGR Plant Transcripts Assemblies (http://plantta.jcvi.org/search.shtml).

To identify St*MKK6* orthologues we searched the UniProt database using the BLASTp algorithm [29]. Multiple-sequence alignments were performed using the MAFFT programme (version 7) [31]. Phylogenetic trees were constructed based on protein sequences using the MEGA 5 software [32], with the neighbour-joining method [33]. Bootstrap values were derived from 1,000 replicates, to quantify the relative support for branches of the inferred phylogenetic tree. To distinguish between *MKK6* orthologues and other *MKKs*, other At*MKKs* were included in the first MAFFT analysis.

Information on the targets of AtMKK6 was collected from the *Arabidopsis* Interactome Network Map [34]. The corresponding targets in potato were assigned by *A. thaliana*-potato orthologue and potato paralogue connections [35].

The location of the St*MKK6* gene (PGSC0003DMG403005720) within the potato genome was identified using the Ensembl Plants portal (http://plants.ensembl.org/index.html). The promoter sequences were obtained from the *Solanum_tuberosum*-PGSC_ DM_v34_superscaffolds database. Promoter sequences were analysed using the PlantCARE web service [36].

Localisation predictions of the *A. thaliana* and potato MKK6 proteins were performed using PredictProtein [37] and the data on the whole *A. thaliana* MKK family using the SUBA3 web services [38].

Metadata analysis

The expression data of the St*MKK* genes in developmental tissues and under stress conditions were collected through the use of the Bio-Analytic Resource for Plant Biology (http://bar. utoronto.ca/welcome.htm) and the Potato eFP browser [39]. The developmental tissues were vegetative and reproductive organs from greenhouse-grown plants. For each *MKK*, the RNA expression signals (FPKM values) based on the RNA sequencing of double monoploid *S. tuberosum* Group Phureja clone DM1-3 (DM) [40] and the genome sequence and analysis of the tuber crop potato [30] were collected. At*MKK6* transcription data from various datasets were analysed using Genevestigator [41] and only two-fold changes were regarded as significant.

We collected data from previously reported microarray expression analyses of the potato HR cv. Rywal and the SA deficient NahG-Rywal infected with PVY[N-Wi] ([24], GEO: GSE46180), for all of the potato *MKKs* and *MAPKs*. Microarray identifiers were connected to the PGSC gene identifiers [35] and only the probes with less than four single nucleotide polymorphisms compared to the PGSC transcripts were used.

Co-expression analysis of microarray expression datasets was carried out with Biolayout Express3D 3.0 software [42], with ratios of PVY/mock normalised signals. The network was constructed using the Pearson correlation coefficient threshold of 0.98, and the graph was clustered using the Markov clustering algorithm with inflation 2.3; the other parameters were set at the default values.

Gene expression analysis

Total RNA from control and SA- treated samples was extracted using MagMAX-96 Total RNA Isolation Kit (Life Technologies) according to manufacturer's instructions. RNA concentration was quantified by UV absorption at 260 nm using a NanoDrop ND1000 spectrophotometer (Nanodrop technologies). Integrity and purity of the RNA samples were determined by agarose gel electrophoresis and OD 260/280 nm absorption ratios. To confirm microarray results the same RNA samples previously analysed in microarray experiments [24] were used. Reverse transcription was performed on 1 µg of total RNA using High Capacity cDNA Reverse Transcription Kit (Applied Biosystems). The expression of St*MKK6* was assayed by real-time PCR (qPCR). The primers and probe (MKK6_F, MKK6_R and MKK6_S) targeting St*MKK6* were designed by Primer Express 2.0 (Applied Biosystems) (Table S1 in File S1). The qPCR was performed using TaqMan chemistry, with Cq values determined as described previously [24]. All reactions were run on a 7500 Fast Real-Time PCR System (Applied Biosystems) in 5 µl volume with 300 nM concentration of primers and 150 nM of probe and performed in duplicate. Linearity (R2) and efficiency (E = 10[−1/slope]) [43] of each reaction were compared to the accepted values. The standard curve method was used for relative gene expression quantification. Target gene accumulation was normalised to two endogenous control genes, cytochrome oxidase (COX) [44] and elongation factor 1 (EF-1) [45]. To determine differentially expressed genes, the Student t-test was performed.

Cloning

The full-length sequences of St*MKK6*, St*MAPK4_1*, St*MAPK4_2*, St*MAPK6* and St*MAPK13* were amplified from potato cv. Rywal cDNA with the primers listed in Table S1 in File S1. The fragments were inserted into the pJET 1.2 blunt cloning vector (Thermo Scientific) and sequenced (GATC Biotech). For the localisation studies, St*MKK6* was cloned into the pENTR/D TOPO vector (Invitrogen), using the LR reaction, following the manufacturer protocol, and recombined into the binary destination vectors pH7YWG2 (YFP) and PH7CWG2 (CFP) [46], to produce proteins with C-terminal YFP/CFP fusion.

The St*MKK6* promoter was amplified using the genomic library from potato cv. Santé as the template, which was constructed using GenomeWalker Universal kit (BD Biosciences Clontech), as described previously [47]. In the genome walking PCR amplifications, the Advantage 2 Polymerase Mix (Clontech) was used with the PCR conditions suggested by the manufacturer. The adaptor primer AP1 and a nested primer AP2, that were provided by the manufacturer were paired with the nested reverse gene-specific primers MKK6_AP1 and MKK6_AP2 for amplification of the region upstream from the St*MKK6* gene. The fragment was inserted into the pJET 1.2 blunt cloning vector (Thermo Scientific) and sequenced. The 200-bp region of the St*MKK6* promoter (identical in all three of the investigated genotypes) was afterwards cloned in front of the St*MKK6* gene in the pH7YWG2 vector, using QuikChange II XL Site-Directed Mutagenesis kit (Agilent Technologies), with the MAPKprom-YFP_F and MKK6prom-YFP_R primers. The St*MKK6* promoter region from potato cv. Rywal was amplified from the genomic DNA using the primers listed in Table S1 in File S1.

For the yeast two-hybrid analysis, the coding regions of the selected genes were cloned into pGBKT7 (St*MKK6*) or pGADT7 (St*MAPK4_1*, - 4_2, - 6, -13) using QuikChange II Site-Directed

Mutagenesis kit (Agilent Technologies). The primers used are listed in Table S1 in File S1.

Hormonal treatment

Two different potato genotypes were used in this study including 1 non-transgenic cv. Rywal and 1 transgenic line NahG, transgenic plants of cv. Rywal (NahG-Rywal) [24]. Potato plants were grown in stem node tissue culture. Two weeks after node segmentation they were transferred to soil and grown as previously described by Baebler et al. 2009 [45].

SA treatment was performed spraying plants with SA-analog 300 µM INA (98% 2,6-Dichloroisonicotinic acid, Aldrich) in distilled water. Control plants were sprayed with distilled water alone. Leaves were harvested after 24 h of treatment and immediately frozen in liquid nitrogen. Two biological replicates per treatment were analysed.

Localisation studies

N. benthamiana plants were grown in a growth chamber at 22/20°C under a 16/8 h light/dark cycle. PVY[NTN] (isolate NIB-NTN, AJ585342) inoculation of *N. benthamiana* was performed as described by Baebler et al. 2009 [45]. Plasmids pH7YWG2 (YFP), pH7CWG2 (CFP), pH7YWG2::StMKK6-YFP, pH7CWG2::StMKK6-CFP under the 35S promoter, and pH7YWG2::StMKK6 under the control of the native promoter, were introduced into *Agrobacterium tumefaciens* strain GV3101 by electroporation. The *A. tumefaciens* cells were cultured, harvested by centrifugation, and re-suspended in 0.2 mM acetosyringone solution, and infiltrated into 4–5-week-old *N. benthamiana* leaves, 8 days after PVY/mock inoculation.

After 48 h to 72 h, YFP/CFP were visualized with a Leica TCS SP5 laser-scanning microscope mounted on a Leica DMI 6000 CS inverted microscope (Leica Microsystems, Germany), with an HC PL FLUOTAR 10.0×0.30 DRY objective. For excitation, the 458 nm (enhanced cyan fluorescent protein) and 514 nm (enhanced yellow fluorescent protein) lines of an Argon laser were used. The enhanced cyan fluorescent protein emission was measured from 475 nm to 495 nm, and the enhanced yellow fluorescent protein from 525 nm to 550 nm. Autofluorescence emission was measured from 690 nm to 750 nm. Differential interference contrast (DIC) images were captured using the transmission light detector of the confocal microscope. All of the images were acquired with the 10× objective or with an additional 3.64× zoom. The acquired images were processed using the Leica LAS AF Lite software (Leica Microsystems). pH7YWG2 and pH7CWG2 without the inserted St*MKK6* were used as the controls.

Yeast two-hybrid assay

The coding regions of selected genes were cloned into pGBKT7 (St*MKK6*) or pGADT7 (St*MAPK4_1*, -4_2, -6, -13). To introduce insertions into plasmids the QuikChange II Site-Directed Mutagenesis Kit (AgilentTechnologies) was used. The primers used are listed in Table S1 in File S1.

Yeast two-hybrid analysis was performed using the Matchmaker GAL4-based two-hybrid assay (Clontech). The vector pGBKT7, containing bait protein fused to Gal DNA-binding domain, was used to transform the yeast strain Y2H Gold through the polyethylene glycol/LiAc-based method (Clontech), according to manufacturer instructions. After selection on SD/-Leu medium, the transformed colonies were used for co-transformation following the protocol described by Clontech, with the pGADT7 vector. It was used as a prey and contained target genes fused to Gal DNA-activation domain. Co-transformants were selected on SD/-

Leu/-Trp (DDO) medium. Positive interaction transformants were selected on SD/-Leu/-Trp/-His/-Ade/x-a-Gal/Aba (QDO/X/A) medium. All of the genes were previously tested for autoactivation and toxicity, plating yeast colonies transformed with pGBKT7_StMKK6 on SD/-Trp/x-a-Gal/Aba and pGADT7_StMAPK4_1, -4_2, -6, -13 on SD/-Leu/x-a-Gal/Aba. Positive (pGBKT7_SV40 large T-antigen -pGADT7_53) and negative (pGBKT7_SV40 large T-antigen -pGADT7_Lam) interaction controls, included in the Matchmaker Gold Yeast Two-Hybrid System (Clontech), were performed in parallel.

Results

The MKK family in potato

Our genome analysis of the potato, tomato and *Nicotiana attenuata* genomes identified five *MKK* genes in each, with four in the tobacco *Nicotiana tabacum* L. genome, and eight in the *N. benthamiana* genome (Figure 1). Based on the phylogenetic tree of *Arabidopsis* and five *Solanaceae* species, for only two of the *A. thaliana* genes (At*MKK3*, At*MKK6*) a specific orthologue could be assigned from potato, tomato or *Nicotiana* spp., while the other *MKK*s form larger orthologue groups. The *MKK* family in potato, tomato and *Nicotiana* spp. can also be divided into four groups: group A (*A. thaliana* and *N. benthamiana*, three genes; potato, tomato, *N. tabacum* and *N. attenuata*, two genes); B (*N. benthamiana*, two genes; the rest, one gene each); C (*A. thaliana* and *N. benthamiana*, two genes; potato, tomato, *N. tabacum* and *N. attenuata*, one gene each); and D (*A. thaliana*, four genes; *Solanaceae* spp., one gene each, except *N. tabacum* which has none). Out of the five examined *Solanaceae* spp., *N. benthamiana* is the only one of these in which duplication of the *MKK* genes occurred after speciation and thus *N. benthamiana* has possible paralogues for *MKK3* (NbS00014007g0022.1, NbS00037566g0012.1), *MKK4/5* (NbS00012713g0030.1, NbS00006609g0002.1) and *MKK6* (NbS00006857g0015.1, NbS00045036g0007.1) (Figure 1).

Regulation of potato MKK family gene expression in developmental processes and stress

We first analysed the gene-expression metadata for the potato *MKK* family for developmental processes and after exposure to different biotic and abiotic stress [40]. The tissue specificities of the *MKK* family members differed between both of the potato varieties studied here, the potato Group Phureja Clone DM, and Group Tuberosum Clone RH, especially regarding the expression in tubers and stolons. In general, this analysis showed that in potato, St*MKK1/2* and St*MKK4/5* are relatively uniformly expressed, with the highest expression in tubers and stolons. St*MKK3* transcripts had the highest abundance in carpels and petals, and St*MKK7/9* in tubers and flowers, and also in roots for Clone RH. Under biotic and abiotic stress conditions, all of the potato *MKK*s except St*MKK4/5*, were down-regulated after wounding. None of the other stress conditions and hormonal treatments influenced the *MKK* family gene expression by more than two-fold compared to controls (File S2).

In a previous transcriptome analysis of potato HR responses to PVY infection [24], we checked the expression profile of identified potato *MKK*s [35]. According to the gene expression profiles obtained for the potato cv. Rywal, three out of five *MKK*s showed differential expression for the HR response to PVY: St*MKK3*, St*MKK6* and St*MKK4/5* (Figure 2). St*MKK6* showed the strongest response of all, thus its expression data were also confirmed by quantitative real-time PCR (qPCR) (Table S2 in File S1). St*MKK6* was strongly up-regulated before lesion formation (3 dpi). In contrast, at the same time, St*MKK3* was down-regulated.

St*MKK4/5* showed two-fold induction at the later time (6 dpi) following inoculation. For the SA-deficient plants (NahG-Rywal), the regulation of these *MKK*s at the gene level was either attenuated (for St*MKK3* and St*MKK4/5*) or delayed (for St*MKK6*). Only St*MKK7/9* was induced earlier and stronger in the NahG-Rywal plants, compared to the non-transgenic plants.

Based on these data we focused on further investigations on the At*MKK6* orthologue in potato, St*MKK6*.

St*MKK6* orthologues across the plant kingdom

To study the sequence diversity and evolution of the *MKK6* gene, we performed a phylogenetic analysis within the plant kingdom. We identified 39 plant *MKK6* orthologues among higher plants. As well as for different families of Angiosperms, one *MKK6* gene was also identified in Gymnosperms, and one in mosses (Figure 3 and File S3). *MKK6* is mostly a single-copy gene, as these 39 *MKK6* orthologues belong to 31 plant species (Figure 3).

Comparison of the *MKK6* sequence from potato cv. Rywal and the published potato genome sequence of *S. tuberosum* Group Phureja, Clone DM showed that the coding domain sequences are different for three nucleotides at different sites, which results in only one amino-acid difference; this is not part of the protein kinase domain.

Regulation of St*MKK6* gene expression in development and stress

To better understand the function of the *MKK6*s, we examined the gene expression data that is available for *A. thaliana MKK6* in Genevestigator. At*MKK6* is up-regulated in developmental processes (callus formation, germination) and after treatments with translation inhibitor cyclohexamide and hormones auxins. At*MKK6* is also up-regulated after flagellin 22 treatment, and under abiotic stress (cold, salt, acidic pH, hypoxia). At*MKK6* is down-regulated after treatments with abscisic acid and combinations of abscisic acid with jasmonate or SA.

The expression profiles of St*MKK6* obtained from the Potato eFP Browser database showed that in the potato Group Phureja Clone DM, St*MKK6* tissue expression was high in the tubers and callus, and very low (below detection levels) in the flowers. In the Group Tuberosum Clone RH under non-stress conditions, St*MKK6* gene expression was strong also in the shoot apex (Figure 4A; [40]). Among all of the stress-related conditions included, compared to the controls, St*MKK6* was down-regulated after wounding of the secondary leaves and after treatment with the fungal elicitor DL-β-amino-n-butyric acid (Figure 4B).

Analysis of the St*MKK6* native promoter

In the Potato Genome Sequencing Consortium (PGSC) Potato Genome Browser, St*MKK6* is assigned to PGSC gene ID PGSC0003DMG403005720, and it is localised to chromosome 3. Based on this information, we also cloned the St*MKK6* promoter sequence from two cultivars of *S. tuberosum* Group Tuberosum, both of which are resistant to PVY infection: cv. Rywal (GenBank accession number KF837128) and cv. Santé (GenBank accession number KF837129). The 790 bp-long promoter isolated from cv. Rywal (Table S3 in File S1) is identical to the promoter of *S. tuberosum* Group Phureja, but different from that of cv. Santé (Table S4 in File S1), due to a 27-bp-long region that is localised 230 bp upstream of St*MKK6*. In general, the most common domains were the TATA-box, CAAT box and domains related to light responsiveness. Based on the development and stress-related domains found, St*MKK6* appears to be involved in

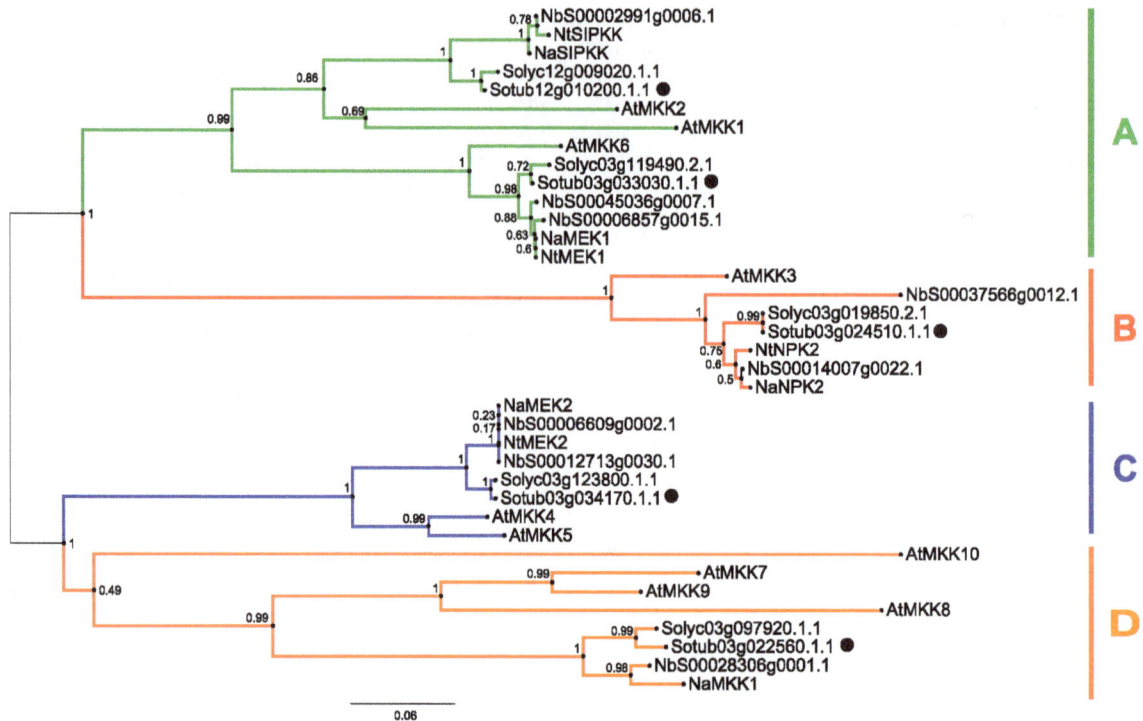

Figure 1. Phylogenetic tree of *MKK* family in *A. thaliana* and five *Solanaceae* species. The species are *A. thaliana* (At), potato (Sotub), tomato (Solyc), *N. benthamiana* (Nb), *N. attenuata* (Na) and *N. tabacum* (Nt). Genes are grouped into 4 groups: A (green), B (red), C (blue) and D (orange) [4]. Potato genes are marked with dots. The numbers on the nodes are percentages from a bootstrap analysis of 1000 replicates. The scale bar indicates the branch length that corresponds to 0.06 substitutions per site.

development of the endosperm and in responses to heat stress and wounding. St*MKK6* is also predicted to be involved in gibberellic acid and salicylic acid signalling, and is under regulation by the circadian clock (Figure 5).

Regulation of St*MKK6* by salicylic acid

Promoter analysis of St*MKK6* has demonstrated potential involvement of St*MKK6* in SA signalling. Therefore, we examined expression of St*MKK6* under the SA treatment. The expression of St*MKK6* was measured by qPCR 24 h after treatment with SA in Rywal and NahG-Rywal plants.

The treatment showed St*MKK6* to be sensitive to the present active form of SA since significant changes of St*MKK6* expression were observed in the SA-deficient genotype NahG-Rywal (Figure 6). There was a substantial, two-fold rise, of the St*MKK6* RNA, compared to the non-treated plants. In the Rywal plants,

that received the same SA treatment, the change in the St*MKK6* expression was not significant (Table S5 in File S1).

We also observed significant difference in the RNA level of the endogenous St*MKK6* in the non-treated Rywal, compared to the NahG-Rywal (Figure 6). In Rywal plants there was at least twice as much of expressed St*MKK6* comparing to the level in the NahG-Rywal (Table S6 in File S1).

Localisation of the StMKK6 protein in healthy and PVY-infected epidermal leaf cells

To learn more about the function of the StMKK6 protein in defence responses against the virus PVY, we analysed StMKK6 localisation in healthy and infected *N. benthamiana* plants. The infected leaves of *N. benthamiana* developed no necrotic lesions due to infection with PVY or overexpression of St*MKK6*. *In-silico* analysis predicted the localisation of StMKK6 as in the nucleus

Name	PGSC gene ID	*A. thaliana* orthologues	Microarray probe ID	Rywal			NahG-Rywal		
				1 dpi	3 dpi	6 dpi	1 dpi	3 dpi	6 dpi
StMKK1/2	PGSC0003DMG400000273	AtMKK1/2	MICRO.6723.C1	-0.16	0.10	0.25	0.17	0.85	0.95
StMKK3	PGSC0003DMG402015209	AtMKK3	MICRO.610.C1	-1.23	-1.49	0.02	-0.37	-0.02	0.77
StMKK4/5	PGSC0003DMG400009183	AtMKK4/5	MICRO.14487.C2	0.50	0.65	0.95	0.39	0.39	0.66
StMKK6	PGSC0003DMG403005720	AtMKK6	MICRO.17148.C1	1.40	1.74	-0.17	0.90	0.77	0.65
StMKK7/9	PGSC0003DMG400033696	AtMKK7/9	MICRO.12818.C1	-0.38	-0.15	0.48	0.27	0.96	1.01

Figure 2. Gene expression pattern of *MKK* family in the HR response against PVY. Cultivar Rywal (HR response, conferred by *Ny-1* gene) and NahG-Rywal (impaired accumulation of SA) were analysed for whole transcriptome response 1, 3 and 6 days after PVY[N-Wi] infection [24]. *A. thaliana* and *S. tuberosum* PGSC orthologues were assigned to each probe. Log$_2$ fold changes of PVY in infected vs. mock-inoculated plants are indicated for each time point (1, 3 and 6 dpi). Statistically significant differences (FDR corrected p<0.05) are in bold. Up-regulated values are in blue and down-regulated values are in yellow.

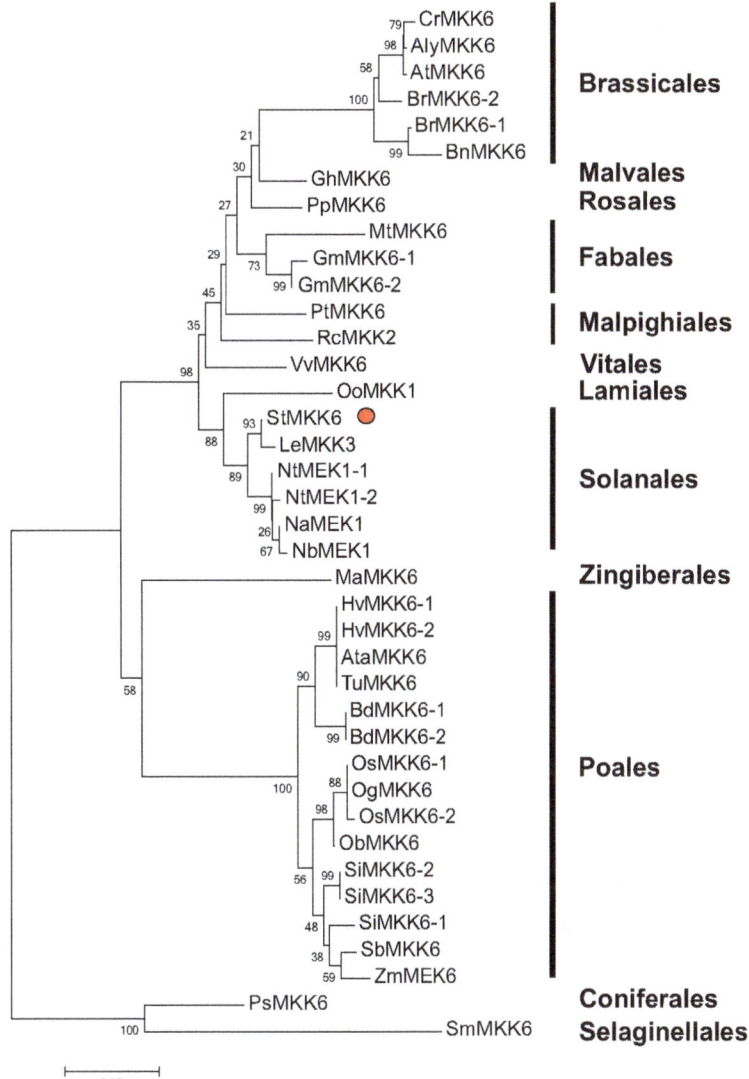

Figure 3. Phylogenetic tree of potato *MKK6* and its orthologues in different plant species, divided into classes. The species of origin for each *MKK* is marked by species acronym before the protein name: Cr, *Capsella rubella*; Aly, *Arabidopsis lyrata*; At, *Arabidopsis thaliana*; Br, *Brassica rapa*; Bn, *Brassica napus*; Gh, *Gossypium hirsutum*; Pp, *Prunus persica*; Mt, *Medicago trunculata*; Gm, *Glycine max*; Pt, *Populus trichocarpa*; Rc, *Ricinus communis*; Vv, *Vitis vinifera*; Oo, *Origanum onites*; St, *Solanum tuberosum*; Le, *Solanum lycopersicum (Lycopersicum esculetum)*; Nt, *Nicotiana tabacum*; Na, *Nicotiana attenuata*; Nb, *Nicotiana benthamiana*; Ma, *Musa acuminata*; Hv, *Hordeum vulgare*, Ata, *Aegilops tauschii*; Tu, *Triticum urartu*; Bd, *Brachypodium distachyon*; Os, *Oryza sativa*, Og, *Oryza glaberrima*; Ob, *Oryza brachyantha*; Si, *Setaria italica*; Sb, *Sorghum bicolor*; Zm, *Zea mays*, Ps, *Picea sitchensis*; Sm, *Selaginella moellendorffii*. Potato *MKK6* is marked with red dot. The numbers on the nodes are percentages from a bootstrap analysis of 1000 replicates. The scale bar indicates the branch length that corresponds to 0.06 substitutions per site.

and cytoplasm. The same results were obtained also for AtMKK6 (Table S7 in File S1). An *in-vivo* study of the subcellular localisation of StMKK6 was performed under the native promoter and in C-terminal translational fusion with YFP in *N. benthamiana* leaves. In the mock-inoculated leaves, no fluorescence of the StMKK6-YFP fusion protein was observed (Figure 7A), while in PVY-infected leaves, there was strong StMKK6-YFP fluorescence in the nucleus and very weak StMKK6-YFP fluorescence in the cytoplasm (Figure 7B).

The fluorescent protein itself did not affect the localisation of StMKK6, as fusions with different fluorescent proteins were observed in the same intracellular compartments. The overexpression of StMKK6, driven by the CaMV 35S promoter, resulted in localisation of the protein in the cytoplasm and nucleus in mock-

inoculated as well as in PVY-infected leaves (Figures S1B and S1C).

Search for St*MKK6* co-regulated genes and potential targets in the MAPK cascade

To further improve our knowledge of St*MKK6* function, we searched for the other components of the MAPK module that is involved in the HR response in potato, the genes that are co-expressed with St*MKK6*, and the possible MAPK targets.

We found 22 co-expressed genes (File S4), for which we analysed the promoter regions. The most common biotic stress-related domains in their promoters were TC-rich repeat motifs (an element involved in defence and stress responsiveness; 13 promoters with the region), TGACG and CGTCA motifs

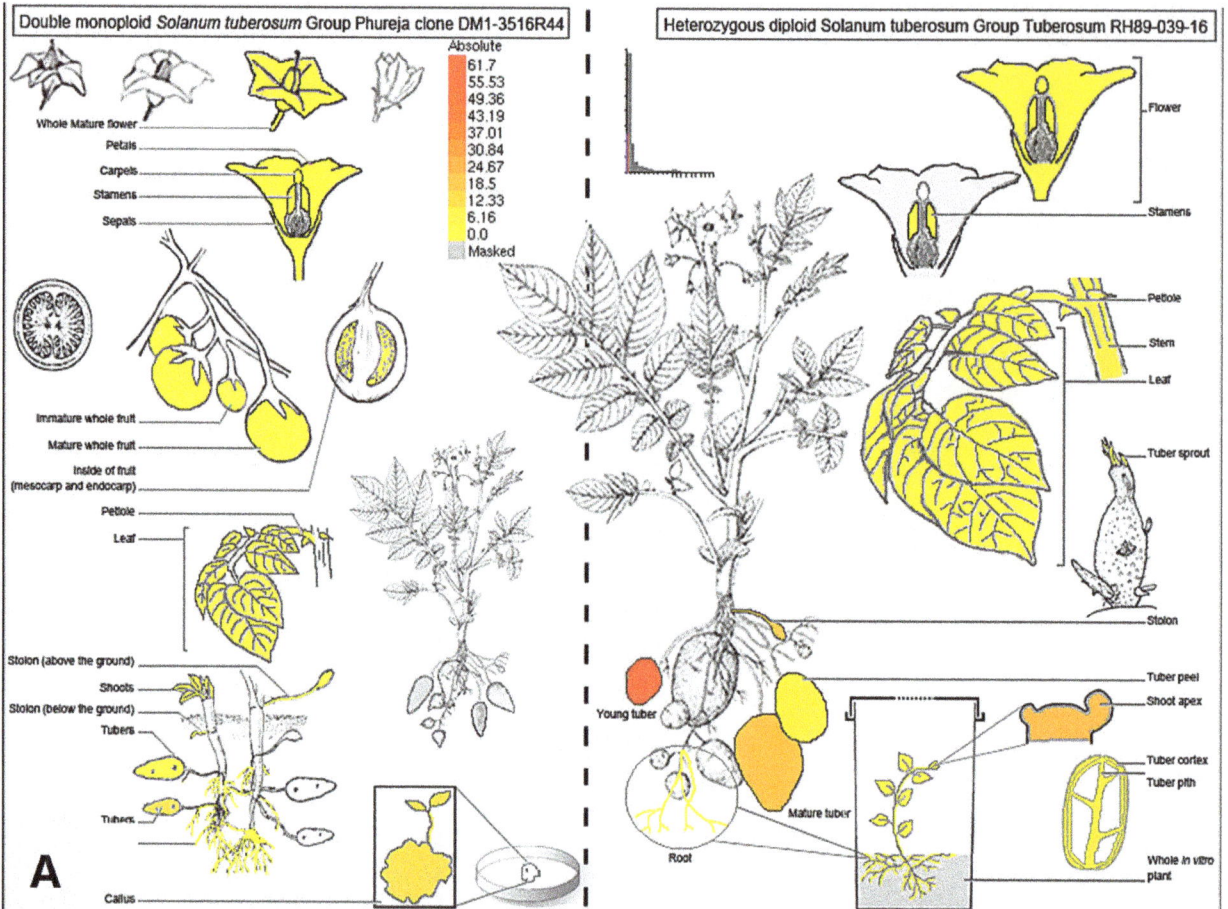

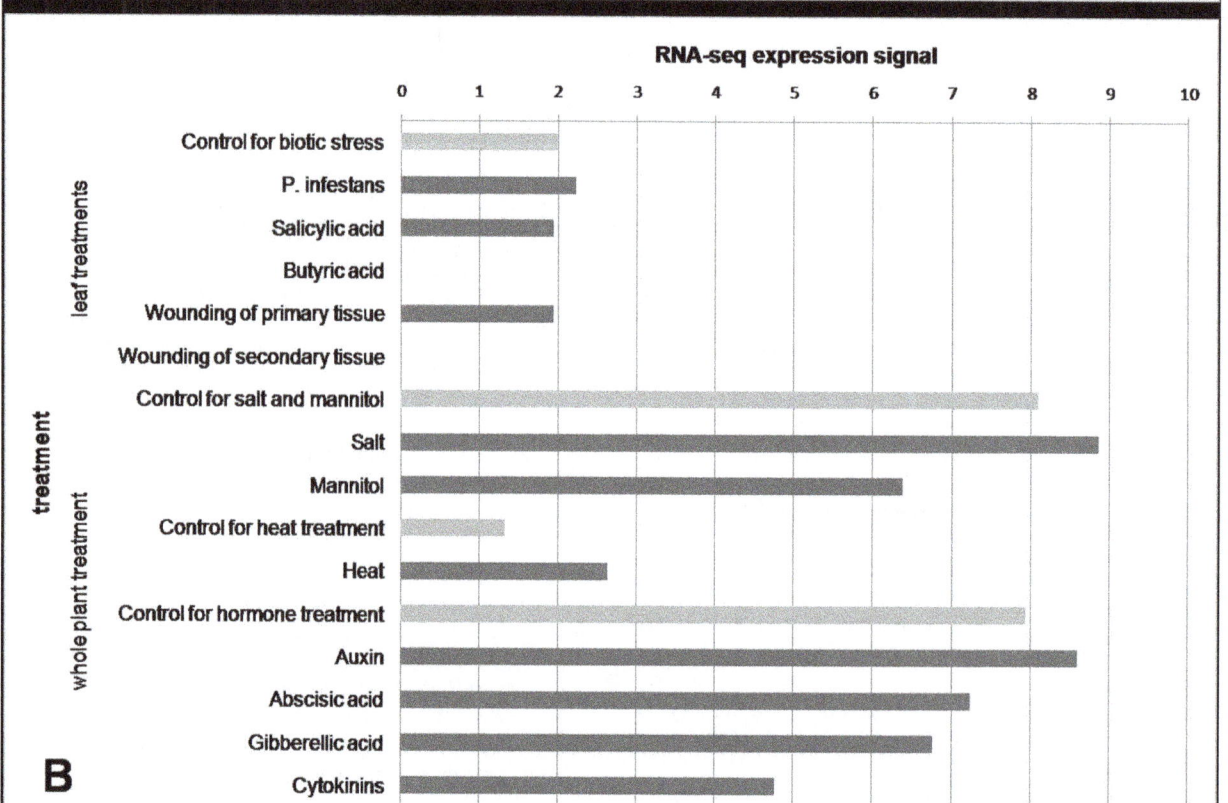

Figure 4. St*MKK6* gene expression in development and stress [40]. Data are obtained from the potato eFP browser [39]. **A.** Tissue and developmental gene expression pattern of St*MKK6* in double monoploid *S. tuberosum* Group Phureja (left) and in heterozygous diploid *S. tuberosum* Group Tuberosum (right). **B.** Changes in St*MKK6* gene expression under biotic stress conditions (treatments of leaves with *P. infestans*, SA analogues acibenzolar-S-methyl and fungal elicitor DL-β-amino-n-butyric acid and wounding) and abiotic stress conditions (treatments of whole plants with salt, heat and hormones cytokinins, 6 benzylaminopurine; gibberellins, GA3; abscisic acid; and auxin, IAA) and control treatments.

(regulatory elements involved in methyl jasmonate responsiveness; 9 promoters with the region), Box-W1 (fungal elicitor responsive element; 8 promoters with the region), and TCA elements (SA responsiveness; 8 promoters with the region). The only region that was common to all 22 genes was the Skn-1 motif, which is related to endosperm expression.

In an *A. thaliana* interactome study [34], AtMKK6 was shown to interact with: AtMAPK4 (AT4G01370), AtMAPK6 (AT2G43790), AtMAPK11 (AT1G01560), AtMAPK12 (AT2G46070), AtMAPK13 (AT1G07880), and the calmodulin-like protein TCH3 (AT2G41100).

We first identified the orthologues of all of the potential interacting MAPKs in potato. AtMAPK4 together with AtMAPK11 and AtMAPK12 represent one orthologue cluster with StMAPK4_1 (PGSC003DMP400037535) and StMAPK4_2 (PGCS003DMP400000144). We also identified specific orthologues of AtMAPK6 in Sotub08g010260.1.1, and for AtMAPK13 in PGSC003DMP400044003 (Figure S2). We further cloned all of the potential MAPK interactors of StMKK6 from potato cv. Rywal. Comparisons of the four MAPKs cloned from potato cv. Rywal and those published in the potato genome sequence of *S. tuberosum* Group Phureja, Clone DM showed Rywal MAPK4_1 to be one-amino-acid different from PGSC0003DMP400037535, while Rywal MAPK 4_2 and MAPK13 are identical to their Phureja orthologues. Rywal MAPK6 had no corresponding proteins in Group Phureja, Clone DM. Yeast two-hybrid assays confirmed the interactions of StMKK6 with three MAPKs: StMAPK4_2, StMAPK6 and StMAPK13 (Figure 8). StMAPK4_1 could not be confirmed as a target of StMKK6 in potato cv. Rywal.

St*MAPK4_2* and St*MAPK13* show differential gene expression patterns after PVY inoculation (Figure 9), and have different expression patterns than St*MKK6* (Figure 2).

```
+ TGTCTTCGAA TGGCTCCAAC CAAATGCATT TCAAGTTTAC AATGAAAGAT CAGCTGGGAG AATAGTCATG
- ACAGAAGCTT ACCGAGGTTG GTTTACGTAA AGTTCAAATG TTACTTTCTA GTCGACCCTC TTATCAGTAC -823

+ CTGTTTAGGC TAGCACTGTT GTTTGGCTTT TCCTTTTGTG TTGTTCGTAA AGATATTGCT AGCACTGTTG
- GACAAATCCG ATCGTGACAA CAAACCGAAA AGGAAAACAC AACAAGCATT TCTATAACGA TCGTGACAAC -753

+ TTTGGCTTTT CAACCAATTG CTTTATATTT GTAAAGATTT GGTAGCTGGA AATTGATTTC AAGCCATCTC
- AAACCGAAAA GTTGGTTAAC GAAATATAAA CATTTCTAAA CCATCGACCT TTAACTAAAG TTCGGTAGAG -683

+ TACTTGAAAC TCAACACGCT TTGGTTTAAA TCCAACTATT TGGGCTGGTC GGATAGAATA TCATGGGTGA
- ATGAACTTTG AGTTGTGCGA AACCAAATTT AGGTTGATAA ACCCGACCAG CCTATCTTAT AGTACCCACT -613

+ ACGGGTTATA GTATTTTCGT GCAATTCAAC TATTTCATTA GTATTTGCTA CATGCATCAA TATAAACATG
- TGCCCAATAT CATAAAAGCA CGTTAAGTTG ATAAAGTAAT CATAAACGAT GTACGTAGTT ATATTTGTAC -543

+ GAATAACTCT TCCTTCGAAA AACTAGACAT CCAAATGAAA AGAAATCACC TCATTTTTAT CGGAAAATTT
- CTTATTGAGA AGGAAGCTTT TTGATCTGTA GGTTTACTTT TCTTTAGTGG AGTAAAAATA GCCTTTTAAA -473

+ CACTATATTA GGAGTAACAA TTAGGTAACA TATGTATGTT CAAGCATTTA GGTGACATAT ATTTATCCCG
- GTGATATAAT CCTCATTGTT AATCCATTGT ATACATACAA GTTCGTAAAT CCACTGTATA TAAATAGGGC -403

+ TTGAGAAGAA AACACAATAT ATATAAGAAA TTTTATATAG GCAAATCTTT TTACAAAGGT AGAATCCATC
- AACTCTTCTT TTGTGTTATA TATATTCTTT AAAATATATC CGTTTAGAAA AATGTTTCCA TCTTAGGTAG -333

+ TCAACGTTAT TCTTTTTTTT TTTTTAAAAA AAAAAAGAGA GGATAAAATA GGGGTATAAT TGGTTAATTT
- AGTTGCAATA AGAAAAAAAA AAAAATTTTT TTTTTTCTCT CCTATTTTAT CCCCATATTA ACCAATTAAA -263

+ TAATAGTTGA AAGATATTTT TAACTTTTCG TTTAAATAAT AAAAAGAATA AAGGATCACT GATTGCTTTG
- ATTATCAACT TTCTATAAAA ATTGAAAAGC AAATTTATTA TTTTTCTTAT TTCCTAGTGA CTAACGAAAC -193

+ TTTTTATTCA ATTTTCAAAA TAAAAATAAA AATGTGTAGG CTAAACATTC AAACCTGTCA GTAATAGAAA
- AAAAATAAGT TAAAAGTTTT ATTTTTATTT TTACACATCC GATTTGTAAG TTTGGACAGT CATTATCTTT -123

+ TAGAATATAA TTTTGTTGGT ATTTGTCTGA GTTTGAAAAT CAGTGGGTAT TTTGGGCGGT TCGTTTCACT
- ATCTTATATT AAAACAACCA TAAACAGACT CAAACTTTTA GTCACCCATA AAACCCGCCA AGCAAAGTGA -53

+ TTTCAAAATC TAAAATCTCC ACTTCCCCTT TCAAACCCTA GTGGATCAAA ATTATG
- AAAGTTTTAG ATTTTAGAGG TGAAGGGGAA AGTTTGGGAT CACCTAGTTT TAATAC
```

Figure 5. Predicted regulatory domains of St*MKK6* native promoter from *S. tuberosum* cv. Rywal, 790-bp-long. The crucial domains that regulate stress responses and development are CCAAT-box (MYBHv1 binding site; yellow), GCN4 motif (endosperm expression; pink), HSE (heat stress; brown), P-box (gibberellins; blue), Skn-1 motif (endosperm expression; purple), TCA element (salicylic acid; green), WUN motif (wounding; red) and circadian motif (control of circadian clock; grey). The arrow indicates the predicted beginning of the 5′ UTR region and the red shaded ATG indicates the start of St*MKK6* coding region. The complete list of St*MKK6* promoter domains from cv. Santé and cv. Rywal with their position, strand and sequence are specified in Tables S3 and S4 in File S1.

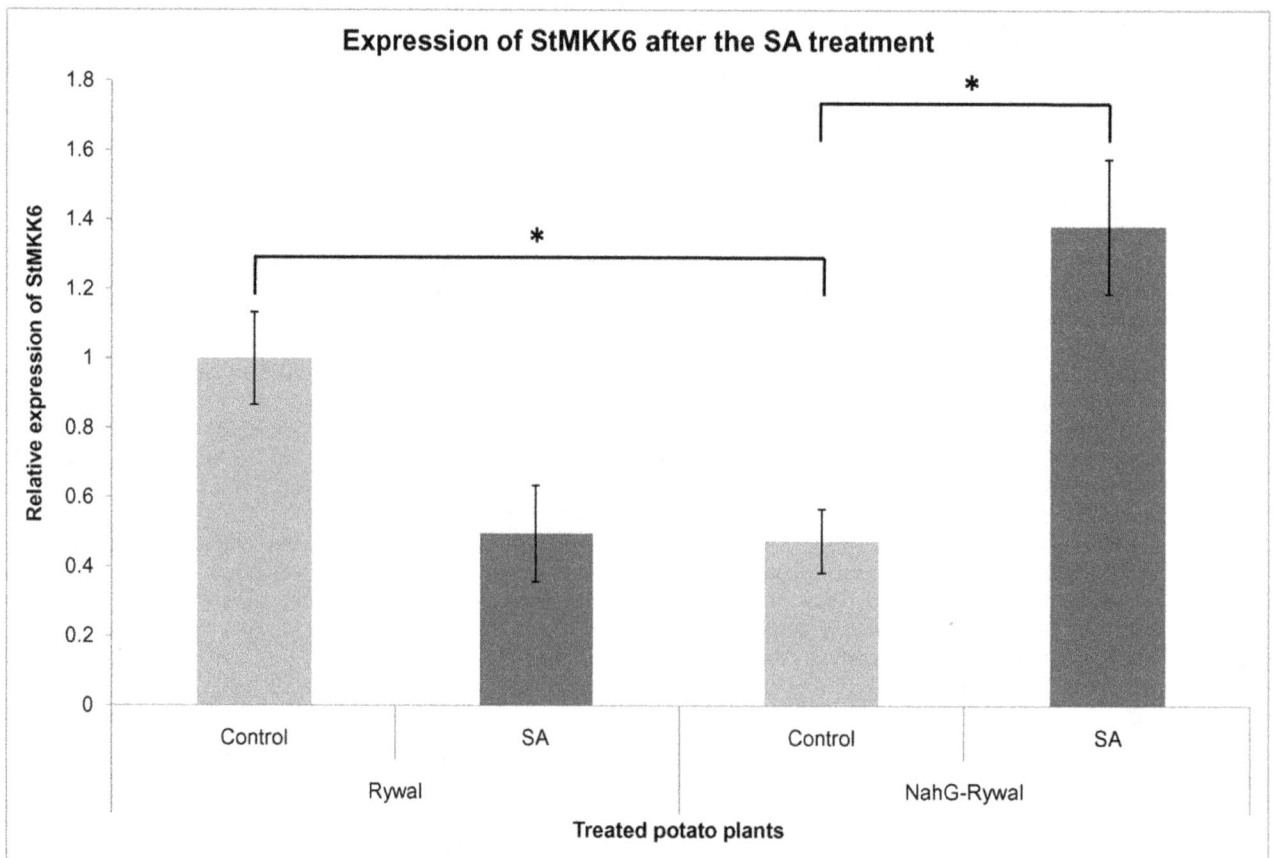

Figure 6. Expression of St*MKK6* in control and SA-treated Rywal and NahG-Rywal potato plants. Expression of the endogenous St*MKK6* in NahG-Rywal is at least two-fold lower than in Rywal. After the SA-treatment the expression of St*MKK6* in NahG-Rywal increased for at least three times, comparing to the non-treated control. Difference in the endogenous St*MKK6* expression between Rywal and NahG-Rywal as well as difference in the St*MKK6* expression between SA-treated and non-treated NahG-Rywal plants are statistically significant (p<0.05).

Discussion

To shed further light on the hypersensitive resistance signalling responses during PVY infection of potato, we analysed the involvement of potato MKKs in this process. The results revealed that the St*MKK6* is regulated by the SA signalling pathway in the HR to PVY infection. The virus induces the expression of St*MKK6* by the protein accumulating preferentially in nucleus.

Potato has five *MKKs*

To date, MKK genes have been analysed in several plants: in *A. thaliana*, 10 have been identified [4], in rice, 8 [48], in *Brachypodium*, 12 [49], in poplar, 10 [50], in apple, 9 [51] and in canola, 7 [52]. In the *Solanaceae* family, the MKK family is smaller than in other plant species. Our analysis showed that the potato and tomato genomes have five MKK genes, and that the *N. benthamiana* genome has eight. Five *MKKs* were identified also in *N. attenuata*, while in *N. tabacum*, an additional Nb*MKK1* orthologue was not found among the currently existing sequences (Figure 1, [53]). The phylogenetic tree of the different MKK genes shows that only *N. benthamiana* has three possible pairs of paralogues, with none found in potato or tomato. Higher numbers of *MKK* family members in other plant species might be the result of gene duplications in the *MKK* ancestors that occurred after divergence into different families. Consequently, it was not possible to assign a single *A. thaliana* orthologue to three out of

five potato *MKKs*. This issue of assigning orthologues was previously discussed by Dóczi et al. 2012 [54], and was attributed to the high expansion and functional diversification of the MAPKs, and to the rapid evolution of MAPK signalling.

It has been speculated that all 10 AtMKKs are fully functional, although there have been doubts expressed for AtMKK8 and AtMKK10. Although AtMKK8 has all of the typical MKK motifs, there is no evidence of its expression, while AtMKK10 lacks a correctly constructed target site in the activation loop [7]. Thus, in plants with lower numbers of MKK genes, the plasticity of the MAPK signalling module is not necessarily also proportionally lower.

StMKK6 preferentially accumulates in the nucleus only after PVY infection

Transcriptomic analysis indicated that the St*MKK6* gene is involved in the HR response against PVY, as the most intensively regulated of the *MKKs* (Figure 2). Thus, we focused on different aspects of St*MKK6* function in this process. Here, meta-analysis of the publicly available datasets (Figure 4) and our *in-planta* localisation study (Figure 7) show that St*MKK6* expression is very low in non-stressed potato leaves. Interestingly, the infection with PVY induces the expression of St*MKK6*, with the StMKK6 protein concentrated predominantly in nucleus.

The information on the intracellular localisation of StMKK6 and its orthologues is currently really limited. The only report on

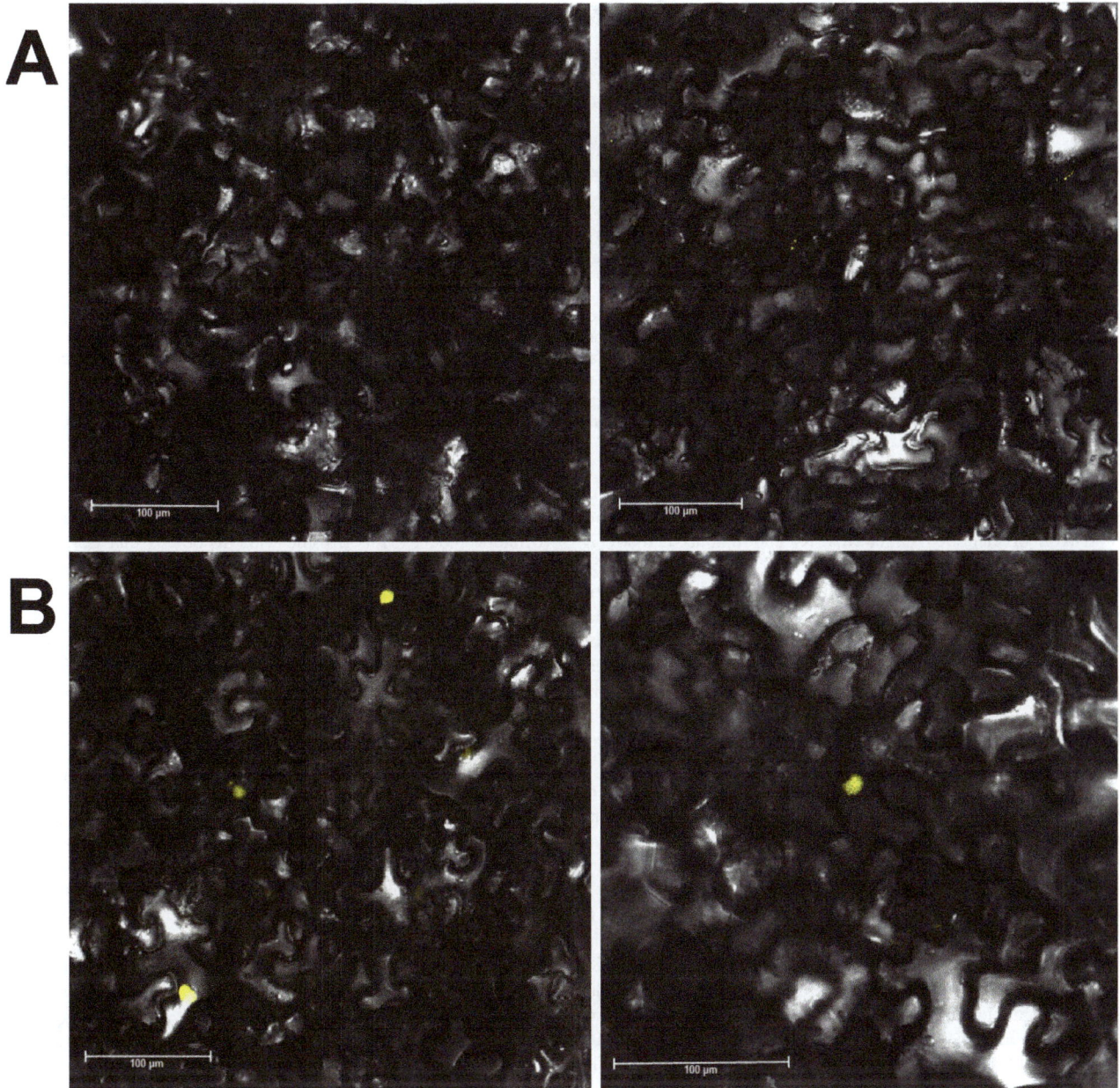

Figure 7. Epidermal cells of *N. benthamiana* expressing translational fusion of StMKK6 with YFP under native promoter. Leaves were agroinfiltrated when the virus has spread uniformly through the inoculated leaves (8 days after inoculation) and observed after 72 h in two independent experiments. Examples from two plants (left and right panels) are shown. Control of transformation (fluorescent marker without StMKK6 fusion) is in the Figure S1A. **A.** Localisation of StMKK6 in mock-inoculated leaves. No fluorescence was observed. **B.** Localisation of StMKK6 in PVY-inoculated leaves, where the protein accumulates predominantly in nucleus. Additional images of StMKK6 localisation under native promoter are in Figures S1D and S1E.

experimental results of MKKs intracellular localisation showed that AtMKK6 accumulated in the equatorial plane of the phragmoplast in dividing root epidermal cells, while all of the other AtMKKs (AtMKK1-5) did not [10]. In *N. benthamiana*, localisation of MKK1 (an orthologue of AtMKK7, -8 and -9) was studied for the HR to *Phytophtora infestans* infection [55,56]. *N. benthamiana* MKK1 preferentially accumulated in the nucleus also under non-stress conditions. Liang et al. 2013 [52], studied the localisation of MKK2, -3 and -4 in *Brassica napus* epidermal leaf cells, and detected the accumulation of these MKKs both in the cytoplasm and the nucleus. On the other hand, other reports

on the localisation of the MKK target MAPKs show preferential accumulation of those proteins in the cytoplasm [52,55]. One however has to note that all above mentioned experiments were studying localisation of MKKs when expressed under strong (e.g. CaMV 35S) promoter, while we have studied localisation under its native promoter (Figure 7). Also in our hands expression of protein under 35S promoter caused ubiquitous cellular localisation (e.g. cytoplasm and nucleus, see Figures S1B and S1C) presumably due to excessive amounts of protein produced, which is known to generate localisation artefacts [57]. Anyhow, our results support the hypothesis that activation of MKK6 targets occurs predom-

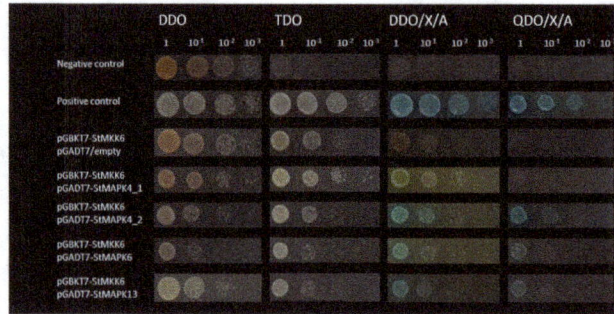

Figure 8. Yeast two-hybrid assays screening StMKK6 interaction partners. StMKK6 protein was fused with Gal DNA-binding domain as bait and StMAPK4_1, StMAPK4_2, StMAPK6 and StMAPK13 were used as prey fused with Gal DNA-activation domain. Interaction pairs P53/SV40 large T-antigen and Lam/SV40 large T-antigen were used as positive and negative controls respectively. StMKK6/empty vector pair was used as a control to discard auto-activation of bait protein. Serial dilutions of interaction pairs were plated on DDO media for co-transformation selection. Interactions of bait and prey proteins were examined by assessing growth on several selective media with different levels of restrictiveness i.e. SD/-Leu/-Trp/-His (TDO), SD/-Leu/-Trp/-x-a-Gal/Aba (DDO/X/A) and QDO/X/A. Only co-transformed colonies growing on the most restrictive media (QDO/X/A) were considered as positive interaction transformants.

inately in the nucleus as indicated also by *in silico* localisation prediction of identified target MAPKs and that its activitiy is regulated on the level of MKK6 transcription.

Several studies have focused on the gene expression profiles of St*MKK6* orthologues in other species. *A. thaliana* At*MKK6* and maize *MKK6* (Zm*MEK1*) are involved in the regulation of cytokinesis [10,58,59] and are required for lateral root formation [60]. Zm*MEK1* is also induced by polyethylene glycol, abscisic acid and SA, and it is down-regulated by NaCl [61]. In rice, Os*MEK1* (referred to as Os*MKK6* in Hamel et al. 2006 [7]) was reported to be involved in signalling of moderately low temperature stress [62,63]. The present investigation into St*MKK6* expression in different tissues of the *S. tuberosum* group Phureja shows that St*MKK6* is highly expressed in tissues with intensive cell proliferation, similar to what has been reported for its orthologues in *A. thaliana*, tobacco and maize [10,58,59,64].

After the SA treatment of the NahG-Rywal, the level of St*MKK6* transcripts three-fold increased, while in non-transgenic Rywal we detected trend of decrease in transcription of StMKK6 gene (Figure 6). Higher basal SA levels in cv. Rywal could explain different regulation of St*MKK6* after exogenous treatment. In fact, our results showed that basal levels of St*MKK6* transcripts in Rywal are at least two-fold higher than in tha SA-deficient NahG-Rywal which indicates that SA is necessary for the maintenance of basal levels of the St*MKK6*. This effect was previously reported for other SA-dependent defence-related genes in cv. NahG-Rywal [24] and cv. NahG-Desirée [23]. The network of SA signalling is complex [23,65–67] and expected to show some nonlinear behaviour. Consequently the response we see might be both time as well as concentration dependent and we will not be able to fully

predict the outcome of SA signalling before its appropriate dynamic model exisits. In line with hypothesis that St*MKK6* is regulated by SA, the analysis of the St*MKK6* promoter revealed motifs that are responsive to SA (Figure 5). To the best of our knowledge, there has been only one study linking St*MKK6* expression and SA signalling for any of the St*MKK6* orthologues (Zm*MEK1*) [61]. In *B. napus*, *MKK1*, *MKK2*, *MKK4* and *MKK9* were induced by SA, while only *MKK3* was not [52], which indicates that the majority of MKKs are targets of the SA signalling network.

The microarray results show that only *MKK3* (group B *MKK*s) responded in the opposite direction; i.e., *MKK3* was repressed after viral infection, contrary to all of the other potato *MKK*s, which were induced. Similarly, it has been shown also for *A. thaliana* and tobacco, that the MKK3 acts as a negative regulator in the JA signalling pathway and response to herbivores [11,16,68].

MAPK signalling network in hypersensitive resistance response to PVY

We identified here three downstream targets of StMKK6 in potato, StMAPK4_2 (orthologue of AtMAPK4, -11 and -12), StMAPK6 (orthologue of AtMAPK6 and tobacco SIPK) and StMAPK13 (orthologue of AtMAPK13 and tobacco NTF6/ NRK1). All of these interactions had already been confirmed in *A. thaliana* [34,69,70]. In rice, Os*MEK1* (or Os*MKK6*) interacts with OsMAPK1, -3, -5 and -6 (orthologues of AtMAPK6, -3 and - 11/4, respectively [48,63]). To date, only one MAPK interactor has been identified for the MKK6 orthologues in *Solanaceae*: in *N. tabacum*, NQK1/NtMEK1 was reported to interact with

Name	PGSC gene ID	*A. thaliana* orthologues	Microarray probe ID	Rywal			NahG-Rywal		
				1 dpi	3 dpi	6 dpi	1 dpi	3 dpi	6 dpi
StMAPK4_2	PGSC0003DMG401000057	AtMAPK4	MICRO.5536.C1	-0.58	0.67	0.52	-0.02	1.57	1.66
StMAPK6	Sotub08g010260*	AtMAPK6	MICRO.3797.C3	0.04	0.11	0.16	0.18	0.26	0.55
StMAPK13	PGSC0003DMG400025366	AtMAPK13	MICRO.1181.C1	-1.63	-1.41	0.11	-0.97	1.02	1.63

Figure 9. Gene expression pattern of confirmed StMKK6 targets in the HR response against PVY. The experimental setup is as in Figure 2. To each probe an *A. thaliana* orthologue, potato orthologue and a PGSC gene ID were assigned. Log2 of gene expression differences between PVY-infected and mock-inoculated plants are indicated for each time point (1, 3 and 6 dpi). Statistically significant differences (FDR corrected p-value <0.05) are given in bold. The star (*) for StMAPK6 PGSC gene ID indicates that the gene was not predicted in PGSC gene model. Up-regulated values are in blue and down-regulated values are in yellow.

NTF6/NRK1 (orthologue of AtMAPK13 [64]). According to this, we can conclude that the MKK6 signalling module is evolutionarily relatively stable, as most of the interactions appear to be conserved between unrelated species. There are, however, some specificities. For example, out of two potato MAPK4/11/12 paralogues StMKK6 interacts with only one of them, while in *A. thaliana* it interacts with all three, and in rice, with two.

Among the StMKK6 downstream targets, the present study shows that only St*MAPK13* expression (Figure 9) is significantly regulated in the HR response to PVY infection, although while St*MKK6* is strongly up-regulated, St*MAPK13* is strongly down-regulated. Interestingly St*MAPK4_2* and St*MAPK13* are significantly induced in SA deficient plants at the same time points as St*MKK6*. All of the target MAPKs are, however, expressed at (substantially) higher levels in both the mock-infected and PVY-infected leaves, compared to St*MKK6* (Figure S3). Therefore, we can hypothesise that the crucial regulation of St*MAPK4_2*, 6 and 13 is not at the gene expression level, but at a later step; e.g., phosphorylation by newly synthesised StMKK6, or any other post-translational process (e.g. protein translocation or degradation).

As well as St*MKK6*, we showed that St*MKK4/5* (also named as St*MEK1* in some studies) is also up-regulated in the HR response of potato to PVY, albeit to a lower extent (slightly below the strict significance cut-off), and showed a delay comparable to St*MKK6* in the NahG plants (Figure 2). In *N. tabacum*, both NtMEK2 (orthologue of StMKK4/5) and NQK1/NtMEK1 (orthologue of StMKK6) are required for N-mediated resistance against tobacco mosaic virus [71–73]. Similarly, the silencing of At*MEK1* and At*MEK2* allowed for higher amplification of *Cucumber mosaic virus* in *A. thaliana* [74]. This indicates the close connectivity and interdependency between the St*MEK1* and St*MEK2* signalling modules in plant defence.

Since an effective HR takes place in cv. Rywal in response to PVY, it is possible that some of the regulated potato MKKs are part of this response. A comprehensive analysis of the different *MKK* functions was performed in Pto-mediated resistance in tomato by Ekengren et al., 2003 [75], and Pedley and Martin, 2004 [76]. Ekengren et al. [75] showed that silencing of either *MEK1* or *MEK2* breaks the resistance against *Pseudomonas syringae*. However, Pedley and Martin [76] showed that Le*MKK2* (orthologue of At*MKK4/5*) and Le*MKK4* (orthologue of At*MKKK7/8/9*) caused programmed cell death when overexpressed, while Le*MKK1* (orthologue of At*MKK1/2*) and Le*MKK3* (orthologue of *MKK6*) did not. In *A. thaliana*, two studies have shown that as well as the MKK4/MKK5 (MEK2) module, another branch of the group A MAPKKs, AtMKK1 and AtMKK2, function together as the second MAPK module involved in the HR response [68,77]. In *N. benthamiana MKK1* (orthologue of At*MKK7/8/9*) is a potent inducer of HR-like cell death in response to *P. infestans* [11,55,78] and the same was shown also for At*MKK7* and *MKK9* in *A. thaliana* [79]. In *A. thaliana*, *MKK5* was shown to have a role in the cascade that triggers the HR response [80]. In the present study, there was strong repression of potato *MKK3*, while this dysregulation was diminished in the NahG-Rywal plants (Figure 2).

The interconnectivity between the NbMEK1 (orthologue of AtMKK6) and NbMEK2 (orthologue of AtMKK4/5) kinase signalling modules was studied mechanistically by del Pozo et al., 2004 [81]. They showed that the silencing of Nb*MEK1* abolished the cell death caused by constitutively active NbMEK2. In *N. attenuata*, *MEK1* (orthologue of St*MKK6*) and *SIPKK* (orthologue of At*MKK1/2*) are involved in the regulation of the accumulation of 12-oxo-phytodienoic acid and JA [53], which from another point, supports the role of SA in MKK6 signalling,

due to the known antagonistic cross-talk between JA and SA in defence pathways [82].

Conclusions

Our results show that StMKK6 is an important player in potato HR response against PVY infection, as shown on the gene-expression level and protein localisation. We identified potential StMKK6 downstream targets and have shown that albeit the signalling network seems to be evolutionary relatively stable the fine-tuned interdependency within the broader MAPK signalling network might be different in different plants, and when a plant is exposed to different pathogens.

Supporting Information

Figure S1 Localisation of StMKK6, under 35S promoter and native promoter, in epidermal cells of *N. benthamiana*. A. Control of transformation. Epidermal cells, transformed with plasmids containing 35S::pH7CWG2-CFP (left) and 35S::pH7YWG2-YFP (right) fusion. The fluorescence of the CFP or YFP alone (without the fusion with StMKK6) is observed only in cytoplasm. **B.** Localisation of StMKK6 fused with CFP (upper panel) or YFP (lower panel) with expression under the CaMV 35S promoter in mock-inoculated epidermal cells. The protein is localised in cytoplasm and nucleus. **C.** Localisation of StMKK6 fused with CFP (upper panel) or YFP (lower panel) with expression under the CaMV 35S promoter in PVY-inoculated epidermal cells. The protein is localised in cytoplasm and nucleus. **D.** Localisation of StMKK6 fused with YFP with expression under native promoter in mock-inoculated epidermal cells, where no fluorescence is observed. **E.** Localisation of StMKK6 fused with YFP with expression under native promoter in PVY-inoculated epidermal cells, where the protein accumulates predominantly in nucleus.

Figure S2 Phylogenetic tree of potato and *A. thaliana* MAPKs from group A and group B. Potato (Sotub, PGSC and St) and *A. thaliana* (At) MAPKs from group A (MAPK3, 6 and 10) and group B (MAPK4, 5, 11, 12 and 13) as was already proposed by Ichimura et al. 2002 [4]. Besides the potato sequences from the PGSC Browser (Sotub and PGSC) the tree also includes four MAPKs from cv. Rywal: StMAPK4_1, StMAPK4_2, StMAPK6 and StMAPK13. The sequences from *Arabidopsis* are AtMAPK3 (AT3G45640.1), AtMAPK4 (AT4G01370.1), AtMAPK5 (AT4G11330.1), AtMAPK6 (AT2G43790.1), AtMAPK10 (AT3G59790.1), AtMAPK11 (AT1G01560.1), AtMAPK12 (AT2G46070.1) and AtMAPK13 (AT1G07880.2). The scale bar indicates the branch length that corresponds to 0.06 substitutions per site.

Figure S3 Expression pattern of St*MKK6* and its confirmed targets St*MAPK4_2*, St*MAPK6* and St*MAPK13* in mock-inoculated plants. Log2 of normalized signals for mock-inoculated plants 1 day post inoculation are shown. The expression is shown for four mock treated Rywal (R) and NahG-Rywal (nah) plants. In both sets of plants the expression of the *MKK6* interacting *MAPKs* is higher than of St*MKK6*. Calculated are differences in the expression between Rywal and NahG-Rywal plants.

File S1 Table S1 in File S1. Oligonucleotides used for fragment isolation, qPCR and cloning. In primers for cloning, the underlined parts of sequences are complementary to the destination plasmids. **Table S2 in File S1. Validation of St*MKK6* microarray (μarray) results by real-time PCR (qPCR).** To validate the microarray results for St*MKK6*, its expression was analysed by qPCR in the same RNA samples as for the microarray analysis. Log_2 of ratio between virus- and mock-inoculated plants in cv. Rywal and NahG-Rywal 1, 3 and 6 days after PVY inoculation for St*MKK6* obtained by both methods are shown. Statistically significant values ($p<0.05$) are marked with bold. **Table S3 in File S1. Predicted regulatory domains of St*MKK6* native promoter from *S. tuberosum*, cv. Rywal.** The promoter sequence is 899 bp-long and analysed with PlantCare software [36]. **Table S4 in File S1. Predicted regulatory domains of St*MKK6* native promoter from *S. tuberosum*, cv. Santé.** The promoter sequence is 247 bp-long and analysed with PlantCare software [36]. **Table S5 in File S1. Expression values of St*MKK6* in control and SA-treated potatoes cv. Rywal and NahG-Rywal.** Two biological replicates per treatment were analysed. Relative expression values and fold-changes (compared to Rywal control) are shown in the table. Differences between control and SA-treated plants were statistically evaluated by t-test. **Table S6 in File S1. Comparison of St*MKK6* basal expression values in Rywal and NahG-Rywal plants.** Two biological replicates per treatment were analysed. Relative expression values and fold-changes (compared to Rywal control) are shown in the table. Differences between Rywal and NahG-Rywal plants were statistically evaluated by t-test. **Table S7 in File S1. Subcellular localisation prediction of StMKK6 and AtMKK6.** Amino acid sequence of AtMKK6 (GenBank accession number NM_125041.2) and StMKK6 (GenBank accession number KF837129.1) proteins were used as query in PredictProtein service. The localisation for each was predicted by three different prediction algorithms: PROSITE, LOCkey and LOCtree. For each subcellular prediction a confidence to the prediction is given.

File S2   Gene expression of potato *MKK* family in different tissues and after several stress treatments. For each *MKK*, the RNA expression signals (FPKM values), based on the RNA sequencing of double monoploid *Solanum tuberosum* Group Phureja clone DM1-3 (DM) [40] and Genome sequence and analysis of the tuber crop potato [30] were collected in the Potato eFP browser [39]. **A.** Tissue expression of MKK gene family in Clone DM and Clone RH. In bold are the expression values that are more than 2-times different from the average expression in all the organs. **B.** Expression of *MKK* gene family after several stress treatments. In bold are the expression values that are more than 2-times different from the controls.

File S3   St*MKK6* orthologs with corresponding UniProt IDs. List of St*MKK6* orthologues across plant kingdom. Listed are the Uniprot IDs for the proteins, names of the proteins as in the phylogenetic tree (Figure 3) and the organism the gene was originally isolated from.

File S4   Expression of genes co-regulated with St*MKK6* and their regulatory domains in promoter regions. Cultivar Rywal (HR response, conferred by Ny-1 gene) and NahG-Rywal (impaired accumulation of SA) were analysed for whole transcriptome response 1, 3 and 6 dpi after PVY infection [24]. Log2 fold changes of PVY in infected vs. mock inoculated plants are indicated for each time point (1, 3 and 6 dpi). Statistically significant differences (FDR corrected $p<0.05$) are in bold. Higher expression values are blue and lower are yellow. The promoter sequences were obtained from the *Solanum_tuberosum*-PGSC_DM_v34_superscaffolds database. 1000 bp-long promoter sequences upstream of the gene were analysed by PlantCARE web service [36]. Biotic stress-related domains are coloured and the colours are explained in the legend on the right. (XLS)

Acknowledgments

The authors would like to acknowledge Tina Demšar, Katja Stare, Tanja Guček and Klavdija Mohorič for all of their help and Chris Berrie for language revision. For the use of the confocal microscope, we thank the Laboratory of Biotechnology of the National Institute of Chemistry, Ljubljana, Slovenia.

Author Contributions

Conceived and designed the experiments: AL AC JŽ KG. Performed the experiments: AL AC DD. Analyzed the data: AL ŠB. Contributed to the writing of the manuscript: AL AC ŠB KG. Contribution to yeast two-hybrid experiments: ABZ.

References

1. Meng X, Zhang S (2013) MAPK cascades in plant disease resistance signaling. Annu Rev Phytopathol 51: 245–266.
2. Rodriguez MCS, Petersen M, Mundy J (2010) Mitogen-activated protein kinase signaling in plants. Annu Rev Plant Biol 61: 621–649.
3. Pitzschke A, Hirt H (2009) Disentangling the complexity of mitogen-activated protein kinases and reactive oxygen species signaling. Plant Physiol 149: 606–615.
4. Ichimura K, Shinozaki K, Tena G, Sheen J (2002) Mitogen-activated protein kinase cascades in plants: a new nomenclature. Trends Plant Sci 7: 301–308.
5. Andreasson E, Ellis B (2010) Convergence and specificity in the Arabidopsis MAPK nexus. Trends Plant Sci 15: 106–113.
6. Rasmussen MW, Roux M, Petersen M, Mundy J (2012) MAP kinase cascades in Arabidopsis innate immunity. Front Plant Sci 3: 169.
7. Hamel L-P, Nicole M-C, Sritubtim S, Morency M-J, Ellis M, et al. (2006) Ancient signals: comparative genomics of plant MAPK and MAPKK gene families. Trends Plant Sci 11: 192–198.
8. Qiu J-L, Zhou L, Yun BW, Nielsen HB, Fiil BK, et al. (2008) Arabidopsis mitogen-activated protein kinase kinases MKK1 and MKK2 have overlapping functions in defense signaling mediated by MEKK1, MPK4, and MKS1. Plant Physiol 148: 212–222.
9. Teige M, Scheikl E, Eulgem T, Dóczi R, Ichimura K, et al. (2004) The MKK2 pathway mediates cold and salt stress signaling in Arabidopsis. Mol Cell 15: 141–152.
10. Takahashi Y, Soyano T, Kosetsu K, Sasabe M, Machida Y (2010) HINKEL kinesin, ANP MAPKKKs and MKK6/ANQ MAPKK, which phosphorylates

and activates MPK4 MAPK, constitute a pathway that is required for cytokinesis in Arabidopsis thaliana. Plant Cell Physiol 51: 1766–1776.
11. Takahashi F, Yoshida R, Ichimura K, Mizoguchi T, Seo S, et al. (2007) The mitogen-activated protein kinase cascade MKK3-MPK6 is an important part of the jasmonate signal transduction pathway in Arabidopsis. Plant Cell 19: 805–818.
12. Dóczi R, Brader G, Pettkó-Szandtner A, Rajh I, Djamei A, et al. (2007) The Arabidopsis mitogen-activated protein kinase kinase MKK3 is upstream of group C mitogen-activated protein kinases and participates in pathogen signaling. Plant Cell 19: 3266–3279.
13. Hwa CM, Yang XC (2008) The AtMKK3 pathway mediates ABA and salt signaling in Arabidopsis. Acta Physiol Plant 30: 277–286.
14. Cho SK, Larue CT, Chevalier D, Wang H, Jinn TL, et al. (2008) Regulation of floral organ abscission in Arabidopsis thaliana. Proc Natl Acad Sci U S A 105: 15629–15634.
15. Asai T, Tena G, Plotnikova J, Willmann MR, Chiu WL, et al. (2002) MAP kinase signalling cascade in Arabidopsis innate immunity. Nature 415: 977–983.
16. Wang H, Ngwenyama N, Liu Y, Walker JC, Zhang S (2007) Stomatal development and patterning are regulated by environmentally responsive mitogen-activated protein kinases in Arabidopsis. Plant Cell 19: 63–73.
17. Yoo S-D, Cho YH, Tena G, Xiong Y, Sheen J (2008) Dual control of nuclear EIN3 by bifurcate MAPK cascades in C2H4 signalling. Nature 451: 789–795.
18. Zhou C, Cai Z, Guo Y, Gan S (2009) An arabidopsis mitogen-activated protein kinase cascade, MKK9-MPK6, plays a role in leaf senescence. Plant Physiol 150: 167–177.

19. Zhang X, Dai Y, Xiong Y, DeFraia C, Li J, et al. (2007) Overexpression of Arabidopsis MAP kinase kinase 7 leads to activation of plant basal and systemic acquired resistance. Plant J 52: 1066–1079.
20. Naseem M, Dandekar T (2012) The role of auxin-cytokinin antagonism in plant-pathogen interactions. PLoS Pathog 8: e1003026.
21. Lewsey M, Palukaitis P, Carr JP (2009) Plant-virus interactions: defence and counter-defence. In: Parker J, editor. Annual Plant Reviews. Oxford: Wiley-Blackwell, Vol. 34. 134–176.
22. Vlot a C, Dempsey DA, Klessig DF (2009) Salicylic acid, a multifaceted hormone to combat disease. Annu Rev Phytopathol 47: 177–206.
23. Baebler Š, Stare K, Kovač M, Blejec A, Prezelj N, et al. (2011) Dynamics of responses in compatible potato-Potato virus Y interaction are modulated by salicylic acid. PLoS One 6: e29009.
24. Baebler Š, Witek K, Petek M, Stare K, Tušek-Žnidarič M, et al. (2014) Salicylic acid is an indispensable component of the Ny-1 resistance-gene-mediated response against Potato virus Y infection in potato. J Exp Bot 65: 1095–1109.
25. Singh DP, Moore CA, Gilliland A, Carr JP (2004) Activation of multiple antiviral defence mechanisms by salicylic acid. Mol Plant Pathol 5: 57–63.
26. Mur LAJ, Kenton P, Lloyd AJ, Ougham H, Prats E (2008) The hypersensitive response; the centenary is upon us but how much do we know? J Exp Bot 59: 501–520.
27. Kogovšek P, Ravnikar M (2013) Physiology of the potato-Potato virus Y interaction. In: Lüttge U, Beyschlag W, Francis D, Cushman J, editors. Progress in Botany. Berlin, Heidelberg: Springer Berlin Heidelberg, Vol. 74. 101–133.
28. Szajko K, Chrzanowska M, Witek K, Strzelczyk-Zyta D, Zagórska H, et al. (2008) The novel gene Ny-1 on potato chromosome IX confers hypersensitive resistance to Potato virus Y and is an alternative to Ry genes in potato breeding for PVY resistance. Theor Appl Genet 116: 297–303.
29. Altschul SF, Gish W, Miller W, Myers EW, Lipman DJ (1990) Basic local alignment search tool. J Mol Biol 215: 403–410.
30. Potato Genome Sequencing Consortium (2011) Genome sequence and analysis of the tuber crop potato. Nature 475: 189–195.
31. Katoh K, Standley DM (2013) MAFFT multiple sequence alignment software version 7: improvements in performance and usability. Mol Biol Evol 30: 772–780.
32. Tamura K, Peterson D, Peterson N, Stecher G, Nei M, et al. (2011) MEGA5: molecular evolutionary genetics analysis using maximum likelihood, evolutionary distance, and maximum parsimony methods. Mol Biol Evol 28: 2731–2739.
33. Saito N, Nei M (1987) The neighbour-joining method: A new method for reconstructing phylogenetic trees. Mol Biol Evol 4: 406–425.
34. Arabidopsis Interactome Mapping Consortium (2011) Evidence for network evolution in an Arabidopsis interactome map. Science 333: 601–607.
35. Ramšak Ž, Baebler Š, Rotter A, Korbar M, Mozetič I, et al. (2014) GoMapMan: integration, consolidation and visualization of plant gene annotations within the MapMan ontology. Nucleic Acids Res 42: D1167–75.
36. Lescot M, Déhais P, Thijs G, Marchal K, Moreau Y, et al. (2002) PlantCARE, a database of plant cis-acting regulatory elements and a portal to tools for in silico analysis of promoter sequences. Nucleic Acids Res 30: 325–327.
37. Rost B, Yachdav G, Liu J (2004) The PredictProtein server. Nucleic Acids Res 32: W321–6.
38. Tanz SK, Castleden I, Hooper CM, Vacher M, Small I, et al. (2013) SUBA3: a database for integrating experimentation and prediction to define the SUBcellular location of proteins in Arabidopsis. Nucleic Acids Res 41: D1185–91.
39. Winter D, Vinegar B, Nahal H, Ammar R, Wilson G V, et al. (2007) An "Electronic Fluorescent Pictograph" browser for exploring and analyzing large-scale biological data sets. PLoS One 2: e718.
40. Massa AN, Childs KL, Lin H, Bryan GJ, Giuliano G, et al. (2011) The transcriptome of the reference potato genome Solanum tuberosum Group Phureja clone DM1-3 516R44. PLoS One 6: e26801.
41. Hruz T, Laule O, Szabo G, Wessendorp F, Bleuler S, et al. (2008) Genevestigator v3: a reference expression database for the meta-analysis of transcriptomes. Adv Bioinformatics 2008: 420747.
42. Theocharidis A, van Dongen S, Enright AJ, Freeman TC (2009) Network visualization and analysis of gene expression data using BioLayout Express(3D). Nat Protoc 4: 1535–1550.
43. Rasmussen R (2001) Quantification on the LightCycler. In: Meuer S, Wittwer C, Nakagawara K, editors. Rapid Cycle Real-Time PCR: Methods and Applications, Heidelberg: Springer Berlin Heidelberg, 21–34.
44. Weller SA, Elphinstone JG, Smith NC, Boonham N, Stead DE (2000) Detection of Ralstonia solanacearum strains with a Quantitative, Multiplex, Real-Time, Fluorogenic PCR (TaqMan) assay. Appl Environ Microbiol 66: 2853–2858.
45. Baebler Š, Krečič-Stres H, Rotter A, Kogovšek P, Cankar K, et al. (2009) PVY NTN elicits a diverse gene expression response in different potato genotypes in the first 12 h after inoculation. Mol Plant Pathol 10: 263–275.
46. Karimi M, De Meyer B, Hilson P (2005) Modular cloning in plant cells. Trends Plant Sci 10: 103–105.
47. Pohleven J, Obermajer N, Sabotič J, Anžlovar S, Sepcić K, et al. (2009) Purification, characterization and cloning of a ricin B-like lectin from mushroom Clitocybe nebularis with antiproliferative activity against human leukemic T cells. Biochim Biophys Acta 1790: 173–181.
48. Singh R, Jwa NS (2013) The rice MAPKK-MAPK interactome: the biological significance of MAPK components in hormone signal transduction. Plant Cell Rep 32: 923–931.
49. Chen L, Hu W, Tan S, Wang M, Ma Z, et al. (2012) Genome-wide identification and analysis of MAPK and MAPKK gene families in Brachypodium distachyon. PLoS One 7: e46744.
50. Nicole MC, Hamel LP, Morency MJ, Beaudoin N, Ellis BE, et al. (2006) MAPping genomic organization and organ-specific expression profiles of poplar MAP kinases and MAP kinase kinases. BMC Genomics 7: 223.
51. Zhang S, Xu R, Luo X, Jiang Z, Shu H (2013) Genome-wide identification and expression analysis of MAPK and MAPKK gene family in Malus domestica. Gene 531: 377–387.
52. Liang W, Yang B, Yu BJ, Zhou Z, Li C, et al. (2013) Identification and analysis of MKK and MPK gene families in canola (Brassica napus L.). BMC Genomics 14: 392.
53. Heinrich M, Baldwin IT, Wu J (2011) Three MAPK kinases, MEK1, SIPKK, and NPK2, are not involved in activation of SIPK after wounding and herbivore feeding but important for accumulation of trypsin proteinase inhibitors. Plant Mol Biol Report 30: 731–740.
54. Dóczi R, Okrész L, Romero AE, Paccanaro A, Bögre L (2012) Exploring the evolutionary path of plant MAPK networks. Trends Plant Sci 17: 518–525.
55. Takahashi Y, Nasir KH Bin, Ito A, Kanzaki H, Matsumura H, et al. (2007) A high-throughput screen of cell-death-inducing factors in Nicotiana benthamiana identifies a novel MAPKK that mediates INF1-induced cell death signaling and non-host resistance to Pseudomonas cichorii. Plant J 49: 1030–1040.
56. Yoshihiro T, Nasir KH Bin, Ito A, Kanzaki H, Matsumura H, et al. (2007) A novel MAPKK involved in cell death and defense dignaling. Plant Signal Behav 2: 396–398.
57. Piston DW, Patterson, George H. Lippincott-Schwartz J, Claxton, Nathan S. Davidson MW (n.d.) Imaging parameters for fluorescent proteins.
58. Hardin SC, Wolniak SM (1998) Molecular cloning and characterization of maize ZmMEK1, a protein kinase with a catalytic domain homologous to mitogen- and stress-activated protein kinase kinases. Planta 206: 577–584.
59. Hardin SC, Wolniak SM (2001) Expression of the mitogen-activated protein kinase kinase ZmMEK1 in the primary root of maize. Planta 213: 916–926.
60. Zeng Q, Sritubtim S, Ellis BE (2011) AtMKK6 and AtMPK13 are required for lateral root formation in Arabidopsis. Plant Signal Behav 6: 1436–1439.
61. Liu Y, Zhou Y, Liu L, Sun L, Zhang M, et al. (2012) Maize ZmMEK1 is a single-copy gene. Mol Biol Rep 39: 2957–2966.
62. Wen J, Oono K, Imai R, Osmek R (2002) Two novel mitogen-activated protein signaling components, OsMEK1 and OsMAP1, are involved in a moderate low-temperature signaling pathway in rice. Plant Physiol 129: 1880–1891.
63. Xie G, Kato H, Imai R (2012) Biochemical identification of the OsMKK6-OsMPK3 signalling pathway for chilling stress tolerance in rice. Biochem J 443: 95–102.
64. Soyano T, Nishihama R, Morikiyo K, Ishikawa M, Machida Y (2003) NQK1/NtMEK1 is a MAPKK that acts in the NPK1 MAPKKK-mediated MAPK cascade and is required for plant cytokinesis. Genes Dev 17: 1055–1067.
65. Naseem M, Philippi N, Hussain A, Wangorsch G, Ahmed N, et al. (2012) Integrated systems view on networking by hormones in Arabidopsis immunity reveals multiple crosstalk for cytokinin. Plant Cell 24: 1793–1814.
66. Glazebrook J, Chen W, Estes B, Chang H-S, Nawrath C, et al. (2003) Topology of the network integrating salicylate and jasmonate signal transduction derived from global expression phenotyping. Plant J 34: 217–228.
67. Miljkovic D, Stare T, Mozetič I, Podpečan V, Petek M, et al. (2012) Signalling network construction for modelling plant defence response. PLoS One 7: e51822.
68. Meng X, Xu J, He Y, Yang K-Y, Mordorski B, et al. (2013) Phosphorylation of an ERF transcription factor by Arabidopsis MPK3/MPK6 regulates plant defense gene induction and fungal resistance. Plant Cell 25: 1126–1142.
69. Melikant B, Giuliani C, Halbmayer-Watzina S, Limmongkon A, Heberle-Bors E, et al. (2004) The Arabidopsis thaliana MEK AtMKK6 activates the MAP kinase AtMPK13. FEBS Lett 576: 5–8. Available: http://www.ncbi.nlm.nih.gov/pubmed/15474000.
70. Lin W-Y, Matsuoka D, Sasayama D, Nanmori T (2010) A splice variant of Arabidopsis mitogen-activated protein kinase and its regulatory function in the MKK6–MPK13 pathway. Plant Sci 178: 245–250.
71. Liu Y, Schiff M, Dinesh-Kumar SP (2004) Involvement of MEK1 MAPKK, NTF6 MAPK, WRKY/MYB transcription factors, COI1 and CTR1 in N-mediated resistance to tobacco mosaic virus. Plant J 38: 800–809.
72. Liu Y, Ren D, Pike S, Pallardy S, Gassmann W, et al. (2007) Chloroplast-generated reactive oxygen species are involved in hypersensitive response-like cell death mediated by a mitogen-activated protein kinase cascade. Plant J 51: 941–954.
73. Jin H, Liu Y, Yang K-Y, Kim C, Baker B, et al. (2003) Function of a mitogen-activated protein kinase pathway in N gene-mediated resistance in tobacco. Plant J 33: 719–731.
74. Shang J, Xi D-H, Xu F, Wang S-D, Cao S, et al. (2011) A broad-spectrum, efficient and nontransgenic approach to control plant viruses by application of salicylic acid and jasmonic acid. Planta 233: 299–308.
75. Ekengren SK, Liu Y, Schiff M, Dinesh-Kumar SP, Martin GB (2003) Two MAPK cascades, NPR1, and TGA transcription factors play a role in Pto-mediated disease resistance in tomato. Plant J 36: 905–917.
76. Pedley KF, Martin GB (2004) Identification of MAPKs and their possible MAPK kinase activators involved in the Pto-mediated defense response of tomato. J Biol Chem 279: 49229–49235.

77. Gao M, Liu J, Bi D, Zhang Z, Cheng F, et al. (2008) MEKK1, MKK1/MKK2 and MPK4 function together in a mitogen-activated protein kinase cascade to regulate innate immunity in plants. Cell Res 18: 1190–1198.

78. Asai S, Ohta K, Yoshioka H (2008) MAPK signaling regulates nitric oxide and NADPH oxidase-dependent oxidative bursts in Nicotiana benthamiana. Plant Cell 20: 1390–1406. doi:10.1105/tpc.107.055855.

79. Popescu SC, Popescu G V, Bachan S, Zhang Z, Gerstein M, et al. (2009) MAPK target networks in Arabidopsis thaliana revealed using functional protein microarrays. Genes Dev 23: 80–92.

80. Liu H, Wang Y, Xu J, Su T, Liu G, et al. (2008) Ethylene signaling is required for the acceleration of cell death induced by the activation of AtMEK5 in Arabidopsis. Cell Res 18: 422–432.

81. Del Pozo O, Pedley KF, Martin GB (2004) MAPKKKalpha is a positive regulator of cell death associated with both plant immunity and disease. EMBO J 23: 3072–3082.

82. Gimenez-Ibanez S, Solano R (2013) Nuclear jasmonate and salicylate signaling and crosstalk in defense against pathogens. Front Plant Sci 4: 72.

10

Using Crowdsourcing to Evaluate Published Scientific Literature: Methods and Example

Andrew W. Brown*, David B. Allison

Office of Energetics, Nutrition Obesity Research Center, School of Public Health, University of Alabama at Birmingham, Birmingham, Alabama, United States of America

Abstract

Systematically evaluating scientific literature is a time consuming endeavor that requires hours of coding and rating. Here, we describe a method to distribute these tasks across a large group through online crowdsourcing. Using Amazon's Mechanical Turk, crowdsourced workers (microworkers) completed four groups of tasks to evaluate the question, "Do nutrition-obesity studies with conclusions concordant with popular opinion receive more attention in the scientific community than do those that are discordant?" 1) Microworkers who passed a qualification test (19% passed) evaluated abstracts to determine if they were about human studies investigating nutrition and obesity. Agreement between the first two raters' conclusions was moderate ($\kappa = 0.586$), with consensus being reached in 96% of abstracts. 2) Microworkers iteratively synthesized free-text answers describing the studied foods into one coherent term. Approximately 84% of foods were agreed upon, with only 4 and 8% of ratings failing manual review in different steps. 3) Microworkers were asked to rate the perceived obesogenicity of the synthesized food terms. Over 99% of responses were complete and usable, and opinions of the microworkers qualitatively matched the authors' expert expectations (e.g., sugar-sweetened beverages were thought to cause obesity and fruits and vegetables were thought to prevent obesity). 4) Microworkers extracted citation counts for each paper through Google Scholar. Microworkers reached consensus or unanimous agreement for all successful searches. To answer the example question, data were aggregated and analyzed, and showed no significant association between popular opinion and attention the paper received as measured by Scimago Journal Rank and citation counts. Direct microworker costs totaled $221.75, (estimated cost at minimum wage: $312.61). We discuss important points to consider to ensure good quality control and appropriate pay for microworkers. With good reliability and low cost, crowdsourcing has potential to evaluate published literature in a cost-effective, quick, and reliable manner using existing, easily accessible resources.

Editor: Vincent Larivière, Université de Montréal, Canada

Funding: This project was supported by National Institutes of Health (http://nih.gov) grants NIDDK P30DK056336 and NIDDK T32DK62710-01A1. The funders had no role in study design, data collection and analysis, decision to publish, or preparation of the manuscript.

Competing Interests: The authors have declared that no competing interests exist.

* Email: awbrown@uab.edu

Introduction

The evaluation of published research literature requires human intelligence to complete (so called Human Intelligence Tasks, or HITs) and is therefore difficult or impossible to automate by computer. Yet, many of the individual tasks that make up a typical evaluation (e.g., extraction of key information, coding of qualitative data) may be amenable to crowdsourcing, the practice of dividing a large task across an often disjointed group of individuals. Crowdsourcing has been used in other contexts to evaluate such complex topics as codifying sleep patterns [1], classifying biomedical topics [2], and creating speech and language data [3]. Herein, our primary purpose is to examine the feasibility of using crowdsourcing to evaluate scientific literature. To do so, we use the following question as an example: "Do nutrition-obesity studies with conclusions concordant with popular opinion receive more attention in the scientific community than do those that are discordant?" If conclusions that are concordant with popular opinion are afforded more attention by the scientific community (i.e., higher citation counts or published in a higher-ranked journal), it may be evidence of distortion of the scientific record. This question requires four kinds of HITs: 1) categorization of

study abstracts; 2) iterative refinement of free-text responses; 3) eliciting subjective opinions; and 4) a simple data extraction of citation counts. We will first describe the crowdsourcing platform, followed by discussing each of these HITs, and then use the data to examine the example question. We conclude the narrative with additional discussion and remarks about crowdsourcing, followed by a more in depth description of the methodology.

Method Narrative and Evaluation

We chose to use Amazon's Mechanical Turk (MTurk; www.mturk.com) crowdsourcing platform. MTurk is a flexible, online, task-based, microwork marketplace in which crowdsourcing workers (microworkers) can choose to accept tasks posted by requesters (in this case, AWB and DBA) for a fee defined by the requester. The requester can set qualification criteria (e.g., self-reported age and location; MTurk aggregated work history; custom-defined qualifications), and choose whether to accept or reject work after it has been submitted before paying. Freely available MTurk interfaces allow the creation of common digital question formats (e.g., list box, multiple choice, free-text) with more formats available through custom coding. Considerations for each step in our workflow (Fig. 1) include: 1) how involved our non-crowdsourced research team (AWB and DBA) would be in evaluating questions; 2) how qualified each microworker needed to be; 3) how to phrase questions and tasks such that HITs could be completed by the targeted microworkers; and 4) how many times each question needed to be evaluated to expect reliable responses.

The first set of HITs (Fig. 1A) was designed to examine the feasibility of employing microworkers to categorize study abstracts. Microworkers were asked to rate 689 abstracts about nutrition and obesity to determine: 1) if the study was about humans, thereby excluding animal studies, *in vitro* studies, commentaries, and reviews (a multiple choice question); 2) whether the study investigated the relation of foods or beverages (hereafter referred to as 'food' for simplicity) with obesity (a multiple choice question); 3) which food was described in the paper (a free-text question); and 4) the conclusions about the relation between the food and obesity

(a multiple choice question). This also limited studies to those comparing one food along a range of exposures or to a control. Microworkers were only allowed to complete these HITs if they satisfied two types of qualification. First, built-in MTurk Qualifications selected only microworkers that were United States residents and at least 18 years old. Second, a custom MTurk Qualification required microworkers to correctly categorize 3 example abstracts (Fig. 1Q and Fig. 2Q).

Agreement was reached in 529 of the 689 abstracts after two microworker assessments (77%, in Fig. 2A). Interrater agreement was moderate from these first two ratings [4] ($\kappa = 0.528$ when all free-text conclusions were categorized as 'other;' $\kappa = 0.586$ when excluding free-text answers). An additional 17% reached consensus after a third rater, and 2% more reached consensus when free-text answers ("Other") were reviewed by AWB. Only 4% of abstracts had to be rated by AWB because of lack of consensus,

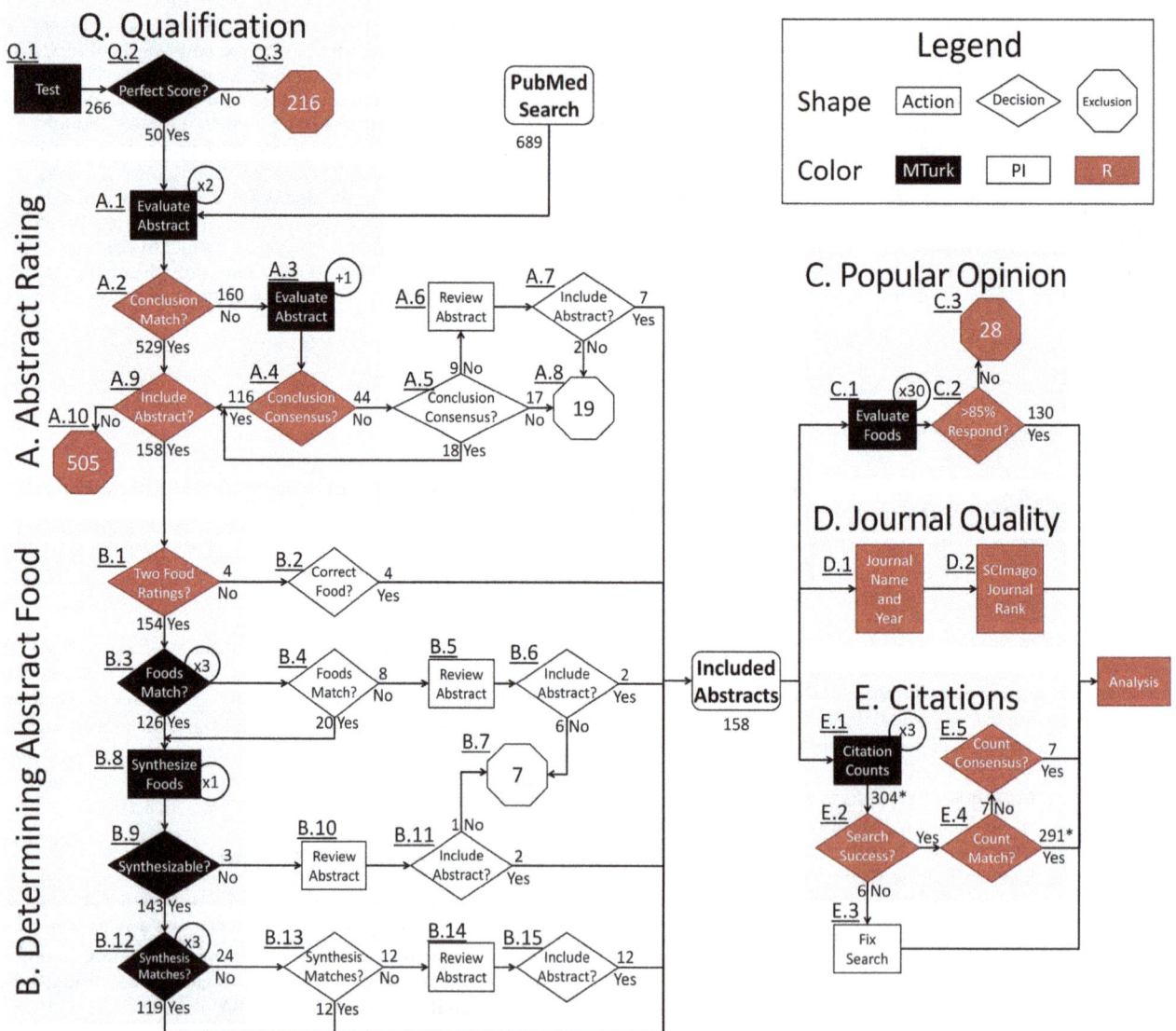

Figure 1. Flow chart of crowdsourcing procedures. Steps are separated into sections based on separate tasks. The more granular of a process that can be made, the more amenable the process is to crowdsourcing. MTurk: process completed on MTurk. PI: process completed by AWB. R: process completed with custom R scripts. Gray boxes include tasks automated through R or MTurk, while steps outside of the gray boxes were manually completed. Circled numbers represent the number of times a task was completed. In box E, 304 preliminarily included abstracts (including the 158 ultimately included) had citation counts extracted while the final abstract ratings were concurrently being completed from box A.

Q. Qualification Test
(n=266)
13%
6%
81%
■ Continued ■ Passed ■ Failed

B.12 Synthesis Matches
(n=143)
8%
9%
83%
■ Consensus ■ MR Yes ■ MR No

A. Abstract Rating
(n=689)
2% 4%
17%
77%
■ Two ■ Three ■ "Other" ■ MR

C.1 Popular Opinion
(n=4740)
0.2%
99.8%
■ Complete ■ Incomplete

Tasks (n)
0 5 10 15 20 25 30 35
Hourly pay/task (USD)

Tasks (n)
0 50 100 150 200 250 300
Completion time (s)

C.2 Known Foods
(n=158)
18%
82%
■ Known ■ Unknown

B.3 Foods Match
(n=154)
11% 4%
85%
■ Crowdsourcing ■ Review ■ Fail

E. Citation Counts
(n=304)
2% 2%
96%
■ Match ■ Consensus ■ Error

Figure 2. Performance metrics of microworkers. In each chart, green = desirable, yellow = acceptable, red = undesirable, blue = issue external to crowdsourcing, and black = failure. The letters preceding each chart title corresponds to the steps in Fig. 1, and the number in parentheses represents the number of units compared. **A)** Two: consensus after two ratings; Three: consensus after three ratings; "Other": consensus after reviewing free-text answers; MR: manually reviewed by AWB. The accompanying histograms show the distributions of hourly pay calculated by completion time, and completion time in seconds for each task. **B.3)** Review: microworkers did not agree the foods matched, AWB reviewed the foods extracted from abstracts, and they matched; Fail: the foods did not match after review from AWB. **B.12)** MR Yes: AWB determined the synthesized term matched the original two foods; MR No: AWB determined the synthesized term did not match the original two foods. **C.1)** Popular opinion questions were either completely answered or not. **C.2)** Known: less than 15% did not know the food; Unknown: greater than 15% did not know the food. **E)** Match: 3 ratings matched; Consensus: 2 of 3 matched; Error: the link provided to the microworkers was faulty.

and only 7 of the 158 abstracts coded to be included (Step A.9) were subsequently rejected in other steps (4% of the 158). The median abstract evaluation time was 38 seconds, which resulted in an estimated median wage of \$6.63/hr USD as calculated from task completion time (Fig. 2A histograms).

The next set of three HITs investigated iterative refinement of free-text answers by synthesizing the free-text food-topic answers into one coherent term (Steps B.1-B.15). Three microworkers confirmed that the two food-topics reported for each abstract (e.g., "Pistachios" and "pistachio nuts") referred to the same food (Fig. 2B.3). For topics that at least two of three microworkers said matched, a separate HIT was posted to synthesize the two foods into one coherent phrase (e.g., "pistachios"). New phrases were compared to the original phrases by two additional microworkers to confirm that they were a reasonable synthesis of the original phrases (Fig. 2B.12). If the phrases failed to match or be synthesized at any point during this quality control, AWB reviewed them (Steps B.2, B.4, and B.13), and if need-be he directly reviewed the abstract (Steps B.5, B.10, and B.14). Each assignment was awarded \$0.01 USD.

In addition to categorization and iterative synthesis, subjective opinions can also be elicited from microworkers. Microworkers were asked to rate the perceived obesogenicity of the foods synthesized above (Step C.1) and to predict the overall US perception of the food using a 7 point, horizontal, multiple choice scale from "prevents obesity" to "causes obesity" (e.g., x-axis in Fig. 3), with an option to indicate they did not know a food. Microworkers were allowed to respond to all 158 foods, but only once for each food. Thirty responses were collected for each food at \$0.01 USD each. Only 8 obesogenicity responses were unusable out of 4740 (<1%; Fig. 2C.1). These survey results are only meant to reflect an estimate of popular opinion, and do not necessarily reflect the actual obesogenicity of any given food.

Qualitative and quantitative review of the opinion answers supports internal and external consistency of microworker opinions. Most foods were known by the microworkers (Fig. 2C.2). Foods categorized as "unknown" (>15% of respondents did not know the food; Fig. S1 and Table S1) included conjugated linoleic acid (n = 5), references to glycemic index (n = 3), and specific foods such as mangosteen juice, Chungkook-jang, and "Street food in Palermo, Italy;" none of these foods would be expected to be widely known in the general US population. Also, "known" and "unknown" foods comprised mutually exclusive lists. In addition, microworkers' predictions of US opinions track well with the aggregated individual opinions (Fig. 3). Finally, popular opinions seem to match our expert expectations: fruits and vegetables were rated as preventing obesity, with less certainty about fried vegetables or fruit juice, and sweetened beverages were rated as causing obesity, with artificial sweeteners considered to be less obesogenic.

A simple data extraction task was also presented to microworkers. Google Scholar search links were presented to microworkers and microworkers reported the number of times the paper was cited. Each of three ratings was awarded \$0.01 USD. Only 7

papers did not have identical reported citations counts among the 304 abstracts searched (2%; Fig. 2E). All 7 of these papers had two of three responses matching, which matched a manual review by AWB.

To answer the question initially posed, the data extracted from these steps were synthesized with a measurement of journal quality (ScImago Journal Rank, SJR). As expected, the number of citations a paper received increased with journal quality (15.13 ± 2.66 citations per unit increase of SJR, p = 6.4196e-8) and time since publication (10.26 ± 1.32 citations per year since publication, p = 1.0187e-12). No models that tested whether opinion-conclusion concordance was associated with scientific attention were statistically significant. Specifically: log citation counts were not predicted by a continuous measurement of concordance when controlling for publication date and SJR (1.09 per unit concordance increase; 95% CI: 0.96,1.24; p = 0.2021), nor was log SJR predicted by continuous concordance (0.96 per unit concordance increase; 95% CI: 0.86,1.08; p = 0.5218); and log citation counts were not different between categorically concordant versus discordant conclusions when controlling for publication date and SJR (1.02; 95% CI: 0.78,1.33; p = 0.8823), nor was log SJR different (0.87; 95% CI: 0.69,1.10; p = 0.2454).

Discussion

The above narrative describes the potential and feasibility of using crowdsourcing to evaluate published literature. Several important considerations must be made regarding the implementation of crowdsourcing for literature analysis or any other complex task, some of which are described in more detail in the methods section below. First and foremost, using crowdsourcing does not preclude the need for the quality control checks that would be implemented in any other data extraction or survey methodology. Indeed, when highly technical information is being acquired through crowdsourcing, quality checks are essential to not only guarantee the validity of the results but also to reassure the reader that competent individuals have assembled the information. It is also important to note that microworkers self-certify their age and location, which is a common limitation of any self-reported results. Using HITs stored outside of MTurk can provide the potential for IP address confirmation, which is a more reliable confirmation of location; we did not use such methodology herein. Investigators must determine what level of confidence they need in the identities of respondents. For technical tasks, such as literature evaluation, qualification tests like the ones we employed can be used to help guarantee competency, which in this example was more important than identity. Finally, it is important for investigators to consider the desired scope of the literature evaluation. In this case-study, our inclusion/exclusion methods may have resulted in papers being excluded by chance early that should have been included, but the final corpus should not have included papers that should have been excluded. It was more important for our example that the final corpus only include studies of interest than to have an exhaustive inclusion; this is

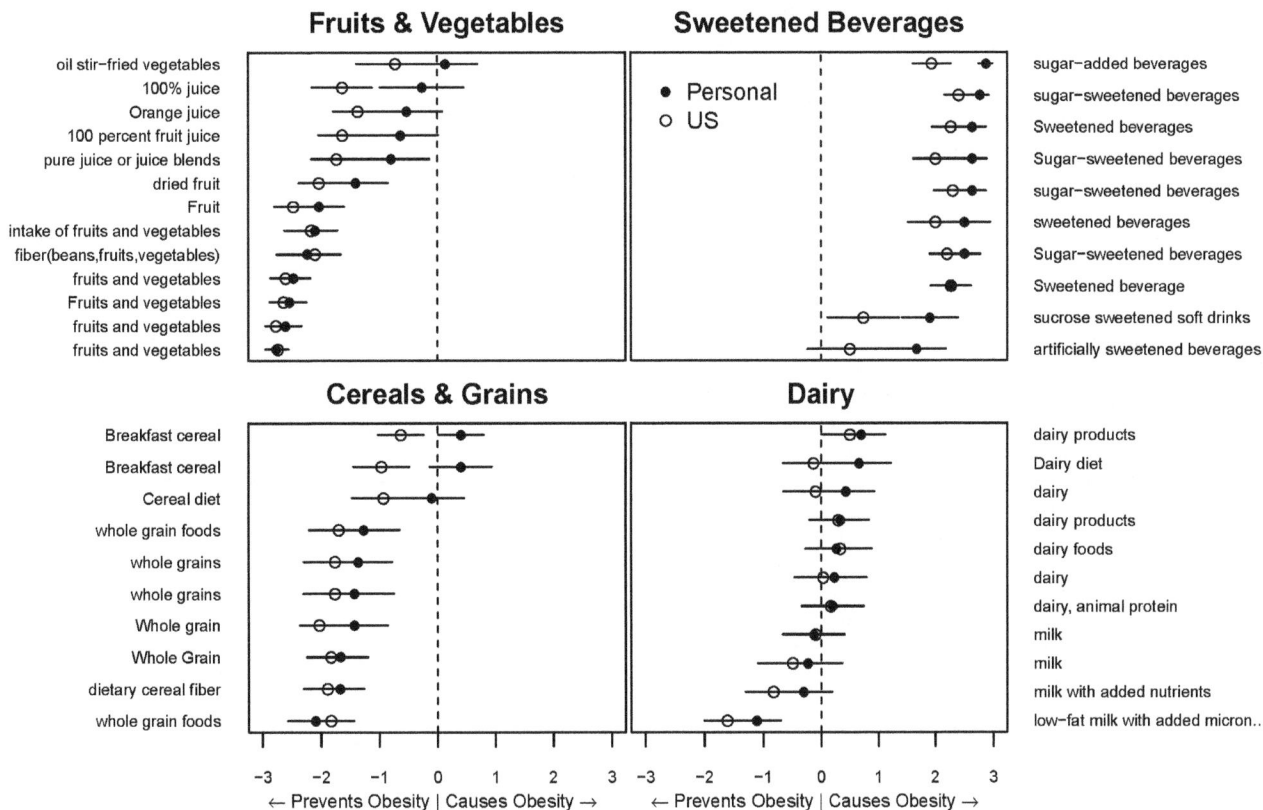

Figure 3. Example of groups of foods and average obesogenicity ratings. Each food listed on the y-axes is shown as synthesized by microworkers in Fig. 1 Step B.9. Foods are ordered within each panel top to bottom with the personal opinion from most to least obesogenic. The vertical dotted line represents the transition from popular opinion indicating the food prevents obesity (left) to causes obesity (right). Each point represents the mean ±95% CI. Note that these results are only meant to reflect an estimate of popular opinion, and do not necessarily reflect the actual obesogenicity of any given food.

clearly not the case for someone interested in conducting a comprehensive systematic review, and thus other controls would need to be included in such cases.

There is also an upfront cost for investigators to become familiar with the infrastructure that is not calculated into the cost of microworkers reported above. For laboratories with sufficient programming knowledge (e.g., SOAP and REST requests, HTML, Java, XML, or command line), adapting the existing MTurk infrastructure should be fairly straightforward. Similarly, for simple data-extraction tasks where an objective, easy-to-understand answer to the task exists (in our example, citation counts), applying quality control to the results is also straightforward. In other circumstances, researchers must weigh the time it takes to become familiar with and implement a new method versus the time it takes to complete the tasks with existing methodology. Time and cost savings will depend on the difficulty of the task, the difficulty in validating microworker responses, the size of the task, and the prospects of continued use of the methods by the research group. We are presently working to address some of the barriers other groups may encounter when attempting to implement crowdsourcing for literature evaluation.

The amount of money paid for each task must also be carefully considered. Because resources such as MTurk are microwork marketplaces, the quality of work, the number of microworkers attracted to a task, and the timeliness of task completion may be related to how much is paid. Just as important is the fairness of pay for the microworkers. The total cost paid to microworkers in these tasks was $221.75 USD. In our abstract evaluations, the extrapolated median hourly pay for the abstract evaluation task was $6.63 USD, which was marginally lower than our target ($7.25 USD, the United States' minimum wage). The overall median hourly pay across all tasks was $5.14 USD (Fig. S2). Although each assignment a microworker accepts is, in effect, a form of microcontract work and is therefore, as we understand it, not subject to minimum wage laws, the ethics of having an individual working at a rate below established minimum pay is questionable. Inflating our $221.75 estimate to have a median hourly pay of $7.25 USD, the expected cost of tasks in this project would have been $312.61. We have identified several areas *post hoc* that we and others may want to consider to better ensure fair pay. 1) Calculate pay rates based on individuals that have completed at least several tasks. We noted that the first few tasks an individual completes are typically slower, so a group of individuals that completes few tasks will inflate the amount of time it takes to complete a task, which will unfairly raise the cost of the tasks for the requester. 2) Calculate pay rates by discarding particularly rapid completers. Although investigators want work completed rapidly and accurately, calculating pay rates based on the most rapid individuals may effectively deflate the pay for average, but otherwise good, microworkers. 3) Estimate time-to-completion *a priori* by having an investigator complete the task in the MTurk Sandbox. This will allow the investigator to include the amount of time it takes for the webpages to load and submissions to be processed, which is included in the calculated time it takes to complete a task. Webpage load time is likely the greatest influence on our average pay rate for the $0.01 tasks because webpages can take several seconds to load. Decreasing the load time for the task can increase an investigator's return on investment and microworkers' pay. 4) If the final rate target was missed, give bonuses to individuals completing some threshold of tasks to at least average minimum wage.

Conclusions

The value of science is dependent on complete and transparent reporting of scientific investigations, but recent examples highlight the existence of incomplete or distorted research reporting [5,6]. Here, we have demonstrated that the careful use of crowdsourcing can be an economical and timesaving means to evaluate large bodies of published literature. Manually and continuously evaluating literature may be manageable for specific subtopics, but may be unwieldy even for a topic as specific as obesity (> 18,000 papers in PubMed indexed in 2012 alone). Crowdsourcing can save calendar time by increasing the number of individuals working on a given task at one time. Because many of the steps in study preparation (e.g., coding sheets; literature searches) and calculations (e.g., interrater reliability; hypothesis testing) are analogous to those of traditional in-house evaluation methods, the time investment from appropriately trained research groups should not be substantially greater. With good reliability and an estimated total microworker cost of $312.61 for the tasks evaluated herein, appropriately constructed and controlled crowdsourcing has potential to help improve the timeliness of research evaluation and synthesis.

Methods

Human subjects approval

This work was approved under Exemption Status by the University of Alabmama at Birmingham's Institutional Review Board (E130319007). Each HIT included a highlighted statement that informed the microworkers that the task was being used as part of a research project, and by accepting the HIT they certified that the microworker was at least 19 years old. Contact information for AWB and the University of Alabama at Birmingham IRB were provided with this statement.

Interfacing with MTurk

HITs were constructed using HTML and Java, using the Mechanical Turk Command Line Tools (version 1.3.1) and the Java API (version 1.6.2).

Search

PubMed was searched for abstracts of human studies about foods and beverages (hereafter referred to simply as food) and obesity that were not reviews: obesity [majr] AND food [majr] NOT review [ptyp] AND humans [mh] AND English [lang], where abstracts were available and date range 2007-01-01 to 2011-12-31. Because public opinion about foods may drift through time, the abstract date range was limited; in addition, ending the search at the end of 2011 allowed time for the articles to be cited. Abstracts from 689 papers were considered.

Microworker Qualifications

We restricted microworkers to United States residents over 18 years of age using built-in MTurk qualification requirements. Because the age of majority is 19 in the state of Alabama, microworkers were informed in the HIT description and the IRB

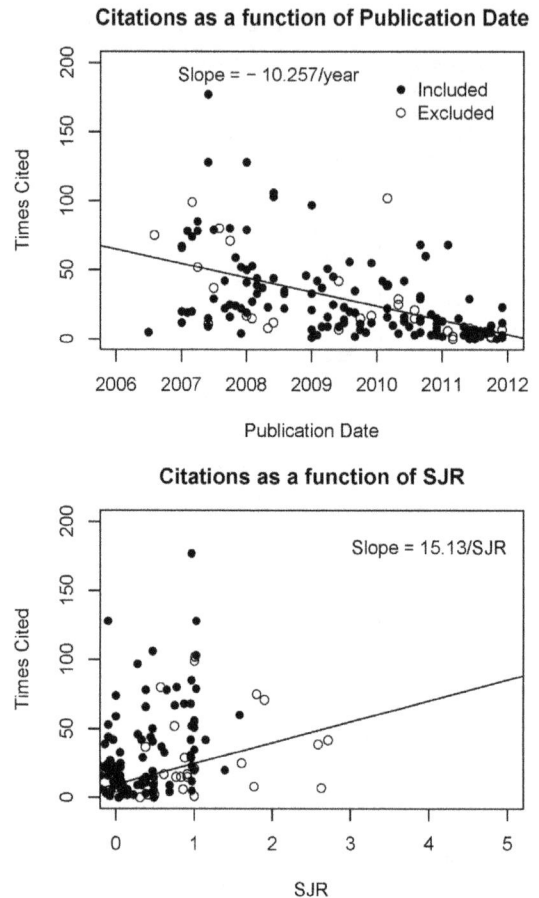

Figure 4. Citations increase with journal quality and time since publication. To confirm that extracted citation counts, journal rank, and publication date conformed to expected patterns, citation counts were fit as a function of publication date controlling for journal quality (SJR; upper panel) and citation counts were fit as a function of journal quality (lower panel). No differences were seen between papers that were "known" (included) and "unknown" (excluded).

header on each HIT that they needed to be older than 19. Age and location restrictions are only confirmed by self-certification. To complete HITs for the initial categorization of abstracts (Fig. 1, Step A.1), microworkers had to complete a qualification test including three example abstracts (Step Q.1). The abstracts used for the qualification test were related to, but not included in, the 689 abstracts to be categorized. Microworkers had to complete the qualification with 100% accuracy to proceed to categorize abstracts, and were not allowed to retake the qualification test (Step Q.2). The other tasks (B, C, and E) were simple and straight forward, so only the residency and age restriction qualifications were enforced.

Abstract Rating

A web form with the PubMed abstract page embedded in an HTML iFrame was presented to microworkers who elected to complete the sorting and successfully completed the qualification test. The response tools available in the Command Line Tools and Java API reflect standard web-based response tools, including multiple choice, checkbox, list selection, and free-text, among others. This HIT used a mixture of these by asking microworkers to determine: 1) if the study was about humans, which was meant

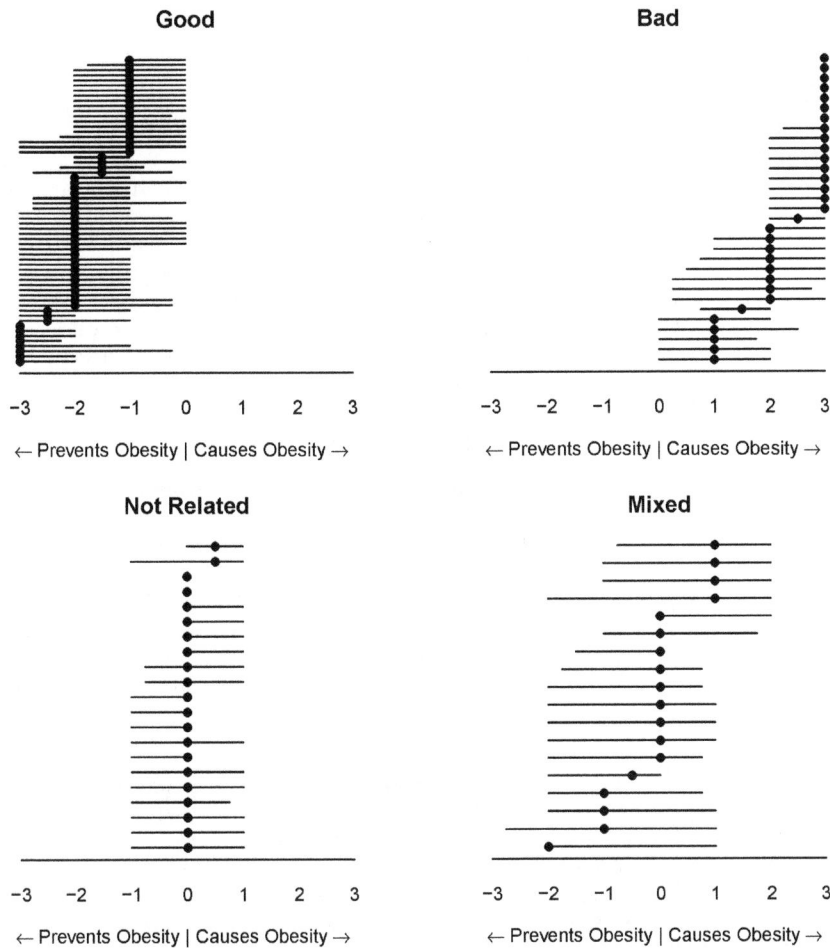

Figure 5. Quartile-based categorization of perceived obesogenicity foods. Foods were categorized dependent on the median and quartiles as follows: Good ("Prevents Obesity"), median ≤ −1 and Q3≤0; Bad ("Causes Obesity"), median ≥1 and Q1≥0; Not Related, median and IQR between −1 and 1; Mixed (depends on who eats the food), median and IQR spanning other categories. Each dot represents the median spanned by Q1 to Q3 for each included food.

to exclude animal studies, *in vitro* studies, commentaries, and reviews (multiple choice); 2) if the study investigated the relation of foods with obesity (multiple choice); 3) what food was described in the paper (free-text answer); and 4) the conclusions about the food and obesity (multiple choice). Responses had to be internally consistent. For example, an abstract could not be rated as being a non-human study and then have conclusions about the food and obesity given. Only 10 of 1538 responses (<1%) had internal consistency issues.

If two microworkers concurred about the conclusions (Step A.2), then the abstract was passed to the next phase (Step A.9). Some rating disagreements were the result of microworkers typing answers into the provided free-text 'other' box rather than selecting an option. When disagreements existed, the abstract was evaluated a third time (Step A.3). The abstract was passed to the next phase (Step A.4-A.9) if at least 2 of 3 ratings agreed. Otherwise, the abstract was manually rated by AWB (Steps A.5-A.8). Each abstract rating received $0.07.

Kappa statistics are typically calculated with raters rating each item in a set. Because microworkers did not rate each item within a task, a kappa-type statistic (κ in the text) was calculated by ordering raters based on the number of items a microworker completed within each task. Within each item, the most prolific

microworker was designated rater 1 and the next most prolific microworker as rater 2. Interrater agreement was calculated by Conger's extension of Cohen's kappa [7] using the 'exact' option of the kappam.fleiss function of R package "irr" version 0.84.

Determining Abstract Food Topic

The iterative synthesis was described in the narrative text.

Popular Opinion About Foods

In Step C.1, microworkers were asked their opinions on the perceived obesogenicity of foods identified in Step B. Opinion questions of this nature are subject to the same concerns as survey or opinion questions evaluated in other settings (e.g., [8]). Microworkers were told to evaluate on a 7 point, horizontal, multiple choice scale whether a food "prevents obesity" or "causes obesity". In an effort to prevent microworkers from overstating their personal beliefs to influence a perceived counter belief of others, they were also asked to predict whether most Americans thought the food "prevents obesity" or "causes obesity". Microworkers were encouraged to click an option indicating they did not know what a food was rather than looking up information or guessing. HITs were posted on 3 different days and times to include a variety of microworkers and microworkers were allowed to respond to all 158 foods, but only once for each food. Thirty responses were collected for each food at $0.01 each.

Foods were categorized as "known" if >85% of respondents ranked the food, and "unknown" if >15% of respondents marked that they did not know the food (Fig. S1 for cutoffs; Table S1 for the list of "known" and "unknown" foods). Known foods were included in subsequent analyses. Figure 3 shows the ratings of related known foods.

Journal Quality

Publication dates were extracted using custom R scripts to first look for an Epub year and month in PubMed (Step D.1); if not, year and month of the journal issue was extracted; if only a year was available, the PD was set as January 1 of that year. SCImago Journal Rank (SJR) was used as a measurement of journal quality (http://www.scimagojr.com/journalrank.php, accessed: 13 Aug 2013]. SJR's were extracted for the year prior to the PD to estimate the quality of the journal around the time the authors would have submitted an article (Step D.2).

Citations

Google Scholar search links were presented to microworkers for each title instead of embedding in iFrames because Google does not allow their material to be placed in iFrames. Links were generated from simple concatenation of "http://scholar.google.com/scholar?q = " and URL-encoded article titles. Microworkers followed the link and reported the number of times the paper was cited on 3 June 2013 (Step E.1). Although Google Scholar includes some citation sources that are not classically considered relevant to academic circles and may miss others, it is a free, stable alternative to commercial sources [9]. 304 papers (including all papers in the final evaluation) were reviewed for citation counts. If the search failed (e.g., because of special characters in the title), citation counts were obtained by AWB (Step E.3). Reported citation counts were tested for verbatim matching (Step E.4) or consensus (Step E.5). Each of three ratings was awarded $0.01 USD.

Model Analysis

Linear models were fitted to confirm the expectation that 1) citation counts decrease with more recent publication date; and 2) citation counts increase with increasing SJR (Fig. 4). Models were also fitted to test whether natural-log-transformed (log) citation counts were associated with abstract conclusions, controlling for publication date and log SJR, and whether log SJR was associated with abstract conclusions. Two separate models were fit to test whether log citation counts or log SJR were associated with conclusions agreeing with popular opinion. In Model 1, the concordance of abstract conclusions with microworker food obesogenicity opinions were rated continuously on a scale of -3 to 3, limited only to papers that concluded the food was beneficial or detrimental for obesity. In Model 2, abstract conclusions were categorized as agreeing or disagreeing with microworker food obesogenicity opinions using quartiles as follows (Fig. 5):

- Beneficial: at least 75% of responses were less than or equal to 0, and at least 50% were less than or equal to -1.
- Detrimental: at least 75% of responses were greater than or equal to 0, and at least 50% were greater than or equal to 0.

- Not Related: the median had to be less than or equal to 1 and greater than or equal to -1, with the interquartile range also within that range.
- Unclearly Related: broad interquartile ranges not included in the definitions above.

These categorizations were then dichotomized as either matching or not matching the conclusions of the abstract. This dichotomization was used to predict log citation counts, correcting for publication date and log SJR; or to predict log SJR directly.

Variables were extracted and compared from the MTurk output using custom scripts in R. Statistical models were calculated using the glm function with default options in R version 3.0.1 using RStudio on a 64-bit Windows 7 machine.

Supporting Information

Figure S1 The proportion of microworkers knowing each food. Complete answers are those where a response was given for both the US and personal opinions, or the respondent indicated they did not know what the food was. Most foods were known and rated by all microworkers as demarked by the large bar at 1.0 on the x-axis. The proportion of people knowing a food decreased until a natural break at 0.85 (vertical dashed line), which was chosen as the cutoff between classifying a food as "known" or "unknown".

Figure S2 Median extrapolated hourly pay for microworkers. Median hourly pay for each microworker on a given task was calculated by dividing 3600 seconds by their median completion time in seconds and multiplying by the reward amount. For task A, this amount was $0.07 USD; for tasks B.3, B.8, B.12, C, and E, the amount was $0.01. The solid horizontal line represents minimum wage ($7.25 USD); the dashed horizontal line represents median extrapolated pay rate across all tasks ($5.14 USD). Each circle represents one microworkers' median extrapolated pay rate; the area of the circle is proportional to the number of assignments a microworker completed within a HIT. The same microworkers did not necessarily work on all HITs.

Acknowledgments

The authors thank G. Pavela and N. Menachemi for providing feedback on manuscript drafts, and J. Dawson for help calculating the kappa statistics.

Author Contributions

Conceived and designed the experiments: AWB DBA. Performed the experiments: AWB. Analyzed the data: AWB. Contributed reagents/materials/analysis tools: AWB DBA. Wrote the paper: AWB DBA. Designed custom R code: AWB. Designed and executed HITs on MTurk: AWB.

References

1. Warby SC, Wendt SL, Welinder P, Munk EG, Carrillo O, et al. (2014) Sleep-spindle detection: crowdsourcing and evaluating performance of experts, non-experts and automated methods. Nat Methods 11: 385–392.

2. Mortensen JM, Musen MA, Noy NF (2013) Crowdsourcing the verification of relationships in biomedical ontologies. AMIA Annu Symp Proc 2013: 1020–1029.

3. NAACL. Workshop on Creating Speech and Language Data with Amazon's Mechanical Turk : Proceedings of the Workshop; 2010;Los Angeles, CA.

4. Landis JR, Koch GG (1977) The measurement of observer agreement for categorical data. Biometrics 33: 159–174.

5. Brown AW, Bohan Brown MM, Allison DB (2013) Belief beyond the evidence: using the proposed effect of breakfast on obesity to show 2 practices that distort scientific evidence. Am J Clin Nutr 98: 1298–1308.

6. Yavchitz A, Boutron I, Bafeta A, Marroun I, Charles P, et al. (2012) Misrepresentation of randomized controlled trials in press releases and news coverage: a cohort study. PLoS Med 9: e1001308.

7. Conger AJ (1980) Integration and generalization of kappas for multiple raters. Psychological Bulletin 88: 322–328.

8. Aday LA, Cornelius LJ (2011) Designing and conducting health surveys: a comprehensive guide : Wiley.com.

9. Harzing AW (2013) A preliminary test of Google Scholar as a source for citation data: A longitudinal study of Nobel prize winners. Scientometrics 94: 1057–1075.

11

Automatic Detection of Regions in Spinach Canopies Responding to Soil Moisture Deficit Using Combined Visible and Thermal Imagery

Shan-e-Ahmed Raza[1]*, Hazel K. Smith[2], Graham J. J. Clarkson[3], Gail Taylor[2], Andrew J. Thompson[4], John Clarkson[5], Nasir M. Rajpoot[1,6]*

1 Department of Computer Science, University of Warwick, Coventry, United Kingdom, **2** Centre for Biological Sciences, Life Sciences, University of Southampton, Southampton, United Kingdom, **3** Vitacress Salads Ltd., Lower Link Farm, St Mary Bourne, Andover, United Kingdom, **4** Soil and Agri-Food Institute, School of Applied Sciences, Cranfield University, Bedford, United Kingdom, **5** School of Life Sciences, University of Warwick, Wellsbourne, United Kingdom, **6** Department of Computer Science and Engineering, Qatar University, Doha, Qatar

Abstract

Thermal imaging has been used in the past for remote detection of regions of canopy showing symptoms of stress, including water deficit stress. Stress indices derived from thermal images have been used as an indicator of canopy water status, but these depend on the choice of reference surfaces and environmental conditions and can be confounded by variations in complex canopy structure. Therefore, in this work, instead of using stress indices, information from thermal and visible light imagery was combined along with machine learning techniques to identify regions of canopy showing a response to soil water deficit. Thermal and visible light images of a spinach canopy with different levels of soil moisture were captured. Statistical measurements from these images were extracted and used to classify between canopies growing in well-watered soil or under soil moisture deficit using Support Vector Machines (SVM) and Gaussian Processes Classifier (GPC) and a combination of both the classifiers. The classification results show a high correlation with soil moisture. We demonstrate that regions of a spinach crop responding to soil water deficit can be identified by using machine learning techniques with a high accuracy of 97%. This method could, in principle, be applied to any crop at a range of scales.

Editor: Roeland M. H. Merks, Centrum Wiskunde & Informatica (CWI) & Netherlands Institute for Systems Biology, Netherlands

Funding: S.E.A.R was funded by the Horticultural Development Company (HDC) and by the Department of Computer Science, University of Warwick. Work on sustainable water use in salad crops in the laboratory of GT is funded by Vitacress Salads Ltd. with the award of a PhD to H.K.S. Research on baby leaf salads in the lab of Gail Taylor is funded by Biotechnology and Biological Sciences Research Council (BBSRC). Field experiments were designed with the help of Vitacress Salads Ltd UK. All the other funders had no role in study design, data collection and analysis, decision to publish, or preparation of the manuscript.

Competing Interests: S.E.A.R was funded by the Horticultural Development Company (HDC) and by the Department of Computer Science, University of Warwick. Work on sustainable water use in salad crops in the laboratory of GT is funded by Vitacress Salads Ltd. with the award of a PhD to H.K.S. Graham Clarkson is an employee of Vitacress Salads Ltd.

* E-mail: s.e.a.raza@warwick.ac.uk (SEAR); n.m.rajpoot@warwick.ac.uk (NMR)

Introduction

Infrared thermometers have been used in the past by researchers to determine temperature differences in both individual plants and their canopies for irrigation scheduling purposes. The development of thermal imagers has extended the opportunities for analysis of thermal properties of plants and canopies [1]. The non-contact, non-destructive nature and repeatability of measurements makes thermal imaging useful in agriculture, the food industry and forestry [2,3]. Imaging has been used as a tool in plants for predicting crop water stress, early disease detection, predicting fruit yield, bruise detection and detection of foreign bodies in food material. Under soil water deficits beyond a critical threshold, plants tend to close their stomata, and the rate of transpiration is reduced. This reduction in transpiration leads to an associated increase in leaf temperature. It also widens the range of temperature variation within the canopy which can be detected using infrared thermometry or by the use of thermal imagers [4]. There has been a lot of work focused on water stress analysis of plants using thermal imaging; however few researchers have

exploited the information from the visible light images for analysis. Most of the work conducted uses stress indices [5,6] and researchers have conducted various experiments to investigate the relationship between different stress indices and temperature values determined by thermal imaging [7,8]. The use of thermal imaging as an indicator of plant stress has also been tested in a number of environmental conditions and the conditions best suited to its successful application have been explored. Leaf energy balance equation was formulated to estimate stomatal conductance [9], but the proposed energy balance equation was dependent on a range of environmental factors and plant variables such as emissivity of the leaf surface, air density and specific heat capacity. The complexity, and associated difficulty of measuring these variables accurately, made it difficult to obtain accurate estimates of stomatal conductance from leaf temperature. Consequently, leaf energy balance equation was rearranged to derive thermal indices based on 'wet' and 'dry' reference surfaces [10,11], using the 'Crop Water Stress Index' (CWSI) [5,6], thus making stomatal conductance more straightforward to calculate from leaf temperatures. There is a debate within the scientific community as

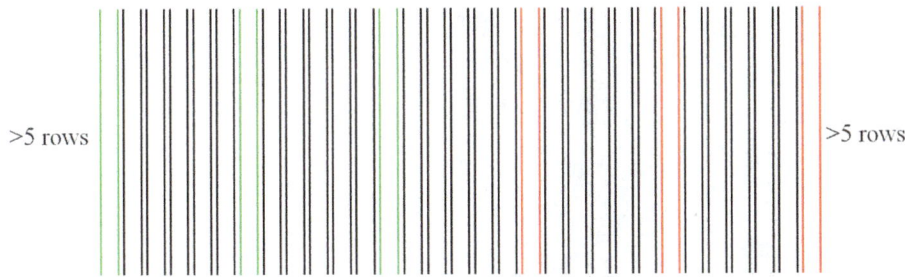

Figure 1. Sampling layout for the collection of thermal images and soil moisture measurements, 2010. The rows represent beds of Spinach (cv. Racoon), with those marked in green showing irrigated sample rows and the red indicating non-irrigated sample rows. Point measurements were made every 20 m for the full length of each bed (n = 54 for each treatment).

to the ideal choice of reference surfaces and much work has been undertaken to find the best choice for reference surfaces and in what conditions they must be used [12].

The robustness, sensitivity and limitations of thermal imaging for detecting changes in stomatal conductance and leaf water status in plants has been analysed by researchers in various conditions [13]. The temperature of surfaces within the canopy is highly dependent on whether they are shaded or in direct sunlight; this variation has been investigated and various options have been suggested to minimise the effect. It was suggested that the average temperature of the canopy was more useful to reduce the effect of leaf angles and other environmental factors when compared to individual leaf temperatures [14]. Researchers have also compared various techniques for image acquisition and have performed experiments to investigate the potential of infrared thermography for irrigation scheduling and to evaluate the consistency and repeatability of measurements under a range of environmental conditions [15]. It was suggested to exclude pixels which are outside the wet-dry threshold range to allow for semi-automated analysis of a large area of canopy. In addition, the authors proposed using thermal data from shaded leaves for improved data consistency, since there is less variability in temperature within an image, and smaller errors resulting from differences in radiation absorbed by reference and transpiring shaded leaves. Variation coefficients of stress indices were found to be of considerable importance and discriminatory powers of the techniques for estimates of stomatal conductance were found to be limited. In a later study, it was proposed that sunlit leaves show a wider range of temperatures because, although natural leaf orientation has little effect on the energy balance of shaded leaves, there is a large effect

on exposed leaves [16]. Based on these observations, the information from temperature distribution can be combined with the leaf orientation for thermal analysis in high resolution images.

Combining information from thermal and visible light images has the potential to provide a better estimate of stress indices and to identify regions in the canopy responding to soil water deficit. The use of thermal and visible imaging has been studied to maintain mild to moderate water stress levels in grapevine [17]. To estimate the canopy temperature, different sections of the canopy were used, including: the whole canopy, all of the sunlit canopy, the centre of the canopy and only sunlit leaves from the centre of the canopy. The best correlation between CWSI and stomatal conductance was calculated from the centre of the canopy measurements (or its sunlit fraction). The authors observed that CWSI computed with wet and dry references was the most robust index and suggested that the fusion of thermal and visible imaging can not only improve the accuracy of remote CWSI determination but also provide precise data on water status and stomatal conductance of grapevine.

Partly automated methods have also been used in the past to study plant stress indices [18]. The authors exploited colour information from visible light images to identify leaf area, as well as sunlit and shaded parts of the canopy. As a pre-processing step, images of constant temperature background were subtracted from the actual image to correct for relative errors in calibration of the camera caused by internal warming. Ground Control Points (GCPs) were manually selected to overlay the thermal image on the visible light image. Different regions in the visible light images were classified, using a supervised classification method, into pixels which represent leaves, other parts of the plant and background.

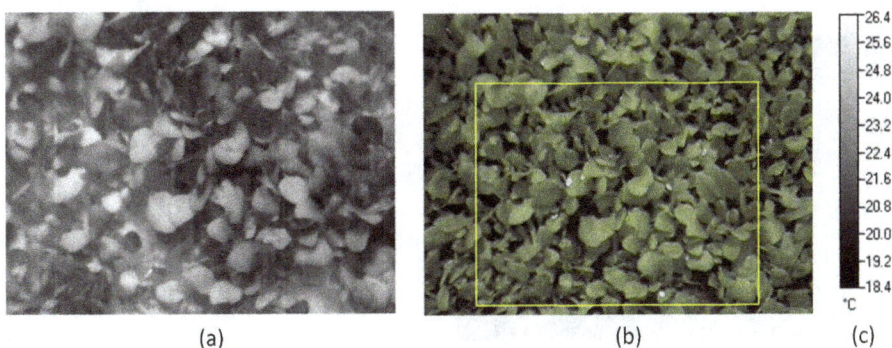

Figure 2. Image(s) obtained using a thermal imaging camera (NEC Thermo TracerTH9100 Pro). (a) thermal image with pixel values ranging from 0–255. (b) Region (rectangle) corresponding to the thermal image in the visible light image. (c) corresponding temperature range.

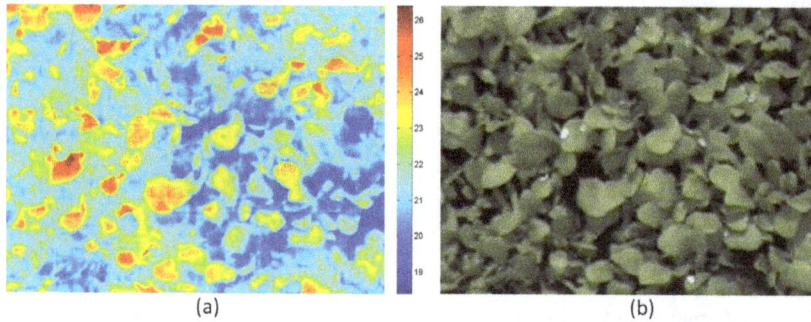

Figure 3. Visible light thermal images of **Figure 2** obtained after pre-processing; (a) the thermal image in **Figure 2**(a) has been replaced by temperature values. (b) visible light image in **Figure 2**(b) has been transformed to match thermal image in a way that same pixel locations correspond to same point located on the plant.

Statistical parameters and stress indices were calculated based on temperature values from the corresponding classified regions of the plant. The results showed that temperature distribution can be used as an indicator of stomatal conductance and plant stress. More recently, researchers have used automated methods to estimate water status using aerial thermal images of palm tree canopies [19]. The authors used watershed segmentation of thermal images to detect the palm trees, and found the detected temperature to be a good indicator of the tree's water status.

Here, we aim to use combined information from thermal and visible light images of a spinach canopy to classify well-watered and water deficient plants. We present a new technique to enhance the 'discriminatory power' of thermal imaging to identify parts of the canopy which have reduced their transpiration rates in response to soil moisture deficit. Instead of using stress indices to identify stress regions, we combine information from visible light and thermal images and use machine learning techniques to

classify between canopies growing in well-watered soil or under soil moisture deficit. Furthermore, we have acquired information about the light intensity and green-ness of the plant from the visible light images. These data are subsequently used, along with statistical information from thermal images, to classify between crop irrigation treatments using 1. Support Vector Machines (SVM), 2. Gaussian Processes Classifier (GPC) and 3. a combination of both classifiers. All three classifiers show promising results with the set of features extracted using combined information from thermal and visible light images.

Materials and Methods

Image Acquisition

Spinach (cv. Racoon) was drilled on 11 March 2010 at Mullens Farm, Wiltshire and was maintained with commercial practice. Permission for this study was given by the farm manager (Graham

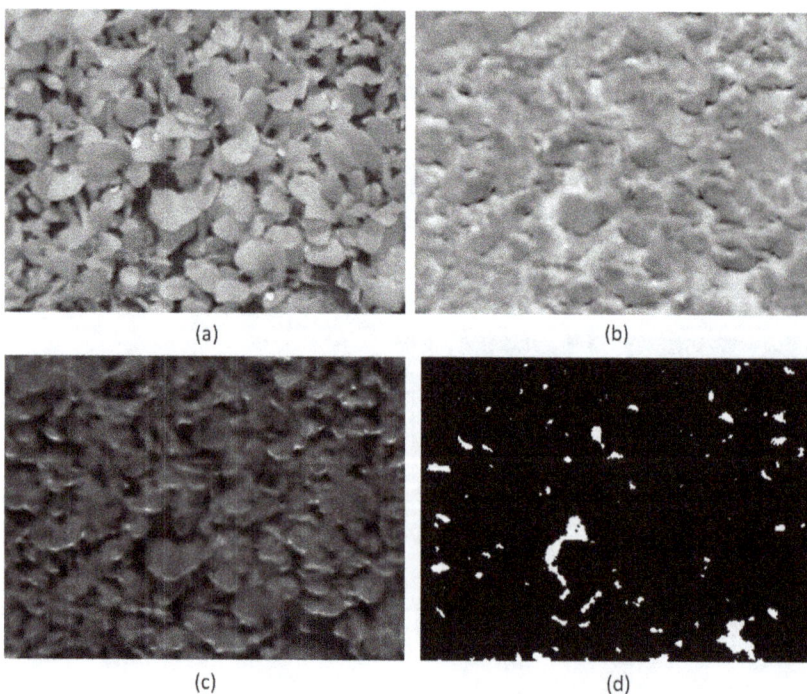

Figure 4. (a), (b) and (c) 'L', 'a' and 'b' channels of the visible light image. (d) thresholded a-channel.

Table 1. Features selected for our experiments.

	Symbol	Description	Type	p-value
1.	μ_{LT}	Luminance has been found to be a major factor which affects the thermal profile of an image [16]. In this work the temperature values were linearly scaled (multiply) with the corresponding L-channel of the colour image so that the effect of light intensity was incorporated into the model. After scaling temperature data with the L-channel, mean temperature values of an image was used as a feature.	C/T	0.154
2.	μ_a	The colour information indicates the amount of area covered by the plants or by other types of region. In Figure 4(b), lower intensities corresponded to green parts of the plant whereas the background shows a higher intensity value. For this reason the mean of the a-channel in our set of features was used.	C	1.92×10^{-07}
3.	μ_b	Similar to Feature 2, in Figure 4 (c) darker regions corresponded to background and hence the mean of b-channel was included in the set of features.	C	1.67×10^{-04}
4.	σ_{nT}	The amount of variation present in an image is also important [30]. Each row of the temperature data was therefore normalised by its median and then the standard deviation of the temperature values employed as a feature, to determine the amount of variation in the canopy region covered by the image.	T	2.89×10^{-19}
5.	μ_{aT}	In Lab colour space, lower values in a-channel corresponded to green regions. The a-channel was thresholded using Otsu's method [31] to find the background regions as represented by white pixels in Figure 4 (d). Temperature values corresponding to the background were discarded and the mean of the temperature values corresponding to the rest of pixels calculated, as a measure of the mean temperature of green parts of the plant.	C/T	1.88×10^{-21}
6.	σ_{aT}	Similar to Feature 5, the temperature values corresponding to background were discarded and the standard deviation of temperature values corresponding to the rest of the pixels calculated as a measure of variation in thermal intensities of green parts of the plant.	C/T	1.024×10^{-04}
7.	μ_T	Mean of temperature values	T	1.46×10^{-21}
8.	σ_T	Standard deviation of temperature values	T	1.12×10^{-04}

Feature type shows that the corresponding feature contains information about colour (C) or thermal (T) data or both (C/T). The rightmost column shows p-values of the features calculated using analysis of variance (ANOVA).

Clarkson) who is also a contributing author to this manuscript. Measurements were taken on 27 April of two treatment areas in bright and clear conditions; well-watered and water-deficient. The former treatment had been irrigated during the preceding week, while the latter had not, and were both harvested the following week for market. Both treatment areas were crops of spinach of the same age and variety and both had reached full canopy cover. Sampling consisted of taking a single image and soil moisture measurement at 20 m intervals for the length of each row. Three rows were sampled per treatment, with five rows separating the sampled rows (Figure 1). Soil moisture measurements were made using a Delta-T ML2x Thetaprobe connected to a HH2 moisture meter (Delta-T Devices, Cambridge, UK), with the probe position being in the centre of the bed at a depth of approximately 7 cm. The infra-red thermal images were taken using a TH9100WR thermal camera (NEC, Metrum) from a fixed distance of approximately 1 m above the crop. The camera operated in the region of 8–14 μm with 0.1°C thermal resolution and a spatial resolution of 320 (V) and 240 (H) pixels. Emissivity was set at 1.0 because it has been reported to induce errors of less than 1°C [20,21]. All measurements were taken between 11:00 and 13:00 hrs on a single day.

Pre-processing

Information from both thermal and visible light images (Figure 2) was used for classification. Thermal images were obtained as images with pixel intensity values ranging from 0 to 255. Initially, the image values were transformed to temperature values. A character recognition algorithm based on cross correlation was used, which automatically recognised the charac-

ters in the temperature bar (Figure 2c) and identified the temperature range for the thermal image [22]. This made it possible to replace the image values, which ranged from 0 to 255, with temperature values. In order to extract useful information from thermal and visible light images, both must be aligned so that the pixel location in both images corresponds to the same physical location with respect to the plant. Since both thermal and visible light images are acquired using a single device, there is a fixed transformation between thermal and visible light images. In order to compute this transformation, the transformation between a single pair of thermal and visible light images was calculated by manually selecting control points. To reduce the amount of noise present in the visible light image, anisotropic diffusion filtering was applied [23]. These pre-processing steps resulted in the images shown in Figure 3 and further calculations were conducted on these images.

Feature Computation

In order to get good classification results, we extracted information from the data in the form of features which carry discriminating information from different treatments and similar information from the same treatment type. Features were selected on the basis of observations made by various researchers [13–18]. Average values and variation in the thermal profile of the canopy were selected and combined with information from the visible light image. As a first step, the colour space of the visible light image from RGB to Lab colour space was transformed (Figure 4). In Lab colour space, instead of Red, Green and Blue channels, an L-channel exists for luminance, as well as 'a' and 'b' channels for the

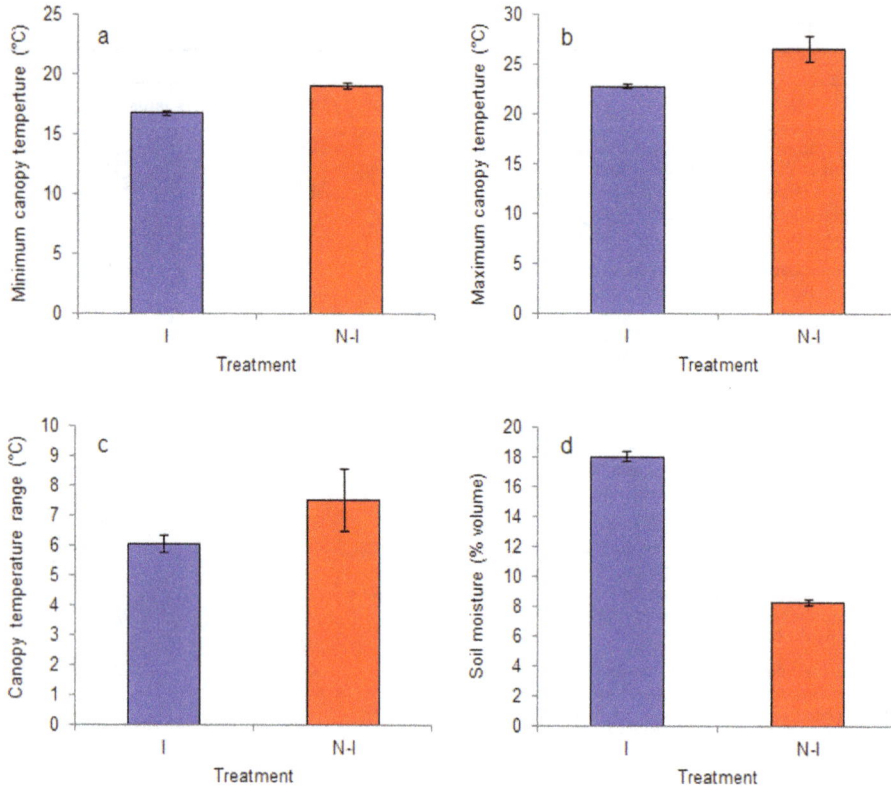

Figure 5. Crop canopy thermal properties (a–c) and soil moisture (d) of irrigated (I) and non-irrigated (N-I) beds of spinach. Crops were grown commercially at Mullens Farm, UK in April 2010. Each bar represents the mean value ± SE n = 3.

colour components. Features selected for experiments are given in Table 1.

Support Vector Machines (SVM)

SVM is a supervised learning method used for classification and regression analysis [24]. SVM constructs a hyperplane in high dimensional space and tries to find the hyperplane which maximises the separation between two classes of training data points. In this work, we used linear SVM which uses the model,

$$y = \mathbf{w}^T \mathbf{x} + b \qquad (1)$$

where $\mathbf{x} = [\mu_{LT}, \mu_a, \mu_b, \sigma_{nT}, \mu_{aT}, \sigma_{aT}, \mu_T, \sigma_T]$ denotes the input feature vector and y denotes the classification output (+1 for plants undergoing water stress, and -1 for well-watered plants). SVM models the parameters b and $\mathbf{w}$ to find the maximum margin hyperplane between data points from two classes.

Gaussian Processes for Classification (GPC)

Gaussian Processes (GP) can be defined as a class of probabilistic models comprised of distributions over functions instead of vectors [25–27]. A Gaussian distribution can be expressed by a mean vector and a covariance matrix. A GP is fully characterised by its mean and covariance functions. In machine learning, GPs have been used for regression analysis and classification. Similar to SVM, GPCs also belong to the class of supervised classification methods. However, instead of giving discriminant function values it produces output with probabilistic interpretation, i.e., a prediction for $p(y = +1|\mathbf{x})$ which denotes the probability of assigning a label (y) value +1 to the input feature

vector $\mathbf{x}$ [28]. GPCs do not calculate this probability directly on the input variables and assume that the probability of belonging to a class is linked to an underlying GP in the form of a latent function. Given a training set $D = \{(\mathbf{x}_i, y_i) | i = 1, 2, \ldots n\}$ consisting of training images of both classes (water deficit and well-watered), with manually assigned labels y_i to the corresponding feature vectors $\mathbf{x}_i$ extracted from those images, GPC makes prediction about the label of the feature vector computed from an unseen image $\mathbf{x}_*$, using posterior probability,

$$p(y_* = +1|D, \mathbf{x}_*) = \int p(y_* = +1|f_*) p(f_*|D, \mathbf{x}_*) df_* \qquad (2)$$

The probability of belonging to a class $y_i = +1$ for an input $\mathbf{x}_i$ (known data point) is related to the value f_i of a latent function f [29]. This relationship is defined with the help of a squashing function. In this case, a Gaussian cumulative distribution function was used as the squashing function.

$$p(y = +1|f_i) = \frac{1}{2}\left[1 + \frac{erf(y_i f_i)}{\sqrt{2}}\right] \qquad (3)$$

where $erf(z)$ is the error function defined as $erf(z) = \frac{2}{\sqrt{\pi}} \int_0^z e^{-t^2} dt$.

The second term in the integral in equation (2) is given by,

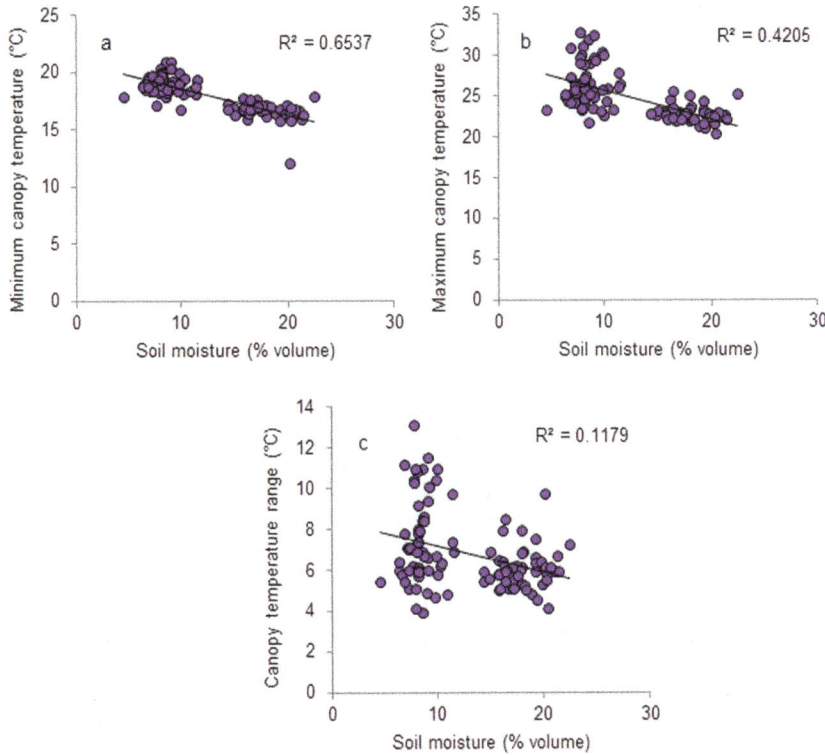

Figure 6. Regressions of crop canopy thermal properties (temperature minimum (a), maximum (b) and range (c)) and soil moisture measurements of irrigated and non-irrigated spinach beds. Crops were grown commercially in April 2010. Trend lines are shown when p< 0.005 and the R^2 value is given.

$$p(f_*|D,\mathbf{x}_*) = \int p(f_*|\mathbf{X},\mathbf{x}_*,\boldsymbol{f})p(\boldsymbol{f}|D)d\boldsymbol{f} \qquad (4)$$

where $\mathbf{X} = [\mathbf{x}_1,\mathbf{x}_2,...,\mathbf{x}_n]$ and $\boldsymbol{f} = [f_1,f_2,...f_n]$, n is the number of samples. $p(\boldsymbol{f}|D)$ can be formulated by the Bayes' rule as follows,

$$p(\boldsymbol{f}|D) = \frac{p(\boldsymbol{f}|\mathbf{X})}{p(\mathbf{y}|\mathbf{X})} \prod_{i=1}^{n} p(y_i|f_i) \qquad (5)$$

and $p(y_i|f_i)$ can be calculated by equation (3) and $p(\boldsymbol{f}|\mathbf{X})$ is the GP prior over latent function. Since a GP is characterised by a mean function and a covariance function, a zero mean was used for symmetry reasons, and a linear covariance function selected which has been found to be effective in classification problems [26]. The normalisation term in the denominator is the marginal likelihood given by,

$$p(\mathbf{y}|\mathbf{X}) = \int p(\boldsymbol{f}|\mathbf{X}) \prod_{i=1}^{n} p(y_i|f_i) \qquad (6)$$

where $\mathbf{y} = \{y_1,y_2,...y_n\}$. The second term in the above equation is not Gaussian and this makes the posterior in equation (5) analytically intractable. However, analytical approximations or Monte Carlo methods can be used. Two commonly used approximation methods are Laplace approximation and Expectation Propagation (EP). EP minimises the local Kullback-Leibler (KL) divergence between the posterior and its approximation and has been found to be more accurate in predicting than Laplace

Table 2. Total variance explained by Principle Component Analysis when both well-watered and droughted spinach crops were measured for their thermal properties (maximum, minimum and range of temperatures) and soil moisture.

Component	Initial Eigenvalues			Extraction Sums of Squared Loadings		
	Total	% of Variance	Cumulative %	Total	% of Variance	Cumulative %
1	2.863	71.567	71.567	2.863	71.567	71.567
2	.830	20.742	92.309			
3	.308	7.691	100.000			
4	6.967E-16	1.742E-14	100.000			

Extraction Method: Principal Component Analysis.

Table 3. Component Matrix[a] from Principle Component Analysis when both irrigated and non-irrigated spinach crops were measured for their thermal properties (maximum, minimum and range of temperatures) and soil moisture.

	Component
	1
Soil moisture	−.750
Minimum temperature	.869
Maximum temperature	.969
Temperature range	.779

approximation and hence EP was used for approximation in these experiments [25,26].

Experiments and Results

Classification using Machine Learning Methods

A total of 108 images of spinach canopies and corresponding soil moisture point measurements were acquired, with 54 images of well-watered beds and 54 images of droughted beds. The thermal images demonstrated significant variation between the two treatments when judged by soil moisture and thermal canopy properties as taken from the primary thermal images (Figure 5).

Well-watered canopies exhibited lower minimum ($F_{1,5} = 59.74$, p = 0.002) and maximum ($F_{1,5} = 8.71$, p<0.05) temperatures than droughted beds. However, the range of temperatures did not differ between treatments when the droughted beds were compared to irrigated spinach plots ($F_{1,5} = 1.80$, p>0.05). Additionally, it was confirmed that soil moisture differed significantly between treatments ($F_{1,5} = 556.19$, p<0.0001). All analyses were conducted using 1-way ANOVA.

Regressions demonstrated a number of relationships linking crop canopy thermal properties, taken from the primary thermal images, to direct soil moisture measurements (Figure 6). Moreover, there was a clear segregation into two clusters, accounting for the two treatments. To establish how these relationships interacted, PCA was performed upon the four traits of: soil moisture, minimum temperature, maximum temperature and range of temperature. Components were extracted when their Eigenvalue exceeded a threshold value of 1. One component was extracted which explained 71.6% of total variance (Table 2). This component measured all four traits thus showing their tight coupling and the need for more complex analysis if they are to be used for the detection of soil water deficits. All thermal properties were strongly, positively related to each other while soil moisture was negatively related to all thermal traits. These results implied that the thermal properties of spinach canopies can be used as an

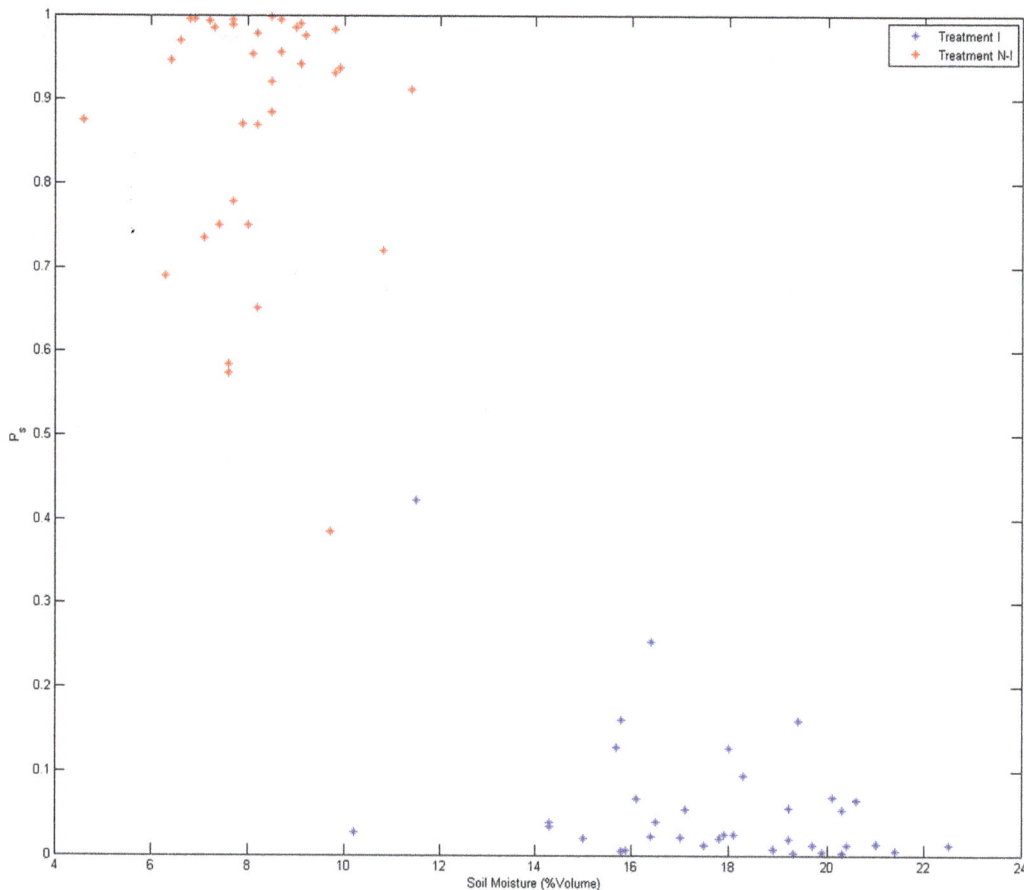

Figure 7. Probability of belonging to treatment N-I (P_s) versus Soil moisture values (correlation value = −0.89, High moisture means less probability of stress). Soil moisture is given as percentage soil water content v/v. Classification accuracy for this particular set of training and testing data was 98.6% as given by GPC.

Table 4. Comparison of average classification results of different classifiers using the proposed set of features.

Feature(s) selected	Classifier	Sensitivity (%)	Specificity (%)	PPV (%)	Accuracy (%)	$\sigma_{accuracy}$
Color only (μ_a, μ_b)	SVM	67.28	70.29	70.98	67.74	3.36
	GPC	80.68	52.96	21.68	56.87	3.55
	Both Classifiers	67.32	70.42	71.11	67.80	3.40
Thermal only (μ_T,σ_T)	SVM	93.35	91.28	90.89	92.14	1.92
	GPC	93.06	80.30	76.67	85.42	2.29
	Both Classifiers	93.35	91.28	90.88	92.14	1.92
Features (1–8) Table 1.	SVM	95.52	96.39	96.30	95.85	1.97
	GPC	96.38	97.39	97.30	96.79	1.56
	Both Classifiers	96.62	96.93	96.84	96.70	1.60
Features (1–6) Table 1.	SVM	95.86	96.86	96.80	96.27	1.58
	GPC	96.53	96.99	96.90	96.68	2.00
	Both Classifiers	**96.97**	**97.38**	**97.31**	**97.12**	**1.52**

indicator of soil water content (Table 3) yet that this approach is not able to accurately detect soil moisture status using primary thermal images. A more complex analysis method is required which is able to utilise both visual and thermal image data to improve soil moisture detection.

The same 108 images of spinach canopies were used for the image processing approach, with the 54 images of well-watered beds being designated treatment I, while the 54 images of the water deficit canopy were designated treatment N-I. The identity of the two treatments was not known during the development of image analysis. After pre-processing, six different features (1–6, Table 1) were obtained from each image. SVM and GPCs were used to classify the test images into water deficient and well-watered. For SVM linear kernel was used and for GPC a zero mean and a linear covariance function were chosen. As discussed before, SVM gives discrete classification results and classifies each image as treatment I or treatment N-I, whereas GPC gives the probability (likelihood) of each image belonging to a particular treatment. Figure 7 shows the probability of an image belonging to treatment N-I (P_s) versus the values of soil moisture for one set of training and testing data. It was clear that the probability (P_s) was highly related to manually calculated soil moisture values (correlation value $= -0.89$ for Figure 7). Based on the probabilities given by GPC, each image was classified as an image from either treatment I or treatment N-I.

Since two different types of classifier were used, disagreement between the results of both the different classifiers could be assessed, which occurred in some cases. This disparity was utilised to further refine the classification results; although this refinement is not very significant, it produces better results. Information from both classification methods was combined to reduce the error from classification. If an image was classified by SVM as treatment I and its probability of belonging to treatment N-I according to GPC was higher than 80% then this image was classified as treatment N-I. On the other hand, if an image was classified as treatment N-I and its probability according to GPC was less than 20%, the image was classified as treatment I. It was found experimentally that the 80–20% threshold gave the best results.

200 iterations were employed to test the accuracy of the classifiers for different pairs of training and testing sets. In each iteration, 36 images were chosen at random (18 from each treatment) for training purposes and the proposed algorithm was tested on the other 72 images. Results showed that GPC demonstrated a higher level of accuracy than the SVM classifier (Table 4); however if information from the results of both of the classifiers was combined, results were improved in terms of sensitivity, specificity, positive predictive value (PPV) and accuracy. An average accuracy of 96.3% was obtained for SVM, 96.7% by using GPC and a slightly higher 97.1% when information from both classifiers was combined. When the results of colour-only and temperature-only features were compared, it was found that combining information from both temperature and colour data increased the accuracy of classification. Furthermore, including mean and standard deviation of temperature values without

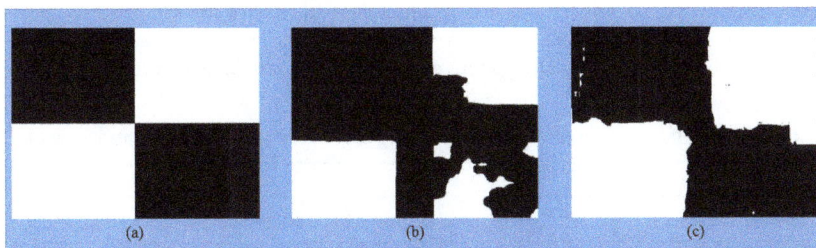

Figure 8. (a) The ground truth pattern for mixed condition mosaicked image. Black colour represents image region corresponding to treatment I and white colour represents the image region which corresponds to treatment N-I. (b) & (c) show classification results obtained using combined classifier with thermal only and proposed feature set respectively.

Figure 9. GPC classification result in terms of confidence score (C_s). Bright shade represents high confidence in classification results and dark shade represents low confidence in the classification. The classifier has higher confidence in the region with image from treatment I or treatment N-I, however the confidence value is low, as depicted by darker shade, around the boundary of two merging images from different treatments.

combining them with colour information diminished the accuracy of results; thus the mean (μ_T) and standard deviation (σ_T) were removed from the set of features.

To further investigate the strength of classifier with the proposed set of features, we created an artificial image with mixed conditions by combining randomly picked thermal and visible light images from Treatment I and Treatment N-I to form a mosaic. The ground truth pattern for the mosaicked image is shown in **Figure 8** (a). Black colour represents image region corresponding to treatment I and white colour represents the image region which corresponds to treatment N-I. A block of size 50×50 pixels was defined at each pixel location in the mosaicked image and the classifier was tested using the features extracted from each of these small blocks (307,200 blocks in total). The classifier for this experiment was trained in a similar way as for the real data (i.e., on randomly selected 36 original images). By using 50×50 blocks to simulate mixed conditions, we reduced the amount of information available, so the accuracy of classification is expected to deteriorate. However, the results show robustness of our proposed feature set when compared to thermal only features. The classification results using the combined classifier with thermal only and the proposed feature set are shown in **Figure 8** (b) & (c) respectively. The classification accuracy using SVM, GPC and the combined classifier was calculated to be **89.1%**, **94.1%** and **92.5%** using the proposed feature set compared to **78.3%**, **54.1%** and **76.3%** when using thermal only features. The classification accuracy for the combined

classifier is less than GPC in the proposed feature set and less than SVM in the thermal only feature set in mixed conditions, however, we still consider this classifier to be important as it gives the best results on real data. **Figure 9** shows GPC classification results using the proposed set of features in terms of the confidence score (C_s). For treatment I, $C_s = 1 - P_s$ and for treatment N-I, $C_s = P_s$, where P_s is the probability of belonging to treatment N-I as given by GPC. The bright shade represents high confidence in classification results and dark shade represents low confidence in the classification. It can be observed that the classifier has higher confidence in the region where the image is from treatment I or treatment N-I, however the confidence value is low, as depicted by low grey values around the boundary of two merging images from different treatments. The mean and standard deviation of C_s was calculated to be **90.5%** and **17.8%** using proposed feature set and **51.1%** and **32.3%** using thermal only features respectively.

Discussion and Conclusions

Our results show that by combining information from thermal and visible light images and using machine learning techniques, canopies which are experiencing water deficits can be identified with high accuracy – more than 97%. Thus we have considerably improved the use of remote images in the detection on canopy stress using this combined approach. The purpose of this study was to test a new dimension of automated classification methods for the detection of regions of a crop canopy that are responding to soil water deficit and to go beyond the restrictions of commonly used statistical approaches. We showed that extraction of a good set of image features can be useful for classifications of this type. In this study, we were able to detect regions of the canopy which were experiencing soil moisture deficit by using a machine learning approach instead of stress indices. Initially, the effect of reflected light and background information was reduced in order to extract features. In the second step these features were classified using SVM, GPC and a combination of both classifiers. The colour information in visible light images provides information about the amount of reflected light intensity from the plant. Using this information, temperature values were scaled on the basis of reflected light. Plant regions can also be identified in the registered thermal image using colour information. This helped to discard temperature values belonging to the background and extract useful information from plant regions in [15]. Based on information from visible light and thermal images, a worthy set of features can be extracted. In these experiments, it was found that scaling with luminance intensity (μ_{LT}) plays an important role in classification. When the luminance intensity scaling feature was removed from our set of features, we found that the accuracy of the classifiers decreased (Table 5). In the case of GPC classification, accuracy fell by up to 7%. This showed that the selection of suitable features is critical when data from thermal images are classified for stress analysis. We have also tested the proposed classifier on an artificially generated mixed condition image. The classification

Table 5. Comparison of average classification results of different classifiers without using light intensity scaling feature (μ_{LT}).

	Sensitivity (%)	Specificity (%)	PPV (%)	Accuracy (%)	$\sigma_{accuracy}$
SVM	94.98	95.01	94.83	94.84	2.01
GPC	88.21	91.84	92.05	89.70	2.61
Both Classifiers	95.28	95.27	95.08	95.12	1.89

results in this image show a significant improvement using the proposed feature set when compared to the thermal only feature set. We found the proposed set of features robust to amount of input information and to mixed-condition images.

In the future, we plan to extend this work to identify canopies under multiple levels of stress. Furthermore, information about leaf angles and distance of the plant from the camera will be used to estimate a more accurate model of the thermal profile, which in this case was linear scaling with light intensity values. For information about depth and leaf angles, a stereo image setup is needed in order to model the effect of leaf angles and distance of leaves from the camera. This model can be combined with more sophisticated machine learning techniques for early water stress detection in crops, and, if automated, could be used to improve irrigation efficiency by optimising the timing and spatial distribution of irrigation events. Other plant stresses such as disease could also potentially be detected rapidly and pre-symptomatically using these methods.

Acknowledgments

The authors would like to thank the staff at Mullens Farm where the experimental field work was conducted.

Author Contributions

Conceived and designed the experiments: SEAR HKS GJJC. Performed the experiments: SEAR HKS. Analyzed the data: SEAR HKS. Contributed reagents/materials/analysis tools: GJJC. Wrote the paper: SEAR HKS NMR GT GJJC JC AJT. Performed computational analysis under supervision of NMR, JC and AJT: SEAR. Performed experiments in the field under supervision of GJJC and GT: HKS.

References

1. Hackl H, Baresel JP, Mistele B, Hu Y, Schmidhalter U (2012) A Comparison of plant temperatures as measured by thermal imaging and infrared thermometry. J Agron Crop Sci 198: 415–429.

2. Eberius M (2011) Automated image based plant phenotyping - Challenges and chances. 2nd Int Plant Phenotyping Symp.

3. Pierce L, Running S, Riggs G (1990) Remote detection of canopy water stress in coniferous forests using the NS 001 Thematic Mapper Simulator and the thermal infrared multispectral scanner. Photogramm Eng Remote Sensing: 579–586.

4. Fuchs M (1990) Infrared Measurement of Canopy Temperature and Detection of Plant Water Stress. Theor Appl Climatol 261: 253–261.

5. Idso SB, Jackson RD, Pinter Jr PJ, Reginato RJ, Hatfield JL (1981) Normalizing the stress-degree-day parameter for environmental variability. Agric Meteorol 24: 45–55. doi:10.1016/0002-1571(81)90032-7.

6. Jackson RD, Idso SB, Reginato RJ, Pinter PJ (1981) Canopy temperature as a crop water-stress indicator. Water Resour Res 17: 1133–1138.

7. Alchanatis V, Cohen Y, Cohen S, Moller M, Sprinstin M, et al. (2009) Evaluation of different approaches for estimating and mapping crop water status in cotton with thermal imaging. Precis Agric 11: 27–41.

8. Reinert S, Bögelein R, Thomas FM (2012) Use of thermal imaging to determine leaf conductance along a canopy gradient in European beech (Fagus sylvatica). Tree Physiol 32: 294–302.

9. Jones HG (1992) Plants and Microclimate: a quantitative approach to environmental plant physiology. 2nd ed. Cambridge: Cambridge University Press.

10. Jones HG (1999) Use of infrared thermometry for estimation of stomatal conductance as a possible aid to irrigation scheduling. Agric For Meteorol 95: 139–149. doi:10.1016/S0168-1923(99)00030-1.

11. Jones HG (1999) Use of thermography for quantitative studies of spatial and temporal variation of stomatal conductance over leaf surfaces. Plant, Cell Environ 22: 1043–1055. doi:10.1046/j.1365-3040.1999.00468.x.

12. Leinonen I, Grant OM, Tagliavia CPP, Chaves MM, Jones HG (2006) Estimating stomatal conductance with thermal imagery. Plant, Cell Environ 29: 1508–1518.

13. Grant OM, Chaves MM, Jones HG (2006) Optimizing thermal imaging as a technique for detecting stomatal closure induced by drought stress under greenhouse conditions. Physiol Plant 127: 507–518. doi:10.1111/j.1399-3054.2006.00686.x.

14. Grant OM, Tronina L, Jones HG, Chaves MM (2007) Exploring thermal imaging variables for the detection of stress responses in grapevine under different irrigation regimes. J Exp Bot 58: 815–825. doi:10.1093/jxb/erl153.

15. Jones HG (2002) Use of infrared thermography for monitoring stomatal closure in the field: application to grapevine. J Exp Bot 53: 2249–2260. doi:10.1093/jxb/erf083.

16. Stoll M, Jones H (2007) Thermal imaging as a viable tool for monitoring plant stress. Int J Vine Wine Sci 41: 77–84.

17. Möller M, Alchanatis V, Cohen Y, Meron M, Tsipris J, et al. (2007) Use of thermal and visible imagery for estimating crop water status of irrigated grapevine. J Exp Bot 58: 827–838. doi:10.1093/jxb/erl115.

18. Leinonen I, Jones HG (2004) Combining thermal and visible imagery for estimating canopy temperature and identifying plant stress. J Exp Bot 55: 1423–1431. doi:10.1093/jxb/erh146.

19. Cohen Y, Alchanatis V, Prigojin A, Levi A, Soroker V (2011) Use of aerial thermal imaging to estimate water status of palm trees. Precis Agric 13: 123–140.

20. Jackson RD (1982) Canopy temperature and crop water stress. Adv Irrig 1: 43–85.

21. López a., Molina-Aiz FD, Valera DL, Peña A (2012) Determining the emissivity of the leaves of nine horticultural crops by means of infrared thermography. Sci Hortic (Amsterdam) 137: 49–58. doi:10.1016/j.scienta.2012.01.022.

22. Eikvil L (1993) Optical Character Recognition. Oslo.

23. Perona P, Malik J (1990) Scale-space and edge detection using anisotropic diffusion. IEEE Trans Pattern Anal Mach Intell 12: 629–639.

24. Cortes C, Vapnik V (1995) Support-vector networks. Mach Learn 20: 273–297.

25. Rasmussen C (2004) Gaussian processes in machine learning. Adv Lect Mach Learn: 63–71.

26. Rasmussen CE, Williams CK (2006) Gaussian processes for machine learning. Cambridge: MIT Press.

27. Haranadh G, Sekhar CC (2008) Hyperparameters of Gaussian process as features for trajectory classification. Neural Networks (IEEE World Congr Comput Intell IEEE Int Jt Conf: 2195–2199.

28. Bazi Y, Melgani F (2010) Gaussian process approach to remote sensing image classification. Geosci Remote Sensing, IEEE Trans 48: 186–197.

29. Ebden M (2008) Gaussian Processes for Regression: A Quick Introduction.

30. Jones HG, Stoll M, Santos T, Sousa C de, Chaves MM, et al. (2002) Use of infrared thermography for monitoring stomatal closure in the field: application to grapevine. J Exp Bot 53: 2249–2260.

31. González RC, Woods RE (2008) Digital image processing. Pearson/Prentice Hall.

12

Quick and Label-Free Detection for Coumaphos by Using Surface Plasmon Resonance Biochip

Ying Li[1,2], Xiao Ma[3], Minglu Zhao[3], Pan Qi[4], Jingang Zhong[1,3]*

1 Key Laboratory of Optoelectronic Information and Sensing Technologies of Guangdong Higher Education Institutes, Jinan University, Guangzhou, Guangdong, China, **2** Pre-university Department, Jinan University, Guangzhou, Guangdong, China, **3** Department of Optoelectronic Engineering, Jinan University, Guangzhou, Guangdong, China, **4** Department of Electronics Engineering, Guangdong Communication Polytechnic, Guangzhou, Guangdong, China

Abstract

Coumaphos is a common organophosphorus pesticide used in agricultural products. It is harmful to human health and has a strictly stipulated maximum residue limit (MRL) on fruits and vegetables. Currently existing methods for detection are complex in execution, require expensive tools and are time consuming and labor intensive. The surface plasmon resonance method has been widely used in biomedicine and many other fields. This study discusses a detection method based on surface plasmon resonance in organophosphorus pesticide residues. As an alternative solution, this study proposes a method to detect Coumaphos. The method, which is based on surface plasmon resonance (SPR) and immune reaction, belongs to the suppression method. A group of samples of Coumaphos was detected by this method. The concentrations of Coumaphos in the samples were 0 µg/L, 50 µg/L, 100 µg/L, 300 µg/L, 500 µg/L, 1000 µg/L, 3000 µg/L and 5000 µg/L, respectively. Through detecting a group of samples, the process of kinetic reactions was analyzed and the corresponding standard curve was obtained. The sensibility is less than 25 µg/L, conforming to the standard of the MRL of Coumaphos stipulated by China. This method is label-free, using an unpurified single antibody only and can continuously test at least 80 groups of samples continuously. It has high sensitivity and specificity. The required equipments are simple, environmental friendly and easy to control. So this method is promised for a large number of samples quick detection on spot and for application prospects.

Editor: Sabato D'Auria, CNR, Italy

Funding: This work was supported by National Natural Science Foundation of China (NSFC) under the Grant No. 41206081, Key Laboratory of Integrated Marine Monitoring, Guangdong Natural Science Foundation under the Grant No. 8451063201000076, and Applied Technologies for Harmful Algal Blooms under the Grant No. MATHAB20120208, and breakthroughs in key areas of Guangdong and Hong Kong tender No. 2005A20501001. The funders had no role in study design, data collection and analysis, decision to publish, or preparation of the manuscript.

Competing Interests: The authors have declared that no competing interests exist.

* Email: tzjg@jnu.edu.cn

Introduction

Organophosphorus pesticide (OPPs) is a type of phosphate with different substituent groups that cause the inhibition of acetylcholinesterase to produce an insecticidal effect [1–3]. China is one of the world's leading users of pesticide, nearly 70% of which is organophosphorus. Organophosphorus pesticide often has high levels of neurotoxicity in human and animals through gradual accumulation and regular intake. Organophosphorus pesticide may lead to symptoms such as neurological disorders, tremors, language disorders and even death [4–15]. Due to its high levels of toxicity, many organophosphorus pesticides have been banned or highly limited in use by most countries. Efforts are being made to revamp restrictions and regulations of organophosphorus pesticide residues in imported fruits and vegetables across America, Europe and Asia, which makes improved detection methods urgent and necessary.

Coumaphos is one of the organophosphorus pesticides. Coumaphos is harmful to human health and has a strictly stipulated maximum residue limit (MRL) on fruits and vegetables. China stipulates that the residues of Coumaphos in vegetables and fruits must be lower than 0.05 mg/kg [16]. Currently, it can be detected through methods of chromatography such as low pressure gas chromatography-mass spectroscopy (LP GC-MS) [17–20], high performance liquid chromatography (HPLC) [21–23] and thin layer chromatography (TLC) [24–25]. Additionally the wave spectrum method [26–28] and enzyme-linked immunosorbent assay (ELISA) [29–30], are also used.

Although these methods are highly sensitive and accurate, the sample pretreatment procedures are time-consuming and labor intensive, complex in execution and require expensive equipment. Furthermore, they do not enable real-time detection and rapid screening of large amount of samples. In contrast, the surface plasmon resonance biochip method is quick, highly sensitive, and label-free, and has been widely used in pharmaceutical analysis, food analysis, environment monitoring and many other fields [31–33]. This study discusses a detection technique based on surface plasmon resonance in organophosphorus pesticide residues, for example Coumaphos. This research adopts the self-developed portable surface plasmon resonance biochip detector by using of the specificity of immunoreaction to study the detection of higher toxicity organophosphorus pesticide-Coumaphos, to analyze the process of kinetic reactions and to establish its standard curve. The technique, which is based on surface plasmon resonance (SPR) and immune reaction, belongs to the suppression method. Comparing with ELISA methods and other methods, this method is label-free and has high specificity, and the sample pretreatment procedure is simple. The utilized equipment is cheap, simple, environmental friendly and easy to control. It achieves the continuousdetection and the quick screening of a large number of samples. This method can be applied in places where real-time quick detection and quality control is needed such as supermarkets, bazaars and factories.

Materials and Methods

Instrument and equipment, materials and reagents

The self-constructed portable SPR biochip detector is a sensitive and accurate angle scanning device. The use of this device is simple and easy, with a precision of up to 0.002°. It can also be controlled easily by LabView and has a user friendly interface [34].

Coumaphos standard (0.02 g/mL, the molecular weight of 362.78) is obtained from Dr. Ehrenstorfer GmbH (Augsburg, Germany). H_{11}-OVA (Coumaphos-ovalbumin, 83 mg/L) and unpurified monoclonal antibody ascites of Coumaphos (5 mg/mL) are produced by the College of Food Science, South China Agricultural University, as described in the previous work [35–36]. $HS(CH_2)_{10}COOH$ (mercapto-undecanoic acid) and $HS(CH_2)_6OH$ (mercapto-hexanoate), N-hydroxysuccinimide (NHS), N-ethyl-N'-(dimethylaminopropyl) carbodiimide (EDC), Ethanolamine (Eth), and Sodium dodecyl sulfate (SDS) are purchased from Sigma, USA. Other reagents are purchased from Beijing Chemical Reagent Company. The PBS buffer (2 mmol/L NaH_2PO_4, 2 mmol/L Na_2HPO_4, 150 mmol/L NaCl, pH 7.4) is used as immune reaction buffer. Coumaphos antigen and monoclonal antibody ascites of Coumaphos are diluted to the appropriate concentration solution with PBS buffer.

Preparation of Biochip

A gold film (thickness of 50 nm) is deposited on the surface of a piece of glass (diameter of 20 mm and thickness of 1 mm). The biochip with gold film is attached to the instrument. The flow cell is installed, and the PBS buffer is injected into it. The biochip will be self-assembledwhen the baseline is stable for a few minutes. The ethanol solution of mercaptoundecanoic acid (0.1 mmol/L $HS(CH_2)_{10}COOH$ and 0.9 mmol/L $HS(CH_2)_6OH$ (Mercapto-hexyl hexanoate)) are injected into the flow cell. The gold film is chemically modified with self-assembled monolayers (SAMs) by the prepared solution for 2 hours. The sulfhydryl terminal of the modified liquid can interact with gold and they bond to each other (S-Au bond), and then adsorb on the gold film stably and orderly. C-terminus can be activated to active ester by EDC/NHS as active groups. Active ester can react with proteins (antibody or antigen) to form amide linkage which will fix protein. After the biochip is modified, the PBS buffer is injected to wash the biochip for forming the baseline of the whole response. The 0.1 mol/L

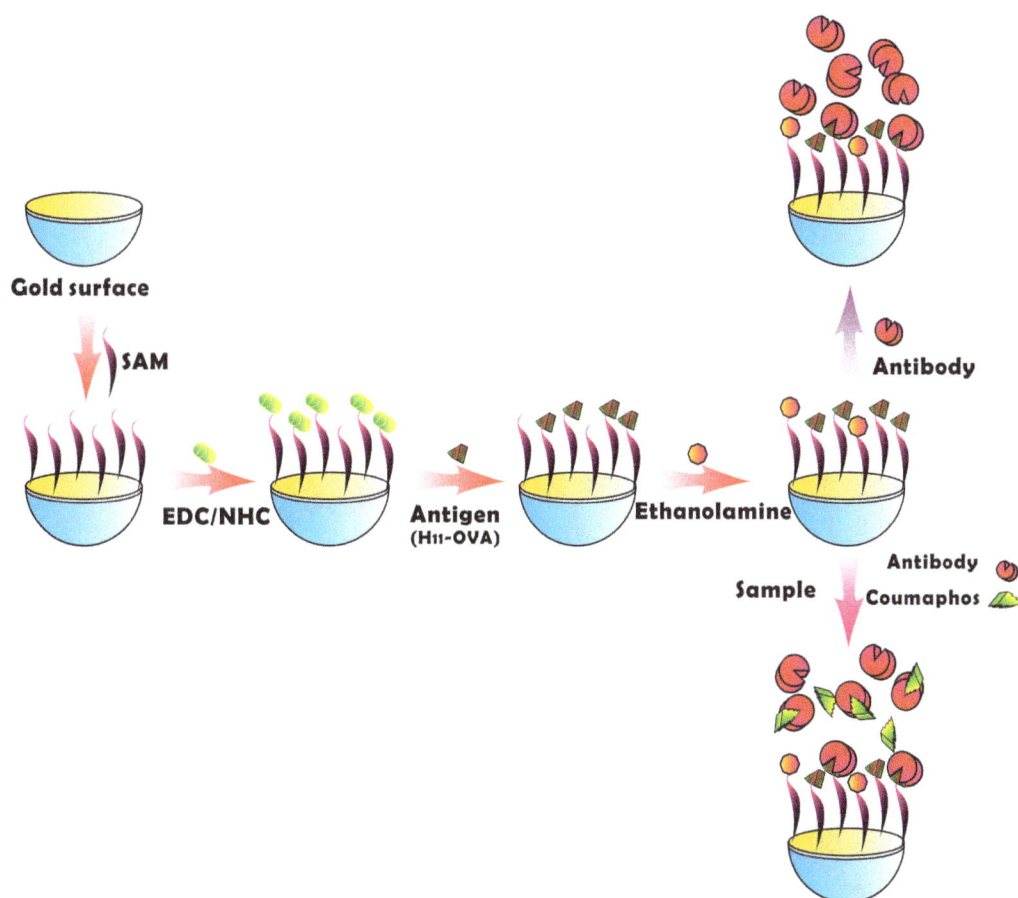

Figure 1. Scheme of Coumaphos immune detection by surface plasmon resonance biochip.

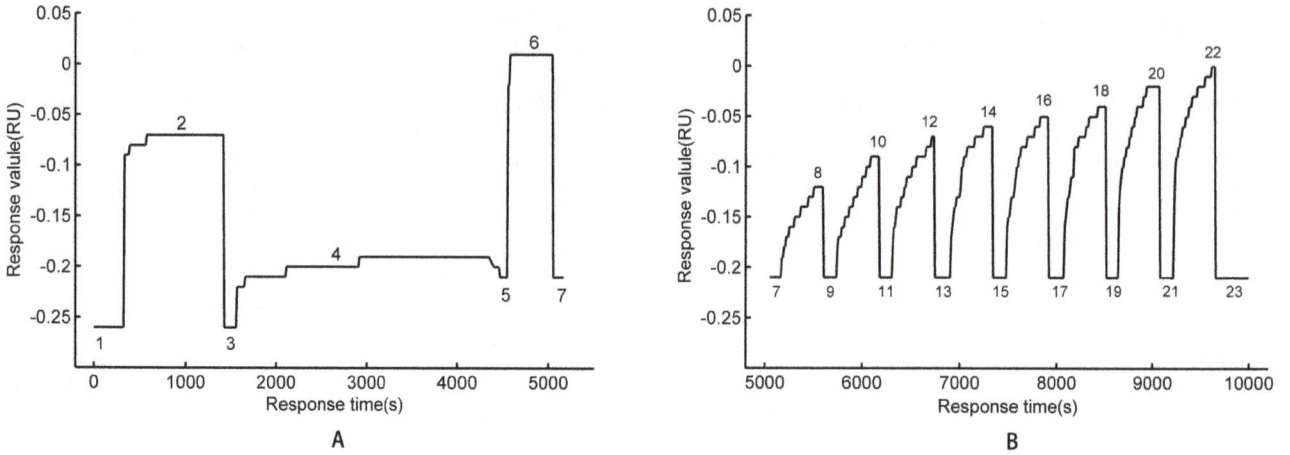

Figure 2. The process of a group of Coumaphos molecules detected by suppression method. A) The preparation of the biochip. 1: PBS; 2: active ester; 3: PBS; 4: The fixation of the bioprobe H_{11}-OVA; 5: PBS; 6: inactivated; 7: PBS. B) Detecting a group of samples by suppression method. 8, 10, 12, 14, 16, 18, 20, 22 plot samples 5000 µg/L, 3000 µg/L, 1000 µg/L, 500 µg/L, 300 µg/L, 100 µg/L, 50 µg/L and 0 µg/L, respectively. 7, 9, 11, 13, 15, 17, 19, 21, 23 are PBS.

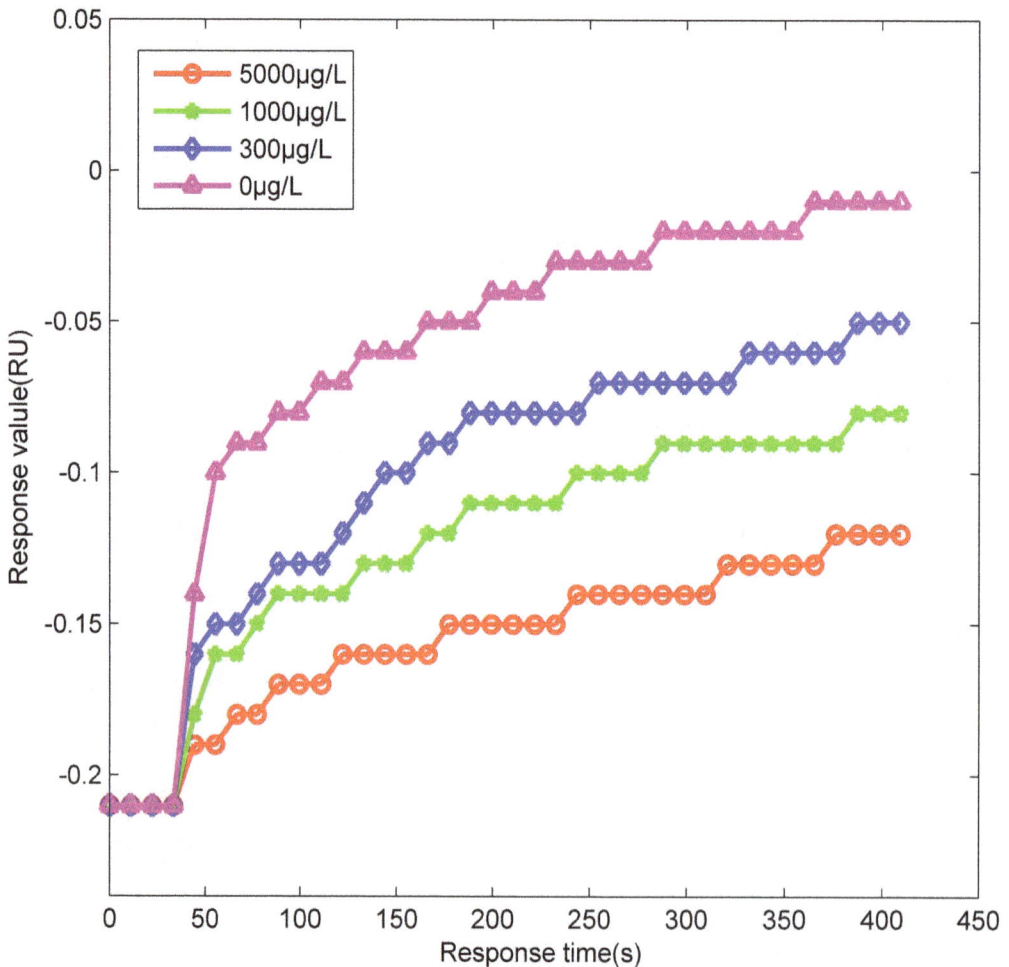

Figure 3. The dynamic curve of Coumaphos molecules detected by suppression method. The concentrations of samples are 0 µg/L, 300 µg/L, 1000 µg/L and 5000 µg/L.

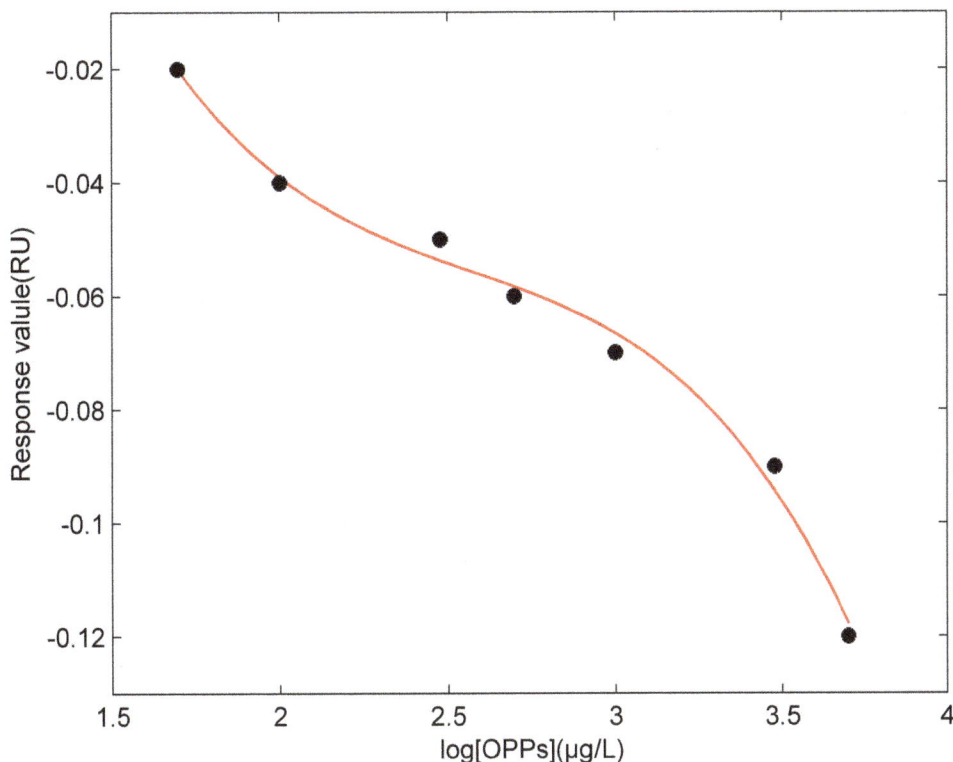

Figure 4. Standard curve of detecting Coumaphos molecules by suppression method. The y-coordinate represents the response value, and the x-coordinate is the concentration of Coumaphos. The curve appears approximately to be an inverted "S". The value of IC_{50} is 1000 µg/L.

NHS and 0.1 mol/L EDC (1:1, V/V) are added on to the surface of the biochip after the baseline being stable to activate the biochip for 15 minutes. The bioprobe H_{11}-OVA is fixed after washed by PBS for 2 minutes. The H_{11}-OVA which has been diluted 15 times as bioprobe is fixed on the surface of the biochip. The response value of the SPR increases significantly. Thirty minutes later, the biochip is washed by PBS for two minutes. The response value of the SPR after fixing is higher than it was before fixing, meaning that the effect of the fixed bioprobe is better. The ethanolamine (pH 8.5, 1 mol/L) is added for the inactivation of the remaining ester bond. The PBS is used to wash the gold film 5–7 minutes after for inactivation. The preparation of the biochip is complete and it can be used for the following detection. The process of Coumaphos immune detection by surface plasmon resonance biochip is shown in Fig. 1.

Suppression Detection

If the probe of biochip is the antibody of Coumaphos, Coumaphos, with a molecular weight of 362.78, can be combined with the antibody that is fixed on the surface of the biochip. This is the direct detection of Coumaphos, but the effect of the refractive index changing around the biochip surface is small. The effect of the direct detection will be unsatisfied because the variation of resonance peak is small. For the trace detection of substances with small molecular weight like Coumaphos, the suppression method was proposed. The bioprobe of the biochip is H_{11}-OVA. 5 minutes after the antibodies were mixed with different concentrations of Coumaphos, the resultant mixtures are added on the surface of the biochip. The SPR effect is detected and the kinetic process can be analyzed. If the concentration of Coumaphos molecules is low, more antibodies of Coumaphos will combine

with the probes H_{11}-OVA on the surface of the biochip, and then the response value of SPR will become larger. The concentration of Coumaphos molecules is inversely proportional to the response value of SPR. That is the meaning of the method of suppression detection.

Results and Discussion

The process of a group of Coumaphos molecules detected by suppression method is shown in Fig. 2. It plots the baseline of PBS in Marking 1 in Figure 2A, the activation of the surface of the biochip, and c-terminus activates into active ester in Marking 2, and the washing by PBS after activation in Marking 3. At that time the response value of the SPR falls back to baseline. The fixation of the bioprobe H_{11}-OVA is shown in Marking 4, and the response value increases drastically. The fixed probe is monitored in real time. The response value no longer increases after about 45 minutes, meaning the bioprobe fixed on the biochip has been saturated. The fixation of the bioprobe is complete and is washed by PBS as shown in the Marking 5. At this time the response value of the SPR decreases slightly but is higher than the baseline in Marking 3, which suggests that the fixation effect is good. In Marking 6, the rest of the ester linkages are inactivated and confined with ethanolamine in order to avoid the formation of nonspecific adsorption. Marking 7 is the cleaning process by PBS to get rid of the substances that are not fixed. With all the operations above, the preparation of the biochip is complete and can be used for the detection of samples.

The antibodies against Coumaphos are diluted with PBS, standing for 5 minutes to ensure sufficient reaction after the antibody was mixed with samples of Coumaphos molecules. The concentration of the antibodies in the mixture is 10 mg/L, the

Table 1. The analysis of the detection of Coumaphos molecules in Coumaphos standard solution by using the SPR biochip in suppression method.

Concentration (µg/L)	concentration measured SPR detector (µg/L) estimated value			average value	standard deviation	absolute deviation	relative deviation
50	50.2	49.7	49.7	49.867	0.289	-0.133	0.58%
100	100.9	100.5	100.4	100.600	0.265	0.600	0.26%
500	501.2	502.1	504.4	502.567	1.650	2.567	0.33%
1000	998.5	1002.4	1004.5	1001.800	3.045	1.800	0.30%

The absolute deviation of the real value compares and the estimated value is smaller, and so is the relative deviation.

concentrations of Coumaphos molecules are 0 µg/L, 50 µg/L, 100 µg/L, 300 µg/L, 500 µg/L, 1000 µg/L, 3000 µg/L and 5000 µg/L, the mixtures are added on the surface of the biochip. Then, the dynamic variation of the SPR is recorded. As Markings 8, 10, 12, 14, 16, 18, 20, 22 plot the dynamic curve of samples in Figure 2B, the concentrations of Coumaphos molecules are 5000 µg/L, 3000 µg/L, 1000 µg/L, 500 µg/L, 300 µg/L, 100 µg/L, 50 µg/L and 0 µg/L, respectively. Thus it can be seen when the concentrations of the samples decrease, the response value of SPR is increasing.

There is a competitive relationship between the Coumaphos molecules in the mixture and the probe H_{11}-OVA on the surface of the biochip, as both of them can be combined with the antibody in the mixture. The Coumaphos molecules suppress the combination between the antibody and the probe H_{11}-OVA on the surface of the biochip. The antibody of Coumaphos is macromolecule, and Coumaphos is micromolecule. The molecular weight of the antibody is larger than Coumaphos, and the weight of H_{11}-OVA-antibody is larger than Coumaphos-antibody. After the combination between the antibody and the probe H_{11}-OVA on the surface of the biochip, the antigen-antibody conjugates on the surface of the biochip increase while the SPR response value also increases. The concentration of Coumaphos molecules is inversely proportional to the response value of SPR. If the concentration of Coumaphos molecules is low more antibodies of Coumaphos will combine with the probes on the surface of the biochip, and then the response value of SPR, namely the resonance angle, will become larger, as is shown in Fig. 2 and Fig. 3. Six minutes after the immune reaction, the SDS-HCl solution is used to elute the combination of the antigen and antibody. The PBS is used to wash the SDS-HCl after two minutes. The response value of SPR can be returned to the baseline, as is shown in Markings 9, 11, 13, 15, 17, 19, 21, 23 of Figure 2B.

Figure. 3 shows the dynamic curve of the partial samples, the concentrations of samples are 0 µg/L, 300 µg/L, 1000 µg/L and 5000 µg/L, respectively. With the concentration of Coumaphos molecules increasing, the response value of SPR declines.

In this case, the biochip with the fixation to the bioprobe can be used in the continuous detection of 80 groups of samples. The response value of SPR declines significantly and the sensitivity decreases after the detection of 80 groups of samples, signaling the bioprobe fixed on the surface of the biochip is damaged. The biochip regeneration can be realized after 0.1 mol/L HCl is added on the surface of the biochip. The sensor biochip can still be used after self-assembling and the fixation of bioprobe.

Figure 4 shows the standard curve of immunoreaction of Coumaphos molecules tested by suppression method with the SPR biochip. The y-coordinate represents the response value of the immune reaction at 5 minutes, and the x-coordinate is the concentration of Coumaphos. The curve appears approximately to be an inverted "S". The standard deviation is less than 10% after the repeated detections of each concentration (3 times). The value of IC_{50} is 1000 µg/L. The detection limit is less than 25 µg/L. After detections of the SPR response value, the concentration of Coumaphos in the sample could be obtained by checking the standard curve. This method could contribute greatly to the quality control of fruits and vegetables and in-field real time detection.

To prepare different concentrations of Coumaphos standard solution, the concentrations of samples are 50 µg/L, 100 µg/L, 500 µg/L, 1000 µg/L, respectively, as the true value of samples. The content of Coumaphos molecules in Coumaphos standard solution is detected by using the SPR biochip in suppression method. And the estimated value of the concentrations of

Coumaphos molecules can be calculated by the combination of the data detected by SPR detector and the standard curve shown in Fig. 4. The concentration of the Coumaphos standard is regarded as the real value and the data calculated by the standard curve is regarded as the estimated value. The real value compares with the estimated value, as is shown in Table 1. The absolute deviation of the real value compares and the estimated value is smaller, and so is the relative deviation. The SPR biochip method is effective, and the experimental method is feasible, and may successfully be used in food quality control and real-time detection.

Conclusions

Organophosphorus, specifically Coumaphos, has become one of the main pesticides for the prevention and control of plant diseases and insect pests. Excessive pesticide residues in fruits and vegetables, however, threaten the health of human beings and animals directly. As the inspection of food imports and exports becomes stricter, it is urgent and necessary to build a simple, effective and cheap detection method of pesticide residue for avoiding pesticide poisoning and protect the health of humans [37].

The detection of Coumaphos by using the biochip of SPR in suppression method mentioned in this paper, has a lot of advantages including inexpensive equipment, label-free detection, high accuracy, specificity and low cost. It also does not require a second antibody, chemical substances, such as fluorescence dye, and use unpurified antibody ascites. It can be used to detect pesticide residues in fruits and vegetables and quickly provide quantitative results. The equipment utilized is inexpensive, portable, environmentally friendly and easy to control. It ensures the achievement of real-time detection and rapid screening of large amount of samples making it ideal for usage in environments such as supermarkets, bazaars and factories.

In this paper, the adopted working concentration of antibody is 10 mg/L, if the working concentration of antibody is reduced, the detection limit of the sample of small Coumaphos molecules can also be reduced.

Acknowledgments

We thank Zong Dai and Yun Chen for the technical assistance.

Author Contributions

Conceived and designed the experiments: YL JZ. Performed the experiments: YL XM. Analyzed the data: YL MZ PQ. Contributed reagents/materials/analysis tools: YL. Contributed to the writing of the manuscript: YL MZ. Designed the software used in analysis: PQ XM.

References

1. Eversole JW, Lilly JH, Shaw FR (1965) Toxicity of droppings from Coumaphos-Fed hens to little house fly larvae. Journal of Economic Entomology 58: 709–710.

2. Pardío VT, Ibarra NDJ, Waliszewski KN, López KM (2007) Effect of coumaphos on cholinesterase activity, hematology, and biochemical blood parameters of bovines in tropical regions of Mexico. Journal of Environmental Science and Health, Part B: Pesticides, Food Contaminants, and Agricultural Wastes 42: 359–366. doi:10.1080/03601230701310500. PubMed: 17474014.

3. Johnson RM, Pollock HS, Berenbaum MR (2009) Synergistic interactions between In-Hive miticides in apis mellifera. Journal of Economic Entomology 102: 474–479. doi:http://dx.doi.org/10.1603/029.102.0202. PubMed: 19449624.

4. Balayiannis G, Balayiannis P (2008) Bee Honey as an environmental bioindicator of pesticides' occurrence in six agricultural areas of Greece. Arch Environ Contam Toxicol 55: 462–470. doi:10.1007/s00244-007-9126-x. PubMed: 18231699.

5. Yavuz H, Guler GO, Aktumsek A, Cakmak YS, Ozparlak H (2010) Determination of some organochlorine pesticide residues in honeys from Konya, Turkey. Environ Monit Assess 168: 277–283. doi:10.1007/s10661-009-1111-6. PubMed: 19685151.

6. Das YK, Kaya S (2009) Organophosphorus Insecticide Residues in Honey Produced in Turkey. Bull Environ Contam Toxicol 83: 378–383. doi:10.1007/s00128-009-9778-5. PubMed: 19452111.

7. Choudhary A, Sharma DC (2008) Pesticide residues in honey samples from Himachal Pradesh (India). Bull Environ Contam Toxicol 80: 417–422. doi:10.1007/s00128-008-9426-5. PubMed: 18506381.

8. Škerl MIS, Bolta ŠV, Česnik HB, Gregorc A (2009) Residues of pesticides in Honeybee (Apis mellifera carnica) bee bread and in pollen loads from treated apple orchards. Bull Environ Contam Toxicol 83: 374–377. doi:10.1007/s00128-009-9762-0. PubMed: 19434347.

9. Weick J, Thorn RS (2002) Effects of acute sublethal exposure to Coumaphos or Diazinon on acquisition and discrimination of odor stimuli in the honey bee (Hymenoptera: Apidae). Journal of Economic Entomology 95: 227–236. doi:http://dx.doi.org/10.1603/0022-0493-95.2.227. PubMed: 12019994.

10. Huang Z (2001) Mite zapper a new and effective method for Varroa Mite control. American Bee Journal: 730–732.

11. Savage EP, Keefe TJ, Mounce LM, Heaton RK, Lewis JA, et al. (1988) Chronic neurological sequelae of acute Organophosphate Pesticide poisoning. Archives of Environmental Health 43: 38–45. doi:10.1080/00039896.1988.9934372. PubMed: 3355242.

12. Popendorf WJ, Leffingwell JT (1982) Regulating OP pesticide residues for farmworker protection. Residue Reviews 82: 125–201. doi:10.1007/978-1-4612-5709-7_3. PubMed: 7051208.

13. Marrs TC (1993) Organophosphate poisoning. Pharmacology & Therapeutics 58: 51–66. doi:0163-7258/9. PubMed: 8415873.

14. Hill EF, Fleming WJ (1982) Anticholinesterase poisoning of birds: Field monitoring and diagnosis of acute poisoning. Environmental Toxicology and Chemistry 1: 27–38. doi:10.1002/etc.5620010105.

15. Eddleston M, Szinicz L, Eyer P, Buckley N (2002) Oximes in acute organophosphorus pesticide poisoning: a systematic review of clinical trials. Oxford Journals 95: 275–283. doi:10.1093/qjmed/95.5.275. PubMed: 11978898.

16. Ministry of Health of the People's Republic of China and the Ministry of Agriculture of the People's Republic of China (2012) National food safety standard Maximum residue limits for pesticides in food. GB 2763–2012.

17. Albero B, Sánchez-Brunete C, Tadeo JL (2004) Analysis of pesticides in honey by Solid-Phase Extraction and Gas Chromatography-Mass Spectrometry. Journal of Agricultural and Food Chemistry 52: 5828–5835. doi:10.1021/jf049470t. PubMed: 15366828.

18. Rissato SR, Galhiane MS, de Almeida MV, Gerenutti M, Apon BM (2007) Multiresidue determination of pesticides in honey samples by gas chromatography–mass spectrometry and application in environmental contamination. Food Chemistry 101: 1719–1726. doi:10.1016/j.foodchem.2005.10.034.

19. Jimenez JJ, Bernal JL, del Nozal MJ, Martin MT, Mayorga AL (1998) Solid-phase microextraction applied to the analysis of pesticide residues in honey using gas chromatography with electron-capture detection. Journal of Chromatography A 829: 269–277. PII: S0021-9673(98)00826-7. PubMed: 9923084.

20. Pang GF, Fan CL, Liu YM, Cao YZ, Zhang JJ, et al. (2006) Multi-residue method for the determination of 450 pesticide residues in honey, fruit juice and wine by double-cartridge solid-phase extraction/gas chromatography-mass spectrometry and liquid chromatography-tandem mass spectrometry. Food Additives and Contaminants 23: 777–810. doi:10.1080/02652030600657997. PubMed: 16807205.

21. Grimalt S, Sancho JV, PozoÓJ, Garca-Baudin JM, Fernndez-Cruz ML, et al. (2006) Analytical study of Trichlorfon residues in Kaki Fruit and Cauliflower aamples by Liquid Chromatography-Electrospray Tandem Mass Spectrometry. Journal of Agricultural and Food Chemistry 54: 1188–1195. doi:10.1021/jf052737j. PubMed: 16478235.

22. Famiglini G, Palma P, Termopoli V, Trufelli H, Cappiello A (2009) Single-step LC/MS method for the simultaneous determination of GC-Amenable Organochlorine and LC-Amenable Phenoxy Acidic Pesticides. Analytical Chemistry 81: 7373–7378. doi:10.1021/ac9008995. PubMed: 19663448.

23. Sannino A, Bolzoni L, Bandini M (2004) Application of liquid chromatography with electrospray tandem mass spectrometry to the determination of a new generation of pesticides in processed fruits and vegetables. Journal of Chromatography A 1036: 161–169. doi:10.1016/j.chroma.2004.02.078. PubMed: 15146917.

24. Lawrence JF (1976) Fluorogenic labeling of organophosphate pesticides with dansyl chloride: Application to residue analysis by high-pressure liquid chromatography and thin-layer chromatography. Journal of Chromatography A 121: 343–351. PubMed: 932151.

25. Butz S, Stan HJ (1995) Screening of 265 pesticides in water by Thin-Layer Chromatography with automated multiple development. Analytical Chemistry 67: 620–630. doi:10.1021/ac00099a021.

26. Janotta M, Karlowatz M, Vogt F, Mizaikoff B (2003) Sol–gel based mid-infrared evanescent wave sensors for detection of organophosphate pesticides in aqueous

solution. Analytica Chimica Acta 496: 339–348. doi:10.1016/S0003-2670(03)01011-0.

27. Tanner PA, Leung KH (1996) Spectral interpretation and qualitative analysis of organophosphorus pesticides using FT-Raman and FT-Infrared Spectroscopy. Applied Spectroscopy 50: 565–571. doi:0003-7028/96/5005-0565.

28. Viveros L, Paliwal S, McCrae D, Wild J, Simonian A (2006) A fluorescence-based biosensor for the detection of organophosphate pesticides and chemical warfare agents. Sensors and Actuators B: Chemical 115: 150–157. doi:10.1016/j.snb.2005.08.032. PubMed: 17616234.

29. Kim YJ, Cho Y, Lee HS, Lee YT (2003) Investigation of the effect of hapten heterology on immunoassay sensitivity and development of an enzyme-linked immunosorbent assay for the organophosphorus insecticide fenthion. Analytica Chimica Acta 494: 29–40. doi:10.1016/j.aca.2003.07.003.

30. Brun EM, Garcés-García M, Puchades R, Maquieira Á (2004) Enzyme-linked immunosorbent assay for the organophosphorus insecticide fenthion. Influence of hapten structure. Journal of Immunological Methods 295: 21–35. doi:10.1016/j.jim.2004.08.014. PubMed: 15627608.

31. Moeller N, Mueller-Seitz E, Scholz O, Hillen W, Petz M, et al. (2007) A new strategy for the analysis of tetracycline residues in foodstuffs by a surface plasmon resonance biosensor. European Food Research and Technology 224: 285–292. doi:10.1007/s00217-006-0392-z.

32. Nogues C, Leh H, Langendorf CG, Law RHP, Buckle AM, et al. (2010) Characterisation of peptide microarrays for studying antibody-antigen binding using surface plasmon resonance imagery. PLoS ONE 5: e12152. doi:10.1371/journal.pone.0012152. PubMed: 20730101.

33. Kuo YC, Ho JH, Yen TJ, Chen HF, Lee OKS (2011) Development of a surface plasmon resonance biosensor for real-time detection of Osteogenic differentiation in live mesenchymal stem cells. PLoS ONE 6: e22382. doi:10.1371/journal.pone.0022382. PubMed: 21818317.

34. Li Y, Qi P, Ma X, Zhong JG (2014) Quick detection technique for clenbuterol hydrochloride by using surface plasmon resonance biosensor. European Food Research and Technology 239: 195–201. doi:10.1007/s00217-014-2201-4.

35. Dai Z, Liu H, Shen YD, Su XP, Xu ZL, et al. (2012) Attomolar determination of Coumaphos by electrochemical displacement immunoassay coupled with oligonucleotide sensing. Analytical Chemistry 84: 8157–8163. doi:10.1021/ac301217s. PubMed: 22934793.

36. Xu ZL, Wang Q, Lei HT, Eremin SA, Shen YD, et al. (2011) A simple, rapid and high-throughput fluorescence polarization immunoassay for simultaneous detection of organophosphorus pesticides in vegetable and environmental water samples. Analytica Chimica Acta 708: 123–129. doi:10.1016/j.aca.2011.09.040.

37. Yin GH, Sun ZN, Liu N, Zhang L, Song YZ, et al. (2009) Production of double-stranded RNA for interference with TMV infection utilizing a bacterial prokaryotic expression system. Applied Microbiology and Biotechnology 84: 323–333. Doi:10.1007/s00253-009-1967-y. PubMed: 19330324.

13

Partial Resistance of Carrot to *Alternaria dauci* Correlates with *In Vitro* Cultured Carrot Cell Resistance to Fungal Exudates

Mickaël Lecomte[1,2,3◉], **Latifa Hamama**[1,2,3◉], **Linda Voisine**[1,2,3], **Julia Gatto**[4], **Jean-Jacques Hélesbeux**[4], **Denis Séraphin**[4], **Luis M. Peña-Rodriguez**[5], **Pascal Richomme**[4], **Cora Boedo**[1,2,3¤], **Claire Yovanopoulos**[1,2,3], **Melvina Gyomlai**[1,2,3], **Mathilde Briard**[1,2,3], **Philippe Simoneau**[1,2,3], **Pascal Poupard**[1,2,3], **Romain Berruyer**[1,2,3*]

1 Agrocampus-Ouest, UMR 1345 IRHS, Angers, France, 2 Université d'Angers, UMR 1345 IRHS, SFR QUASAV, Angers, France, 3 INRA, UMR 1345 IRHS, Angers, France, 4 Université d'Angers, UPRES EA921SONAS, SFR 4207 QUASAV, Angers, France, 5 Unidad de Biotecnología, Centro de Investigación Científica de Yucatán, Mérida, Yucatán, Mexico

Abstract

Although different mechanisms have been proposed in the recent years, plant pathogen partial resistance is still poorly understood. Components of the chemical warfare, including the production of plant defense compounds and plant resistance to pathogen-produced toxins, are likely to play a role. Toxins are indeed recognized as important determinants of pathogenicity in necrotrophic fungi. Partial resistance based on quantitative resistance loci and linked to a pathogen-produced toxin has never been fully described. We tested this hypothesis using the *Alternaria dauci* – carrot pathosystem. *Alternaria dauci*, causing carrot leaf blight, is a necrotrophic fungus known to produce zinniol, a compound described as a non-host selective toxin. Embryogenic cellular cultures from carrot genotypes varying in resistance against *A. dauci* were confronted with zinniol at different concentrations or to fungal exudates (raw, organic or aqueous extracts). The plant response was analyzed through the measurement of cytoplasmic esterase activity, as a marker of cell viability, and the differentiation of somatic embryos in cellular cultures. A differential response to toxicity was demonstrated between susceptible and partially resistant genotypes, with a good correlation noted between the resistance to the fungus at the whole plant level and resistance at the cellular level to fungal exudates from raw and organic extracts. No toxic reaction of embryogenic cultures was observed after treatment with the aqueous extract or zinniol used at physiological concentration. Moreover, we did not detect zinniol in toxic fungal extracts by UHPLC analysis. These results suggest that strong phytotoxic compounds are present in the organic extract and remain to be characterized. Our results clearly show that carrot tolerance to *A. dauci* toxins is one component of its partial resistance.

Editor: Richard A. Wilson, University of Nebraska-Lincoln, United States of America

Funding: M. Lecomte was granted a doctoral fellowship by SFR 4207 QUASAV. The funder had no role in study design, data collection and analysis, decision to publish, or preparation of the manuscript.

Competing Interests: The authors have declared that no competing interests exist.

* Email: romain.berruyer@univ-angers.fr

◉ These authors contributed equally to this work.

¤ Current address: INRA, UMR 1095 GDEC, Clermont-Ferrand, France

Introduction

Partial or quantitative resistance of plants to pests and diseases has been intensively studied among crops. The prospect of developing a sustainable control method has fostered a tremendous amount of work geared towards identifying the genetic factors determining this resistance (known as Quantitative Resistance Loci, or QRLs) to numerous plant diseases or pests. As a snapshot of this activity, in 2011 alone, 41 papers were published on this topic in *Theoretical and Applied Genetics*, dissecting the determinism of partial resistance to 27 distinct pest species amongst 14 crops. Papers have been published on the subject in that journal every year since 1993, with a peak in 2004 (51 articles). On the other hand, as there is much less data addressing the mechanisms involved in plant pathogen partial resistance, these mechanisms are not clearly understood.

Several reviews on Quantitative Disease Resistance (QDR) have recently been published ([1], [2], [3], [4], [5]). A comprehensive survey of disease resistance mechanisms is presented in some of these reviews. A comparison of major types of plant immune responses (Pathogen Associated Molecular Pattern Triggered Immunity, or PAMP Triggered Immunity or PTI *vs* Effector Triggered Immunity or ETI) suggests that molecular mechanisms of plant-pathogen interactions linked to PTI (basal resistance) and ETI (race specific resistance) share common signaling networks. Similarly, it is quite possible that PTI and ETI share common mechanisms with QDR. With this possibility in mind, Kushalappa and Gunnaiah [6] defined quantitative resistance as the ability of a plant to produce resistance-related metabolites and proteins (also referred to as resistance-related biochemicals) to mitigate the action of pathogenicity factors (enzymes, toxins). The genetic basis of plant resistance is complicated by the existence of different pathogen lifestyles, e.g. necrotrophic, hemibiotrophic and biotrophic agents have been described amongst fungi. Recently, significant progress has been achieved in the understanding of the host response to necrotrophic pathogens, including *Alternaria* species [7]. Plant immunity processes are now better explained through the identification of virulence effectors from fungal necrotrophs and their host cellular targets.

In an excellent review, Poland et al. [5] propose for the first time a classification of the possible mechanisms underlying QDR. Six categories of possible QDR mechanisms underlying observed QRLs were distinguished: (i) QRLs could be linked to genes regulating morphological and developmental traits, (ii) mutations or allelic changes in genes involved in basal defense could have an effect on QDR, e.g. chitin receptor kinase 1 in the *Arabidopsis thaliana-Alternaria brassicicola* pathosystem [8], (iii) allelic forms of genes involved in the regulation of signaling pathways, such as the transcription factor WRKY33 in *Arabidopsis* [9], might correspond to QRLs that could modulate resistance levels against necrotrophic or biotrophic pathogens, (iv) QRLs could represent weak forms of major resistance genes (R-genes) or QRLs may colocalize with R-genes (numerous examples, including several plant species in contact with fungal pathogens, are reported in the literature), (v) loci or genes that confer QDR could be components of chemical warfare between the plant host and its pathogen, or (vi) QRLs might represent novel classes of genes, that were not previously described as defense genes supporting resistance mechanisms. Two examples could be mentioned in this latter category: the loss of function of the proline-rich protein Pi 21 is responsible for non-race specific QDR of rice to the hemibiotrophic fungus *Magnaporthe grisea* [10]; and rice indole-3-acetic acid -amido synthetase GH3-2 mediates broad-spectrum partial resistance against two pathogenic bacteria and *M. grisea* by suppressing pathogen-induced auxin production [11].

Since the review of Poland et al. [5], recent advances on determining the mechanisms underlying QDR have been reported in studies involving cultivated monocots of high economic importance. In these studies, specific genes conferring partial resistance to bacterial or fungal pathogens were described: the wheat kinase start protein WKS1 towards the stripe rust pathogen, *Puccinia striiformis* f. sp. *tritici* [12], the wheat serine/threonine protein kinase Stpk-v towards the powdery mildew pathogen *Blumeria graminis* f. sp. *tritici* [13], and the rice putative receptor like cytoplasmic kinase BSR1 towards *Xanthomonas oryzae* pv. *oryzae* and *M. grisea* [14]. In the barley genome, hotspots of non-race specific disease resistance to *Blumeria graminis* were identified with candidate genes encoding components of PAMP-triggered immunity, such as receptor-like protein kinases, factors of vesicle transport and secreted class III peroxidases [15]. In the present paper, QDR will be considered through the involvement of chemical warfare components in the host-pathogen system, as previously suggested by Poland et al. [5]. The production of plant defense compounds in a quantitative or qualitative manner (see for example [16], [17]), or the mechanisms deployed by the plant against pathogen-produced phytotoxins, might contribute to higher partial resistance.

Toxins produced by necrotrophic pathogens, such as *Alternaria* species, have been recognized as important compounds responsible for plant disease, through host cellular death [18]. The capacity of the plant host to resist pathogen-produced toxins via different modes, including detoxification and metabolic bypass, has been extensively described in two pathosystems (*Cochliobolus carbonum*/maize [19]; *Alternaria alternata* f.sp. *lycopercisi*/tomato [20]). In these two examples, toxin resistance mechanisms were described however with respect to qualitative resistance mechanisms. Another example of toxin resistance was reported in the study of Walz et al. [21] using transgenic tomato lines. The introduction of a wheat oxalate oxidase gene in tomato reduced disease symptoms in plants infected by *Botrytis cinerea* or *Sclerotinia sclerotiorum*, two necrotrophic fungi producing oxalic acid, a toxin that is considered to be an important factor determining pathogenicity. In the same line of thought, a correlation between partial resistance and toxin resistance has been found in two other plant-necrotrophic fungal pathogen interactions: *Allium sativum-Stemphylium solani* [22] and *Hevea brasiliensis-Corynespora cassiicola* (V. Pujade-Renaud, personnal communication). To our knowledge, the discovery of partial resistance mechanisms based on QRLs and linked to a pathogen-produced toxin has never been published. This latter hypothesis is tested in the present paper based on the carrot-*Alternaria dauci* pathosystem.

Phytotoxins produced by necrotrophic fungal pathogens were classified as non-host selective (NHST) and host-selective (HST) toxins. These two toxin categories are respectively related to quantitative and qualitative pathogenicity components [23], but their potential contribution, as aggressiveness factors or factors contributing to the host range, is probably more complex, especially when considering the role of NHST in infection processes. Plant pathogens belonging to the *Alternaria* genus are well-known producers of both types of toxins, most of which are described in different *A. alternata* pathotypes [18]. The necrotrophic pathogen *Alternaria dauci* causes leaf blight, one of the most destructive foliar diseases in cultivated carrot. Brown lesions formed on leaves are often surrounded by a chlorotic halo probably due to the action of one or several toxins. This fungus may produce NHST and HST, but literature concerning the toxin produced by this species is relatively scarce. Papers concerning this pathosystem are mainly focused on the characterization of zinniol, which is assimilated as an NHST. Zinniol could exert its phytotoxic activity through disturbance of membrane due to its effect on calcium channels [24], [25]. It was previously demonstrated that different plant pathogen species of *Alternaria* (generally species exhibiting large conidia with a long beak) and the sunflower pathogen *Phoma macdonaldii* can produce zinniol [26], [27]. In a recent study dealing with the *Alternaria tagetica*-marigold (*Tagetes erecta*) pathosystem, the classification of zinniol as a phytotoxin was however controversial [28]. By comparison to other NHSTs, high zinniol concentrations are indeed required to obtain phytotoxicity in *T. erecta* cell cultures.

Other secondary metabolites synthesized by *A. dauci* have been described, such as alternariol or alternariol monomethyl ether [29], [30], which are also known as mycotoxins synthesized by *Alternaria* species in tomato [31]. It was suggested that alternariol produced by *A. alternata* acts as a tomato tissue colonization factor [32]. Among secondary metabolites of *A. dauci*, four unknown species-specific compounds were reported [29]. The phytotoxin role of these unknown compounds was not specified and remains to be clarified. In a previous paper, we showed that, in greenhouse conditions, the studied host range of *A. dauci* was not restricted to cultivated carrot [33]. Lesions varying in severity and extent were indeed observed on wild *Daucus* species, different cultivated Apiaceae species, and also on all tested dicotyledonous species, such as tomato or radish. Thus, *A. dauci* can exhibit a broad host range in controlled conditions, which suggests that HST production does not have an important role in the biology of this species.

The aims of the present work were to (i) determine if the partial resistance of carrot to *A. dauci* could at least partly be based on resistance mechanisms against toxic metabolites produced by the fungus and (ii) better characterize those metabolites. Carrot has been used as a model plant for somatic embryogenesis studies since the discovery of this regeneration pathway [34]. Carrot is thus very well adapted for *in vitro* studies using plant cells and tissues [35], [36]. Embryogenic cellular cultures were obtained from carrot genotypes with varying degrees of resistance to *A. dauci* and were confronted with fungal exudates. Two levels of response were analyzed: (i) cytoplasmic esterase activity which was previously used as a marker of cell growth and viability [37] and (ii) the differentiation of embryogenic cells to somatic embryos (globular, heart-shaped and torpedo-shaped embryos) in auxin-depleted culture medium. We also confronted these cultures with synthetic zinniol at different concentrations, aqueous and organic fungal extracts. Moreover, zinniol concentrations in fungal extracts, and its chemical stability in our experimental conditions were evaluated. Our results suggest that carrot tolerance to *A. dauci* toxic metabolites is one important component of the partial resistance in this pathosystem. It was also demonstrated that the phytotoxic activity is not caused by zinniol, but instead is linked to the organic phase obtained from the fungal exudates.

Materials and Methods

Plant and fungal material, inoculation and symptom scoring

The *Daucus carota* genotypes used in this study were Bolero, Presto, K3, I2, H4 and H1. Bolero and Presto are Nantaise type hybrid cultivars used as standards for resistance and susceptibility, respectively, as in [17,38], while K3, I2, H4 and H1 are breeding material. H1 plants were obtained by self-pollinating a single plant of a susceptible S3 line obtained from French genetic background at Vilmorin (France). I2 and K3 were obtained in the same fashion from two partially resistant Asiatic lines both developed at Agrocampus Ouest (Angers, France). I2 and K3 are genetically different according to preliminary molecular studies (Le Clerc et al., submitted). H4 was obtained from a partially resistant South American cultivar. All fungal material used in this study was from the *A. dauci* reference strain FRA017, which was also used in previous studies [33,38,39]. This strain was isolated in 2000 from naturally infected carrot leaves collected in Gironde, France.

All plant cultivation and inoculation procedures have already been described in detail in [39] (plant cultivation) and [38] (fungus cultivation, inoculum production, drop inoculation). Briefly, plants were grown in greenhouse conditions in boxes containing peat moss/sand mixture for 6 weeks. *Alternaria dauci* was grown in petri dishes on V8 agar, incubated at 24°C in darkness for 7 days, and then exposed to near-ultraviolet light for 12 h/day for 10–15 days for conidia production. The conidia suspension concentration was adjusted to 200 conidia mL^{-1} in 0.05% Tween 20. Individual L3 leaves were inserted in an incubation chamber without being detached from the plant, and forty 5 μL drops of inoculum were applied using a micropipette. The symptom number was evaluated at 7, 9 and 13 dpi and is expressed as the number of symptoms per conidia. The areas under the disease progression curve (AUDPC) were calculated from these data. Leaves were then harvested for qPCR analysis. qPCR evaluation of *A. dauci* in carrot leaves has already been described [38]. Briefly, fungus genome copy numbers (N_f), evaluated by qPCR from 25 ng DNA samples, were used to calculate infection ratios $I = 100 \times N_f/N_p$, as described in Berruyer

et al. [40], where N_p stands for carrot genome copy number. For each genotype, the experiment was repeated four to five times, with each repetition consisting of four inoculated leaves.

Fungal extract preparation

Fungal extracts were prepared from liquid cultures. Erlenmeyer flasks (250 mL) containing 100 mL of liquid carrot juice medium [Joker 100% pure carrot juice (Eckes-Granini Group GmbH, Nieder-Olm, Germany): 20% v/v, CaCO$_3$: 3 g L^{-1}; pH 6.8; H$_2$O: q.s.p. 1 L] were inoculated with a conidial suspension to reach a final concentration of 5.10^3 conidia mL^{-1}. The fungus culture was grown in the dark for 48 h at 24°C on an orbital shaker set at 125 rpm. Liquid phase (raw *Alternaria* extract, rA) was recovered by filtration through Sefar Nitex (Sephar AG, Heiden, Switzerland) nylon membranes of the following decreasing porosities: 200 μm, 11 μm and 1 μm. Organic compounds were derived from the raw extract by liquid-liquid extraction. pH was adjusted to 7 and one volume of ethyl acetate was added to one volume of raw extract. The mixture was strongly agitated, left to rest, and the phases were separated in a separating funnel. The operation was repeated thrice; the organic phases were pooled and labeled organic *Alternaria* extract (oA). The remaining aqueous phase (aqueous *Alternaria* extract, aA) and the raw extract were freeze dried, weighed and stored in a dessiccator. The organic phase was dried over sodium sulfate, filtered and evaporated under reduced pressure using a rotary evaporator (Rotavapor Büchi Labortechnik AG, Flawil, Switzerland) with a water bath at 25°C, weighed and stored at −20°C. Typical yields were of 13 mg mL^{-1} for the raw extract and aqueous phase, and 30 μg mL^{-1} for the organic phase. Mock extracts (raw, organic and aqueous, respectively labeled rM, oM and aM) were obtained with similar yields from mock cultures incubated in the same conditions. Fungal extracts were also prepared from cultures grown in liquid V8 medium for four days in similar conditions, or in anoxic conditions (12 days at 24°C without shaking).

Zinniol synthesis and conservation

We wanted to develop a safer zinniol synthesis procedure by reducing the use of toxic reagents such as zinc cyanide and hydrogen chloride gas during the formylation step. Unfortunately, we were unable to modify the previously reported strategy under any of the tested experimental conditions. Therefore the synthetic zinniol samples used in this study were prepared using the approach developed by Martin and Vogel [41]. All the spectroscopic data were in accordance with those reported in that paper. Proton Nuclear Magnetic Resonance (^{1}H NMR) analyses were performed in deuterated solvents or a mixture of solvents (Deuterated chloroform, or CDCl$_3$; Dimethyl sulfoxide, or DMSO; deuterium oxide, or D$_2$O) using a JEOL GSX270WB spectrometer. Stability of zinniol was studied in CDCl$_3$, deuterated DMSO-aqueous buffer at pH 5.6, and B5 Gamborg medium [42]. For stability in B5 Gamborg medium, the solutions were sampled at different times, and the samples were stored at −80°C. High Pressure Liquid Chromatography (HPLC) analyses were performed on a Waters 2695 separation module coupled to a Waters 2996 Photodiode Array (PDA) Detector using the Empower software package. A QK Uptisphere 3ODB RP18 column (150×4.6 mm, 3 μm, Interchrom) was used for organic extract analysis with the following gradient: initial mobile phase MeOH/H$_2$O 10/90 reaching 100/0 (v/v) in 25 min, with a 0.7 mL min^{-1} flow rate.

Figure 1. Range of symptoms observed on leaves 13 days after inoculation. The symptom number was assessed at 7, 9 and 13 dpi. At 13 dpi, leaves were detached, imaged using a desktop image scanner, and then subjected to DNA extraction and qPCR for fungal biomass evaluation (see Table 1). The leaves shown here show a symptom severity representative of the plant partial resistance level. **A**: H1, **B**: Presto, **C**: K3, **D**: H4, **E**: Bolero, **F**: I2. H1, K3, H4 and I2 are breeding lines, while Presto and Bolero are widely cultivated Nantaise type carrot cultivars.

Zinniol detection

Tandem ultra high-performance liquid chromatography- mass spectroscopy (UHPLC-ESI-MS) analyses allowed us to determine the detection level and the amounts of zinniol in different *A. dauci* cultures extracts. Dried extracts of *A. dauci* cultures were extemporaneously dissolved in ethyl acetate/methanol (50:50, v/v) at a working concentration of 6.67 mg mL^{-1} and filtered through a 0.2 μm nylon membrane prior to immediate analysis by UHPLC. These analyses were performed using an Accela High Speed LC System (ThermoFisher Scientific) consisting of a quaternary pump with an online degasser, autosampler, PDA detector and a TSQ Quantum Access MAX triple stage quadrupole mass spectrometer with an ESI interface. The chromatographic analysis was achieved on a Agilent Zorbax Eclipse Plus C$_{18}$ reversed-phase analytical column (2.1×100 mm×1.8 μm). An elution gradient of water (Milli-Q

quality) and acetonitrile (LC–MS grade) was used. Two microlitres of each *A. dauci* culture extract or standard zinniol solution were injected using the partial loop injection mode (10 μL loop size). The PDA detector was set in the 200–500 nm wavelength range with two selected channels at 210 and 233 nm. Data were acquired and processed using the Xcalibur 2.0 software package (ThermoFisher Scientific). Standard zinniol solutions were freshly prepared to obtain five concentrations in the 0.05–5 mg mL^{-1} range.

In vitro culture methods

Plants were grown in greenhouse conditions for 2 months as previously described. For callogenesis induction, petiole explants (10 cm) were surface disinfected for 5 min with ethanol at 70% (v/v), followed by immersion in a 25% (v/v) commercial bleach solution for 20 min and subsequently washed three times with

Table 1. Comparison of two different carrot *A. dauci* colonization evaluation methods, symptom number assessment and qPCR-based fungal biomass evaluation.

| genotype | log (AUDPC) | | log(I+1) | |
	mean	homogeneity groups[1]	mean	homogeneity groups[1]
H1	3.09	a	0.79	ab
Presto	2.92	b	0.91	a
H4	2.71	c	0.48	c
I2	2.56	cd	0.51	bc
Bolero	2.53	d	0.39	c
K3	2.46	d	0.36	c

Carrot plants of six different genotypes were tested for *Alternaria dauci* resistance using two different methods simultaneously. Plants were grown in greenhouse conditions. The third leaf was inoculated after it was isolated in an incubation chamber without detaching it from the plant. The symptom number was assessed at 7, 9 and 13 dpi. At 13 dpi, leaves were detached and then subjected to DNA extraction and qPCR for fungal biomass evaluation. Log(AUDPC) was calculated from the visual assessments, log(I+1) from the qPCR experiments. Both were subjected to variance analysis followed by a Waller-Duncan multiple comparison. As could be expected, the two parameters were closely correlated (r^2 = 0.793). Interestingly, log(AUDPC) seemed to show a higher resolution, as the homogeneity groups appeared to be more numerous (4 vs 2).
[1]Homogeneity goups were calculated using the Waller-Duncan multiple comparison following an ANOVA analysis.

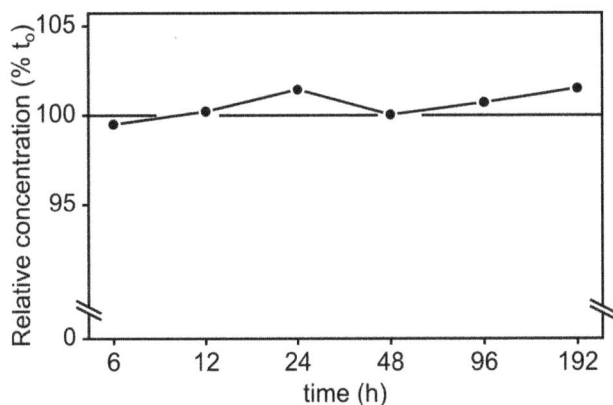

Figure 2. Stability of zinniol over time. Synthetic zinniol was added to Gamborg medium in order to check its stability over time under our experimental conditions (dark, 22°C, shaking). HPLC was used to measure variations in the zinniol concentration over a time course. Three different HPLC analyses were performed for each time. Zinniol concentrations were divided by the initial zinniol concentration in the medium, giving a relative zinniol concentration (noted % t_0). Except for small (less than 2%) random variations, the zinniol concentration did not vary over time, indicating stability. Standard errors are not represented because they were smaller than the dots we used.

sterilized twice distilled water. Petioles were sectioned (1 cm) and placed in Petri dishes containing solidified B5 Gamborg medium [42] supplemented with 30 g L^{-1} sucrose, and 0.5 mg L^{-1} 2,4-dichlorophenoxyacetic acid (2,4-D) and 7 g L^{-1} agar. The pH was adjusted to 5.8. The cultures were maintained at 23°C (16 h) and 19°C (8 h) in the dark. In order to induce embryogenic callus development, calli were separated from the original petiole material and propagated by subculturing every 6 weeks in solidified B5 Gamborg medium (macronutrients diluted for ¾) supplemented with 0.1 mg L^{-1} 2,4-D.

For the embryogenic suspension cell cultures, 1 g of friable calli was transferred to a Corning flask containing 25 mL of B5 Gamborg liquid medium (hereafter called "B5 medium"). The medium was supplemented with 0.25 mg L^{-1} 2,4-D and 0.05 mg L^{-1} kinetin to maintain cells in a dedifferentiated state. The cultures were maintained under continuous agitation (125 rpm) at 22°C in the dark. After 3 weeks, cells were separated from calli by sieving through 450 μm mesh pore sieves (Laboratory sieves Ø45 mm; Saulas, France). Cells were retained on nylon membrane (50 μm pore diameter: Sephar Nitex) and transferred on the same fresh medium for 2 weeks of culture. For somatic embryo development, cells were sieved through 200 μm mesh pore sieves. Cells retained on nylon membrane (25 μm pore diameter) were transferred onto 12.5 mL of the B5 medium without growth regulators. In the absence of growth regulation factor, carrot cells spontaneously undergo embryogenesis.

Embryogenic cell treatments

Lyophilized raw and aqueous fractions were resuspended in growth regulator-free B5 liquid medium in the same proportion (w/v) prior to lyophilization. Organic fractions were resuspended in DMSO and then diluted in growth regulator-free B5 liquid medium. For all fractions, after the pH was adjusted to 5.8, solutions were filter sterilized and kept at −20°C until use. Zinniol (2 mM) was prepared in DMSO (0.4%) and growth regulator-free B5 liquid medium. Cells in the 25–200 μm size range were recovered by filtration and allowed to recover overnight at 22°C in

the dark, under shaking at 125 rpm, in growth regulator-free B5 liquid medium. One mL of cell suspension was distributed into each well of enzyme-linked immunosorbent assay (ELISA) plates, and then one mL of fungal extract solution in growth regulator-free B5 liquid medium was added in order to reach final concentrations of 25% (v/v) of the original culture medium in which the fungus had been grown. After adding the extracts, cell incubation was continued under continuous agitation (125 rpm) at 22°C in the dark. When needed, cells were transferred weekly into fresh growth regulator-free B5 medium containing the same extracts. Cell treatments with zinniol at 0.025 μM (z1), 10 μM (z2) and 500 μM (z3) were performed the same way. DMSO 0.4% in growth regulator-free B5 liquid medium was used as mock extracts. The whole experiment was repeated at least three times per condition.

Fluorimetric evaluation of cell esterase activity

Enzymatic assays were conducted following protein extraction performed according to Vitecek et al. [43] with some modifications. For each condition, 1 mL of cultured cells was collected and centrifuged at 1 800 g for 10 min at 22°C. The supernatant was removed and replaced with 500 μL of 50 mM potassium phosphate buffer (pH 8.75). After centrifugation at 7 200 g for 10 min at 22°C, the pellet was resuspended in a 2 mL microtube in 100 μL of 250 mM potassium phosphate buffer (pH 8.75) containing 1 mM dithiothreitol. Then a thin spatula tip of Fontainebleau sand and one 4-mm diameter stainless steel ball were added. Each sample was frozen in liquid nitrogen, and then ground twice in a Retsch MM301 laboratory ball mill for 30 s at 30 Hz. After grinding, 100 μL of 250 mM potassium phosphate buffer was added to each sample. The homogenate was then centrifuged at 4°C for 15 min at 10 000 g. The supernatant (200 μL) was collected, frozen in liquid nitrogen and stored at − 80°C until further use.

The enzymatic assays were performed at a final volume of 300 μL in 96 well ELISA plates. In each well, 20 μL of supernatant was added to 200 μL of 1 M potassium phosphate buffer at pH 8.75. The reaction was started by adding 80 μL of buffer supplemented with fluorescein diacetate (FDA) at 5 μM final concentration from a 1 mg mL^{-1} stock solution of FDA in acetone stored at −80°C. Twenty μL of extraction buffer was used as a blank. The enzymatic reaction kinetics were recorded using a FLUOstar Omega (BMG Labtech) plate spectrofluorometer set to detect fluorescein fluorescence (excitation wavelength: 485 nm, emission wavelength: 520 nm) for 90 min at 45°C. The fluorescein concentration was calculated by comparing the fluorescence data with a standard curve as in Green et al. [44]. Enzyme activity was expressed in nmol fluoresceine min^{-1} and specific activity in nmol fluorescein min^{-1} mg protein^{-1}. The protein concentration in samples was measured by the method of Bradford [45] with a commercial protein assay kit (Sigma-Aldrich). In the case of cultivar Presto, protein concentrations were too low to accurately calculate specific activity.

Microscopic evaluation of cell viability and embryogenesis ability

The ability of cells to differentiate and develop somatic embryos was monitored for 3 weeks after treatments. Proembryogenic masses and somatic embryo formation were visually checked under a stereo microscope (Olympus SZ61TR) fitted with a digital camera (Olympus DP20). Membrane integrity and cell viability were evaluated by a modified double staining method [43] using fluorescein diacetate (FDA) and propidium iodide (IP). In living cells, FDA is degraded into fluorescein, a green fluorescent

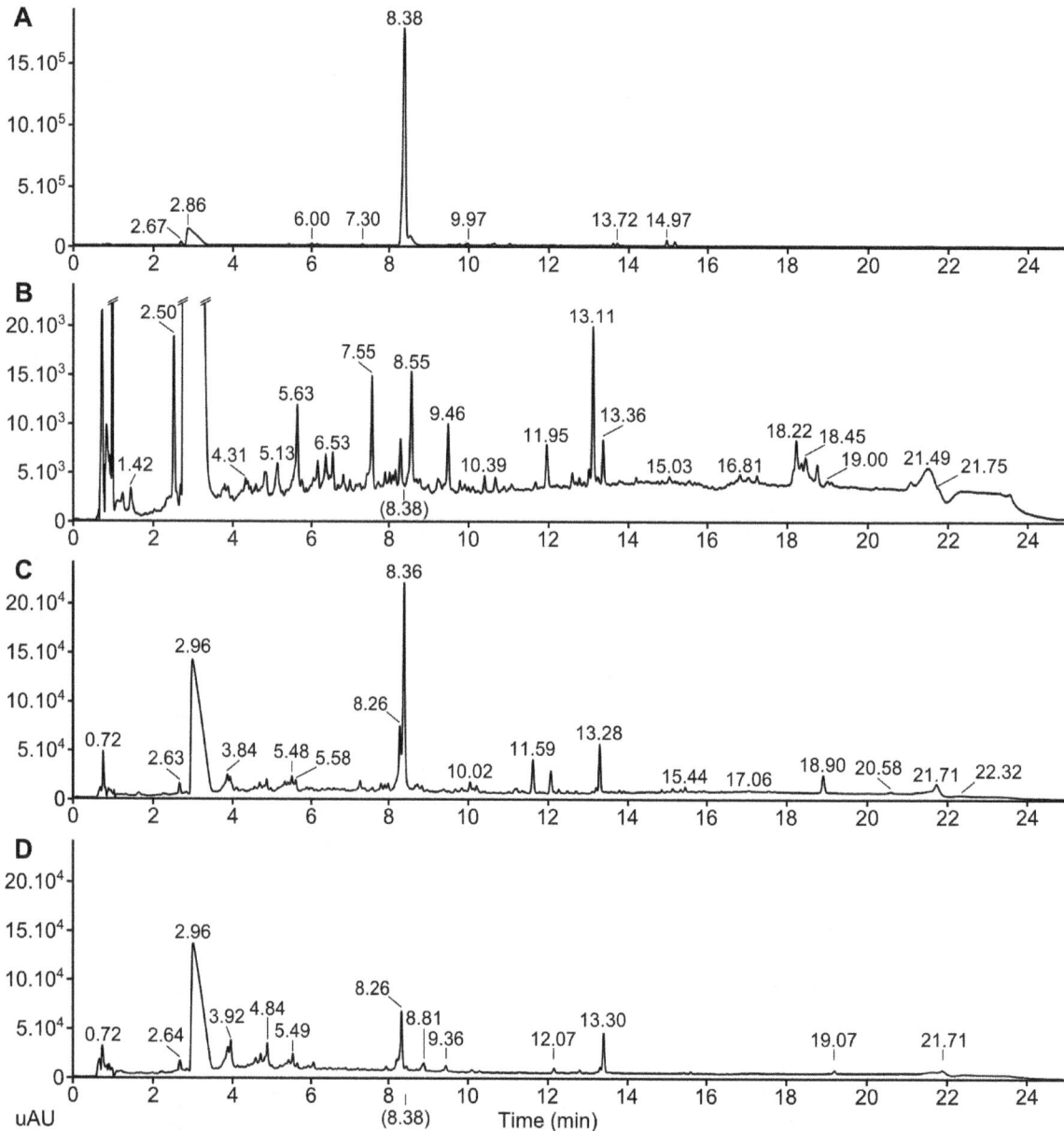

Figure 3. UHPLC detection of zinniol in fungal extracts. UHPLC chromatograms were obtained from different FRA017 *Alternaria dauci* fungal extracts and compared with an UHPLC chromatogram of pure synthetic zinniol. Retention times corresponding to main peaks are indicated **A**: UHPLC chromatogram of 10 μg synthetic zinniol. Observed zinniol retention time is 8.38 minutes **B**: UHPLC chromatogram of 13.4 μg organic extract of an *A. dauci* culture after 48 h under shaking conditions in carrot juice medium. Zinniol expected retention time of 8.38 minutes is indicated. **C**: UHPLC chromatogram of 13.4 μg organic extract of an *A. dauci* culture after 12 days without shaking (anoxic conditions) in V8 medium. A strong peak is visible, corresponding to zinniol retention time. **D**: UHPLC chromatogram of 13.4 μg organic extract of an *A. dauci* culture after 48 h under shaking conditions in V8 medium. Zinniol expected retention time of 8.38 minutes is indicated. Chromatograms C and D have the same scale. uAU: micro Absorption Units (optical density) at 233 nm.

compound that cannot escape the cell. IP can only enter dead or dying cells through damaged plasma membranes. An FDA stock solution (1 mg mL^{-1} in acetone) was maintained at −80°C, and was extemporarily diluted 10-fold in bi-distilled water (working solution). IP 0.15% was prepared in a phosphate buffered saline solution and maintained at 4°C in the dark. One drop of the cell suspension was placed on a microscope slide and 15 μL of IP and FDA working solutions were added. After 5 min incubation in the dark at Room Temperature (RT), stained cells were observed under a fluorescence microscope (Leica DMR HC) fitted with a

digital camera (Qimaging, Retiga 2000R) and monitored using Image Pro Express 6.0 software. Green and red fluorescence indicated viable and dead cells, respectively.

Statistical analysis

All statistical analyses were performed using R-2.6.1 software (R Development Core Team, 2005). Symptom scoring and qPCR data were analyzed as in [38]. Briefly, log(AUDPC) and log(I+1) were subjected to analysis of variance (ANOVA) and Waller-Duncan multiple comparison procedures. Specific activity data

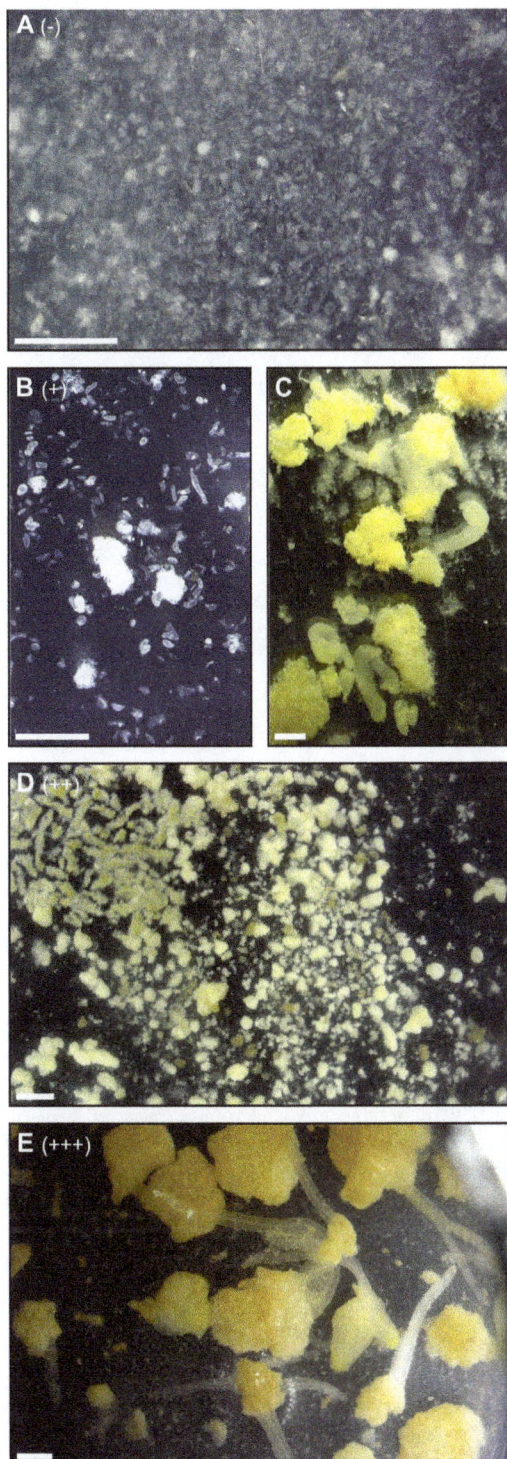

Figure 4. Range of embryogenic activity observed in cell suspensions 3 weeks after treatment. In order to assess carrot cell resistance to fungal toxins, carrot cell suspensions were tested for embryogenesis in the presence of fungal extracts and toxins. Embryogenesis was assessed 3 weeks after treatment, and compared to negative controls. Four levels of embryogenic activity were noted. **A**: (−) no embryogenesis was visible, cells were damaged, **B**: (+) early-stage embryogenic masses were visible, **C**: same as B, but after 6 weeks. **D**: (++) embryos were present, and **E**: (+++) embryogenesis was profuse.

were analyzed as follows: first, the whole dataset was subjected to ANOVA followed by multiple comparisons. The specific activities revealed homoscedasticity (residual vs fitted plot), but a cultivar effect on the residual distribution was observed (residuals vs cultivar box plot). Thus, a separate ANOVA followed by multiple comparisons were also performed for each cultivar. Regardless of the method used, in some instances, mock extracts and DMSO revealed significant effects compared to control. In order to isolate the *A. dauci* exudate and toxin effects from the fungal growth medium and solvent effects, specific activity ratios (rA/rM, oA/oM, aA/aM, z1/DMSO, z2/DMSO and z3/DMSO) were calculated for each independent experiment. For each cultivar × treatment combination, 6–12 figures were calculated from independent repetitions. These results were analyzed by ANOVA using the cultivar × treatment combination as a factor. A 95% confidence interval was calculated in order to check for significant activity variations. When 1 was not included in the interval, the variation was considered significant. Correlations between relative activities were calculated by comparing mean ratios for each cultivar.

Results

Evaluation of plant resistance to fungal disease

Six carrot genotypes representative of a broad spectrum of levels of resistance to *Alternaria dauci* were used in this experiment. They included Presto and Bolero, standard cultivars used respectively for susceptibility and resistance towards *A. dauci*. In previous greenhouse and field resistance tests (Le Clerc et al., submitted), Presto and H1 were found to be susceptible to *A. dauci*, while Bolero, I2 and K3 were found to be more resistant. H4 showed intermediate resistance levels. These genotypes were challenged with *A. dauci* using the drop inoculation method as in [38]. The log(AUDPC) was calculated via visual scoring, and log(I+1) by qPCR evaluation of the fungal biomass. As could be expected, there was a close correlation between the two parameters ($r^2 = 0.793$, see Table 1). Interestingly, the log(AUDPC) seemed to show a higher resolution, as the homogeneity groups appeared to be more numerous (4 vs 2). The resistance classification obtained in this experiment was similar to the findings of previous field and greenhouse experiments. H1 was found to be significantly more susceptible than Presto. H4 was found to be significantly more resistant than Presto, but significantly more susceptible than Bolero. K3 and I2 did not show any significant difference in resistance level with Bolero (Table 1, Fig. 1).

Zinniol synthesis, stability and concentration in fungal extracts

In our hands, the NMR samples of zinniol in CDCl$_3$ proved to be rapidly degraded at room temperature after a few days (Fig. S1). This major stability issue encountered during its analysis raised questions on its storage and extraction from fungal culture filtrates. Many papers have reported the use of chloroform as solvent to both extract and store zinniol produced by *Alternaria* fungi [41], [46]. We suspected that the potential residual acidity of this solvent was the main factor explaining this pattern. As the culture medium used for *in vitro* cultures (B5 Gamborg medium) is about pH 5.8, we aimed to determine the stability of zinniol in these conditions. HPLC analysis proved that zinniol was stable at pH 5.6 in a deuterated DMSO-aqueous buffer solution after 1 week at RT (Fig. S1). We then used HPLC to determine the stability of zinniol in the *in vitro* culture medium over 7 days (Fig. S1 and Fig 2). As no significant zinniol variations were observed

Table 2. Influence of cultivar, fungal exudate fractions and zinniol on cell suspension integrity and somatic embryogenesis.

Treatment	Carrot genotype					
	Bolero	**H1**	**H4**	**I2**	**K3**	**Presto**
rA[1]	++[2]	−	−	+++	++	−
rM	++	+	+	+++	++	+
aA	++	++	+	+++	++	+
aM	++	++	+	+++	++	+
oA	+	−	−	++	++	−
oM	++	+	+	++	++	+
C	++	++	+	+++	++	+
DMSO	++	+	+	+++	++	+
z1	++	+	+	+++	++	+
z2	++	+	+/−	+++	++	+
z3	−	−	−	+	+/−	−

Carrot cell suspensions with six different genotypes were tested for embryogenesis in the presence of fungal extracts and toxins. Embryogenesis was assessed 3 weeks after treatment.
[1]Treatments were as follows: rA: *Alternaria dauci* (strain FRA017) fungal culture raw extract; rM: uninoculated medium raw extract; aA: *A. dauci* fungal culture aqueous extract; aM: uninoculated medium aqueous extract; oA: *A. dauci* fungal culture organic extract; oM: uninoculated medium organic extract; DMSO: DMSO solution, at a concentration corresponding to oM, z1, z2 and z3 treatments; z1: 0.025 µM zinniol; z2: 10 µM zinniol; z3: 500 µM zinniol; C: no treatment.
[2]The signs are as follows: (−) no embryogenesis was visible and cells were damaged, (+) early-stage embryogenic masses were visible, (++) embryos were present, (+++) embryogenesis was profuse. +/− early-stage embryogenic masses were visible, or no embryogenesis was visible depending on the repetition.

between samples, we concluded that the compound was stable in the culture medium conditions used in this study.

Zinniol concentrations in fungal organic extracts were evaluated by UHPLC-MS. No significant amounts of zinniol were found (Fig. 3B compared to Fig. 3A). Based on the injected quantity of the fungal organic fractions, we concluded that the zinniol concentration was below 0.075% w/w in these fractions, which corresponded to 100 nM zinniol in the fungal growth medium. In order to check if the absence of zinniol was due to the genetic background of fungal strain FRA017 or to the culture conditions, FRA017 was grown in V8 liquid medium in the same conditions, and once again, no significant amounts of zinniol were found (Fig. 3D). Furthermore, the fungus was grown in V8 liquid medium for 12 days in anoxic conditions. In the corresponding organic extract, a zinniol concentration of about 4% w/w (corresponding to roughly 5 µM) was detected (Fig. 3C). The detection of significant amount of zinniol in the organic extract is thus dependent on the fungal culture condition: anoxic conditions seem to be needed.

Plant cell resistance to Alternaria exudates and zinniol: cell somatic embryogenic ability

Bolero, Presto, I2, K3, H1 and H4 cultured cells were challenged with various fungal, zinniol, and carrot juice medium extracts. Treatments with DMSO (0.1%), fungal growth medium raw (rM), aqueous (aM) and organic (oM) extracts yielded similar results: as untreated cultures (control), and regardless of the genetic background, cells survived well after treatment and underwent embryogenesis 3 weeks later (Fig. 4, Table 2).

Three weeks after adding the fungal extract (rA), H1, Presto and H4 cells showed marked damage, with the presence of a high quantity of cell debris, while Bolero, I2 and K3 cells formed embryos in a fashion that could not be distinguished from the controls (Fig. 4, Table 2). Similar results were also obtained 3 weeks after treatment with the fungal organic fraction (oA): Bolero, I2 and K3 cell suspensions underwent embryogenesis, H1, Presto

and H4 cell suspensions did not undergo embryogenesis, and showed substantial amounts of cell debris. Conversely, no effects were observed when cell suspensions were treated with fungal aqueous fractions (aA): 3 weeks after treatment, no difference was noted between the treatments and controls (Fig. 4, Table 2). Concerning zinniol, no cultivar differential effect was observed. When 0.025 µM or 10 µM zinniol was added (treatments z1 and z2), no difference was noted between the treated cells and controls, irrespective of the genetic background. At 500 µM zinniol (z3), cell suspensions formed debris, and no embryogenesis was observed 3 weeks after treatment. Both susceptible and resistant cultivars were affected (Fig. 4, Table 2). Plant cells (H1 and K3 genotypes) were also challenged with organic extracts from *A. dauci* growing in various conditions. The results were similar to those obtained previously after treatment with rA or oA: cells from the susceptible H1 cultivar did not undergo embryogenesis, while the resistant K3 cells did (Table S1). These extracts included a 5 µM zinniol-containing organic extract obtained from a fungal culture grown 12 days in anoxic conditions.

Plant cell reaction to Alternaria dauci exudates and zinniol: cytoplasmic esterase activity

Cell suspensions underwent the same set of treatments as in the cell somatic embryogenic ability experiment. Esterase activity was measured 48 h after treatment of cell suspensions. In a first step, the activity was modeled using ANOVA followed by least significant difference (LSD) multiple comparison (Table 3). Two different ANOVAs were performed, one based on the whole dataset while taking cultivar × treatment combinations as a factor (h1), and another whereby the activity was modeled separately in each cultivar while taking treatments as a factor (h2). Classically, ANOVA on whole dataset are preferred, but overall variance was influenced by the cultivar, thus breaching homoscedasticity assumptions. Both methods yielded very similar results, as presented in Table 3. Where not explicitly indicated, only

Table 3. Influence of cultivar, fungal exudates fractions and zinniol on cell suspension esterase activity.

Treatment	Bolero ESA[1]	Bolero h1[2]	H1 h2[3]	H1 ESA	H1 h1	H4 h2	H4 ESA	H4 h1	I2 h2	I2 ESA	I2 h1	K3 h2	K3 ESA	K3 h1
rA[4]	602	b	b	229	nopqrs	cde	211	opqrs	cd	415	cdefghi	b	315	ijklmno
rM	388	cdefghijk	c	288	klmnopq	bc	190	qrs	cd	343	fghijklm	cd	284	klmnopq
aA	484	c	c	256	mnopqrs	bcde	168	rs	cd	486	c	a	402	cdefghij
aM	422	cdefghi	c	300	jklmnop	bc	149	s	d	346	efghijklm	cd	373	defghijkl
oA	770	a	a	185	qrs	e	455	cde	b	282	klmnopq	de	221	opqrs
oM	430	cdefgh	c	193	pqrs	de	446	cdefg	b	224	opqrs	e	218	opqrs
C	436	cdefgh	c	314	ijklmno	b	339	ghijklm	bc	283	klmnopq	de	285	klmnopq
DMSO	482	c	c	279	lmnopq	bc	481	c	b	283	klmnopq	de	256	mnopqrs
z1	771	a	a	449	cdef	a	720	a	a	440	cdefgh	ab	340	fghijklm
z2	602	b	b	262	mnopqr	bcde	676	ab	a	350	efghijklm	c	351	efghijklm
z3	463	cd	c	265	lmnopqr	bcd	490	c	b	337	hijklmn	cd	359	defghijklm

Carrot cell suspensions with six different genotypes were tested for esterase specific activity in the presence of fungal extracts and toxins.

[1]ESA is for Esterase Specific Activity, expressed in nmol min^{-1} mg(prot)$^{-1}$.

[2]Activities with the same letter are not significantly different. h1 homogeneity groups were obtained by a single ANOVA analysis of all the results followed by LSD multiple comparisons.

[3]Activities with the same letter are not significantly different. h2 homogeneity groups were obtained by a separate ANOVA analysis of the results for each cultivar followed by LSD multiple comparisons.

[4]The treatments were as follows: rA: *Alternaria dauci* (strain FRA017) fungal culture raw extract; rM: uninoculated medium raw extract; aA: *A. dauci* fungal culture aqueous extract; aM: uninoculated medium aqueous extract; oA: *A. dauci* fungal culture organic extract; oM: uninoculated medium organic extract; DMSO: DMSO solution at a concentration corresponding to oM, z1, z2 and z3 treatments; z1: 0.025 µM zinniol; z2: 10 µM zinniol; z3: 500 µM zinniol. C: no treatment.

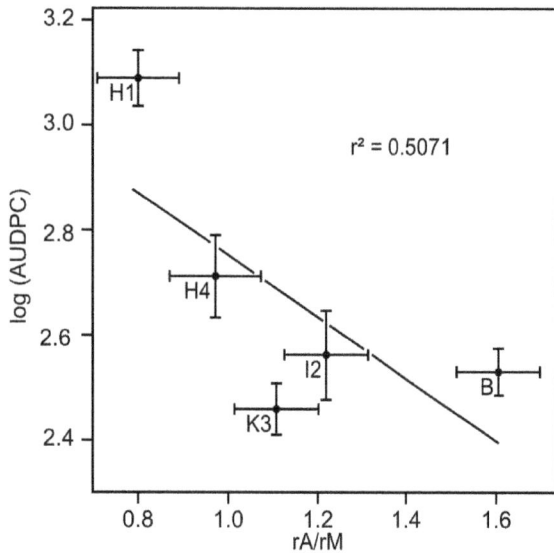

Figure 5. Correlation between cell suspension reactions to *Alternaria dauci* raw extracts and whole plant resistance to the *A. dauci* fungus. log(AUDPC) data, calculated from visual assessments, are the same than in Table 1. The same genotypes were also tested for esterase activity in the presence of fungal (rA) or uninoculated medium (rM) raw extracts. rA/rM denotes esterase activity variations due to the presence of fungal extracts. A negative correlation coefficient (r = − 0.7221, $r^2 = 0.5071$) was noted between rA/rM and log(AUDPC).

homogeneous groups obtained using separate ANOVAs are discussed here.

Cell suspensions treated with uninoculated fungal medium raw extract (rM) did not show significant variations in esterase activity as compared to untreated cells. In contrast, the raw fungal extracts (rA) had significant effects. When compared with the untreated control, esterase activity was significantly lower in the susceptible cultivar H1, and significantly higher in the resistant cultivars Bolero and I2. Non-significant variations were observed in K3 (resistant, rise) and H4 (intermediate, drop). Similar trends were observed when comparing rA and rM, except that the decreased activity in H1 was not significant. Cell suspensions treated with uninoculated fungal medium organic extract (oM) did show any significant variation in esterase activity as compared to untreated cells. The oA effects were thus compared with oM. In these conditions, a significant increase was observed in the resistant cultivar Bolero. Non significant variations were noted in the other cultivars: an increase for I2, and minute variations for genotypes K3, H1 and H4. Cell suspensions treated with aM showed significantly different activities than control in H4 (drop) and K3 (rise). Compared to aM, the aA effects were as follows: a significant increase in the resistant genotype I2, non-significant increases in genotypes K3, H4 and Bolero, and non-significant decrease in the susceptible genotype H1. DMSO treated cell suspensions did not show significant esterase activity variations, with the exception of H4, where a sharp increase was observed. This variation was not significant in the H4 separate ANOVA, but was significant when ANOVA was performed on the whole dataset. The z1 treatment led to a highly significant increase in esterase activity, irrespective of the cultivar considered. The z2 treatment led to a significant increase in specific esterase activity, except for H1, where no significant variation was observed. The z3 treatment led to non-significant variations in esterase activity, except for K3 (significant increase).

Specific activity ratios were calculated in order to isolate the *A. dauci* exudate and toxin effects from the fungal growth medium and solvent effects: *A. dauci* exudates versus uninoculated medium (rA/rM, oA/oM and aA/aM), or zinniol versus DMSO (z1/DMSO, z2/DMSO and z3/DMSO). Correlations between these ratios and between rA/rM and AUDPC were investigated. As expected, a negative correlation (r = −0.7121, $r^2 = 0.5071$) was obtained between rA/rM and AUDPC. Indeed, a trend was noted when AUDPC was plotted against rA/rM (Fig. 5): the susceptible cultivar H1 showed the highest AUDPC and the lowest rA/rM ratio, while the resistant cultivar Bolero combined the highest rA/rM ratio with a very low AUDPC. H4 and I2 seemed intermediate between these two extremes. One of the resistant genotypes (K3) was out of line with the main trend: although it was quite resistant towards *A. dauci*, it did not show strong esterase relative activity in the presence of rA. When K3 was removed, the r^2 increased to 0.7038. When the relative enzymatic activities were compared for the different fungal fractions or toxin concentrations tested, only three revealed a statistically significant correlation (Table 4), and the highest correlation was between oA/oM and z1/DMSO ($r^2 = 0.9736$, p = 0.184%). Similarly, rA/rM was closely correlated with z1/DMSO ($r^2 = 0.8905$, p = 1.59%) and oA/oM ($r^2 = 0.8779$, p = 1.88%). When plotted against each other, these ratios showed a close correlation (Fig. 6). Once again, K3 seemed to be slightly out of line with the main trend, with z1/DMSO and oA/oM values lower than expected in comparison to the rA/rM values.

These results partially confirmed the data obtained in the embryogenesis experiment: a negative correlation was found between infected plant disease extent and relative esterase activity in the presence of fungal raw extracts. Reactions to fungal raw and organic extracts were almost the same. Nevertheless, there were several marked differences between these two datasets. First, there was a strong correlation between the low zinniol concentration effect and the raw or organic extract effect. Second, although raw and organic extracts did effectively block embryogenesis amongst susceptible cultivars, esterase activity was not markedly affected by these extracts in the susceptible cultivar H1 after 48 h of exposure.

In order to investigate this apparent discrepancy, we used microscopy to assess H1 and K3 cell survival and esterase activity rates after 7 and 14 days of exposure to either fungus (oA) or the uninoculated medium (oM) organic phase (Fig. 7). oA treated K3 cell esterase activity, survival and embryogenesis could not be differentiated from oM treated cells. At 7 days, mortality was somewhat higher and esterase activity lower in oA- as compared to oM-treated H1 cells. The much greater differences observed at 14 days followed a similar trend. High mortality was noted amongst oA treated H1 cells as compared to oM-treated cells. Moreover, proembryogenic masses were visible in oM-treated H1 cultures, and not in oA treated cultures.

Discussion

The aim of this study was to investigate the role of fungal toxins in both pathogenicity and resistance in the carrot-*A. dauci* interaction. Since *A. dauci* toxins are not fully known [29], we opted to confront *in vitro* cultured carrot cells with raw fungal extracts. A differential response to phytotoxicity was clearly demonstrated between susceptible and partially resistant carrot genotypes after fungal exudate treatment of plant embryogenic cultures. A close correlation was noted between the resistance to *A. dauci* at the whole plant level and resistance to fungal exudates at the cellular level.

The toxicity of raw and organic fungal extracts was clearly noted, while no toxic reaction of embryogenic cultures was

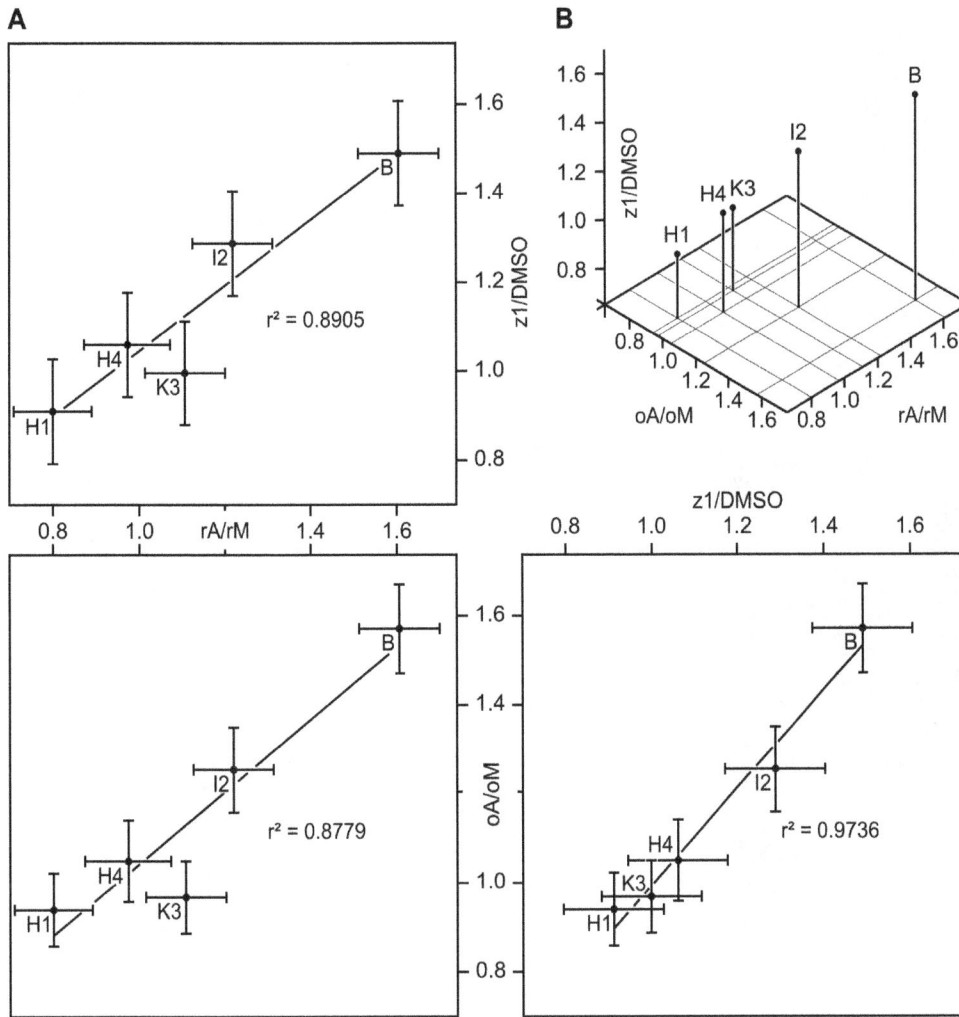

Figure 6. Correlations between cell suspension reactions to _Alternaria dauci_ raw extracts, organic extracts and low zinniol concentrations. Five carrot genotypes were tested for their metabolic activity when _A. dauci_ raw (rA) or organic (oA) extract was added to the plant culture medium. The same experiments were conducted while adding uninoculated medium raw (rM) or organic (oM) extract and 0.025 μM zinniol to DMSO (z1) or DMSO. rA/rM denotes plant cell esterase activity variations due to the presence of fungal raw extracts, oA/oM denotes plant cell esterase activity variations due to the presence of fungal organic extracts, and z1/DMSO denotes plant cell esterase activity variations due to the presence of 0.025 μM zinniol in the medium. A: correlation plots of rA/rM, oA/oM and z1/DMSO by pairs. Bars represent standard errors. The three paired correlated activity indices presented here correspond to the most significant r^2 values (see Table 4). B: 3D correlation plot of rA/rM, oA/oM and z1/DMSO.

Table 4. Correlation coefficients for esterase activity ratios.

	z3/DMSO	z2/DMSO	z1/DMSO	oA/oM	aA/aM
rA/rM	0.0217	0.0026	**0.8906**	**0.8779**	0.4037
aA/aM	0.1610	0.0560	0.5327	0.3723	
oA/oM	0.0463	0.0195	**0.9736**		
z1/DMSO	0.0122	0.0240			
z2/DMSO	0.5734				

Carrot cell suspensions with five different genotypes were tested for esterase relative specific activity in the presence of fungal extracts and toxins. The treatments were as follows: rA: _A. dauci_ (strain FRA017) fungal culture raw extract; rM: uninoculated medium raw extract; aA: _A. dauci_ fungal culture aqueous extract; aM, uninoculated medium aqueous extract; oA: _A.dauci_ fungal culture organic extract; oM: uninoculated medium organic extract; DMSO: DMSO solution at a concentration corresponding to oM, z1, z2 and z3 treatments; z1: 0.025 μM zinniol; z2: 10 μM zinniol; z3: 500 μM zinniol. Correlation coefficients corresponding to significant ($\alpha = 0.05$) linear regressions are in bold.

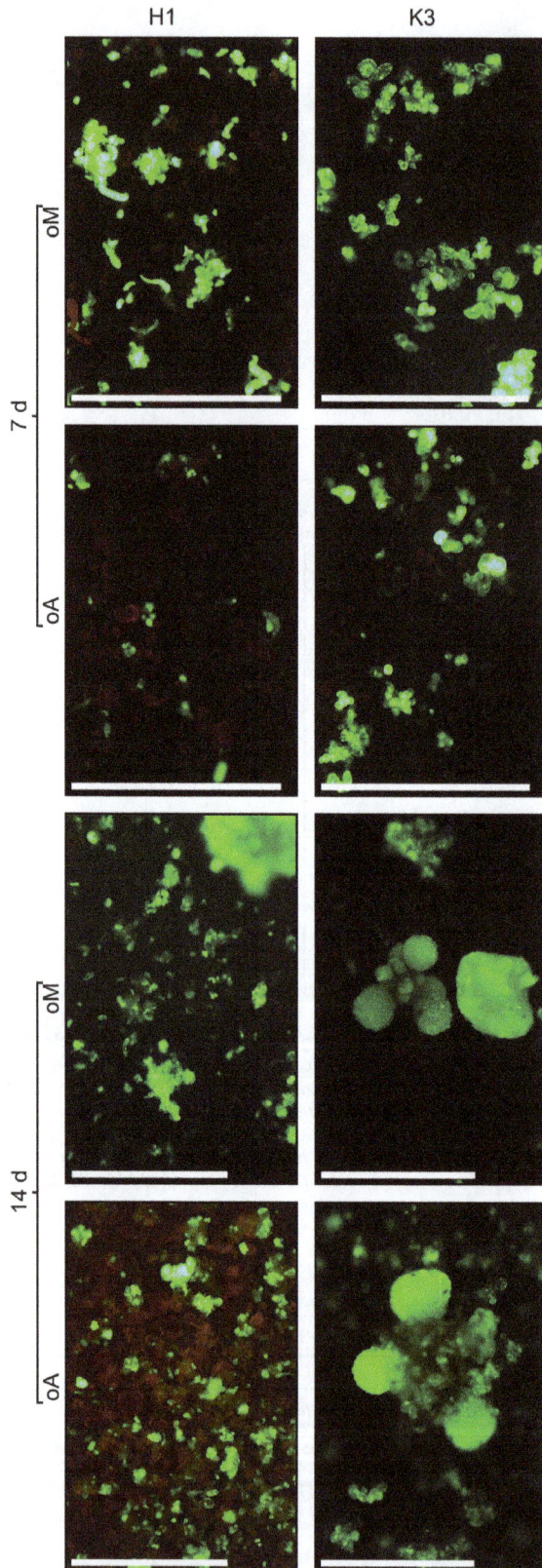

organic extract (oM). Seven and 14 days after adding extracts, membrane integrity and cell viability were evaluated by microscopy using a double staining method with fluoresceine diacetate (FDA) and propidium iodide (IP). The images shown are representative of results obtained from three independent experiments. oA treated K3 cell esterase activity, survival and embryogenesis could not be differentiated from oM treated cells. At 7 days, mortality was somewhat higher and esterase activity lower in oA- than oM-treated H1 cells. At 14 days, much greater observed differences followed a similar trend. High mortality was visible in oA treated H1 cells compared to oM-treated cells. Moreover, proembryogenic masses were visible in oM-treated H1 cultures, and not in oA treated cultures.

obtained after treatment with the aqueous extract. If toxic metabolites were present in this aqueous extract, their concentrations were likely too low to induce toxic effects. Peptidic HSTs have previously been described in other pathosystems involving fungi: *Stagonospora nodorum*, *Pyrenophora tritici-repentis* and two *Alternaria* species (AB-toxin in *A. brassicicola*, AP-toxin in *A. panax*, for a review, see [47]). The production of such toxic peptides in *A. dauci* exudates has, to our knowledge, never been reported. In our study, zinniol, a putative NHST used at physiological concentrations (10 μM), did not exhibit toxicity towards carrot embryogenic cultures. Moreover, zinniol was not detected in exudates collected from 48 h fungal cultures. Five μM zinniol was produced by the same fungal strain in exudates from a 12 day culture under anoxic conditions, which is in line with the findings of Barash et al. [24]. This highly suggests that: (i) zinniol was not responsible for the phytotoxic reactions observed after treatment with the organic extract, and (ii) one or several unknown toxic hydrophobic metabolites were produced by *A. dauci*.

In previous studies, zinniol toxicity was evaluated by direct application of this compound on leaves of different plant species (including carrot) at relatively high concentrations ranging from 150 μM to 1 mM [24,26,48]. Application of 500 μM zinniol to *Tagetes erecta* cell suspensions was deleterious for the cultures [28]. Using the same zinniol concentration, we observed a similar response from carrot cell suspensions irrespective of the plant genetic background. However, due to the high zinniol concentrations used in the papers cited above, the results obtained at whole plant or cellular levels probably overestimated the role of zinniol as a phytotoxin. This bias seems to be absent in papers investigating the activity of other phytotoxins. For example, in the *Stemphylium solani- Allium sativum* pathosystem [22], necrotic lesions were observed on leaves of a susceptible garlic genotype using a purified toxin (SS-toxin) at 11 μg mL^{-1} concentration from a 21 day fungal culture filtrate. In the present study, no phytotoxic reactions were observed using 10 μM zinniol (3 μg mL^{-1}). Consequently, zinniol is probably not a phytotoxin as previously suggested by Qui et al. [28]. By comparison, we obtained toxic effects on carrot embryogenic cultures treated with the organic extract at 7.5 μg mL^{-1} concentration (25% of the original fungal culture medium). Moreover, the HPLC spectra indicated that no dominant hydrophobic metabolite was present. These two combined results suggest the production by *A. dauci* of hydrophobic compounds at least 5-fold more toxic than zinniol to carrot cells.

In this study, carrot *in vitro* cell suspensions from several genotypes were challenged with fungal extracts and zinniol. Compound toxicity and genotype resistance were evaluated on the basis of cell viability and embryogenic ability. In several other studies, plant cell reactions to compounds produced by pathogenic fungi were investigated using *in vitro* cell suspension cultures. Cultured grapevine cell defense-related compound production was enhanced by adding autoclaved *Phaeomoniella clamydospora* biomass

Figure 7. Toxicity and resistance evaluations using fluorescence microscopy. Liquid cell cultures from two carrot genotypes were tested for mortality and metabolic activity when *Alternaria dauci* organic extract (oA) was added to the plant culture medium. The same experiments were conducted while adding uninoculated medium

[49]. Similarly, when challenged with two distinct *Botrytis cinerea* elicitors (botrycin and cinerein), cultured grapevine cells showed defense reactions that differed depending on the tested elicitor [50]. Conversely, fungal toxins from *Rhizoctonia solani* and *Sarocladium oryzae* were shown to inhibit defense-related compounds in rice cell suspensions [51]. A link between plant pathogen partial resistance and toxin resistance has been suggested in the *Allium sativum –Stemphylium solani* pathosystem [22] but, to our knowledge, the present study is the first example where *in vitro* cell viability and embryogenic ability were used as an indicator of fungal toxin plant resistance. In our study, we adapted cell viability measurement methods based on measuring esterase activity using FDA as a substrate to carrot cell suspension cultures. Cell viability is classically measured using counting methods in which viable and nonviable cell numbers are compared. Nevertheless these methods lack accuracy because of the weight and clumpiness of cultured plant cells. Since FDA fluorescence was proposed by Rotman and Papermaster [52] as a way of measuring esterase activity, this procedure has been very widely used to measure cell viability and activity, e.g. in *Medicago truncatula* [37] or soil microorganisms [44]. We also adapted the microscopy techniques proposed by Vitecek *et al.* [43] using both FDA green fluorescence and propidium iodide red fluorescence in damaged cells as a good indicator of viability.

Although this study was not aimed at assessing the kinetics of the effects of fungal toxins on carrot cells, observations were performed at different times: esterase activity quantification was performed 48 h after adding extracts. Microscopic observations were performed after 7 days and 14 days of exposure, while embryogenesis was observed after 21 days of exposure. Overall, these results suggest a long-term effect of the fungal extract: at 48 h, the average esterase activity of susceptible H1 cells relative to that of unexposed cells was 80% (Fig. 5). Differences with respect to the negative control were noted after 7 days, and they were more clearcut after 14 days (Fig. 7). Nevertheless, some cells were still alive. At 21 days, no embryogenesis was visible, and only debris was observed (Fig. 4A). Since no further variations were noted after several more weeks, we assumed that no more living cells were present. This should perhaps be considered in the light of the fact that, even under very favorable conditions (24°C, 100% RH, in susceptible cultivars such as Presto), the first symptoms were only visible 7 days after inoculation. In favorable conditions, other plant fungal pathogens cause visible symptoms earlier (often within 72 h, e.g. with *Magnaporthe grisea*, Fig. 3 in [40], *Botrytis cinerea* (see Fig. 1 in [53]), or *Alternaria brassicicola* (see Fig. 8 in [54]).

Amongst resistant cultivars I2 and Bolero, esterase activity was enhanced after 48 h of exposure to fungal raw extract, organic extract, and low zinniol concentrations. These results surprised us as we expected to detect toxicity through a drop in esterase activity, and resistance through the absence of such a drop in resistant cultivars, as was observed in resistant cultivar K3. As an afterthought, a rise in esterase activity could perhaps be interpreted as a plant resistance reaction. FDA enters plant cells where it can be hydrolyzed by various enzymes, including proteases, lipases and esterases [52]. Such hydrolytic enzymes can be linked with plant defense mechanisms through mobilization of the primary energy metabolism, reducing ability and carbon skeleton for defense [55]. Under that hypothesis, the higher metabolic activity of Bolero and I2 cells could be explained by the fact that, in these cultivars, plant cells are able to detect hydrophobic compounds produced by fungi that include zinniol. The data presented here are not out of line with this interpretation. A high correlation was found between esterase activities in the presence of organic fungal exudates and the low zinniol concentration (Fig. 6). As these effects were measured 48 h

after plant cell exposure to zinniol or organic extract, the low concentrations of zinniol produced by *A. dauci* might be involved in the plant response right after the onset of the plant-pathogen interaction. Zinniol was not found in our fungal exudates, but its presence at very low concentrations could not be ruled out. Besides, zinniol was detected in infected plant tissues in at least two different pathosystems at early stages of plant infection: 2 days after sunflower infection by *Phoma macdonaldii* [27] and 12 h after carrot infection by *A. dauci* [56]. More generally, elicitors are often described as small secreted proteins or polymers, but there seem to be other cases where fungal secondary metabolites [57], or more generally small molecules [58] play such a role.

In conclusion, three main insights emerged from the presented data: (i) strongly phytotoxic compounds are present in the organic phase of *A. dauci* exudates, (ii) zinniol is not the main phytotoxic compound produced by *A. dauci*, and (iii) carrot resistance to *A. dauci* involves cellular resistance to these compounds. Our study also raised new questions, especially concerning the nature of the hydrophobic toxic compounds present in the organic phase. *Alternaria dauci* aggressiveness varies strongly depending on the strain [33]. It would be interesting to determine if these variations are linked with quantitative or qualitative variations in the production of those compounds in fungal strains. Moreover, the role of zinniol in the carrot-*A. dauci* interaction should be redefined.

Supporting Information

Figure S1   HPLC analysis of zinniol stability in different solutions. HPLC chromatograms were obtained from different 10 µg zinniol samples after an incubation of one week at room temperature. **A**: Zinniol incubated in a deuterated DMSO-aqueous buffer at pH 5.6. **B**: Zinniol incubated in B5 Gamborg medium (0.4% DMSO, pH 5.8). **C**: Zinniol incubated in $CDCl_3$. In A and B, one strong peak is visible at 7.015 minutes, corresponding to zinniol expected retention time. In C, a small peak is visible at the same retention time. Other peaks are visible at 9.63, 14.33, 14.55, 15.30 and 16.40 minutes retention time. AU: Absorption Units (optical density) at 233 nm.

Table S1   Influence of culture medium and anoxia on fungal exudates organic extracts toxicity. Carrot cell suspensions with two different genotypes were tested for embryogenesis in the presence of fungal extracts. Embryogenesis was assessed 4 weeks after treatment. [1]Treatments were as follows: C: no treatment, DMSO: DMSO solution at the same concentration than in organic extracts. Organic extracts from *Alternaria dauci* (strain FRA017) fungal culture grown in the following conditions: oA: 48 h shaking in carrot juice medium, oA4d: 96 h shaking in carrot juice medium, oAV: 72 h shaking in V8 medium, oAVa: 12 days no shaking (anoxia) in V8 medium, oC uninoculated carrot medium. [2]The signs are as follows: (−) no embryogenesis was visible and cells were damaged, (+) early-stage embryogenic masses were visible, (++) embryos were present, (+++) embryogenesis was profuse.

Acknowledgments

We would like to thank Nicolas Turnbull and Cédric Dumont for their participation in the preliminary phases of the *in vitro* testing experiments. We also would like to thank Aurélia Rolland and David Macherel who provided access to the spectrofluorometer in their lab, and our colleagues at IRHS (INEM team) for their help in the greenhouse experiments. Anita Suel, Sébastien Huet and Valérie Le Clerc are also gratefully acknowledged for the selection and maintenance of carrot genotypes. We also thank Bruno Hamon for his assistance in the microbiology experiments and fungal strain maintenance, Karlina García-Sosa and Landy Uc-Cen for their technical help in anoxic fungal cultures. We are grateful to Tristan Boureau and Séverine Gagné whose insight helped us in the redaction of the manuscript, Piétrick Hudhomme (Université d'Angers, MOLTECH Anjou, UMR CNRS 6200) for advice on fungal extract preparation and David Manley for revising the English in the manuscript.

References

1. Boyd LA, Ridout C, O'Sullivan DM, Leach JE, Leung H (2013) Plant-pathogen interactions: disease resistance in modern agriculture. Trends in Genetics 29: 233–240.
2. Kou Y, Wang S (2010) Broad-spectrum and durability: understanding of quantitative disease resistance. Current Opinion in Plant Biology 13: 1–5.
3. St.Clair DA (2010) Quantitative disease resistance and quantitative resistance loci in breeding. Annual Review of Phytopathology 48: 247–268.
4. Zhang Y, Lubberstedt T, Xu M (2013) The genetic and molecular basis of plant resistance to pathogens. Journal of Genetics and Genomics 40: 23–35.
5. Poland JA, Balint-Kurti PJ, Wisser RJ, Pratt RC, Nelson RJ (2009) Shades of gray: the world of quantitative disease resistance. Trends in Plant Science 14: 21–29.
6. Kushalappa AC, Gunnaiah R (2013) Metabolo-proteomics to discover plant biotic stress resistance genes. Trends in Plant Science 18: 522–531.
7. Lai Z, Mengiste T (2013) Genetic and cellular mechanisms regulating plant responses to necrotrophic pathogens. Current Opinion in Plant Biology 16: 505–512.
8. Miya A, Albert P, Shinya T, Desaki Y, Ichimura K, et al. (2007) CERK1, a LysM receptor kinase, is esssential for chitin elicitor signaling in Arabidopsis. Proceedings of the National Academy of Sciences of the USA 104: 19613–19618.
9. Zheng Z, Qamar SA, Chen Z, Mengiste T (2006) Arabidopsis WRKY33 transcription factor is required for resistance to necrotrophic fungal pathogens. The Plant Journal 48: 592–605.
10. Fukuoka S, Saka N, Koga H, Ono K, Shimizu T, et al. (2009) Loss of function of a proline-containing protein confers durable disease resistance in rice. Science 325: 998–1001.
11. Fu J, Liu H, Li Y, Yu H, Li X, et al. (2011) Manipulating broad-spectrum disease resistance by suppressing pathogen-induced auxin accumulation in rice. Plant Physiology 155: 589–602.
12. Fu D, Uauy C, Distelfeld A, Blechl A, Epstein L, et al. (2009) A kinase-start gene confers temperature dependant resistance to wheat stripe rust. Science 323: 1357–1360.
13. Cao A, Xing L, Wang X, Yang X, Wang W, et al. (2011) Serine/threonine kinase gene Stpk-V, a key member of powdery mildew resistance gene Pm21, confers powdery mildew resistance in wheat. Proceedings of the National Academy of Sciences of the USA 108: 7727–7732.
14. Dubouzet JG, Maeda S, Sugano S, Ohtake M, Hayashi N, et al. (2011) Screening for resistance against Pseudomonas syringae in rice-FOX Arabidopsis lines identified a putative receptor-like cytoplasmic kinase gene that confers resistance to major bacterial and fungal pathogens in Arabidopsis and rice. Plant Biotechnology Journal 9: 466–485.
15. Schweizer P, Stein N (2011) Large-scale data integration reveals colocalization of gene functional groups with meta-QTL for multiple disease resistance in barley. Molecular Plant-Microbe Interactions 12: 1492–1501.
16. Kliebenstein DJ, Rowe HC, Denby KJ (2005) Secondary metabolites influence Arabidopsis/Botrytis interactions: variation in host production and pathogen sensitivity. The Plant Journal 44: 25–36.
17. Lecomte M, Berruyer R, Hamama L, Boedo C, Hudhomme P, et al. (2012) Inhibitory effects of the carrot metabolites 6-methoxymellein and falcarindiol on development of the fungal leaf blight pathogen Alternaria dauci. Physiological and Molecular Plant Pathology 80: 58–67.
18. Thomma BPHJ (2003) Alternaria spp.: from general saprophyte to specific parasite. Molecular Plant Pathology 4: 225–236.
19. Johal GS, Briggs SP (1992) Reductase activity encoded by the HM1 disease resistance gene in maize. Science 258: 985–987.
20. Spassieva SD, Markham JE, Hille J (2002) The plant disease resistance gene Asc-1 prevents disruption of sphingolipid metabolism during AAL-toxin-induced programmed cell death. The Plant Journal 32: 561–572.
21. Walz A, Zingen-Sell I, Loeffler M, Sauer M (2008) Expression of an oxalate oxidase gene in tomato and severity of disease caused by Botrytis cinerea and Sclerotinia sclerotiorum. Plant Pathology 57: 453–458.
22. Zheng L, Lv R, Huang J, Jiang D, Hsiang T (2010) Isolation, purification, and biological activity of a phytotoxin produced by Stemphylium solani. Plant Disease 94: 1231–1237.
23. Walton JD (1996) Host-selective toxins: agents of compatibilty. The Plant Cell 8: 1723–1733.
24. Barash I, Mor H, Netzer D, Kashman Y (1981) Production of zinniol by Alternaria dauci and its phytotoxic effect on carrot. Physiological Plant Pathology 19: 7–16.
25. Thuleau P, Graziana A, Rossignol M, Kauss H, Auriol P, et al. (1988) Binding of the phytotoxin zinniol stimulates the entry of calcium into plant protoplasts. Proceedings of the National Academy of Science of the USA 85: 5932–5935.
26. Cotty PJ, Misaghi IJ (1984) Zinniol production by Alternaria species. Phytopathology 74: 785–788.
27. Sugawara F, Strobel G (1986) zinniol, a phytotoxin, is produced by Phoma macdonaldii. Plant Science 43: 19–23.
28. Qui JA, Castro-Concha LA, Garcia-Sosa K, Miranda-Ham ML, Peña-Rodriguez LM (2010) Is zinniol a true phytotoxin? Evaluation of its activity at the cellular level against Tagetes erecta. Journal of General Plant Pathology 76: 94–101.
29. Andersen B, Dongo A, Pryor BM (2008) Secondary metabolite profiling of Alternaria dauci, A. porri, A.solani, and A. tomatophila. Mycological Research 112: 241–250.
30. Montemurro N, Visconti A (1992) Alternaria metabolites - chemical and biological data. In: Chelkovsky J, Visconti A, editors. Alternaria biology, plant diseases and metabolites. Amsterdam: Elsevier. pp. 449–557.
31. Somma S, Pose G, Pardo A, Mulè G, Fernandez Pinto V, et al. (2011) AFLP variability, toxin production, and pathogenicity of Alternaria species from Argentinean tomato fruits and puree. International Journal of Food Microbiology 145: 414–419.
32. Graf E, Schmidt-Heydt M, Geisen R (2012) HOG MAP kinase regulation of alternariol biosynthesis in Alternaria alternata is important for substrate colonization. International Journal of Food Microbiology 157: 353–359.
33. Boedo C, Benichou S, Berruyer R, Bersihand S, Dongo A, et al. (2012) Evaluating aggressiveness and host range of Alternaria dauci in a controlled environment. Plant Pathology 121: 55–66.
34. Reinert J (1958) Morphogenese und ihre Kontrolle an Gewebekulturen aus Carotten. Naturwissenschaften 45: 344–345.
35. Steward FC, Mapes MO, Mears K (1958) Growth and organized development of cultured cells. II. Organization in cultures grown from freely suspended cells. American Journal of Botany 45: 705–708.
36. George EF, Hall MA, Klerk GJD, editors (2008) Plant propagation by tissue culture: volume 1. the background. Dordrecht: Springer. 508 p.
37. Steward N, Martin R, Engasser JM, Georgen JL (1999) A new methodology for plant cell viability assessment using intracellular esterase activity. Plant Cell Reports 19: 171–176.
38. Boedo C, Berruyer R, Lecomte M, Bersihand S, Briard M, et al. (2010) Evaluation of different methods for the characterization of carrot resistance to the alternaria leaf blight pathogen (Alternaria dauci) revealed two qualitatively different resistances. Plant Pathology 59: 368–375.
39. Boedo C, Le Clerc V, Briard M, Simoneau P, Chevalier M, et al. (2008) Impact of carrot resistance on development of the Alternaria leaf blight pathogen (Alternaria dauci). European Journal of Plant Pathology 121: 55–66.
40. Berruyer R, Poussier S, Kankanala P, Mosquera G, Valent B (2006) Quantitative and qualitative influence of inoculation methods on in planta growth of rice blast fungus. Phytopathology 96: 346–355.
41. Martin JA, Vogel E (1980) The synthesis of zinniol. Tetrahedron 36: 791–794.
42. Gamborg OL, Murashige T, Thorpe TA, Vasil IK (1976) Plant tissue culture media. In Vitro 12: 473–478.
43. Vitecek J, Petrlova J, Adam V, Havel L, Kramer KJ, et al. (2007) A fluorimetric sensor for detection of living cell. Sensors 7: 222–238.
44. Green VS, Stott DE, Diack M (2006) Assay for fluorescein diacetate hydrolytic activity: optimisation for soil samples. Soil Biology and Biochemistry 38: 693–701.
45. Bradford MM (1976) A rapid and sensitive method for the quantitation of microgram quantities of protein utilizing the principle of protein-dye binding. Analytical Biochemistry 72: 248–254.
46. Cotty PJ, Misaghi IJ, Hine RB (1983) Production of zinniol by Alternaria tagetica and its phytotoxic effects on Tagetes erecta. Phytopathology 73: 1326–1328.
47. Horbach R, Navarro-Quesada AR, Knogge W, Deising HB (2011) When and how to kill a plant cell: infection strategies of plant pathogenic fungi. Journal of Plant Physiology 168: 51–62.
48. Berestetskii AO, Yuzikhin OS, Katkova AS, Dobrodumov AV, Sivogrivov DE, et al. (2010) Isolation, identification, and characteristics of the phytotoxin produced by the fungus Alternaria cirsinoxia. Applied Biochemistry and Microbiology 46: 75–79.
49. Lima MRM, Ferreres F, Dias ACP (2011) Response of Vitis vinifera cell cultures to Phaeomoniella chlamydospora: changes in phenolic production, oxidative state and expression of defence-related genes. European Journal of Plant Pathology 132: 133–146.

Author Contributions

Conceived and designed the experiments: ML LH JJH DS PP RB. Performed the experiments: ML LH LV JG JJH LPR CB CY MG. Analyzed the data: LH JJH DS PR RB. Contributed reagents/materials/analysis tools: MB. Wrote the paper: ML LH JJH MB PS PP RB.

50. Repka V (2006) Early defence responses induced by two distinct elicitors derived from *Botrytis cinerea* in grapevine leaves and cell suspensions. Biologia Plantarum 50: 94–106.

51. Bithell A, Hsu T, Kandanearatchi A, Landau S, Everall IP, et al. (2010) Expression of the Rap1 guanine nucleotide exchange factor, MR-GEF, is altered in individuals with bipolar disorder. PLoS One 5: e10392.

52. Rotman B, Papermaster BW (1966) Membrane properties of living mammalian cells as studied by enzymatic hydrolysis of fluoregenic esters. Proceedings of the National Academy of Sciences of the USA 55: 134–141.

53. Bessire M, Chassot C, Jacquat AC, Humphry M, Borel S, et al. (2007) A permeable cuticle in *Arabidopsis* leads to a strong resistance to *Botrytis cinerea*. The EMBO Journal 26: 2158–2168.

54. Calmes B, Guillemette T, Teyssier L, Siegler B, Pigné S, et al. (2013) Role of mannitol metabolism in the pathogenicity of the necrotrophic fungus *Alternaria brassicicola*. Frontiers in Plant Science 4: 131.

55. Bolton MD (2009) Primary metabolism and plant defense - Fuel for the fire. Molecular Plant-Microbe Interactions 22: 487–497.

56. Montillet JL (1986) Dosage radioimmunologique du zinniol. Application à l'étude de cette toxine dans l'Alternariose de la carotte. Toulouse: Paul Sabatier 73 p.

57. Böhnert HU, Fudal I, Dioh W, Tharreau D, Notteghem JL, et al. (2004) A putative polyketide synthase/peptide synthetase from *Magnaporthe grisea* signals pathogen attack to resistant rice. The Plant Cell 16: 2499–2513.

58. Sinha AK, Hofmann MG, Römer U, Köckenberger W, Elling L, et al. (2002) Metabolizable and non-metabolizable sugars activate different signal transduction pathways in tomato. Plant Physiology 128: 1480–1489.

14

Characterization of Centromeric Histone H3 (CENH3) Variants in Cultivated and Wild Carrots (*Daucus* sp.)

Frank Dunemann[1]*, Otto Schrader[1], Holger Budahn[1], Andreas Houben[2]

1 Julius Kühn-Institut (JKI) - Federal Research Centre for Cultivated Plants, Institute for Breeding Research on Horticultural Crops, Quedlinburg, Germany, 2 Leibniz-Institute of Plant Genetics and Crop Plant Research (IPK), Chromosome Structure and Function Laboratory, Gatersleben, Germany

Abstract

In eukaryotes, centromeres are the assembly sites for the kinetochore, a multi-protein complex to which spindle microtubules are attached at mitosis and meiosis, thereby ensuring segregation of chromosomes during cell division. They are specified by incorporation of CENH3, a centromere specific histone H3 variant which replaces canonical histone H3 in the nucleosomes of functional centromeres. To lay a first foundation of a putative alternative haploidization strategy based on centromere-mediated genome elimination in cultivated carrots, in the presented research we aimed at the identification and cloning of functional CENH3 genes in *Daucus carota* and three distantly related wild species of genus *Daucus* varying in basic chromosome numbers. Based on mining the carrot transcriptome followed by a subsequent PCR-based cloning, homologous coding sequences for CENH3s of the four *Daucus* species were identified. The ORFs of the CENH3 variants were very similar, and an amino acid sequence length of 146 aa was found in three out of the four species. Comparison of *Daucus* CENH3 amino acid sequences with those of other plant CENH3s as well as their phylogenetic arrangement among other dicot CENH3s suggest that the identified genes are authentic CENH3 homologs. To verify the location of the CENH3 protein in the kinetochore regions of the *Daucus* chromosomes, a polyclonal antibody based on a peptide corresponding to the N-terminus of *DcCENH3* was developed and used for anti-CENH3 immunostaining of mitotic root cells. The chromosomal location of CENH3 proteins in the centromere regions of the chromosomes could be confirmed. For genetic localization of the CENH3 gene in the carrot genome, a previously constructed linkage map for carrot was used for mapping a CENH3-specific simple sequence repeat (SSR) marker, and the CENH3 locus was mapped on the carrot chromosome 9.

Editor: Yamini Dalal, National Cancer Institute, United States of America

Funding: This research was supported by the Federal Ministry of Agriculture and Food(BMEL)in frame of an implemented project at the Julius Kühn-Institut, Quedlinburg, Germany. The funder had no role in study design, data collection and analysis, decision to publish, or preparation of the manuscript.

Competing Interests: The authors have declared that no competing interests exist.

* E-mail: frank.dunemann@jki.bund.de

Introduction

The cultivated carrot (*Daucus carota*) is one of the most important vegetable plants in the world. With a current annual world production of more than 30 million tons and a total growing area of about 1.5 million hectares (FAOSTAT 2012) it ranks among the top ten vegetable crops. Carrot is the most widely grown species of the genus *Daucus*, a member of the large and complex Apiaceae plant family. The genus *Daucus* includes around 25 species and was subdivided taxonomically into five [1], and later into seven sections [2], but both classification systems are not yet fully congruent with molecular phylogenetic studies [3]. *Daucus* species are widespread in the temperate areas of the northern hemisphere, but few species exist also in South America and Australia [3]. *D. carota* is a diploid outcrossing species with nine chromosome pairs ($2n = 2x = 18$). *D. capillifolius*, *D. sahariensis* and *D. syrticus* are the other members of the genus with $2n = 18$ chromosomes, whereas *D. muricatus* ($2n = 20$) and *D. pusillus* ($2n = 22$) have a slightly higher chromosome number. It is assumed that $x = 11$ is the basic chromosome number in Apiaceae family, and $x = 10$ and $x = 9$ are its derivatives [4]. However, a few polyploid species as for example *D. glochidiatus* ($2n = 4x = 44$) and *D. montanus* ($2n = 6x = 66$) also exist.

The haploid genome size of carrot has been estimated at 473 Mbp [5], which is similar to rice. First carrot linkage maps have been developed based on several types of molecular markers [6,7], and a BAC library of the carrot genome has been created [8]. Furthermore, the carrot transcriptome has been revealed recently by next generation sequencing (NGS) technology [9]. Carrot is also well known as a model species for gene transfer using both genetic modifications by vector and non-vector methods, which is a major prerequisite for functional gene studies [10].

Despite all these progressed molecular and biotechnological developments comparatively limited work has been done on the cytological and molecular-cytogenetic characterization of the carrot genome. Individual carrot chromosomes are small and uniform in shape and length [11] and are therefore a difficult object for cytogenetic research. Using rDNA genes as probes for fluorescence *in situ* hybridization (FISH) analysis, chromosomal karyotypes were developed for cultivated carrots and other Apiaceae species [11,12]. Carrot BAC clones were used to integrate genetic and physical maps based on pachytene chromosomes of *D. carota*, and mitotic chromosomes of two further 22-chromosome *Daucus* species as well [13].

As a cross-pollinated species suffering from inbreeding depression carrot provides some challenges in plant (hybrid) breeding. Due to the biannual nature of carrots and the difficulties to produce sufficient amounts of seed from selfings, the generation of genetically homogeneous genotypes with a high degree of homozygosity is a long lasting and inefficient task in carrot breeding programs. As an alternative and/or supplement to traditional inbred line production in carrots, double-haploid plants might be produced by *in vitro*-regeneration of plants through anther or microspore culture. However, haploid production by tissue culture techniques is generally highly genotype-dependent and has been reported to be very inefficient in Apiaceae species [14]. The generation of doubled haploids using naturally occurring mechanisms of uniparental genome elimination induced by interspecific hybridization has not yet been reported for *Daucus*.

Recently, a breakthrough technology has been presented by Ravi and Chan [15,16], which uses centromere-mediated genome elimination processes for the generation of haploid and double-haploid plants. It was demonstrated for the first time in *Arabidopsis thaliana*, that haploids can be generated through manual cross-fertilizations after manipulating a single centromere protein, the centromere-specific histone H3 variant CENH3, in one of the parents designated as 'haploid inducer' [15]. Uniparental genome elimination using this strategy was suggested to function in any (crop) plant due to the universal centromere mechanism based on CENH3 function [15,16].

In eukaryotes, centromeres are the assembly sites for the kinetochore, a multi-protein complex to which spindle microtubules are attached at mitosis and meiosis, thereby ensuring segregation of chromosomes during cell division [17]. They are specified by incorporation of CENH3, which replaces canonical histone H3 in the nucleosomes of functional centromeres [18]. Modifications in CENH3 gene transcription or translation could affect the ability to assemble intact CENH3 chromatin and might result in the loss of CENH3 from the centromere region and a loss of proper centromere function. Contrary to canonical histone H3, which is extremely conserved in eukaryotes, CENH3 shows considerable variability between species and shows some signs of adaptive evolution [19]. Presently, investigations on structure and function of plant CENH3s have been reported for a variety of species originating from at least 20 different plant genera. Among them there are most important cereals such as *Zea mays* [20], *Oryza sativa* [21,22], *Saccharum officinarum* [23], *Hordeum* species [24] and a

few other monocots including vegetable *Allium* species [25]. Besides, CENH3s have been intensively studied in the model dicot species *Nicotiana tabacum* [26], some *Brassica* species [27] and several members of the Leguminosae family including soybean, common bean, and peas [28–31]. To our knowledge, no investigation on CENH3s from Apiaceae species has been reported up to now.

To lay a first foundation of a putative alternative haploidization strategy based on centromere-mediated genome elimination in cultivated carrots, the major aim of the present study was to identify functional *Daucus* CENH3 genes and to verify the location of the CENH3 protein in the kinetochore regions of the *Daucus* chromosomes. Complementary coding sequences of CENH3s of four *Daucus* species were identified and phylogenetically compared with previously reported plant CENH3s. A generated polyclonal CENH3 antibody confirmed the centromeric location of CENH3 proteins, and the CENH3 locus was genetically mapped on the carrot chromosome 9.

Materials and Methods

Plant Material and Isolation of Genomic DNA and cDNA

The carrot (*D. carota* subsp. *sativus*) cultivar 'Deep Purple' (DP) and one accession each from the Mediterranean wild species *D. muricatus* ($2n = 2x = 20$, accession W243/06), the South American species *D. pusillus* ($2n = 2x = 22$, accession 989/92–3) and the Australian species *D. glochidiatus* ($2n = 4x = 44$, accession DAL 341/00) were used for this study. Seeds of wild species were originally received from Hortus Botanicus Coimbra, Portugal (*D. muricatus*), Plant Science Laboratory, University of Reading, U.K. (*D. pusillus*) and Warwick Genetic Resources Unit, Warwick University, Wellesbourne, U.K. (*D. glochidiatus*), and have been kindly provided by T. Nothnagel (Julius Kühn-Institut, Quedlinburg, Germany). Plants obtained from seeds were grown in pots in a greenhouse for DNA and RNA isolations. Total genomic DNA from young leaf tissue of individual plants was extracted using the Qiagen DNeasy Plant Mini kit (Qiagen, Hilden, Germany) following the manufacturer's instructions. For RNA isolation, small leaflets were immediately frozen in liquid nitrogen and ground to fine powder by using a swing mill. Total RNA was

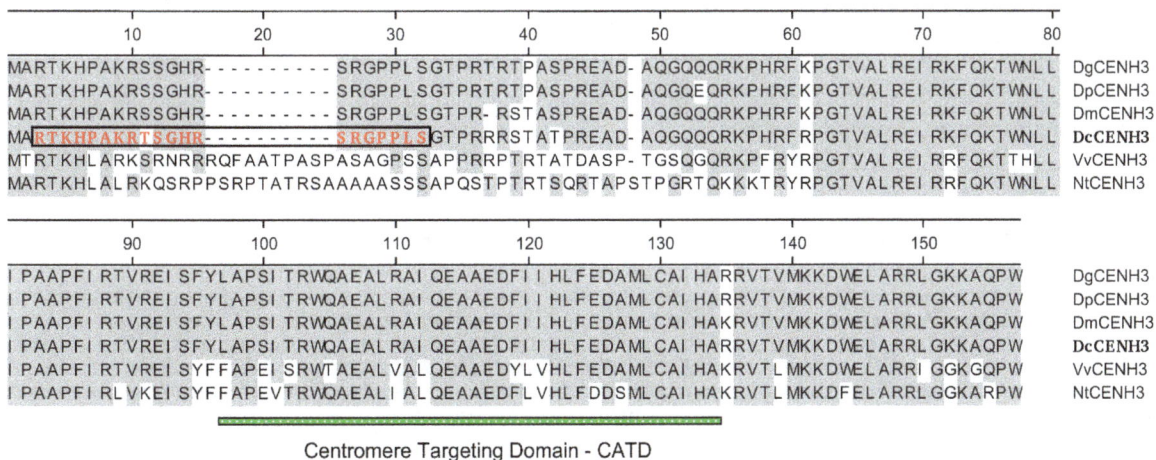

Figure 1. Multiple sequence alignment of the deduced *Daucus* CENH3 proteins and comparison with CENH3 sequences from *Nicotiana tabacum* (GenBank accession number BAH03515) and *Vitis vinifera* (XP_002281073) showing the highest similarity to *Daucus* CENH3s after multiple alignment of various plant CENH3 proteins (see Figure 2). Sequences were compared by ClustalW (Lasergene). The putative centromere targeting domain (CATD) spanning loop 1 and α-2 helix is marked by a crossbar. The position used for construction of a peptide antibody against *DcCENH3* is boxed.

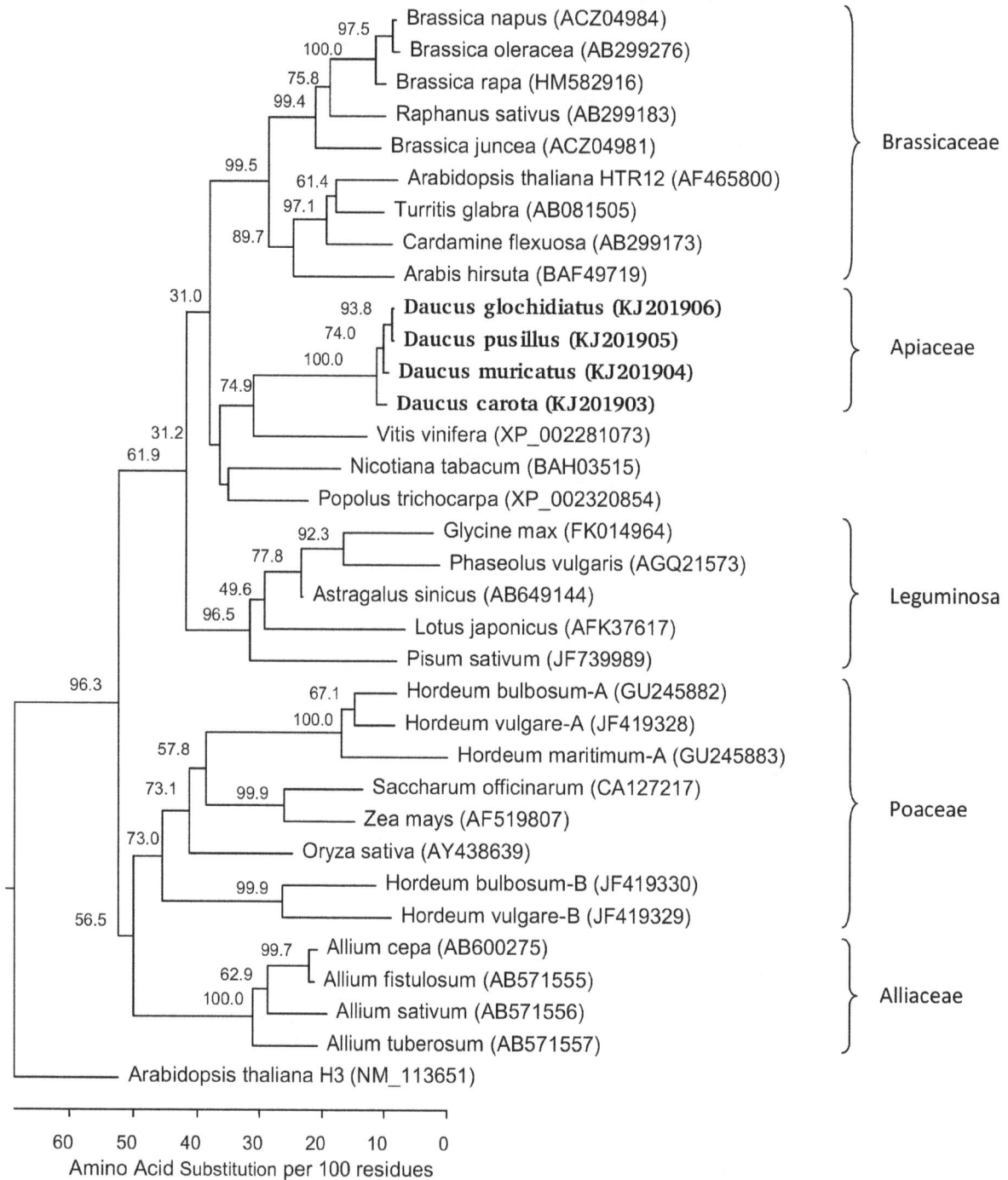

Figure 2. Phylogenetic tree of the deduced *Daucus* CENH3 proteins (printed in bold letters) and a selection of plant CENH3 proteins representing monocot (Alliaceae, Poaceae) and various dicot plant families including Leguminosae and Brassicaceae. Canonical histone H3 of *A. thaliana* was used as an outgroup. For each amino acid sequence, the NCBI accession number is indicated in parentheses. Multiple sequence alignment was performed by ClustalW using the Lasergene (DNASTAR) software package. A phylogenetic tree was constructed using the Kimura distance formula to calculate distance values and bootstrap analysis (10,000 replicates). Numbers indicate bootstrap replication, and branch length is scaled below the tree indicating the number of amino acid substitutions per 100 amino acids.

isolated by using the Qiagen RNeasy Plant Mini kit. An additional DNAse step (Qiagen) was included in this procedure. The qualitatively and quantitatively checked RNA solution was then used to synthesize cDNA with the RevertAid First Strand cDNA Synthesis Kit (Thermo Fisher Scientific, St. Leon-Rot, Germany).

Identification and Cloning of CENH3 Genes

To identify *D. carota* CENH3 orthologous sequences, the assembled carrot transcriptome [9] was used for *in silico* gene mining. A Fasta file containing 58,751 sequences was loaded into the software BioEdit version 7.0.5.3. [32] and screened by NCBI

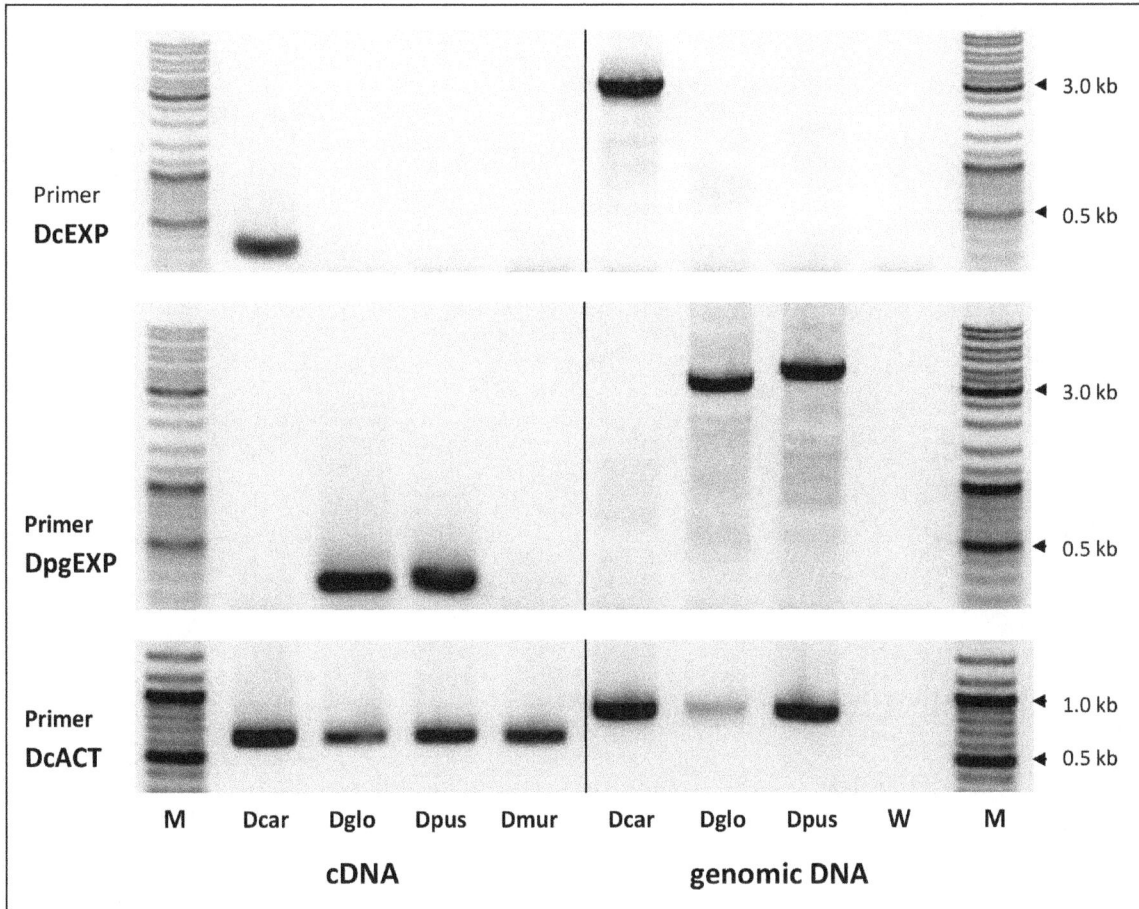

Figure 3. RT-PCR-based transcriptional analysis of CENH3 in *D. carota* **(Dcar),** *D. glochidiatus* **(Dglo),** *D. pusillus* **(Dpus) and** *D. muricatus* **(Dmur) with gene-specific primer pairs designed for** *D. carota* **CENH3 (DcEXP) and CENH3s of** *D. pusillus/D. glochidiatus* **(DpgEXP).** For details, see text, and for position of primers, see Figure S1. For the reference gene *β-actin* the primer pair DcACT was used. Positive control is genomic DNA of *D. carota*, *D. glochidiatus* and *D. pusillus,* and negative control is water (W). Size standard (M) is the Gene Ruler DNA ladder Mix (Thermo Fisher Scientific).

Local BLAST [33] using the tBlastn search option and the translated amino acid sequence of *Nicotiana tabacum* CENH3 (GenBank number BAH03515, [26]) as a query. Since the beginning of the gene was not found, a degenerate PCR forward primer (DCEN1-F: 5'- atg gcg aga acn aar cay) based on the first six amino acids at the N-terminal region of CENH3s of four different dicot species (*A. thaliana*, *Brassica rapa*, *N. tabacum* and *Glycine max*) and an internal gene-specific reverse primer designed from the contig representing the last part of the putative carrot CDS (DCEN1-R: 5'- acg gag cag cag gaa tta ga) were designed and used for PCR-based cloning of the missing carrot CENH3 CDS region. DNA fragments with the expected size of approximately 240 bp obtained after PCR with cDNA templates of 'DP' were excised from the gel, purified using the MinElute Gel Extraction Kit (Qiagen, Hilden, Germany) and cloned by the pGEM-T Easy Vector System (Promega, Madison, USA). Plasmid inserts of selected clones were sequenced by Eurofins-MWG-Operon (Ebersberg, Germany). Based on the sequences obtained a second round of screening the *Daucus* transcriptome was performed with the tBlastx program, and two additional short contigs representing the beginning of the CENH3 gene were detected including a part of the 5'-UTR region in one of the contigs, which was used to design a new PCR primer (DCEN2-F:

5'- ccg tta gaa atc acg gtc atc a). Using this primer together with a newly created reverse primer exactly fitting the last nucleotides of the CDS (DCEN2-R: 5' - acc agg gct gcg ctt tct) we were able to amplify the complete *Daucus* CENH3 coding region as well as its full-length genomic sequence. For latter approach, Long Range PCR (LR-PCR) based on the 'Long PCR Enzyme Mix' (Thermo Scientific) was carried out with genomic DNA of *D. carota* cv. 'DP'. PCR fragments with a size of about 4.5 kb were cloned using the pGEM-T Easy Vector System (Promega). Clones were sequenced by Eurofins-MWG-Operon using the sequencing primers M13uni(-21) and M13rev(-49) for sequencing the beginning and the end of the cloned inserts. To obtain the full-length nucleotide sequence a primer walking approach based on five intermediate primers was used. At least two replications of plasmid insert sequencing were performed for each clone to provide a sufficient reading confidence required for accurate manual assembling of a consensus gene sequence.

For sequence alignment and phylogenetic analysis, putative amino acid sequences were deduced from the determined *Daucus* cDNA sequences and compared with a selection of published CENH3 proteins from other plant species and canonical histone H3 of *A. thaliana* as an outgroup. Multiple sequence alignment (MSA) of CENH3 proteins was performed by ClustalW using the

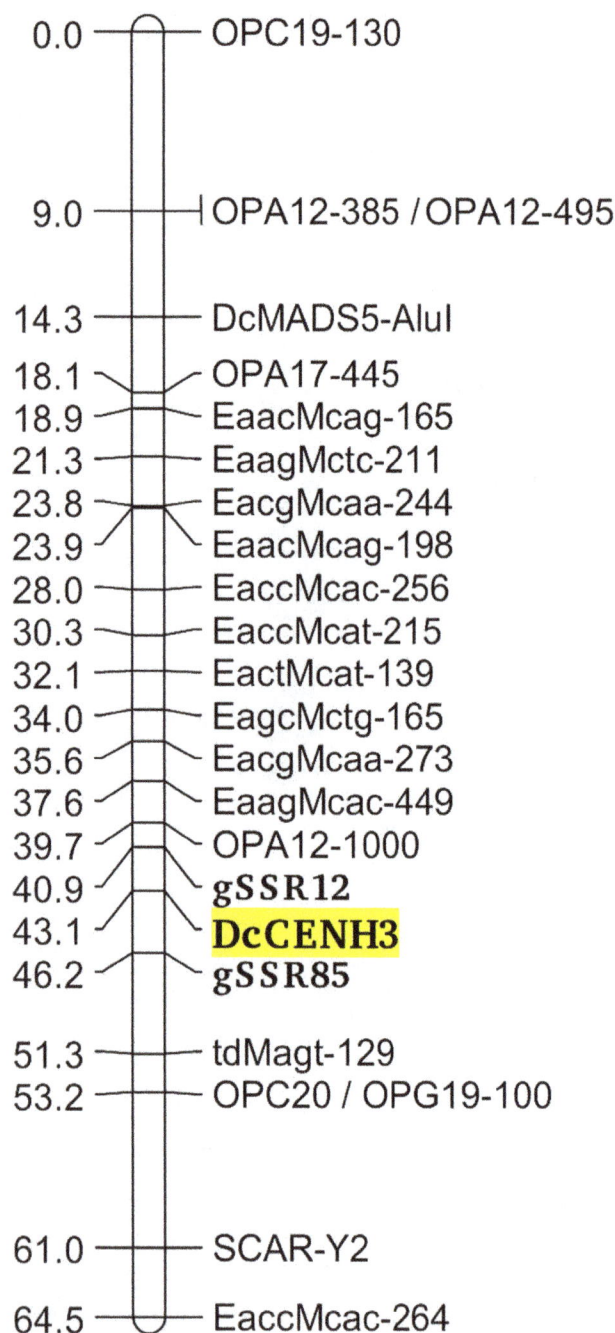

Figure 4. Genetic map of the carrot chromosome 9 (corresponding to linkage group 7) with the calculated position of the *DcCENH3* gene mapped through the DCEN-SSR marker. Scale: centiMorgan (cM).

Lasergene software package (DNASTAR, Madison, WI, USA). A phylogenetic tree was constructed using the Kimura distance formula to calculate distance values and bootstrap analysis (10,000 replicates).

Transcriptional Analysis of CENH3 Gene Expression

For the development of species-specific PCR primers, final CENH3 CDS nucleotide sequences were aligned by ClustalW (Lasergene), and based on sequence differences detected among *D.*

carota and *D. glochidiatus*, a set of two PCR primer pairs (named as DcEXP and DpgEXP) was developed for testing transcription activity of parental CENH3s by reverse transcription (RT) - PCR (for primer sequences, see Figure S3A). As reference gene for RT-PCR the constitutive (house-keeping) gene *β-actin* was chosen, and the following primers were used: DcACT-F: 5′- aca ctg gtg tga tgg ttg ga; DcACT-R: 5′-tgg tga taa ctt gcc cat ca [34]. RT-PCR was carried out in a total volume of 25 µl containing 1 µl of the synthesized cDNA solution, 1 U of 'DreamTaq' DNA polymerase (Thermo Fisher Scientific), 1x *Taq* polymerase buffer with MgCl$_2$ (Thermo Fisher Scientific), 0.2 µM of each primer and 0.2 mM of each dNTP. Amplification conditions were as follows: 1 cycle of 3 min at 94°C; 35 cycles of 94°C for 30 sec, 53°C (DcEXP, DpgEXP) or 57°C (DcACT) for 45 sec, 72°C for 1 min; final extension of 72°C for 5 min. A positive (genomic DNA) and a negative control (water) were included into RT-PCR.

Linkage Mapping

Based on a SSR (simple sequence repeat) sequence found within an intron of the cloned genomic *D. carota* CENH3 sequence, a PCR primer pair (DCEN-SSR-F: 5′- ggt ctc tct ccc tca cac act t; DCEN-SSR-R: 5′- cgt ctc gga gtt ccc tgt ata a) was designed and used for linkage mapping. For chromosomal location of the carrot CENH3 gene a genetic map constructed previously for the carrot progeny DM19 was used [Budahn, unpublished]. DM19 was developed from an initial cross of two parental *D. carota* leaf mutants ('Yellow' and 'Cola'). Selected F$_1$ plants were self-pollinated to produce the F$_2$ generation used for linkage mapping. The genetic map has been constructed on a basis of 161 individual DM19 plants and includes 285 molecular markers located on nine linkage groups [Budahn, unpublished]. SSR analysis was carried out according to the PCR conditions published for carrot SSRs [6] using a LI-COR 4300 DNA analyzer (LI-COR Biosciences, Lincoln, NE, USA). DNA fragments polymorphic for the parents were scored in the DM19 progeny, and marker scores were converted to the segregation type codes required for linkage mapping with the JoinMap version 4.0 software [35]. Linked loci were grouped using LOD thresholds from 5.0 to 10.0 in steps of 0.2 and recombination frequency ≤0.4. The jump threshold was set to 5.0 and the third mapping round was carried out. Map distances in centi-Morgan (cM) were calculated using the Kosambi function.

Immunostaining

Based on a peptide corresponding to the N-terminus of *DcCENH3* (NH$_2$-RTKHPAKRTSGHRSRGPPLS-CONH$_2$; amino acids 3–22) polyclonal IgG antibodies were generated. Peptide synthesis, immunization of three rabbits and affinity purification of *Daucus* CENH3-antiserum on sepharose columns was performed by Pineda Antikörper-Service (Berlin, Germany). In addition, a commercially available mouse antibody to α-tubulin (clone DM 1A, Sigma) was used. A Cy3-conjugated anti-rabbit IgG (Dianova) and an anti-mouse Alexa 488 antibody (Molecular Probes) were used as secondary antibodies. Immunostaining was performed on slides prepared from root tips of *D. carota* and *D. glochidiatus*. Seeds were germinated on moist filter paper at room temperature in dark for 3 days. Prior to incubation with either antibody root tips (1.5–2 cm) were fixed 5 min under mild vacuum at room temperature and 25 min on ice in freshly prepared 3.7% paraformaldehyde solution (PFA) containing phosphate-buffered saline (1xPBS, pH 7.3) and then washed three times for 5 min in 1x PBS on ice. For immunostaining with anti- α-tubulin antibody, material was fixed in 3.7% PFA solution containing microtubules stabilizing buffer (1xMTSB prepared with 50 mM Pipes, 2 mM EGTA,

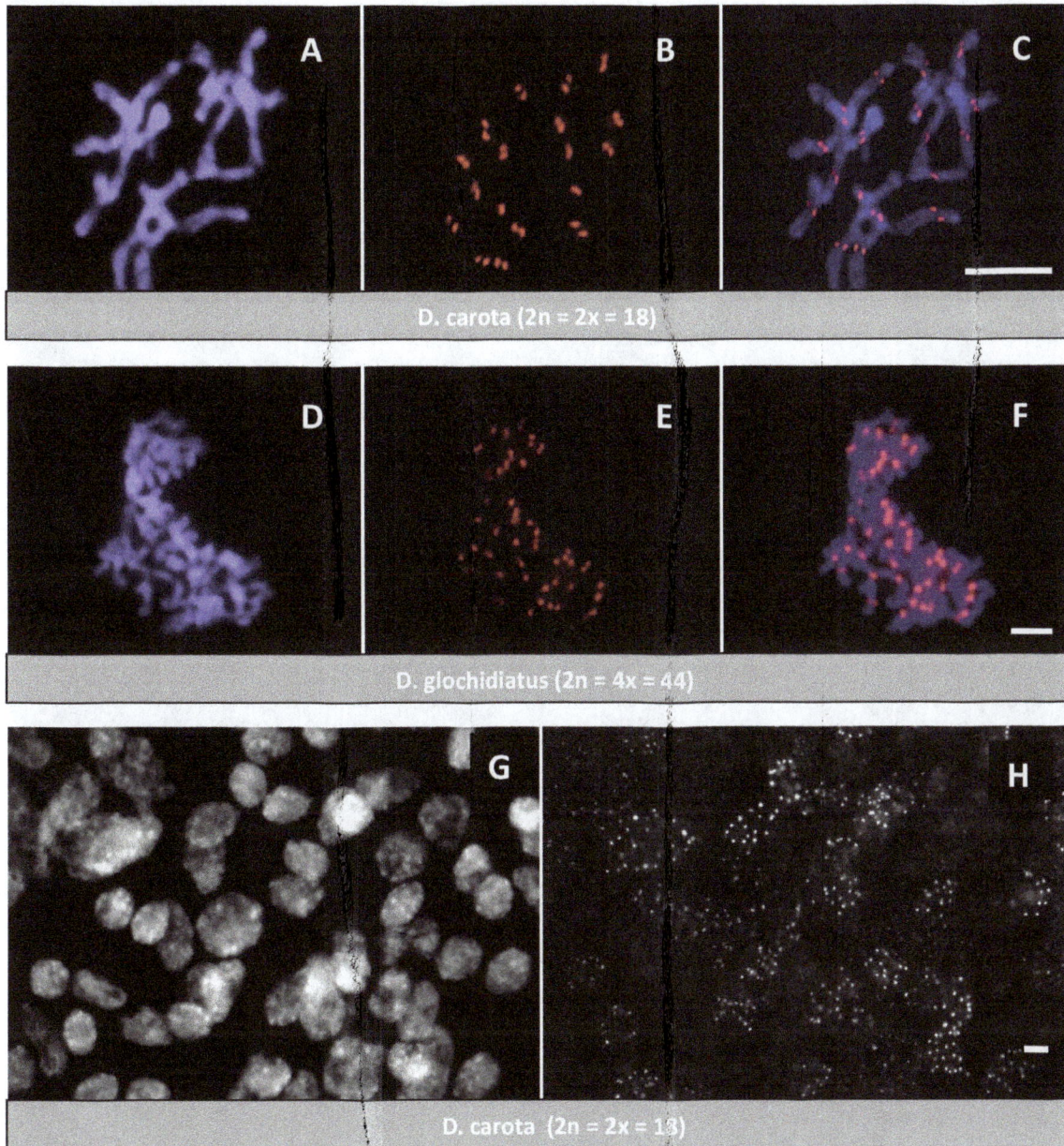

Figure 5. Immunostaining of *Daucus* root tip cells using anti-*DcCENH3* antibody. (**A–C**) *D. carota* (2n = 2x = 18) metaphase chromosomes, (**D–F**) *D. glochidiatus* (2n = 4x = 44) metaphase chromosomes, (**G, H**) interphase nuclei of *D. carota*. A, D and also G are DAPI-stained chromosomes, B, E and also H are CENH3 immunosignals, C and F are merged images. Scale bar 5 μm.

2 mM MgSO$_4$). Meristematic regions of root tips were digested by treating with an enzyme mix (2 vol enzyme mixture: 0.7% cellulase (Calbiochem), 0.7% cellulase R10 (Duchefa), 1% pectolyase (Sigma), 1% cytohelicase (Sigma) plus 1 vol 1x PBS/MTSB, pH 7.5) at 37°C until the material became soft (about 30–40 minutes). The macerated material was shortly washed and then squashed in PBS or MTSB on a slide. Coverslips were removed using liquid nitrogen and slides were immersed in 1x PBS/MTSB and further processed on the same day or the day after. The slides were incubated for 1 h at 37°C in a moisture chamber with blocking solution (3% BSA in 1x PBS/8% BSA in 1xMTSB, 0.1% Tween 20), followed by an incubation at 10°C overnight with the primary antibody diluted in 1xPBS/MTSB supplemented with 1% BSA. Dilutions were 1:500 for anti-CENH3, and 1:100 for

antibody to α-tubulin. Following three washes in 1xPBS/MTSB for 5 min the secondary antibody (anti-rabbit-Cy3 diluted 1:300 in 1xPBS/MTSB supplemented with 1% BSA or anti-mouse-Alexa 488 diluted 1:200 was applied for 45 min at 37°C. After 3 final washes with PBS buffer 5 min each time, the slides were counterstained with 4′,6-diamino-2-phenylindole (DAPI) and mounted in Vectashield mounting medium (Vector Laboratories, Burlingame, CA, USA). In double immunostaining experiments (CENH3 and α-tubulin) the two primary or secondary antibodies were incubated together.

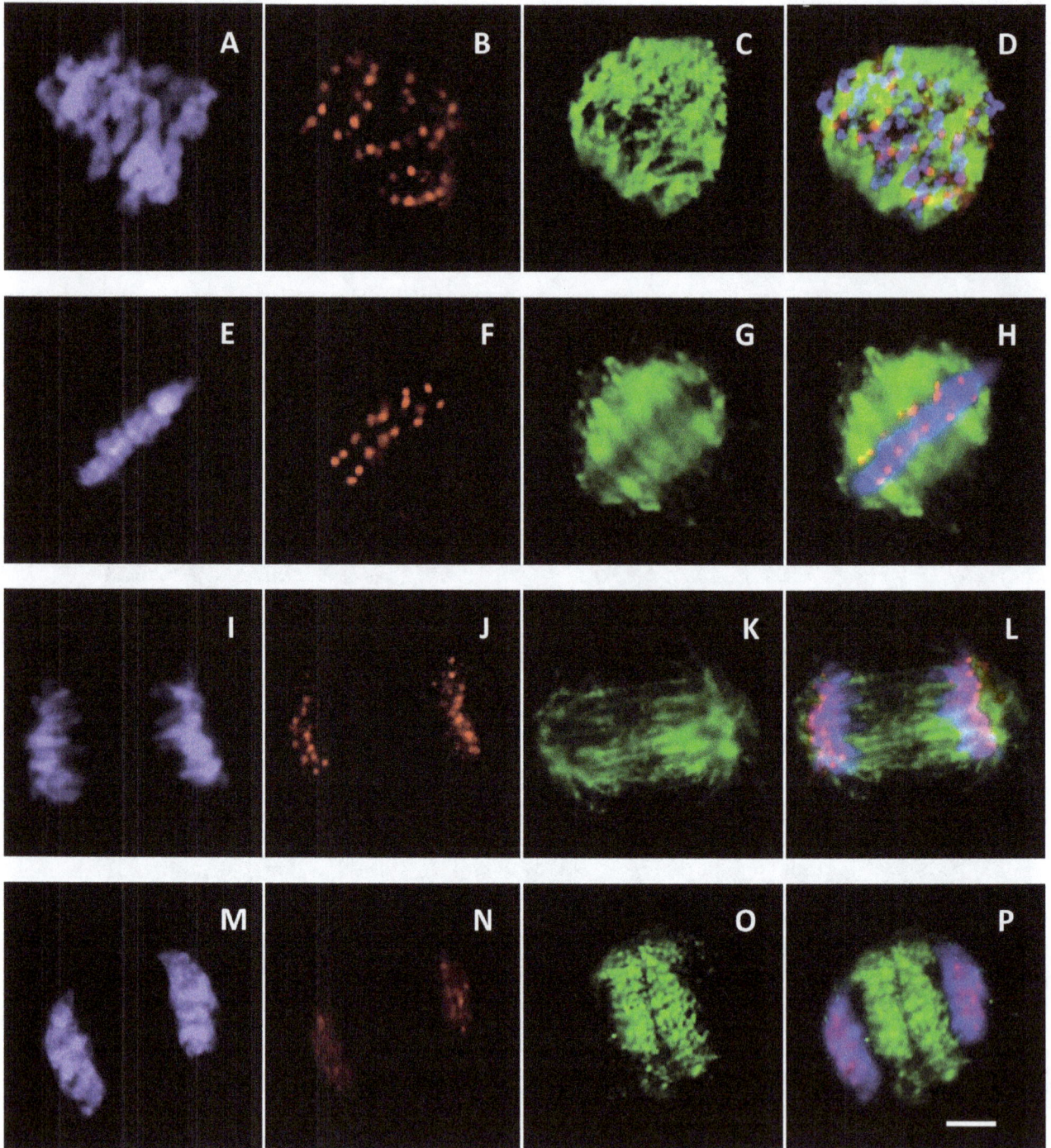

Figure 6. Double-immunostaining of root-tip cells of carrot (*D. carota*) at different stages of mitosis with antibodies against carrot CENH3 (in red) and α-tubulin (in green). Chromosomes are counterstained with DAPI (in blue). (A–D) prophase, (E–H) metaphase, (I–L) anaphase, (M–P) telophase. Scale bar 10 μm.

Results and Discussion

Identification of *Daucus* CENH3 and Phylogenetic Analysis

The amino acid sequence of *N. tabacum* CENH3 [26] was used as a query in a tBlastn search against the assembled carrot transcriptome [9]. After bioinformatic CENH3 mining, two overlapping contigs were identified which represented about 85% of the whole putative CENH3 coding sequence (CDS). Since the highly variable N-terminal region was not found, an intermediate PCR-based cloning step was performed. Finally, four overlapping contigs were found in the *D. carota* transcriptome representing the whole CENH3 coding sequence including a part of the 5′-UTR region. Based on the assembled sequence, specific primers were designed for PCR-based cloning of the complete *Daucus* CENH3 coding region. The *D. carota* homolog of CENH3 was named *DcCENH3* (GenBank number KJ201903) and has been identified with a nucleotide sequence length of 438 bp encoding a 146 amino acid (aa) protein, which is one of the shortest plant CENH3s known so far. Similarly, the ORFs of *D. pusillus* (*DpCENH3*, KJ201905), *D. glochidiatus* (*DgCENH3*, KJ201906) and *D. muricatus* (*DmCENH3*, KJ201904) were isolated. *DpCENH3* and *DgCENH3* also showed a DNA size of 438 bp, whereas *DmCENH3* cDNA was 3 bp shorter (435 bp, 145 aa). To our knowledge, the cloned genes from different carrot species are the first CENH3s isolated from the large Apiaceae plant family.

A multiple sequence alignment of the nucleotide sequences is shown in Figure S1, and the amino acid sequences deduced from the ORFs are shown in Figure 1, respectively. Except for the unique feature of the missing triplet in *D. muricatus* the CENH3 sequences of the four species differed only by a few nucleotides. The highest similarity was found between *DpCENH3* and *DgCENH3* with 98.4% nucleotide identity (Figure S2) resulting in an exchange of a single amino acid (Figure 1). In each comparison among the *Daucus* species the homology was higher than 95% identity on a nucleotide level, with a maximum of six amino acid changes between *D. carota* and *D. glochidiatus*. With regard to the putative centromere targeted domain (CATD), the protein sequences were identical (Figure 1). The nearly identical CENH3 variants in the Australian accession of *D. glochidiatus* (2n = 4x = 44) and the American representative of *D. pusillus* (2n = 2x = 22), which both have the basic chromosome number of x = 11, indicate the putatively close phylogenetic relationship between these two species. Molecular taxonomic studies have placed both species in the same *Daucus II* subclade [36], but it is unknown if polyploid *D. glochidiatus* is the result of a recent hybridization with any of the diploid species investigated in this study. Because of the fragmentary knowledge on systematics and phylogeny of the genus there is no indication yet either for a potential *Daucus* ancestor or a hypothetical ancestral karyotype [12].

To analyse the intron/exon structure of *Daucus* CENH3, the full-length genomic sequence of *D. carota* CENH3 was amplified with the same primer pair (DCEN2) used for the cDNAs, cloned into plasmids and sequenced by a primer walking approach. A single sequence was obtained, with a total length of 4,515 bp. Alignment of the *D. carota* CENH3 cDNA sequence with the genomic sequence resulted in a gene structure consisting of 7 exons and 6 introns of very different sizes (Figures S3A and S3B). A similar structure of 7 exons and 6 introns was observed for rice CENH3 genes [22], and 7 exons were also reported for *Brassica nigra* [27]. Exon 2 of carrot CENH3 was found to be extremely short (14 bp), whereas intron 5 displayed a sequence length of 2,354 bp, which is more than 50% of the whole gene. Exon 3 of *DcCENH3* contains the 3 nucleotides C-G-A (coding for arginine), which are missing in the *DmCENH3* CDS, but their position inside exon 3 indicates, that no alternative splicing has caused the loss of this single triplet in *D. muricatus*.

When the *Daucus* CENH3 proteins were aligned for phylogenetic analysis with those from various other monocot and dicot plant species, and *A. thaliana* canonical histone H3 as an outgroup, the carrot CENH3s formed a specific *Daucus* (Apiaceae) clade which was relatively closely located to CENH3s from grape, tobacco and poplar (Figure 2). The nucleotide identity values of the comparisons to *V. vinifera* CENH3 coding sequence were in a range of 66.4% and 67.4% (Figure S2), and the amino acid identity was about 67% to 69% depending on the *Daucus* species (not shown). Most characteristic for this comparison was the lack of ten consecutive amino acids in the hypervariable N-terminal tail domain of *Daucus* CENH3s, whereas the putative centromere

targeting domain (CATD) was exactly of the same length (Figure 1). The CATD is composed of the loop 1 linker and α-2 helix of the histone fold domain of the C terminal part of CENH3 and is important for binding of CENH3 to centromeric DNA [37]. Its role has been documented also for higher plants like *A. thaliana* [38]. A low degree of amino acid identity of *Daucus* CENH3s to *A. thaliana* H3 was found, and also to the identical *D. carota* canonical H3 sequence identified by bioinformatic mining in the *Daucus* transcriptome (result not shown). Overall, these finding as well as the phylogenetic arrangement of the sequences among some other dicot species suggest, that the deduced *Daucus* CENH3s are authentic CENH3 homologs.

Transcriptional Analysis of *Daucus* CENH3 Variants

The sizes of the PCR products obtained after cDNA-PCR with the single universal primer pair DCEN2 appeared to be similar among the different genotypes, and also the clones obtained from individual accessions of each species did not indicate so far the possibility that alternative splicing might have been occurred. However, we wanted to include in this study some transcriptional analyses of different *Daucus* CENH3 sequences, to examine the possible existence of additional transcribed alleles, which have not been revealed by the cloning procedure used. Therefore, a set of species-specific internal PCR primers was developed for RT-PCR analysis of each CENH3 variant. Using sequence differences present at nucleotide positions 86 and 88, forward SNP primers were designed for *D. carota*, *D. glochidiatus* and *D. pusillus* (Figure S1). In *D. muricatus* it was not yet possible to develop a SNP-specific primer pair. Due to the high similarity of *DgCENH3* and *DpCENH3* the same primer pair (DpgEXP) was chosen. As shown in Figure 3, the *D. carota* - specific primers (DcEXP) produced a single fragment of the expected size of 315 bp only in *D. carota*, but not in any of the other species. Vice versa, the RT-PCR with the DpgEXP primers displayed species-specific transcription in both *D. glochidiatus* and *D. pusillus*, but not in *D. carota* and *D. muricatus*. We can therefore exclude, that tetraploid *D. glochidiatus* cells contain the same expressed CENH3 variant of *D. carota*. Although we assume the existence of a single transcript of the CENH3 gene in this polyploid species, interpretations regarding the number of putative alleles and transcripts should still be done cautiously. Only a relatively small number of five individual clones have been randomly selected and sequenced after PCR-based cloning, and the presence of additional CENH3 alleles is possible. Hirsch et al. [22] identified two distinct CENH3 transcripts in allotetraploid *Oryza* species and were able to trace their origin back to diploid rice species known as putative progenitors. In *Brassica*, where the situation is more complex, up to four distinct CENH3 cDNAs were identified in individuals of each of the diploid species *B. rapa*, *B. oleracea*, and *B. nigra*, and the presence of multiple isoforms in allotetraploids derived from them suggest multiple CENH3 loci in *Brassica* [27]. In natural allopolyploids of wild rice and tobacco obviously all CENH3s from each genome retain their expression, whereas in soybean with its putative polyploid genome structure only a single transcribed homolog of CENH3 was found [28]. In *Daucus* it would be interesting to conduct interspecific crosses i.e. *D. carota* x *D. glochidiatus* followed by transcriptional analyses and immunostaining experiments on chromosomes of hybrid embryos.

Chromosome 9 Encodes CENH3 of *D. carota*

For localization of the CENH3 gene in the carrot genome through linkage mapping, a previously constructed genetic map of the carrot progeny DM19 was used. This well-saturated map has been constructed on the basis of 285 molecular markers and has already been used for mapping of several genes involved in flowering characteristics [Budahn, unpublished]. Based on the compound dinucleotide SSR motif $(CT)_{14}$ CCC $(CT)_3$ TT $(CT)_6$ present in the second intron of the genomic sequence, a specific PCR primer pair was developed (DCEN-SSR, Figure S3A). The CENH3-specific fragments segregated in DM19 progeny as a co-dominant marker (segregation type 'hk×hk' according the Join-Map format [35]), and the CENH3 locus was mapped on the carrot linkage group 7, which has been designated as chromosome 9 after the integration of genetic and physical maps of *D. carota* [13]. As shown in Figure 4, the location of the CENH3 gene was calculated between two anonymous genomic SSR markers (gSSR12, gSSR85) in the bottom part of the chromosome. According to marker information of the dense carrot linkage map presented by Cavagnaro et al. [6] there might be a tight genetic linkage of DCEN-SSR to a structural key gene involved in carotenoid biosynthesis (ζ-*carotene desaturase*, *ZDS2* [39]) located in the middle of the interval between the markers gSSR12 and gSSR85. This finding suggest that the repeat motif within *DcCENH3* might also be useful as a highly informative molecular marker for association studies targeted to carotenoid biosynthesis in carrots.

Visualization of Centromeres with a Carrot CENH3-specific Antibody

To verify that *DcCENH3* proteins localize to *Daucus* centromeres, immunofluorescence experiments were performed on mitotic chromosome preparations. Anti-*DcCENH3* antibody staining in root-tip cells of *D. carota* showed, that signals were exclusively located at the centromere regions of all 18 metaphase chromosomes (Figure 5A–C) providing direct *in situ* evidence for centromeric localization. As shown in Figure 5D–F, also tetraploid nuclei of *D. glochidiatus* appeared to be stained at the centromeric regions, indicating the cross-reactivity of the *D. carota* antibody with CENH3 of other *Daucus* species. Signals were also visible during interphase (Figures 5G and 5H) of carrot mitotic cells. After double immunostaining with anti-CENH3 and anti-α-tubulin, CENH3 signals were present in all stages of *D. carota* root cell mitosis (Figure 6). During anaphase, the antibody signals were located mainly at the tip of microtubule bundles attaching on the leading portions of the chromosomes in opposite orientations (Figure 6I–L). At telophase, the α-tubulin signals were mainly located in the equatorial plane, although the CENH3 signals remained at the two cell poles (Figure 6M–P). The cross-reactivity of the carrot CENH3 antibody with centromeres of distantly related *Daucus* species was not unexpected considering the very small sequence difference of a single amino acid in the N-terminal sequence of the deduced CENH3 protein used to create an antiserum. In several cases, anti-CENH3 antibodies were used to recognize CENH3s of more or less closely related species. The wide cross reactivity of antibodies raised against rice CENH3 [21] has been demonstrated in other *Oryza* species [40] and several other *Poaceae* species such as barley [41], wheat [42], and rye [43].

Wide cross-reactivity was also observed between different *Brassica* species [26] and several *Allium* species [25]. From our results with *Daucus* species it can be assumed that the *DcCENH3* antibody might be eventually also useful for the characterization of CENH3s of members of other genera of the Apiaceae plant family such as fennel (*Foeniculum*), celery (*Apium*), or parsley (*Petroselinum*). Work is in progress to confirm this assumption, and to clone the involved genes for further functional studies.

Supporting Information

Figure S1 Nucleotide sequence alignment (ClustalW, Lasergene) of the CENH3 coding sequences of *D. carota* (Dc), *D. glochidiatus* (Dg), *D. pusillus* (Dp), and *D. muricatus* (Dm). Sequences of PCR forward primers used for species-specific RT-PCR are labeled by red- (*D. carota*) or blue-edged boxes (*D. pusillus*, *D. glochidiatus*), and (identical) reverse primer sequences are marked by a green-edged box.

Figure S2 Phylogenetic tree built on the basis of cDNA nucleotide sequences of CENH3 variants identified in the four *Daucus* species of this study and two published CENH3 sequences (*N. tabacum*, NCBI acc. No. BAH03515; *V. vinifera*, XP_002281073) showing the highest similarity to *Daucus* CENH3s after multiple alignment of various plant CENH3 proteins (see Figure 2). Sequences were compared by ClustalW (Lasergene). Branch length is scaled as number of substitutions per 100 nucleotides. In the table below the nucleotide sequence identity is shown (%) for the six sequences of the dendrogram shown above.

Figure S3 (A) Result of the alignment of the *D. carota* CENH3 coding region (cDNA sequence) with the genomic DNA (gDNA sequence) showing the intron-exon-structure of the *DcCENH3* gene. The position of a PCR primer pair designed for genetic mapping of *DcCENH3* (DCEN-SSR-F/-R) is also shown. **(B)** *DcCENH3* cDNA sequence and the deduced amino acid sequence. The positions of introns are marked by a red arrow.

Acknowledgments

We wish to thank Antje Krüger, Nicole Schäfer and Karla Müller for excellent technical assistance, and Katrin Kumke for technical support and advices in immunostaining. The authors also thank Dr. Thomas Nothnagel for providing seeds of *Daucus* wild species and helpful discussions on carrot breeding and *Daucus* genetic resources.

Author Contributions

Conceived and designed the experiments: FD AH. Performed the experiments: FD OS HB. Analyzed the data: FD AH. Contributed reagents/materials/analysis tools: OS HB AH. Wrote the paper: FD.

References

1. Sáenz Lan C (1981) Research on *Daucus* L. (Umbelliferae). Anal Jard Bot Madrid 37: 481–534.
2. Heywood VH (1983) Relationships and evolution in the *Daucus carota* complex. Isr J Bot 32: 51–65.
3. Grzebelus D (2011) *Daucus*. In: C. Kole (ed.), Wild Crop Relatives: Genomic and Breeding Resources, Vegetables. Springer-Verlag Berlin Heidelberg, 91–113.
4. Pimenov MG, Vasileva MG, Lenov MV, Dauschkcevich JV (2003) Karyotaxonomical analysis in the Umbelliferae. Science Publishers, Enfield, New Hampshire, USA.
5. Arumuganathan K, Earle ED (1991) Nuclear DNA content of some important plant species. Plant Mol Biol Rep 9: 208–218.
6. Cavagnaro PF, Chung S-M, Manin S, Yildiz M, Ali A, et al. (2011) Microsatellite isolation and marker development in carrot - genomic distribution, linkage mapping, genetic diversity analysis and marker transferability across Apiaceae. BMC Genomics 12: 386.
7. Alessandro MS, Galmarini CR, Iorizzo M, Simon PW (2013) Molecular mapping of vernalization requirement and fertility restoration genes in carrot. Theor Appl Genetics 126: 415–423.
8. Cavagnaro PF, Chung SM, Szklarczyk M, Grzebelus D, Senalik D, et al. (2009) Characterization of a deep-coverage carrot (*Daucus carota* L.) BAC library and initial analysis of BAC-end sequences. Mol Genet Genom 281: 273–288.

9. Iorizzo M, Senalik DA, Grzebelus D, Bowman M, Cavagnaro PF, et al. (2011) *De novo* assembly and characterization of the carrot transcriptome reveals novel genes, new markers, and genetic diversity. BMC Genomics 12: 389.

10. Baranski R (2008) Genetic transformation of carrot (*Daucus carota*) and other Apiaceae species. Transgenic Plant Journal 2: 18–38.

11. Schrader O, Ahne R, Fuchs J (2003) Karyoptype analysis of *Daucus carota* L. using Giemsa C-banding and FISH of 5S and 18S- 25S rRNA specific genes. Caryologia 56: 149–154.

12. Iovene M, Grzebelus E, Carputo D, Jiang J, Simon PW (2008) Major cytogenetic landmarks and karyotype analysis in *Daucus carota* and other Apiaceae. Amer J Bot 95: 793–804.

13. Iovene M, Cavagnaro PF, Senalik D, Buell CR, Jiang J, et al. (2011) Comparative FISH mapping of *Daucus* species (Apiaceae family). Chromosome Res 19: 493–506.

14. Ferrie AMR, Bethune TD, Mykytyshyn M (2011) Microspore embryogenesis in Apiaceae. Plant Cell Tiss Organ Cult 104: 399–406.

15. Ravi M, Chan SWL (2010) Haploid plants produced by centromere-mediated genome elimination. Nature 464: 615–619.

16. Ravi M, Chan SWL (2013) Centromere-mediated generation of haploid plants. In: Jiang J, Birchler JA (eds) Plant Centromere Biology, John Wiley & Sons, 169–181.

17. Jiang J, Birchler JA, Parrot WA, Dawe RK (2003) A molecular view of plant centromeres. Trends Plant Sci 8: 570–575.

18. Houben A, Schubert I (2003) DNA and proteins of plant centromeres. Curr Opin Plant Biol 6: 554–560.

19. Malik HS, Henikoff S (2009) Major evolutionary transitions in centromere complexity. Cell 138: 1067–1082.

20. Zhong CX, Marshall JB, Topp C, Mroczek R, Kato A, et al. (2002) Centromeric retroelements and satellites interact with Maize kinetochore protein CENH3. Plant Cell 14: 2825–2836.

21. Nagaki K, Cheng Z, Ouyang S, Talbert PB, Kim M, et al. (2004) Sequencing of a rice centromere uncovers active genes. Nat Genet 36: 138–145.

22. Hirsch CD, Wu YF, Yan HH, Jiang JM (2009) Lineage-specific adaptive evolution of the centromeric protein CENH3 in diploid and allotetraploid *Oryza* species. Mol Biol Evol 26: 2877–2885.

23. Nagaki K, Murata M (2005) Characterization of CENH3 and centromere-associated DNA sequence in sugarcane. Chromosome Res 13: 195–203.

24. Sanei M, Pickering R, Kumke K, Nasuda S, Houben A (2011) Loss of centromeric histone H3 (CENH3) from centromeres precedes uniparental chromosome elimination in interspecific barley hybrids. Proc Natl Acad Sci USA 108: E498–E505.

25. Nagaki K, Yamamoto M, Yamaji N, Mukai Y, Murata M (2012) Chromosome dynamics visualized with an anti-centromeric histone H3 antibody in *Allium*. PLOS ONE 7: e51315.

26. Nagaki K, Kashihara K, Murata M (2009) A centromeric DNA sequence colocalized with and centromere-specific histone H3 in tobacco. Chromosoma 118: 249–257.

27. Wang G, He Q, Liu F, Cheng Z, Talber PB, et al. (2011) Chracterization of CENH3 proteins and centromere-associated DNA sequences in diploid and allotetraploid Brassica species. Chromosoma 120: 353–365.

28. Tek AL, Kashihara K, Murata M, Nagaki K (2010) Functional centromeres in soybean include two distinct tandem repeats and a retrotransposon. Chromosome Res 18: 337–347.

29. Tek AL, Kashihara K, Murata M, Nagaki K (2011) Functional centromeres in *Astragalus sinicus* include a compact centromere-specific histone H3 and a 20-bp tandem repeat. Chromosome Res 19: 969–978.

30. Neumann P, Navratilová A, Schroeder-Reiter E, Koblížková A, Steinbauerova V, et al. (2012) Stretching the rules: monocentric chromosomes with multiple centromere domains. PLOS Genetics 8: e1002777.

31. Iwata A, Tek AL, Richard MMS, Abernathy B, Fonseca A, et al. (2013) Identification and characterization of functional centromeres of the common bean. Plant J 76: 47–60.

32. Hall TA (1999) Bioedit: a user-friendly biological sequence alignment editor and analysis program for Windows 95/98/NT. Nucl Acids Symp Ser 41: 95–98.

33. Altschul SF, Madden TL, Schäffer AA, Zhang J, Zhang Z, et al. (1997) Gapped BLAST and PSI-BLAST: a new generation of protein database search programs. Nucleic Acids Res 25: 3389–3402.

34. Wally O, Jayaraj J, Punja ZK (2009) Broad-spectrum disease resistance to necrotrophic and biotrophic pathogens in transgenic carrots (*Daucus carota* L.) expressing an Arabidopsis *NPR1* gene. Planta 231: 131–141.

35. Van Ooijen J (2006) JoinMap 4. Software for the calculation of genetic linkage maps in experimental populations. Kyazma BV, Wageningen, Netherlands.

36. Spalik K, Downie SR (2007) Intercontinental disjunctions in *Cryptotaenia* (Apiaceae, Oenantheae): An appraisal using molecular data. J Biogeogr 34: 2039–2054.

37. Black BE, Foltz DR, Chakravarthy S, Luger K, Woods VL, et al. (2004) Structural determinants for generating centromeric chromatin. Nature 29: 578–582.

38. Lermontova I, Schubert V, Fuchs J, Klatte S, Macas J, et al. (2006) Loading of *Arabidopsis* centromeric histone CENH3 occurs mainly during G2 and requires the presence of the histone fold domain. Plant Cell 18: 2443–2451.

39. Just BJ, Santos CAF, Fonseca MEN, Boiteux LS, Oloizia BB, et al. (2007) Carotenoid biosynthesis structural genes in carrot (*Daucus carota*): isolation, sequence-characterization, single nucleotide polymorphism (SNP) markers and genome mapping. Theor Appl Genet 114: 693–704.

40. Lee HR, Zhang W, Langdon T, Jin W, Yan H, et al. (2005) Chromatin immunoprecipitation cloning reveals rapid evolutionary patterns of centromeric DNA in *Oryza* species. Proc Natl Acad Sci USA 102: 11793–11798.

41. Houben A, Schroeder-Reiter E, Nagaki K, Nasuda S, Wanner G, et al. (2007) CENH3 interacts with the centromeric retrotransposon cereba and GC-rich satellites and locates to centromeric substructures in barley. Chromosoma 116: 275–284.

42. Liu Z, Yue W, Li DY, Wang RRC, Kong XY, et al. (2008) Structure and dynamics of retrotransposons at wheat centromeres and pericentromeres. Chromosoma 117: 445–456.

43. Houben A, Kumke K, Nagaki K, Hause G (2011) CENH3 distribution and differential chromatin modifications during pollen development in rye (*Secale cereale* L.). Chromosome Res 19: 471–480.

15

Soil Type Dependent Rhizosphere Competence and Biocontrol of Two Bacterial Inoculant Strains and Their Effects on the Rhizosphere Microbial Community of Field-Grown Lettuce

Susanne Schreiter[1,2], **Martin Sandmann**[2], **Kornelia Smalla**[1], **Rita Grosch**[2]*

1 Julius Kühn-Institut – Federal Research Centre for Cultivated Plants (JKI), Institute for Epidemiology and Pathogen Diagnostics, Braunschweig, Germany, 2 Leibniz Institute of Vegetable and Ornamental Crops Großbeeren/Erfurt e.V., Department Plant Health, Großbeeren, Germany

Abstract

Rhizosphere competence of bacterial inoculants is assumed to be important for successful biocontrol. Knowledge of factors influencing rhizosphere competence under field conditions is largely lacking. The present study is aimed to unravel the effects of soil types on the rhizosphere competence and biocontrol activity of the two inoculant strains *Pseudomonas jessenii* RU47 and *Serratia plymuthica* 3Re4-18 in field-grown lettuce in soils inoculated with *Rhizoctonia solani* AG1-IB or not. Two independent experiments were carried out in 2011 on an experimental plot system with three soil types sharing the same cropping history and weather conditions for more than 10 years. Rifampicin resistant mutants of the inoculants were used to evaluate their colonization in the rhizosphere of lettuce. The rhizosphere bacterial community structure was analyzed by denaturing gradient gel electrophoresis of 16S rRNA gene fragments amplified from total community DNA to get insights into the effects of the inoculants and *R. solani* on the indigenous rhizosphere bacterial communities. Both inoculants showed a good colonization ability of the rhizosphere of lettuce with more than 10^6 colony forming units per g root dry mass two weeks after planting. An effect of the soil type on rhizosphere competence was observed for 3Re4-18 but not for RU47. In both experiments a comparable rhizosphere competence was observed and in the presence of the inoculants disease symptoms were either significantly reduced, or at least a non-significant trend was shown. Disease severity was highest in diluvial sand followed by alluvial loam and loess loam suggesting that the soil types differed in their conduciveness for bottom rot disease. Compared to effect of the soil type of the rhizosphere bacterial communities, the effects of the pathogen and the inoculants were less pronounced. The soil types had a surprisingly low influence on rhizosphere competence and biocontrol activity while they significantly affected the bottom rot disease severity.

Editor: Martha E. Trujillo, Universidad de Salamanca, Spain

Funding: The study was funded by 'German Research Foundation' (DFG), SM 59/11-1; GR 1729/8-1: KS RG. The funders had no role in study design, data collection and analysis, decision to publish, or preparation of the manuscript.

Competing Interests: The authors have declared that no competing interests exist.

* Email: grosch@igzev.de

Introduction

Plant pathogens are a limiting factor in crop productivity worldwide and responsible for yield losses [1]. Crop rotation, use of resistant cultivars and application of chemicals are strategies to minimize disease incidence and severity in Integrated Pest Management (IPM). However, resistant cultivars and effective fungicides for the control of diseases caused by soil-borne pathogens such as *Rhizoctonia solani* (Kühn) are often not available. Moreover, adverse eco-toxicological effects of chemical fungicides urge the development of alternative strategies to combat fungal diseases on crops [2–4]. In terms of disease control, it is well-documented that microbial inoculants as part of IPM can contribute to the reduction of adverse environmental effects caused by the exclusive reliance on fungicides [5-7] and thus represent a promising strategy for more sustainable agriculture [8]. Currently, the worldwide bio-pesticide market offers products including 60 bacterial and 60 fungal species [9]. Nonetheless, the exploitation of microbial inoculants as biocontrol agents in agriculture has been hampered by inconsistent results at the field scale [10,11]. The inconsistency observed in biocontrol effects limits the attraction of microbial inoculants for growers but the reasons for this variability remain largely unexplored. Variation in the colonization ability of bacterial inoculants in the rhizosphere (rhizosphere competence) is assumed to be one of the factors contributing to this inconsistency. Several studies showed that the expression of genes, responsible for the capability of a biocontrol strain to suppress a disease, is often regulated in a cell density dependent manner [12,13]. Therefore, the ability of inoculants to colonize the rhizosphere at sufficiently high numbers for an extended period was identified as a prerequisite for their beneficial effect on plants [11,14,15]. Thanks to the application of advanced genomics and microscopy methods the understanding of factors contributing to the biocontrol activity of bacterial strains has clearly improved [11,15–18], especially for *Pseudomonas* strains such as *P. fluorescens* CHA0 or Pf-5. The complex regulation of genes involved in biocontrol and plant-microbe interaction has been studied in more detail [19–23]. Although the mode of action differs from strain to strain, numerous studies supported the assumption that biocontrol activity most likely results from multi-factorial processes such as antibiosis, production of cell wall degrading enzymes, surfactants, volatile substances or sidero-phores, competition for nutrients and space and/or the enhance-ment of plant innate defense responses [24,25]. Several properties of bacterial inoculants such as motility [26], attachment [27], growth [16], production of antifungal metabolites or siderophores [24] and uptake and catabolism of root exudates [17,28] have been shown to be linked to rhizosphere competence.

However, knowledge of factors influencing rhizosphere compe-tence of bacterial inoculants under field conditions is largely lacking. Only in a few studies efforts were made to quantify the inoculant densities in the rhizosphere of field-grown crops or to evaluate the influence of inoculants and/or pathogens on the indigenous rhizosphere microbial community [7,29,30]. In agri-culture, crops are cultivated under various ecological conditions. Therefore, a better understanding of the complex relationships among inoculant, pathogen, plant, and ecological factors such as the soil type is a prerequisite for improved and reliable biocontrol effects. So the goal of the present study was to investigate the influence of soil types on the rhizosphere competence and biocontrol activity of bacterial inoculants and their effects on the indigenous soil bacterial community.

The strains *Pseudomonas jessenii* RU47 [31] and *Serratia plymuthica* 3Re4-18 [32], which revealed remarkably good control effects against bottom rot in previous experiments [7,32], were selected for this study. The causal agent of bottom rot on lettuce, the soil-borne fungus *R. solani* AG1-IB whose genome was recently sequenced [33] was used as model pathogen and lettuce as model host plant. The experimental plot system with three soil types under the same cropping history at the same field site enabled us to study the effects of different soil types on the rhizosphere competence and the biocontrol activity of the bacterial inoculants for the first time. We hypothesized that the soil types influence the rhizosphere competence and the biocontrol activity of the inoculant strains applied. Furthermore, we hypothesized that both the inoculants and the presence of the pathogen (*R. solani* AG1-IB) also influences the structural diversity of microbial communities in the rhizosphere of lettuce, and that the extent of this effect depends on the soil type.

Materials and Methods

Bacterial inoculants

The bacterial inoculant *P. jessenii* RU47 was isolated from a disease-suppressive soil [31,34] and the strain *S. plymuthica* 3Re4-18 originated from the endorhiza of potato [35]. To monitor the survival of inoculants in the rhizosphere spontaneous rifampicin resistant mutants were used [34]. Both strains were stored at − 80°C in Luria-Bertani broth (Carl Roth GmbH & Co. KG, Karlsruhe, Germany) with 20% glycerol.

Design of field experiments

To evaluate the effect of soil types on the rhizosphere competence and disease suppression of the bacterial inoculants *P. jessenii* RU47 and *S. plymuthica* 3Re4-18 without and with *R. solani* inoculation, two independent field experiments were carried out in a unique experimental plot system at the Leibniz Institute of

Vegetable and Ornamental Crops (Großbeeren, Germany, 52° 33′ N, 13° 22′ E). The first experiment was performed in unit 5, with planting on 8 June, harvest on 18 July 2011, the second experiment in unit 6, with planting on 27 July and harvest on 5 September 2011. Each unit was comprised out of three blocks (one block per soil type). The three soil types were characterized as Arenic-Luvisol (diluvial sand, DS), Gleyic-Fluvisol (alluvial loam, AL) and Luvic-Phaeozem (loess loam, LL) [36,37]. Each block consisted of 24 plots of 2 m×2 m in size and a depth of 75 cm. In unit 5, the following crops were cultivated from 2000 to 2010: pumpkin, nasturtium, nasturtium, phacelia, amaranth, wheat, pumpkin, nasturtium, wheat, broccoli, wheat, Teltow turnip and lettuce, and in unit 6 pumpkin, nasturtium, pumpkin, amaranth, wheat, wheat, pumpkin, nasturtium, wheat, wheat and lettuce.

Lettuce seeds (cv. Tizian, Syngenta, Bad Salzuflen, Germany) were sown in seedling trays filled with the respective soil type and incubated at 12°C for 48 h and then grown in the greenhouse at approximately 20/15°C (day/night). To maintain the substrate moisture all seedling trays were watered daily and fertilized weekly (0.2% Wuxal TOP N, Wilhelm Haug GmbH & Co. KG, Düsseldorf, Germany). Lettuce seedlings were transplanted at the 3–4-leaf stage (BBCH 14) in the experimental plot system. Plants were placed in a within-row and intra-row distance of 30 cm (36 plants per plot), and lettuce plantlets were overhead irrigated based on the computer program 'BEREST' [38]. The daily soil water content in the rooted soil layer using the water holding capacity of the soil under consideration of the plant growth stage and the potential evapotranspiration were the input variables for the irrigation program. Irrigation decisions were made on the basis of the calculated soil water content and the expected evapotrans-piration and precipitation of the next five days. The soil temperature (reflectometer PT100b1/3 DIN, Messtechnik Gera-berg GmbH, Martinroda, Germany) and the matric potential (CS616-L water content reflectometer, Campbell Scientific, North Logan, Utah, USA) were recorded by data logger (4MbSRAm data logger, Campbell Scientific). Both reflectometers determine an average value of a 20 cm top soil layer. The fertilizer was added to each plot based on a chemical analysis of soils before planting, done according to the certified protocols of Agricultural Tests and Research Institutions Association (VdLUFA, Germany). Each soil type was adjusted to the same amount of nitrogen (162 mg/100 g) by fertilization with Kalkamon (27% N, TDG mbh Lommatzsch, Germany). Lettuce was harvested six weeks after planting (6WAP, BBCH 49) in both experiments. The lettuce shoot dry mass (SDM) of each plant and the disease severity of bottom rot were scored at harvest. For assessment of SDM each lettuce head was cut in four portions and dried at 80°C until a constant dry mass was achieved. The disease severity was rated in four categories: 1–without bottom rot symptoms; 3–symptoms only on first lower leaves and small brown spots on the underside of leaf midribs; 5–brown spots on leaf midribs on lower and next upper leaf layer and 7–severe disease symptoms on upper leaf layers and beginning of head rot to total head rot according to Grosch et al. [39].

The following treatments of lettuce were investigated for each soil type: no treatment with inoculants without (control) and with *R. solani* (*Rs*) inoculation (control+*Rs*), plants treated with inoculants without (RU47; 3Re4-18) and with *R. solani* inocula-tion (RU47+*Rs*; 3Re4-18+*Rs*). Each treatment included four replicates with 36 plants per replicate.

Preparation of pathogen inoculum and inoculation

The *R. solani* AG1-IB isolate 7/3 from the strain collection of the Leibniz Institute of Vegetable and Ornamental Crops (Großbeeren) was used in the present study. The inoculum was

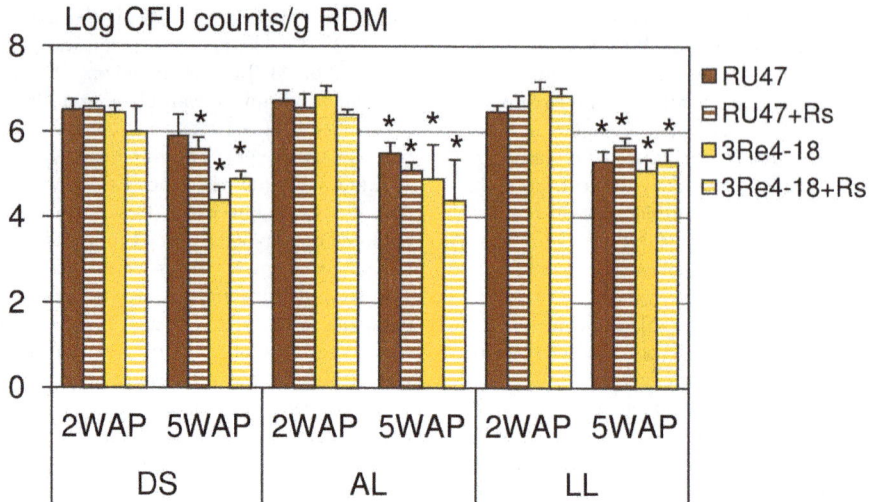

A)

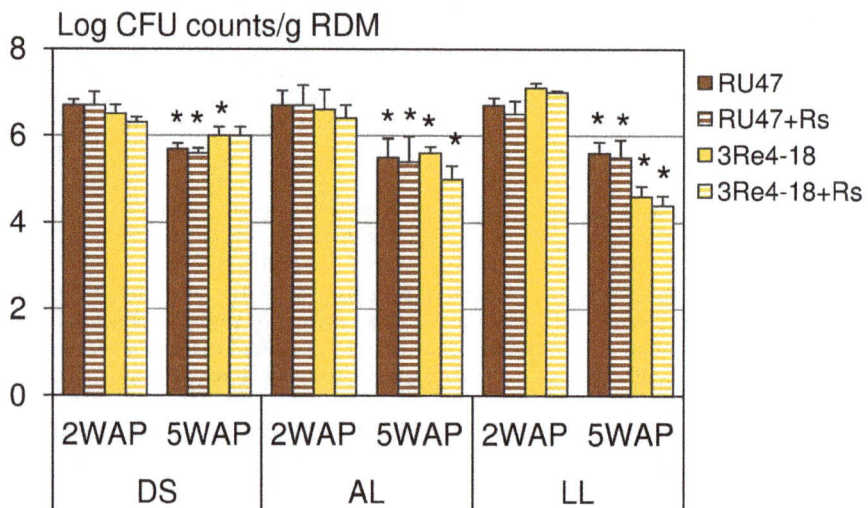

B)

Figure 1. CFU counts of *Pseudomonas jessenii* RU47 and *Serratia plymuthica* 3Re4-18 per gram of root dry mass (RDM) without (RU47, 3Re4-18) and with *Rhizoctonia solani* inoculation (RU47+*Rs*; 3Re4-18+*Rs*) in two experiments, A) and B), two and five weeks after planting (2WAP, 5WAP) in the 2011-season. Plants were grown in three soil types (DS, AL, LL) at the same field site. An asterisk indicates significant differences in CFU counts of RU47 or 3Re4-18 between 2WAP and 5WAP for each soil type (Tukey post-hoc test, *P*<0.05). The bars show the standard deviation.

Table 1. ANOVA results: Factor [soil type, plant growth development stage (PGDS), pathogen] dependent *P*-values for CFU counts of *Pseudomonas jessenii* RU47 and *Serratia plymuthica* 3Re4-18 (*P*<0.05).

	Experiment 1		Experiment 2	
Factor	RU47	3Re4-18	RU47	3Re4-18
Soil type	0.079	0.0007	0.544	<0.0001
PGDS	<0.0001	<0.0001	<0.0001	<0.0001
Pathogen	0.623	0.2609	0.3872	0.006

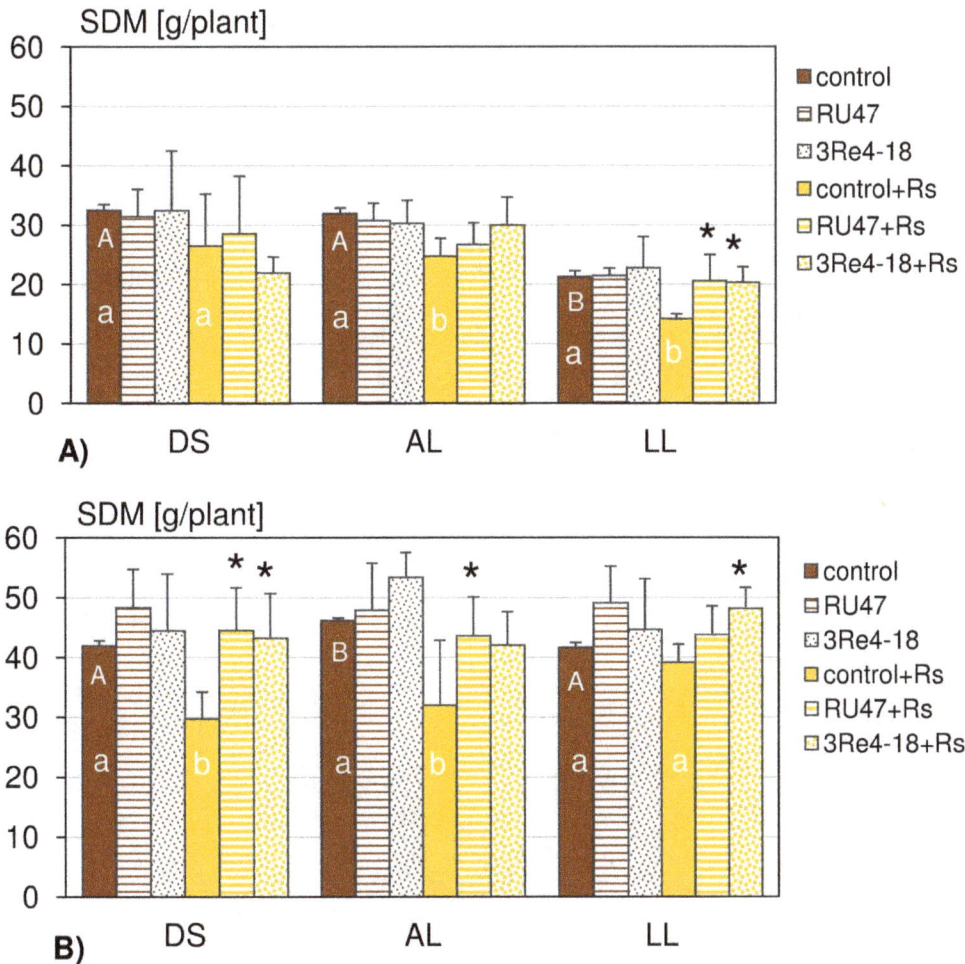

Figure 2. Shoot dry mass (SDM) of lettuce (cv. Tizian) determined for the following treatments: control, RU47, 3Re4-18, control+*Rs*, RU47+*Rs*, 3Re4-18+*Rs*) of two experiments, A) and B), in the 2011-season. Plants were grown in three soil types (DS, AL, LL) for six weeks each, at the same field site. Different capital letters denote significant differences in SDM of lettuce in controls or control+*Rs* between soil types (ANOVA; *P*<0.1). Different lower-case letters indicate significant differences in SDM between control and control+*Rs* within each soil type (Tukey post-hoc test, *P*<0.1). An asterisk denotes significant effects of the inoculants RU47 and 3Re4-18 on SDM to the corresponding control or control+*Rs* in each soil type. The bars show the standard deviation.

prepared as described by Schneider et al. [40] on barley kernels. To ensure a higher pathogen pressure the following inoculation procedure was applied: 36 lettuce plants were planted in each of the 24 plots as described above; after a cultivation time of three weeks they were evenly incorporated into the top soil (10 cm) by means of a rotary hoe, together with 40 g of barley kernels without or with *R. solani* infestation. The experiment started two weeks later assuming a decomposition of incorporated infested or non-infested lettuce plant material.

Preparation of bacterial inocula and application mode

For seed treatment King's B agar (Merck KGaA, Darmstadt, Germany) supplemented with rifampicin (75 µg/ml) were inoculated with the *P. jessenii* RU47 of *S. plymuthica* 3Re4-18 and incubated overnight at 29°C. The bacterial cells were harvested from the Petri dishes by resuspension in 15 ml sterile 0.3% NaCl and the concentration was adjusted in a spectrophotometer to a density of 10^8 colony forming units (CFU)/ml. A total of 200 lettuce seeds (cv. Tizian) were coated with 500 µl of a bacterial cell suspension dripping on the seed during vortexing in a 50 ml Falcon tube.

To prepare the inoculum for the treatment of young plants the inoculant strains were grown in nutrient broth (NB II, SIFIN GmbH, Berlin, Germany) amended with rifampicin (75 µg/ml) on a rotary shaker (90 rpm) at 29°C. After a cultivation time of 16 h the overnight culture was centrifuged at 13,000 g for 5 min, the supernatant discarded and the pellet was resuspended in sterile 0.3% NaCl solution. The cell density was adjusted to 10^7 CFU/ml or 10^8 CFU/ml for the drenching before and after planting, respectively. Lettuce plants were treated by drenching with 20 ml bacterial cell suspension per plant at the 3-leaf stage one week before planting in the field. A second treatment of young plants with 30 ml bacterial cell suspension 10^8 CFU/ml per plant was carried out at the 4-leaf stage two days after planting. The control plants were drenched with 20 ml or 30 ml of 0.3% NaCl solution, respectively, instead of bacterial suspension.

Sampling and sample processing

Rhizosphere samples were collected two and five weeks after planting (2WAP and 5WAP; BBCH 19 and BBCH 49) the lettuce in the experimental plot system. For each treatment and sampling time the roots of three plants per replicate (plot) were combined as

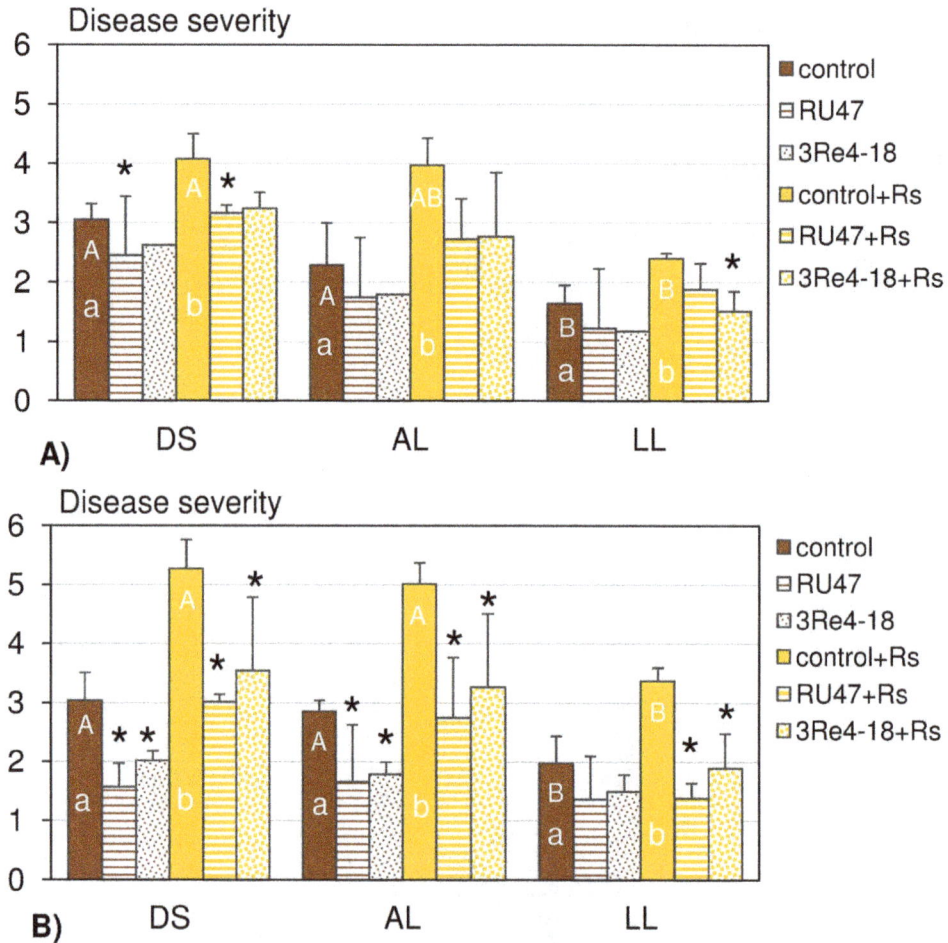

Figure 3. Disease severity of bottom rot on lettuce (cv. Tizian) determined for the following treatments: control, RU47, 3Re4-18, control+*Rs*, RU47+*Rs* and 3Re4-18+*Rs* the in two experiments, A) and B) in the 2011-season. Plants were grown in three soil types (DS, AL, LL) for six weeks each, at the same field site. Different capital letters indicate significant differences in disease severity of bottom rot in controls or control+*Rs* between soil types and different lower-case letters indicate significant differences between control and control+*Rs* within each soil type (Mann-Whitney U-test; $P<0.1$). An asterisk denotes significant differences in disease severity of the RU47 or 3Re4-18 treatment compared to the control and of RU47+*Rs* or 3Re4-18+*Rs* treatments compared to the control+*Rs* in each soil type. The bars show the standard deviation.

a composite sample and considered as one replicate; four replicates were used per treatment. Adhering soil was removed by washing the roots with sterile tap water before microbial cells were extracted as follows: the roots were cut into pieces of approximately 1 cm length and carefully mixed. Five gram of roots were placed in sterile Stomacher bags and treated by a Stomacher 400 Circulator (Seward Ltd, Worthing, UK) for 30 s at high speed after adding 15 ml of sterile 0.3% NaCl. The Stomacher blending step was repeated three times and followed by centrifugation steps as described by Schreiter et al. [37].

Analysis of rhizosphere competence of the bacterial inoculants

The ability of *P. jessenii* RU47 and *S. plymuthica* 3Re4-18 to colonize the rhizosphere of lettuce grown in the three soil types was determined 2WAP and 5WAP. Aliquots of the rhizosphere microbial cell suspension resulting from the combined supernatants of three Stomacher blending steps were immediately processed to determine the inoculant CFU counts by plating serial dilutions onto King's B agar supplemented with rifampicin (75 µg/ml) and cycloheximide (100 µg/ml) and incubated at 29°C

for 48 h. The CFU counts were calculated per gram of root dry mass (RDM). For all soil types Stomacher supernatants obtained from the control plots were plated as well to determine the background of the rifampicin resistant indigenous bacteria.

Data analysis

Data of SDM, disease severity and inoculant plate counts were analyzed with the STATISTICA program (StatSoft Inc., Tulsa, OK, USA). The impact of the soil type, the pathogen and the inoculants on SDM was determined using three-way ANOVA ($P<0.1$) combined with Tukey post-hoc test ($P<0.1$). The data of disease severity was evaluated using the nonparametric Kruskal Wallis test followed by Mann-Whitney U-test ($P<0.1$). The determined inoculants density (CFU counts/g RDM) was calculated and logarithmically (Log_{10}) converted before the impact of the soil type, plant growth development stage, and the presence of the pathogen *R. solani* on the plate counts of each inoculant strain was analyzed using three-way ANOVA ($P<0.05$) combined with Tukey post-hoc test ($P<0.05$). Average values for soil temperature of each soil type were analyzed by Tukey post-hoc test using

Table 2. Treatment-dependent differences (d-values) of bacterial communities in the rhizosphere of lettuce, grown in three soil types (DS, AL, LL), in the 2011-season.

| Soil type | Experiment | Figure | Differences between control and | | | | |
			control+Rs	RU47	3Re4-18	RU47+Rs	3Re4-18+Rs
DS	1	S1	6.1*	2.6*	6.2*	3.6*	10.3*
AL	1	S2	1.9*	13.6*	20.6*	12.3*	13.2*
LL	1	S3	1.7*	6.8*	10.1*	7.8*	10.8*
DS	2	S4	4.9*	4.4*	8.8*	5.7*	7.6*
AL	2	4a	2.9*	2.8	10.0*	9.0*	17.3*
LL	2	S5	3.5	14.6*	16.1*	16.2*	21.8*

The asterisks indicate the significant differences ($P<0.05$) between the untreated control and the respective treatment.

standard errors of difference values and calculation of variance ($P<0.05$).

Analysis of 16S rRNA gene fragments PCR amplified from total community DNA by denaturing gradient gel electrophoresis (DGGE)

Total community DNA (TC-DNA) was extracted from the microbial pellets using the FastDNA SPIN Kit (MP Biomedicals, Heidelberg, Germany) as described by the manufacturer after a harsh lysis step with the FastPrep-24 Instrument (MP Biomedicals, Heidelberg, Germany). The TC-DNA was purified with GENE-CLEAN SPIN Kit (MP Biomedicals, Heidelberg, Germany) according to the manufacturer. The purified TC-DNA was diluted 1:10 with 10 mM Tris HCl before use.

For amplification of 16S rRNA gene fragments, PCR reactions were performed with TC-DNA obtained from rhizosphere samples with the primers F984-GC and R1378 as described by Heuer [41] using Taq DNA polymerase (Stoffel fragment, ABI, Darmstadt, Germany). The PCR products were analyzed by DGGE approach as described by Weinert et al. [42].

Bacterial fingerprints were evaluated with GELCOMPAR II version 6.5 (Applied Maths, Sint-Martens-Latern, Belgium) as described by Schreiter et al. [37]. The obtained Pearson similarity matrices were used for construction of a dendrogram by an Unweighted Pair-Group Method with Arithmetic mean (UPGMA) as well as of statistical analysis by the permutation test, calculating the d-value from the average overall correlation coefficients within the groups minus the average overall correlation coefficients between samples from treatments compared as suggested by Kropf et al. [43].

Results

Soil parameters of both field experiments

The concentration of N, P and K measured for each soil type before planting revealed only minor variations between both experiments (Table S1). In the second experiment, the temperature measured for all three soils in the 20 cm top layer was approximately 0.6°C below that of the first experiment (Table S2).

In contrast to the temperature the volumetric soil water content (VWC) in the top soils varied significantly among the soil types and experiments (except for AL and LL soil in the first experiment; Table S3). The lowest percentage of VWC was recorded in DS, and the highest in AL soil in both experiments. The calculated percentages of VWC in the first experiment were 18.9%, 39.6% and 39.1% in DS, AL and LL soil, and 17.0%, 35.7% and 27.4% in DS, AL and LL soil, in the second experiment. In the first experiment the VWC of AL and LL soil did not differ significantly. In both experiments VWC values for DS soil were significantly lower than those for AL and LL (Tukey post-hoc tests and confidence limits in Table S3).

Effect of the soil types on rhizosphere competence of the bacterial inoculants

In both experiments the inoculants were able to colonize the rhizosphere of lettuce grown in the different soil types at comparable CFU counts (Figure 1). Three-way ANOVA (soil type, plant growth development stage, pathogen) revealed that in both experiments the soil type had no significant effect on the CFU of RU47 in the rhizosphere of lettuce ($P>0.05$) but on the CFU of 3Re4-18 ($P<0.05$) (Table 1). In the first and second experiment 2WAP the CFU counts of 3Re4-18 were significantly higher in LL soil than in DS and AL soil. With increasing plant

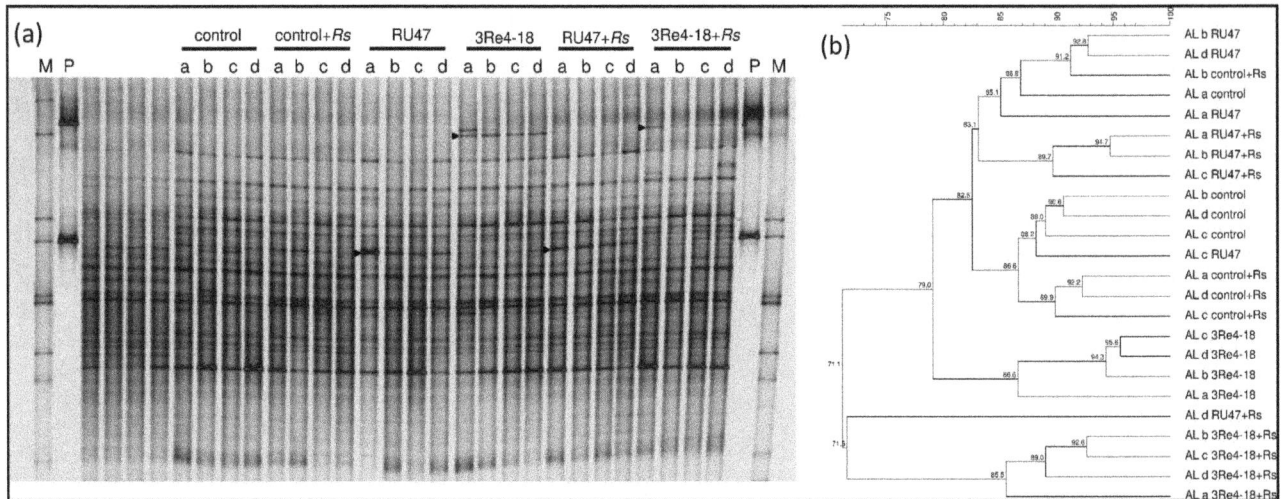

Figure 4. DGGE fingerprints (a) of bacterial 16Sr-RNA gene fragments from community DNA extracts were obtained from lettuce rhizosphere 2WAP of lettuce into the AL soil (second experiment). The corresponding UPGMA dendrogram (b) of these DGGE fingerprints based on the Pearson similarity matrix. The four replicates of the treatments: control, control+*Rs*, RU47, 3Re4-18, RU47+*Rs*, 3Re4-18+*Rs* were indicated by a–d. M: marker [57].

age the CFU counts of both inoculants decreased in the rhizosphere of lettuce grown in all three soils and in both experiments ($P<0.05$) except for the treatment with RU47 in DS soil in the first experiment and the treatment with 3Re4-18+*Rs* in DS soil in the second experiment (Figure 1). No significant effect of *R. solani* on CFU counts of both inoculants was revealed ($P>0.05$) in both experiments. The natural background of the rifampicin mutation was very low in all soil types.

Effects of the soil type, the bacterial inoculants and the pathogen on the lettuce growth

The effects of the soil type, the inoculants and *R. solani* on lettuce growth assessed by comparing the SDM of lettuce plants harvested 6WAP revealed lower lettuce growth in the first experiment compared with the second experiment in particular for LL soil (Figure 2). A significant effect of the soil type on lettuce growth was observed in both experiments. However, the results differed between the experiments. In the first experiment plants grown in LL soil showed the lowest SDM (21.3 g/plant) compared to SDM of plants from DS (32.5 g/plant) and AL soil (31.9 g/plant). In the second experiment the SDM was highest for plants grown in AL soil (46.2 g/plant) while the SDM of plants from DS (41.9 g/plant) and LL soil (41.6 g/plant) were comparable. In contrast to the first experiment an improved lettuce growth was observed in the treatments with the inoculants RU47 and 3Re4-18 in all three soils in the second experiment. However, the inoculant effects on lettuce growth were not significant (Figure 2).

In general the pathogen *R. solani* had a negative effect on lettuce growth in all three soils and in both experiments. The effect was significant in AL soil in both experiments, in LL soil only in the first experiment, and in DS soil in the second experiment (Figure 2). The treatment of lettuce with the inoculants resulted in an improved lettuce growth in RU47+*Rs* and 3Re4-18+*Rs* in all three soils and in both experiments but significant effects were only recorded in the first experiment for the treatments RU47+*Rs* in DS soil and 3Re4-18+*Rs* in LL soil, and in the second experiment for RU47+*Rs* and 3Re4-18+*Rs* in DS soil, for RU47+*Rs* in AL soil and for 3Re4-18+*Rs* in LL soil (Figure 2).

Effect of the soil type on the biocontrol activity of the bacterial inoculants

Bottom rot symptoms were also recorded in the untreated controls in each soil and in both experiments (Figure 3). The inoculation of *R. solani* (control+*Rs*) resulted in a significantly increased disease severity in all three soils and in both experiments. More severe bottom rot symptoms were scored in the second compared to the first experiment based on the disease severity in the pathogen controls in all three soils. The soil type influenced the disease severity of bottom rot in both experiments (Figure 3). The lowest disease severity was always recorded on lettuce grown in LL soil (control and control+*Rs*) compared to disease severity observed for lettuce grown in DS and AL soil.

RU47 and 3Re4-18 were able to reduce severity of bottom rot on lettuce in the treatments without and with *R. solani* inoculation in all three soils and in both experiments (Figure 3). Significant biocontrol effects were observed by the inoculant RU47 in DS soil (RU47; RU47+*Rs*) in both experiments and in AL (RU47; RU47+*Rs*) and LL (RU47+*Rs*) soil in the second experiment only. The inoculant 3Re4-18 showed a significant control effect in LL (3Re4-18+*Rs*) in the first experiment and in all soils in the second experiment (3Re4-18; 3Re4-18+*Rs*) except for 3Re4-18 in LL soil (Figure 3).

Effects of the pathogen and the bacterial inoculants on the bacterial community composition in the rhizosphere of lettuce

The effect of *R. solani* on the bacterial community composition were assessed by comparing the bacterial DGGE fingerprints of the control treatments (control) with those of the pathogen control (control+*Rs*) for each of the soils. In both experiments, the inoculation of *R. solani* had low but an significant effect in all soils except LL in the second experiment, the highest d-values were observed in DS soil (Table 2).

The effects of the inoculants on the bacterial community composition in the rhizosphere of lettuce analyzed by DGGE varied between the two experiments. The bacterial 16S rRNA gene fragments, amplified from TC-DNA of all treatments, were analyzed for each soil type separately and the DGGE gel for AL soil of the second experiment that had high d-values (Figure 4a). The DGGE fingerprint (Figure 4a) revealed that 2WAP the inoculant strain 3Re4-18 belonged to the dominant members of the bacterial community in the rhizosphere as bands with the electrophoretic mobility of 3Re4-18 were detected in the corresponding treatments. However, bands with electrophoretic mobility of the inoculant RU47 could not be detected as it co-migrated with bands of the indigenous bacterial community. Thus RU47 could only be identified in the second experiment in DS and AL soil, while 3Re4-18 was detected in AL and LL soil in the first, and in DS and AL soils in the second experiment (Figures S1–S5). Furthermore, the analysis of DGGE fingerprints by UPGMA revealed that the bacterial community composition of the different treatments shared approximately 70% similarity for all soils (Figure 4b). Although treatment dependent clusters were not always observed, the treatment effects were significant based on the permutation test ($P<0.05$) for all treatments with the inoculants RU47 or 3Re4-18 except for RU47 in AL soil in the second field experiment. However, the differences between the fingerprints of the controls and the treatments given by the calculated d-values varied for both experiments (Table 2). In the first experiment highest d-values were observed for AL soil for both inoculants while in the second experiment the highest d-values were observed for LL soil (Table 2). In both experiments and in all soils higher d-values for the treatments with 3Re4-18 indicated a stronger effect of this inoculant on the indigenous rhizosphere bacterial community compared to RU47 (Table 2). The effects of both inoculants on the bacterial community composition were typically increased for the treatments with *R. solani* inoculation in both experiments (except for RU47 and 3Re4-18 in AL soil in the first experiment).

Discussion

The availability of three soil types differing in their properties but sharing the same cropping history and weather conditions at the same field site enabled us to investigate the effect of the soil type on the rhizosphere competence of the two bacterial inoculants *P. jessenii* RU47 and *S. plymuthica* 3Re4-18 without and with *R. solani* inoculation for the first time. We hypothesized that the soil types with their distinct microbial community composition and physico-chemical properties but also the soil type dependent root exudates of the model plant lettuce [44] might affect the rhizosphere competence of the inoculants and in consequence their biocontrol efficacy. The composition of root exudates collected from lettuce grown in DS, AL and LL soil was recently shown to differ only quantitatively as similar compounds were detected as independent of the soil type [44]. Several studies showed that root exudates are important drivers of the rhizosphere microbial community composition [45,46] and the metabolic activity of the bacterial community including inoculant strains [47]. Recently, we could show by means of DGGE and pyrosequence analysis of 16S rRNA gene fragments, amplified from TC-DNA of bulk soil samples taken from the experimental unit 6 in 2010, that the three soil types indeed harbored a distinct bacterial community structure indicating the importance of the mineral composition and the soil organic matter for shaping the bacterial community composition [37]. Interestingly, in the rhizosphere of lettuce grown in the different soils numerous similar genera were increased in relative abundance, and a high proportion of dominant operational taxonomic units were shared among the rhizosphere samples from different soils [37]. Among the genera, significantly increased in the rhizosphere, was the genus *Pseudomonas* which might explain the negligible effect of

soil types on the ability of RU47 to colonize the lettuce rhizosphere in both experiments performed in 2011. Remarkably, similar CFU counts of RU47 were observed 2WAP in all soil types and in both experiments. The decrease of RU47 CFU counts in the rhizosphere observed 5WAP in both experiments might well be a decrease in its relative abundance per gram of RDM as the rhizosphere pellet was obtained from the complete root system due to changes in the rhizoplane surface/RDM ratio or/and possibly newly emerging roots were not colonized by RU47 as previously reported for *gfp*-tagged inoculants [48]. A decrease in inoculant CFU densities with increasing plant age of lettuce confirmed previous observations for the inoculants used in this study [34,49] and other biocontrol strains [24]. Nonetheless, our results showed an influence of the soil types on the rhizosphere competence of 3Re4-18. While higher CFU counts were observed in LL soil in both experiments 2WAP in comparison to DS and AL soil, a pronounced decline of 3Re4-18 counts was recorded in LL soil in the second experiment 5WAP. Indeed, in the second experiment the SDM was almost doubled for lettuce grown in LL soil compared to the first experiment which might have resulted in a decreased relative abundance of 3Re4-18. Neumann et al. [44] reported that the lettuce growth and the root morphology were influenced by soil type which might have contributed to the soil type dependent differences in rhizosphere competence of 3Re4-18. The differences in lettuce growth between the first and the second experiment in particular in LL soil were not unusual for agricultural practice. Factors that might have contributed to these differences range from slightly lower VWC, temperatures in the field directly after planting to differences in sun light intensity. In addition, the slightly different cropping history in both experimental units could have affected the soil microbiome with potential effects on plant growth. Plant growth in turn is assumed to influence the quantity of root exudates [50]. Several studies have underlined a clear relationship between the level of disease suppression and inoculant densities in the rhizosphere of an introduced biocontrol strain [17,51,52]. Inconsistent results in biological control were often assumed to be associated with inefficient root colonization at the field scale [11,53]. However, despite the good colonization of the lettuce rhizosphere by both inoculants in the present study, the biocontrol effects varied in both experiments. Although in the presence of the inoculant strains the disease severity was reduced. Compared to the control and the pathogen control (control+*Rs*), the biocontrol effects were not always significant. In fact, in the first experiment significant biocontrol effects were only observed for RU47 and RU47+*Rs* in DS soil and 3Re4-18+*Rs* in LL soil while biocontrol effects were found to be much more pronounced and significant in almost all treatments in the second experiment. The improved biocontrol effect seems to be correlated with an increased lettuce growth in the second experiment. It is tempting to speculate that an improved plant growth might be linked to an increased amount of photosynthates released via the roots. The increased root exudation in turn might have not only affected the metabolic activity of the inoculants but also facilitated the interaction of *R. solani* with the lettuce plant. Therefore, more severe bottom rot symptoms in the control and the control+*Rs* treatments were recorded in the second experiment. Furthermore, the disease severity in the control (natural background of *R. solani*) and in the pathogen control (control+*Rs*) was highest in DS soil and lowest in LL soil in both experiments. The higher conduciveness of the DS soil may be caused by higher oxygen availability because of the bigger pore sizes in sandy soils but might also be due to differences in the microbial community structure. Also, Steinberg et al. [54] reported that the capacity of a soil to antagonize the action of a plant pathogen is related to the structure and function of the microbial communities in soil.

Molecular fingerprinting techniques confirmed that the inoculants were indeed dominant members of the bacterial communities in the rhizosphere of lettuce. The additional DGGE bands of the inoculants strains 3Re4-18 and the increased intensity of a band co-migrating with RU47 which did not occur in the control treatments likely contributed to the effects of the inoculants on the rhizosphere bacterial community composition determined by the permutation test (Table 2). An approach to prevent the amplification of the inoculant strains is the use of taxon-specific primers which would exclude the taxonomic group to which the inoculant belongs, as proposed by Gomes et al. [55]. The actinobacterial fingerprint, done for the samples from 2WAP of the second experiment still revealed a significant effect except for RU47 in DS soil (Schreiter et al. unpublished data). The low effects of *R. solani* on the bacterial community composition can most likely be explained by the pathogenesis of *R. solani* AG1-IB which is primarily not a root pathogen but attacks the lower leaves of lettuce [56]. In conclusion, in the present study which was based on two independent field experiments performed in the 2011-season we could not confirm our hypothesis that the soil type influences the rhizosphere competence of biocontrol strains RU47 and 3Re4-18. We are aware that this finding cannot be generalized and can be different for other plants–inoculant combinations. The soil type independent enrichment of *Gamma-proteobacteria* such as *Pseudomonas,* as observed by Schreiter et al. [37], in the rhizosphere of lettuce may be an explanation for the fact that the rhizosphere competence of RU47 was not influenced by soil type. Only minor changes in the bacterial rhizosphere composition were observed due to the inoculation of RU47, 3Re4-18 and *R. solani*, and these did not depend on the soil type. However, the plant growth and the disease severity of bottom rot were influenced by the soil type which in turn also influenced the biocontrol effects. The present study is unique as the rhizosphere competence, biocontrol effects, plant growth and treatment effects on the indigenous rhizosphere bacterial community were assessed in three soil types at the same site in two independent field experiments. The present study showed that the often reported inconsistency of biocontrol is likely more due to plants which in turn are influenced by weather conditions. Thus we conclude that the multitrophic interaction between plant, inoculant, pathogen and indigenous microbial community deserves far more attention in the future.

Supporting Information

Figure S1 *Bacteria* **DGGE fingerprint of rhizosphere-samples were obtained from DS soil 2WAP of the lettuce (first experiment).** Lanes a–d replicates of each of the treatments: untreated control (control), control inoculated with *Rhizoctonia solani* (control+*Rs*), inoculation with *Pseudomonas jessenii* RU47 (RU47), inoculation with *Serratia plymuthica* 3Re4-18 (3Re4-18), inoculation with RU47 and *R. solani* (RU47+*Rs*), inoculation with 3Re4-18 and *R. solani* (3Re4-18+*Rs*); M: marker [57].

Figure S2 *Bacteria* **DGGE fingerprint of rhizosphere-samples were obtained from AL soil 2WAP of the lettuce (first experiment).** Lanes a–d replicates of each of the treatments: untreated control (control), control inoculated with *Rhizoctonia solani* (control+*Rs*), inoculation with *Pseudomonas jessenii* RU47 (RU47), inoculation with *Serratia plymuthica* 3Re4-18 (3Re4-18), inoculation with RU47 and *R. solani* (RU47+*Rs*),

inoculation with 3Re4-18 and *R. solani* (3Re4-18+*Rs*); P: inoculants (upper band 3Re4-18; lower band RU47); M: marker [57].

Figure S3 *Bacteria* **DGGE fingerprint of rhizosphere-samples were obtained from LL soil 2WAP of the lettuce (first experiment).** Lanes a–d replicates of each of the treatments: untreated control (control), control inoculated with *Rhizoctonia solani* (control+*Rs*), inoculation with *Pseudomonas jessenii* RU47 (RU47), inoculation with *Serratia plymuthica* 3Re4-18 (3Re4-18), inoculation with RU47 and *R. solani* (RU47+*Rs*), inoculation with 3Re4-18 and *R. solani* (3Re4-18+*Rs*); P: inoculants (upper band 3Re4-18; lower band RU47); M: marker [57].

Figure S4 *Bacteria* **DGGE fingerprint of rhizosphere-samples were obtained from DS soil 2WAP of the lettuce (second experiment).** Lanes a–d replicates of each of the treatments: untreated control (control), control inoculated with *Rhizoctonia solani* (control+*Rs*), inoculation with *Pseudomonas jessenii* RU47 (RU47), inoculation with *Serratia plymuthica* 3Re4-18 (3Re4-18), inoculation with RU47 and *R. solani* (RU47+*Rs*), inoculation with 3Re4-18 and *R. solani* (3Re4-18+*Rs*); P: inoculants (upper band 3Re4-18; lower band RU47); M: marker [57].

Figure S5 *Bacteria* **DGGE fingerprint of rhizosphere-samples were obtained from LL soil 2WAP of the lettuce (second experiment).** Lanes a–d replicates of each of the treatments: untreated control (control), control inoculated with *Rhizoctonia solani* (control+*Rs*), inoculation with *Pseudomonas jessenii* RU47 (RU47), inoculation with *Serratia plymuthica* 3Re4-18 (3Re4-18), inoculation with RU47 and *R. solani* (RU47+*Rs*), inoculation with 3Re4-18 and *R. solani* (3Re4-18+*Rs*); P: inoculants (upper band 3Re4-18; lower band RU47); M: marker [57].

Acknowledgments

We would also like to thank Petra Zocher, Ute Zimmerling, Sabine Breitkopf and Angelika Fandrey for their skilled technical assistance. We also thank Christin Zachow for providing the *Serratia plymuthica* 3Re4-18 strain and Ilse-Marie Jungkurth for helpful comments on the manuscript.

Author Contributions

Conceived and designed the experiments: KS RG. Performed the experiments: SS RG KS. Analyzed the data: SS RG MS. Contributed reagents/materials/analysis tools: KS RG SS MS. Contributed to the writing of the manuscript: SS RG KS. Designed software used in analysis of volumetric soil water content: MS.

References

1. Oerke EC (2006) Crop losses to pests. J Agr Sci 144: 31–43.
2. Alabouvette C, Olivain C, Steinberg C (2006) Biological control of plant diseases: the European situation. Eur J Plant Pathol 114: 329–341.
3. Leistra M, Matser AM (2004) Adsorption, transformation, and bioavailability of the fungicides carbendazim and iprodione in soil, alone and in combination. J Environ Sci Heal B 39: 1–17.
4. Wang YS, Wen CY, Chiu TC, Yen JH (2004) Effect of fungicide iprodione on soil bacterial community. Ecotox Environ Safe 59: 127–132.
5. Weller DM, Raaijmakers JM, Gardener BBM, Thomashow LS (2002) Microbial populations responsible for specific soil suppressiveness to plant pathogens. Annu Rev Phytopathol 40: 309–348.
6. Kazempour MN (2004) Biological control of Rhizoctonia solani, the causal agent of rice sheath blight by antagonistics bacteria in greenhouse and field conditions. Plant Pathol J 3: 88–96.
7. Scherwinski K, Grosch R, Berg G (2008) Effect of bacterial antagonists on lettuce: active biocontrol of Rhizoctonia solani and negligible, short-term effects on nontarget microorganisms. FEMS Microbiol Ecol 64: 106–116.
8. Tikhonovich IA, Provorov NA (2011) Microbiology is the basis of sustainable agriculture: an opinion. Ann Appl Biol 159: 155–168.
9. Quinlan RJ, Lisansky SG (2010) North America: Biopesticides Market. CPL Business Consultants, Oxfordshire, UK.
10. Mark GL, Morrissey JP, Higgins P, O'Gara F (2006) Molecular-based strategies to exploit Pseudomonas biocontrol strains for environmental biotechnology applications. FEMS Microbiol Ecol 56: 167–177.
11. Barret M, Morrissey JP, O'Gara F (2011) Functional genomics analysis of plant growth-promoting rhizobacterial traits involved in rhizosphere competence. Biol Fert Soils 47: 729–743.
12. Pierson III LS, Pierson EA (1996) Phenazine antibiotic production in Pseudomonas aureofaciens: role in rhizosphere ecology and pathogen suppression. FEMS Microbiol Lett 136: 101–108.
13. Steidle A, Allesen-Holm M, Riedel K, Berg G, Givskov M, et al. (2002) Identification and characterization of an N-acylhomoserine lactone-dependent quorum-sensing system in Pseudomonas putida strain IsoF. Appl Environ Microbiol 68: 6371–6382.
14. De Bellis P, Ercolani GL (2001) Growth interactions during bacterial colonization of seedling rootlets. Appl Environ Microbiol 67: 1945–1948.
15. Ghirardi S, Dessaint F, Mazurier S, Corberand T, Raaijmakers JM, et al. (2012) Identification of traits shared by rhizosphere-competent strains of fluorescent Pseudomonads. Microb Ecol 64: 725–737.
16. Bloemberg GV, Lugtenberg BJJ (2001) Molecular basis of plant growth promotion and biocontrol by rhizobacteria. Curr Opin Plant Biol 4: 343–350.
17. Lugtenberg BJJ, Dekkers L, Bloemberg GV (2001) Molecular determinants of rhizosphere colonization by Pseudomonas. Annu Rev Phytopathol 39: 461–493.
18. Persello-Cartieaux F, Nussaume L, Robaglia C (2003) Tales from the underground: molecular plant-rhizobacteria interactions. Plant Cell Enviro 26: 189–199.
19. Schnider U, Keel C, Blumer C, Troxler J, Defago G, et al. (1995) Amplification of the housekeeping sigma-factor in Pseudomonas fluorescens CHA0 enhances antibiotic production and improves biocontrol abilities. J Bacteriol 177: 5387–5392.
20. Dekkers LC, Mulders IHM, Phoelich CC, Chin-A-Woeng TFC, Wijfjes AHM, et al. (2000) The sss colonization gene of the tomato-Fusarium oxysporum f. sp radicis-lycopersici biocontrol strain Pseudomonas fluorescens WCS365 can improve root colonization of other wild-type Pseudomonas spp. bacteria. Mol Plant Microbe In 13: 1177–1183.
21. Patten CL, Glick BR (2002) Role of Pseudomonas putida indoleacetic acid in development of the host plant root system. Appl Environ Microbiol 68: 3795–3801.
22. Loper JE, Hassan KA, Mavrodi DV, Davis EW, II, Lim C, et al. (2012) Comparative genomics of plant-associated Pseudomonas spp.: insights into diversity and inheritance of traits involved in multitrophic interactions. PLoS Genet 8: e1002784–e1002784.
23. Raaijmakers JM, Mazzola M (2012) Diversity and natural functions of antibiotics produced by beneficial and plant pathogenic bacteria. Annu Rev Phytopathol 50: 403–424.
24. Haas D, Defago G (2005) Biological control of soil-borne pathogens by fluorescent Pseudomonads. Nat Rev Microbiol 3: 307–319.
25. Raaijmakers JM, Paulitz TC, Steinberg C, Alabouvette C, Moënne-Loccoz Y (2009) The rhizosphere: a playground and battlefield for soilborne pathogens and beneficial microorganisms. Plant Soil 321: 341–361.
26. Capdevila S, Martinez-Granero FM, Sanchez-Contreras M, Rivilla R, Martin M (2004) Analysis of Pseudomonas fluorescens F113 genes implicated in flagellar filament synthesis and their role in competitive root colonization. Microbiol-Uk 150: 3889–3897.
27. Rodriguez-Navarro DN, Dardanelli MS, Ruiz-Sainz JE (2007) Attachment of bacteria to the roots of higher plants. FEMS Microbiol Lett 272: 127–136.
28. Mavrodi DV, Joe A, Mavrodi OV, Hassan KA, Weller DM, et al. (2011) Structural and functional analysis of the type III secretion system from Pseudomonas fluorescens q8r1-96. J Bacteriol 193: 177–189.
29. Lottmann J, Heuer H, de Vries J, Mahn A, Düring K, et al. (2000) Establishment of introduced antagonistic bacteria in the rhizosphere of transgenic potatoes and their effect on the bacterial community. FEMS Microbiol Ecol 33: 41–49.
30. Chowdhury SP, Dietel K, Randler M, Schmid M, Junge H, et al. (2013) Effects of Bacillus amyloliquefaciens FZB42 on lettuce growth and health under pathogen pressure and its impact on the rhizosphere bacterial community. PLoS ONE 8: e68818–e68818. doi:10.1371/journal.pone.0068818.
31. Adesina MF, Lembke A, Costa R, Speksnijder A, Smalla K (2007) Screening of bacterial isolates from various European soils for in vitro antagonistic activity towards Rhizoctonia solani and Fusarium oxysporum: Site-dependent composition and diversity revealed. Soil Biol Biochem 39: 2818–2828.
32. Grosch R, Faltin F, Lottmann J, Kofoet A, Berg G (2005) Effectiveness of 3 antagonistic bacterial isolates to control Rhizoctonia solani Kühn on lettuce and potato. Can J Microbiol 51: 345–353.
33. Wibberg D, Jelonek L, Rupp O, Hennig M, Eikmeyer F, et al. (2013) Establishment and interpretation of the genome sequence of the phytopathogenic fungus Rhizoctonia solani AG1-IB isolate 7/3/14. J Biotechnol 167: 142–155.
34. Adesina MF, Grosch R, Lembke A, Vatchev TD, Smalla K (2009) In vitro antagonists of Rhizoctonia solani tested on lettuce: rhizosphere competence, biocontrol efficiency and rhizosphere microbial community response. FEMS Microbiol Ecol 69: 62–74.
35. Berg G, Krechel A, Ditz M, Sikora RA, Ulrich A, et al. (2005) Endophytic and ectophytic potato-associated bacterial communities differ in structure and antagonistic function against plant pathogenic fungi. FEMS Microbiol Ecol 51: 215–229.
36. Rühlmann J, Ruppel S (2005) Effects of organic amendments on soil carbon content and microbial biomass-results of the long-term box plot experiment in Grossbeeren. Arch Agron Soil Sci 51: 163–170.
37. Schreiter S, Ding G-C, Heuer H, Neumann G, Sandmann M, et al. (2014) Effect of the soil type on the microbiome in the rhizosphere of field-grown lettuce. Front Microbiol 5: 144.
38. Gutezeit B, Herzog FN, Wenkel KO (1993) Das Beregnungsbedarfssystem für Freilandgemüse. Gemüse 29: 106–108.
39. Grosch R, Schneider JHM, Kofoet A, Feller C (2011) Impact of continuous cropping of lettuce on the disease dynamics of bottom rot and genotypic diversity of Rhizoctonia solani AG 1-IB. J Phytopathol 159: 35–44.
40. Schneider JHM, Schilder MT, Dijst G (1997) Characterization of Rhizoctonia solani AG 2 isolates causing bare patch in field grown tulips in the Netherlands. Eur J Plant Pathol 103: 265–279.
41. Heuer H, Krsek M, Baker P, Smalla K, Wellington EMH (1997) Analysis of actinomycete communities by specific amplification of genes encoding 16S rRNA and gel-electrophoretic separation in denaturing gradients. Appl Environ Microbiol 63: 3233–3241.
42. Weinert N, Meincke R, Gottwald C, Heuer H, Gomes NCM, et al. (2009) Rhizosphere communities of genetically modified zeaxanthin-accumulating potato plants and their parent cultivar differ less than those of different potato cultivars. Appl Environ Microbiol 75: 3859–3865.
43. Kropf S, Heuer H, Grüning M, Smalla K (2004) Significance test for comparing complex microbial community fingerprints using pairwise similarity measures. J Microbiol Meth 57: 187–195.
44. Neumann G, Bott S, Ohler M, Mock H, Lippman R, et al. (2014) Root exudation and root development of lettuce (Lactuca sativa L. cv. Tizian) as affected by different soils. Front Microbiol 5: 2.
45. Paterson E, Gebbing T, Abel C, Sim A, Telfer G (2007) Rhizodeposition shapes rhizosphere microbial community structure in organic soil. New Phytol 173: 600–610.
46. Henry S, Texier S, Hallet S, Bru D, Dambreville C, et al. (2008) Disentangling the rhizosphere effect on nitrate reducers and denitrifiers: insight into the role of root exudates. Environ Microbiol 10: 3082–3092.
47. Benizri E, Nguyen C, Piutti S, Slezack-Deschaumes S, Philippot L (2007) Additions of maize root mucilage to soil changed the structure of the bacterial community. Soil Biol Biochem 39: 1230–1233.
48. Götz M, Gomes NCM, Dratwinski A, Costa R, Berg G, et al. (2006) Survival of gfp-tagged antagonistic bacteria in the rhizosphere of tomato plants and their effects on the indigenous bacterial community. FEMS Microbiol Ecol 56: 207–218.
49. Grosch R, Dealtry S, Schreiter S, Berg G, Mendonça-Hagler L, et al. (2012) Biocontrol of Rhizoctonia solani: complex interaction of biocontrol strains, pathogen and indigenous microbial community in the rhizosphere of lettuce shown by molecular methods. Plant Soil 361: 343–357.

50. Baudoin E, Benizri E, Guckert A (2002) Impact of growth stage on the bacterial community structure along maize roots, as determined by metabolic and genetic fingerprinting. Appl Soil Ecol 19: 135–145.

51. Bull CT, Weller DM, Thomashow LS (1991) Relationship between root colonization and suppression of *Gaeumannomyces graminis* var. tritici by *Pseudomonas fluorescens* strain 2-79. Phytopathology 81: 954–959.

52. Raaijmakers JM, Weller DM (2001) Exploiting genotypic diversity of 2,4-diacetylphloroglucinol-producing *Pseudomonas* spp.: Characterization of superior root-colonizing *P. fluorescens* strain Q8r1-96. Appl Environ Microbiol 67: 2545–2554.

53. Lemanceau P, Alabouvette C (1993) Suppression of fusarium-wilt by fluorescent pseudomonads: Mechanisms and applications. Biocontrol Sci Techn 3: 219–234.

54. Steinberg C, Edel-Hermann V, Alabouvette C, Lemanceau P (2007) Soil suppressiveness to plant diseases. Modern Soil Microbiology: 455–478.

55. Gomes NCM, Kosheleva IA, Abraham WR, Smalla K (2005) Effects of the inoculant strain *Pseudomonas putida* KT2442 (pNF142) and of naphthalene contamination on the soil bacterial community. FEMS Microbiol Ecol 54: 21–33.

56. Davis R, Subbarao K, Raid R, Kurtz E (1997) Compendium of Lettuce Diseases. APS Press: 15–16.

57. Heuer H, Wieland G, Schönfeld J, Schönwalder A, Gomes NCM, et al. (2001) Bacterial community profiling using DGGE or TGGE analysis; Rochelle PA, editor. 177–190 p.

16

Exploring the Polyadenylated RNA Virome of Sweet Potato through High-Throughput Sequencing

Ying-Hong Gu[1][9], Xiang Tao[2][9], Xian-Jun Lai[1], Hai-Yan Wang[1], Yi-Zheng Zhang[1]*

1 College of Life Sciences, Sichuan University, Key Laboratory of Bio-resources and Eco-environment, Ministry of Education, Sichuan Key Laboratory of Molecular Biology and Biotechnology, Center for Functional Genomics and Bioinformatics, Chengdu, Sichuan, People's Republic of China, 2 Chengdu Institute of Biology, Chinese Academy of Sciences, Chengdu, Sichuan, People's Republic of China

Abstract

Background: Viral diseases are the second most significant biotic stress for sweet potato, with yield losses reaching 20% to 40%. Over 30 viruses have been reported to infect sweet potato around the world, and 11 of these have been detected in China. Most of these viruses were detected by traditional detection approaches that show disadvantages in detection throughput. Next-generation sequencing technology provides a novel, high sensitive method for virus detection and diagnosis.

Methodology/Principal Findings: We report the polyadenylated RNA virome of three sweet potato cultivars using a high throughput RNA sequencing approach. Transcripts of 15 different viruses were detected, 11 of which were detected in cultivar Xushu18, whilst 11 and 4 viruses were detected in Guangshu 87 and Jingshu 6, respectively. Four were detected in sweet potato for the first time, and 4 were found for the first time in China. The most prevalent virus was SPFMV, which constituted 88% of the total viral sequence reads. Virus transcripts with extremely low expression levels were also detected, such as transcripts of SPLCV, CMV and CymMV. Digital gene expression (DGE) and reverse transcription polymerase chain reaction (RT-PCR) analyses showed that the highest viral transcript expression levels were found in fibrous and tuberous roots, which suggest that these tissues should be optimum samples for virus detection.

Conclusions/Significance: A total of 15 viruses were presumed to present in three sweet potato cultivars growing in China. This is the first insight into the sweet potato polyadenylated RNA virome. These results can serve as a basis for further work to investigate whether some of the 'new' viruses infecting sweet potato are pathogenic.

Editor: Darren P. Martin, Institute of Infectious Disease and Molecular Medicine, South Africa

Funding: This work was financially supported by the National Science & Technology Pillar Program of China (No. 2007BAD78B03) and the "Eleven-Five" Key Project of Sichuan Province (No. 07SG111-003-1). The funders had no role in study design, data collection and analysis, decision to publish, or preparation of the manuscript.

Competing Interests: The authors have declared that no competing interests exist.

* E-mail: yizzhang@scu.edu.cn

[9] These authors contributed equally to this work.

Introduction

The sweet potato [*Ipomoea batatas* L. (Lam.)] originated in South America and was transported across the pacific by Polynesians [1]. It has been cultivated by humans for up to 8,000 years, and today it is widely grown around the world due to its strong adaptability, easy management, rich nutrient content and multiple usages. Sweet potato is the fifth most important food crop in developing countries. About 130 million metric tons of tuberous roots are produced globally each year on about 9 million hectares of land [2,3]. China is the biggest producer in the world, accounting for 80% of the global sweet potato production [4]. Compared to other staple food crops sweet potato needs fewer inputs, but produces more biomass [5]. A few researchers have shown interest in sweet potato mainly because of its complex hexaploid inheritance [6]. Recently, the growing awareness of health benefits attributed to sweet potato has stimulated renewed interest in this crop [3].

Viral diseases are the second most significant biotic stress for sweet potato after the sweet potato weevil [7]. Usually, sweet potato viruses will co-infect the plants and severely limit root production [8]. Yield losses caused by these viral diseases reach 20% to 40%, but this can reach near 100% in some African countries [9–12]. Over 30 viruses have been reported to infect sweet potato worldwide, but most of them are asymptomatic [3,13]. Eleven of these viruses have been detected in China [14], including *Sweet potato C6 virus* (SPC6V) [15], *Sweet potato chlorotic fleck virus* (SPCFV) [15], *Sweet potato chlorotic stunt virus* (SPCSV) [15–17], *Sweet potato collusive virus* (SPCV, synonym Sweet potato caulimo-like virus) [10], *Sweet potato feathery mottle virus* (SPFMV) [15,18–20], *Sweet potato leaf curl virus* (SPLCV) [21], *Sweet potato latent virus* (SwPLV) [19,20], *Sweet potato mild mottle virus* (SPMMV) [15], *Sweet potato mild speckling virus* (SPMSV) [15], *Sweet potato virus G* (SPVG) [20,22] and *Sweet potato virus 2* [SPV2, synonym Ipomoea vein mosaic virus (IVMV) or Sweet potato virus Y (SPVY)] [23,24]. SPFMV, SwPLV and SPCFV were recognized to be the most commonly occurring and damaging viruses in China [4,14]. Infection rates of these three viruses in major production regions of China range from 21% to 100% [15,17], resulting in annual

Table 1. Summary of RNA-Seq and DGE data used in this study.

Accession No.	Cultivar	Planting location	Tissue types	No. of tags or read pairs	References	Note
SRA043582	Xushu 18	Chengdu	Leaves, stems and roots	48,716,884	[34]	Sequencing directly
SRA043584	Xushu 18	Chengdu	Flowers	41,533,336	[35]	Sequencing directly
SRA022988	Guangshu 87	Guangzhou	Roots	59,233,468	[32]	Re-analyze
SRA044884	Jingshu 6	Beijing	Roots	25,888,888	[33]	Re-analyze
GSE35929	Xushu 18	Chengdu	Young leaves	3,352,753	[34]	Sequencing directly
GSE35929	Xushu 18	Chengdu	Mature leaves	3,429,018	[34]	Sequencing directly
GSE35929	Xushu 18	Chengdu	Stems	3,453,654	[34]	Sequencing directly
GSE35929	Xushu 18	Chengdu	Fibrous roots	3,583,907	[34]	Sequencing directly
GSE35929	Xushu 18	Chengdu	Initial tuberous roots	3,630,619	[34]	Sequencing directly
GSE35929	Xushu 18	Chengdu	Expanding tuberous roots	3,566,630	[34]	Sequencing directly
GSE35929	Xushu 18	Chengdu	Harvested tuberous roots	3,514,272	[34]	Sequencing directly

economic losses of about $639 million to the Chinese sweet potato industry [25].

Up until 1995, most of the work on sweet potato virus focused on SPFMV, but in the past 18 years, due to the advent of molecular biology, various comprehensive studies on virus composition and the effects of viral diseases were reported [3,26–28]. The development of next-generation sequencing (NGS) technology provides a highly sensitive method for virus detection and diagnosis [29–31]. In this study, we analyzed the NGS data of eight sweet potato tissues and re-analyzed those of the other two published studies [32,33] to identify RNA virus sequences.

Materials and Methods

Plant Material

Sweet potato plants (*I. batatas* cv. Xushu 18) were planted in an experimental field at Sichuan University under natural conditions. All of the following samples were collected from symptomless plants: Fibrous roots (FR) at one month after planting; young leaves (YL), mature leaves (ML), stems and initial tuberous roots (ITR) at 1.5 months; expanding tuberous roots at 3 months; harvested tuberous roots at 5 months; newly opened flowers were collected from symptomless drought-treated plants at 4 months.

High-throughput RNA-sequencing

Total RNA was extracted using the TRIzol Reagent (Invitrogen), and genomic DNA removed with DNase I (Fermentas, Burlington, Ontario, Canada) according to the manufacturer's instructions. Then the purity, concentration and RNA integrity number (RIN) of total RNA were measured with a SMA3000 and/or Agilent 2100 Bioanalyzer. The assessed total RNA was submitted to the Beijing Genomics Institute (BGI)-Shenzhen, Shenzhen, China (http://www.genomics.cn) for mRNA purification and RNA sequencing (RNA-Seq) with Illumina Hiseq 2000.

Viral Sequence Mining and Expression Pattern Analyses

To investigate the polyadenylated RNA virome of sweet potato, viral sequences and expression patterns were mined from the vegetative transcriptome of Xushu 18 according to the annotation information and Digital Gene Expression (DGE) profiling results [34]. We also extracted total RNA from floral organs of sweet potato cultivar Xushu 18 and submitted it to Illumina HiSeq 2000

for RNA-Seq analysis [35]. By using Bowtie [36] under default parameters except seed length of 40 and mismatches of 3, the 90 bp paired-end (PE) reads of the floral organs were mapped to the vegetative transcriptome, that has been known to contain some viral sequences. Moreover, the 75 bp PE reads of Guangshu 87 [32] and Jingshu 6 [33] retrieved from the NCBI's Sequence Read Archive database (http://www.ncbi.nlm.nih.gov/Traces/sra) (Table 1) were re-analyzed by using Bowtie [36] to align them to the vegetative transcriptome. The number of mapped read pairs or tags was counted according to the mapping results. RPKM (Reads Per Kilobase per Million mapped reads) [37] and TPM (Transcripts Per Million clean tags) [38] were calculated and used for quantifying each viral transcript in different sweet potato samples.

Reverse Transcription Polymerase Chain Reaction (RT-PCR) Verification

Equal RNA extracted from FR, YL, ML, ITR and stems were reversely transcribed with Moloney murine leukemia virus (MMLV) reverse transcriptase (Invitrogen, Carlsbad, California, CA) using Oligo(dT) as primer. The resulting cDNA was subjected to viral sequence amplification and viral gene expression level analysis.

Fourteen pairs of primers were designed according to the assembled viral transcripts (Table 2) using Primer Premier 5.0 (PREMIER Biosoft. International, CA, USA) (Table 3), and sequence amplification was implemented using KOD-FX (TOYOBO, Osaka, Japan). The purified PCR products were sequenced with an ABI 3730 instrument to confirm the amplified sequences.

Results

Virus Identification *via* Next-generation Sequencing

Seven vegetative tissues were collected from sweet potato cv. Xushu 18 and equal RNA of each tissue sample was pooled together for RNA-Sequencing. A total of 48,716,884 PE reads were generated by Illumina/Solexa Genome Analyzer II. The *de novo* assembly and sequence annotation information were deposited at the Center for Functional Genomics and Bioinformatics of Sichuan University (http://cfgbi.scu.edu.cn/index.html). All of the results described above have been published in 2012 [34]. Sequences of nine viruses were detected in the vegetative

Table 2. Statistics of viruses found in sweet potato transcriptome annotations, estimated gene expression abundance.

Virus name	Abbreviation	Genus	No. of Sequences	Total length (bp)	Identity (%)	Average expression levels (RPKM)	No. of read pairs
Sweet potato feathery mottle virus	SPFMV	Potyvirus	47	22,302	100	50.99	46,434
Sweet potato virus G	SPVG	Potyvirus	5	2,781	100	14.79	1,680
Northern cereal mosaic virus	NCMV-like	Cytorhabdovirus	1	1,134	46	11.58	536
Sweet potato virus B2	SPVB2	Potyvirus	6	987	97	3.85	155
Sweet potato latent virus	SwPLV	Potyvirus	3	806	100	1.73	57
Sweetpotato badnavirus B	SPBV-B	Badnavirus	4	575	94	1.11	26
Sweet potato virus 2	SPV2	Potyvirus	3	419	100	1.40	24
Sweetpotato badnavirus A	SPBV-A	Badnavirus	2	250	97	1.57	16
Mikania micrantha mosaic virus	MMMV	Fabavirus	5	736	100	1.25	42
Yam mosaic virus	YMV	Potyvirus	1	230	94	5.11	48
Turnip mosaic virus	TuMV-like	Potyvirus	1	217	69	0.68	6
Sunflower mosaic virus	SuMV	Potyvirus	1	198	96	1.36	11

No. of Sequences means the viral transcripts detected in the vegetative transcriptome [34]. Total length means the sum of detected viral transcripts of each virus. Identity refers to sequence identities when similarity BLASTX search was conducted between assembled viral transcripts and NR database. YMV was further confirmed to be SPVG, and TuMV-like and SuMV were further confirmed to be SwPLV.

organs of this cultivar (Table 2, Table S1). Among these viruses, two belonged to the *Badnavirus* genus: SPBV-A (sweet potato badnavirus A) and SPBV-B (sweet potato badnavirus B) which were suggested to be Sweet potato pakakuy virus (SPPV) (International Committee on Taxonomy of Viruses, ICTV, http://www.ictvonline.org). The others were all RNA viruses, in which SPFMV, SPVG, SwPLV, SPV2 and SPVB2 (sweet potato virus B2), YMV (*Yam mosaic virus*), TuMV-like (*Turnip mosaic virus*) and SuMV (*Sunflower mosaic virus*) are from the *Potyvirus* genus; and NCMV-like (*Northern cereal mosaic virus*) and MMMV (Mikania micrantha mosaic virus) are from *Cytorhabdovirus* and *Fabavirus*, respectively. MMMV is also known as Mikania micrantha wilt virus (MMWV) as it was first discovered in *Mikania micrantha* [39]. Furthermore, the results demonstrated that SPFMV and SPVG had the longest total sequence length, the highest mapped reads number and the highest average expression levels (Table 2). Except for SPFMV, SPVG, SwPLV and SPV2, the others were reported in sweet potato in China for the first time.

Sequence alignment analyses demonstrated that the SPFMV transcripts belonged to at least three SPFMV strains in this sweet potato cultivar, including the severe, common and ordinary strains [40]. The common strain had been renamed as *Sweet potato virus C* (SPVC) (ICTV, http://www.ictvonline.org). Furthermore, at least two distinct transcripts related to SPVG strains were identified (Table 4). According to the NGS annotation information, there were 3 short sequences s (230 bp, 217 bp and 198 bp, respectively) been annotated as YMV, TuMV-like and SuMV (Table 2). Further studies of the recent released genome sequences of SPVG and SwPLV have confirmed that the 3 short sequences were indeed from SPVG and SwPLV.

Flowers of this cultivar were also collected and submitted to the NGS platform for RNA-Seq study. A total of 41,533,336 PE reads were generated [35]. By mapping these 90 bp PE reads to viral sequences retrieved from NCBI, and assembled viral sequences described above using Bowtie [36], we found four different virus-related transcripts belonging to SPFMV, SPVG, SPLCV and Cymbidium mosaic virus (CymMV). However, the mapped read number for them was only 6, 22, 3 and 20, respectively. These results indicated that fewer viral sequences presented in flowers than the vegetative organs. Of these, CymMV was found for the first time in sweet potato.

Sequence Amplification by RT-PCR

RT-PCR analysis was conducted to verify whether all of these viral sequences existed in sweet potato cultivar Xushu 18. Fourteen pairs of primers were designed according to the assembled sequences (Table 3). Except MMMV, SPLCV and CymMV, 11 virus fragments of the expected sizes were successfully amplified from Xushu 18 (Figure 1). All amplified fragments were re-sequenced by the Sanger method and then were submitted for a sequence similarity search by BLASTN or TBLASTN. The results showed that all fragments had a high identity of ≥95% with the assembled sequences (Table 4). These indicated that deep sequencing technology could provide a reliable method to identify viral sequences. Two of the three SPFMV sequences identified by NGS showed 99% sequence identity with SPVC and the severe strain of SPFMV. Two SPVG sequences showed 99% and 78% identity with two different SPVG strains.

Comparing the re-sequenced fragments with the reference sequences retrieved from NCBI, the identities decreased for most of these amplified fragments, especially for the NCMV–like fragment (Table 4). The SPBV-B fragment shared 77% sequence identity with the reference sequence, and SPBV-A shared 94% identity with the reference in a short segment. These results

Table 3. Primers used for virus fragment amplification.

No.	Virus name	Primer name	Primer sequence (5'–3')	Product length (bp)
1	SPVC	SPVC-CP1F	TCGGTGTATCATCAATCTGGC	560
		SPVC-CP1R	CCATCCATCATCGTCCAAAC	
2	SPFMV	SPFMV-SP1F	CTCCACCACCCACAATAACTG	402
		SPFMV-SP1R	TCCCCATTCCTGTATCGTCA	
3	SPVB2	SPVB2-P3F	GAGACAGCAGAAACAGCAGTGATA	471
		SPVB2-P3R	GGCATCACAATAAACCCATCCT	
4	SPVG	SPVG-1F	ACAACGTGCATCATCAGTCT	459
		SPVG-1R	CATTTGCCATTGGTGCTCTT	
5	SPVB2	SPVB2-P1F	AAGCATGTGGTGAAAGGAAAGTG	361
		SPVB2-P1R	TTGCTTGTTCATCCATTCCCTC	
6	SPVG	SPVG-2F	CGCCAACTAATAGCGAACTCT	549
		SPVG-2R	ACTATACGTCCATTCGCCATC	
7	SPBV-B	SPBV-C2PF	CAGGATTCACTCAGCAGACG	352
		SPBV-C2PR	ATGTCATGAAGGCACCTTCC	
8	SPBV-A	SPBV-C1PF	CAGCTTTGGTTGCTCTGCTATTT	477
		SPBV-C1PR	AAGACGGTTGGCCCATTGATAT	
9	SPV2	IVMV-P1F	TGCTGAAATGGGCATACTCC	489
		IVMV-P1R	TGCACACCTCTCATTCCTAACA	
10	SwPLV	SwPLV-P1F	CGAAGTGGATGACCAGCAGAT	500
		SwPLV-P1R	GGATTCCACGCATTCCAAGTAG	
11	NCMV-like	IbRNLS-PF	TCACCACAGAGGTACAAAGGAAA	1317
		IbRNLS-PR	ACCATGATTTACATCTCTGTCGG	
12	MMMV	MMWV2-P1F	ATGGTTGAATGCTCCCAAGACA	499
		MMWV2-P1R	CTCTCCATCCAATTCCCACCTAT	
13	SPLCV	SPLCV-P1F	GAAGCTATGTCCCGGTTTCAAGAG	300
		SPLCV-P1R	GCCTTCTGTCACGAATCAACCA	
14	CymMV	CymMV-P1F	CCTGAGCCCTTCTGTACCATA	775
		CymMV-P1R	GTGTTGGTGGAGCCAAGATG	

indicated that SPBV-A and SPBV-B could perhaps be new sweet potato viruses. In this study we tentatively named them as Sweet potato badnavirus C (SPBV-C). Interestingly, the NCMV-like fragment failed to find a homologous nucleotide sequence from NCBI by BLASTN, so TBLASTN was employed to blast the deduced protein sequences to the translated nucleotide database. The results showed that it was homologous with Rhabdovirus N-like sequences (RNLSs) [41] and should be a new sweet potato virus. So we tentatively named it as *Ipomoea batatas* Rhabdovirus N-like sequences, IbRNLS.

Confirmation of Four New Sweet Potato Viruses

To identify whether these four new virus-related sequences are present in other sweet potato cultivars in China, we collected eight different sweet potato tuberous root samples from different regions in Sichuan Province, China (Table 5). All of these eight sweet potato cultivars were cultivated by farmers under natural conditions. Total RNA was extracted for RT-PCR analysis. The results demonstrated that the SPBV-C1 viral sequence was amplified from three cultivars, SPBV-C2 was amplified from four cultivars, the NCMV-like presented itself in all eight tuberous root samples, and CymMV presented itself in four samples (Figure 2).

These results confirmed that four new virus-related sequences are present in most of the sweet potato cultivars in this region.

Expression Patterns among Different Tissues

DGE provides a new expression analysis method showing major advances in robustness, resolution and inter-lab portability over microarray and quantitative RT-PCR [42]. For this technology, 21 bp tags were sequenced for each mRNA; the tag number of each transcript gave a digital signal to characterize the expression patterns. To study the gene expression patterns of each virus transcript, all DGE tags from the seven vegetative tissues [34] (Table 1) were used for expression profiling. It was found that there were 16 transcripts containing a *Nla*III recognition site (CATG), which is the motif of DGE tags. These transcripts belonged to SPVC (5 transcripts), SPFMV (5 transcripts), SPVG (3 transcripts), SPV2 (2 transcripts), SwPLV (1 transcript) and NCMV-like (1 transcript). DGE quantification results showed that different viral transcripts had different expression levels (Figure 3). Transcripts of SPVC, SPFMV and SPVG had very high expression levels, which were about 100 times higher than that of the SwPLV transcript. These are consistent with the findings described above and illustrate that SPVC, SPFMV and SPVG may be the most prevalent viruses in China. Furthermore, it was

Table 4. Sequencing results of amplified viral fragments.

#	Length (bp)	Identities (%)[a]		Coverage[b]		E-value[c]	Viral names	GenBank Accession Numbers
		A	B	(%)	B			
1	497	97.99	99	99	99	0	SPVC	JQ902097
2	358	98.88	99	100	99	3.00E-175	SPFMV	JQ902098
3	436	99.54	83	83	83	5.00E-97	SPVB2	JQ902099
4	394	98.48	99	100	99	0	SPVG-1	JQ902100
5	315	99.68	77	98	77	3.00E-60	SPVB2	JQ902101
6	510	98.24	78	99	78	4.00E-112	SPVG-2	JQ902102
7	308	99.25	77	99	77	7.00E-48	SPBV-B	JQ902103
8	438	99.58	94	23	94	0.004	SPBV-A	JQ902104
9	428	99.77	99	100	99	0	SPV2	JQ902105
10	462	95.64	96	98	96	0	SwPLV	JQ902106
11	981	99.80	39	88	39	4.00E-76	* IbRNLS	JQ902107

Sequence numbers are corresponded to Table 3;

[a]Sequence identities (A) were calculated between assembled and Sanger sequenced fragments; Sequence identities (B) were calculated between Sanger sequenced fragments and reference sequences from NCBI;

[b], [c]Indicate the query coverage and E-value of Sanger sequenced segments BLAST against the reference sequences from NCBI;

*This is a Cytorhabdovirus-like sequence; the corresponding sequence identity (B), query coverage and E-value were obtained by TBLASTN searches.

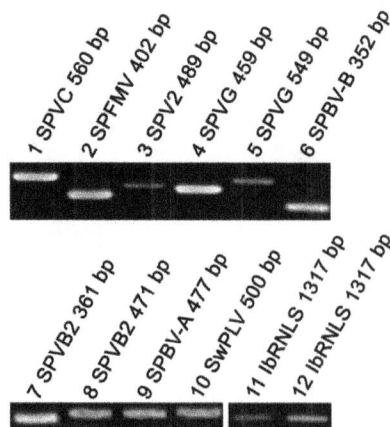

Figure 1. Electrophoresis of PCR products of amplified viral sequences. 1~10 correspond to sequence numbers in Table 3 of 1 (560 bp), 2 (402 bp), 9 (489 bp), 4 (459 bp), 6 (549 bp), 7 (352 bp), 5 (361 bp), 3 (471 bp), 8 (477 bp) and 10 (500 bp), respectively. 11 and 12 correspond to sequence numbers in Table 3 of 11 (1317 bp) amplified from cDNA and genomic DNA, respectively.

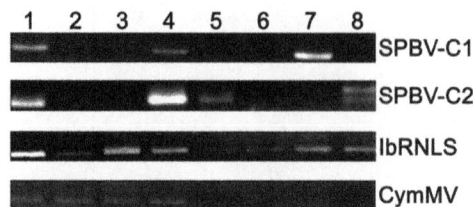

Figure 2. Amplification of four novel viruses from different sweet potato cultivars. 1~8 correspond to sample numbers in Table 5.

found that all these viral transcripts were unevenly distributed in different tissues (Figure 3). Five of these six viruses possessed the highest expression level in fibrous roots, while the remaining one, in expanding tuberous roots. Initial tuberous roots also had a comparably high expression level for most of them, but young leaves, mature leaves and stem had lower expression levels. For example, SPFMV (Transcript_11) had an expression level of 107.97 TPM in fibrous roots and 31.12 TPM in initial tuberous roots, but only 1.75 TPM in mature leaves and no expression in young leaves. The highest expression levels of SPVC (Transcript_859) was also observed in fibrous roots (688.35 TPM), followed by stem (230.48 TPM) and initial tuberous roots (219.25 TPM).

By using sweet potato beta-actin as an internal control, expression levels of SPVC, SPFMV, SPVG, SPV2 and SwPLV transcripts were analyzed by semi-quantitative RT-PCR (Figure 4). For SPVC, fibrous and initial tuberous roots had almost equal expression levels, while young and mature leaves had the lowest levels. For SPFMV, highest expression levels were detected in fibrous roots, followed by initial tuberous roots, and no expression was observed in leaves. Similar expression patterns were also found for SPVG. But SPV2 and SwPLV had different expression patterns. The highest expression level was observed in stems for

SPV2 and young leaves for SwPLV. However, for these two viruses, fibrous or initial tuberous roots also had relatively high expression abundance. There are some slightly differences of the relative expression levels among different tissues quantified by DGE and RT-PCR. But the reason for this discrepancy of the results between these two methods is unknown.

Virus Identification in other Sweet Potato Cultivars

To investigate the polyadenylated RNA virome of the Guangshu 87 and Jingshu 6 cultivars, all the PE reads of their transcriptomes [32,33] were aligned with viral sequences retrieved from NCBI, and the assembled sweet potato transcriptome. Results demonstrated that sequences of 11 virus species were found in cultivar Guangshu 87, including SPFMV, SPVG, SwPLV, SPLCV, SPCFV, SPVB1, SPVB2, SPVB3, SPBV-A, SPBV-B and CymMV. Sequences of 4 viruses were found in Jingshu 6, including SPFMV, SPVG, SPCFV and Cucumber mosaic virus (CMV). SPFMV and SPVG had the highest expression levels amongst all the viruses in these two cultivars (Table 6). Combining together all viruses identified from the three cultivars, we detected a total of 15 viruses, most of which were reported for the first time in China. Among these viruses, SPVC, SPFMV and SPVG had the highest expression levels in all three cultivars.

Discussion

Sweet potato virus is usually detected using indicator plants such as *Ipomoea setosa*, *Ipomoea nil* and *Chenopodium quinoa* [10,43], and electron microscopic observation [44], while molecular diagnosis is conducted using enzyme-linked immunosorbent assay (ELISA) [43,45] or RT-PCR [18,26,46–48]. During the last decades, over 30 sweet potato viruses were detected in the world [3,13], and 11 of these have been reported in China [14]. SPFMV, SPLV and

Table 5. Sweet potato cultivars collected from different regions.

Sample No.	Cultivar	Source
1	Chuanshu 34	Sichuan academy of agricultural sciences, Chengdu City
2	Liyuan 1	Dayi County, Chengdu City
3	Nanshu 007	Dayi County, Chengdu City
4	Nanshu 88	Dayi County, Chengdu City
5	Unknown	Tianquan County, Ya'an City
6	Xushu 18	Weiyuan County, Neijiang City
7	Yusu 303	Dayi County, Chengdu City
8	Xushu 18	Zizhong County, Neijiang City

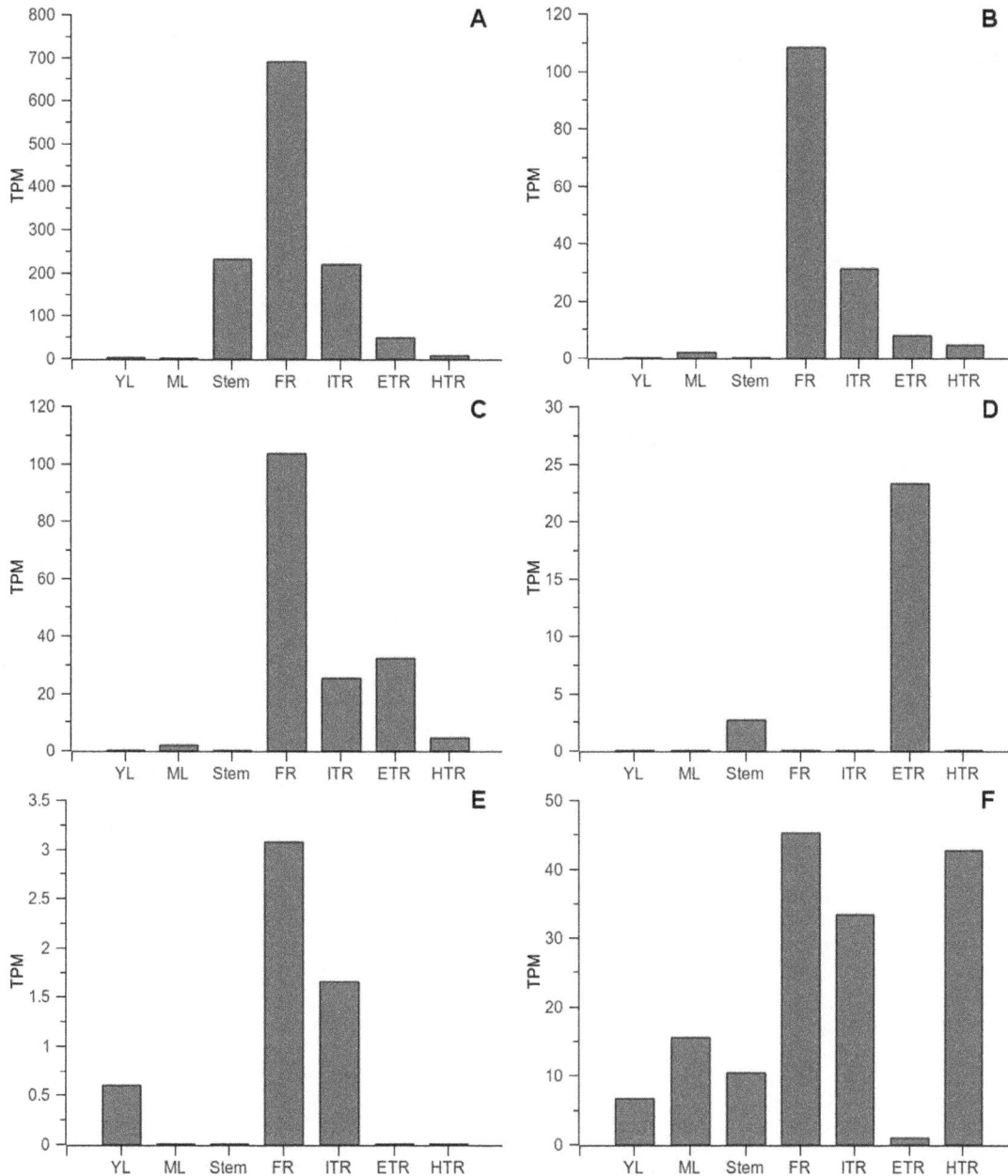

Figure 3. Digital gene expression analyses of different viral transcripts. TPM: Transcripts Per Million clean tags (Morrissy AS, 2009); YL, ML, Stem, FR, ITR, ETR and HTR indicate young leaves, mature leaves, stems, fibrous roots, initial tuberous roots, expanding tuberous roots and harvest tuberous roots, respectively; A~F: indicate SPVC (Transcript_859), SPFMV (Transcript_11), SPVG (Transcript_5373), SPV2 (Transcript_81617), SwPLV (Transcript_95916) and NCMV-like (Transcript_5902), respectively.

SPCFV are considered as the major viruses in China [4,14]. The advent of high-throughput sequencing technology offers a new and powerful approach for characterization of viruses. This methodology shows major advances in robustness, resolution and inter-lab portability [42]. It not only identifies known viruses, but also can identify low-titer and novel viral species without any prior knowledge [31].

In recent years, there were several groups investigating viral infection agents using high-throughput sequencing technology. For example, Kreuze *et al* successfully detected novel viruses from infected sweet potato and constructed complete viral genomic sequences by *de novo* assembling of 21 and 22 bp NGS reads [31],

and Coetzee *et al* characterized the virome of a diseased South African vineyard [30]. In this study, transcripts of 11 virus species were identified in cultivar Xushu 18 through NGS data mining. For the vegetative transcriptome, 88% of the mapped viral PE reads were aligned to SPFMV or SPVC, and 87% of the rest were aligned to SPVG, these illustrated that SPFMV, SPVC and SPVG may be the most prevalent viruses in this cultivar. However, transcripts of only four virus species were detected in the flowers of Xushu 18, all of which had very low expression levels, which may indicate that viruses primarily accumulate in vegetative organs than in floral ones. Totally, transcripts of 15 viruses were identified from three sweet potato cultivars growing in China, four of which

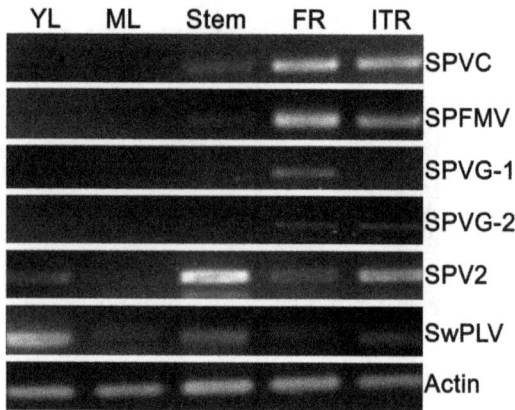

Figure 4. Virus gene expression levels among different tissues. YL, ML, Stem, FR, ITR, ETR and HTR indicate young leaves, mature leaves, stems, fibrous roots, initial tuberous roots, expanding tuberous roots and harvest tuberous roots, respectively.

are novel sweet potato viruses, and several of which are reported in China for the first time.

Of the reported sweet potato viruses, most are associated with symptomless infections in sweet potato and in some cases even in the indicator plant. Some are synergized by SPCSV, the mediator of severe virus diseases in sweet potato, while others apparently are not [3]. Otherwise, sweet potato cultivars differ greatly in their reaction to the viruses, with some being symptomlessly infected, while others apparently immune [49]. The most common virus infecting sweet potato worldwide, SPFMV, can be symptomless, at least in some varieties [45,50]. Previous research showed that nearly 70% of the symptomless plants were SPFMV-infected in a virus survey in Kenya [51]. In this study, although transcripts of 15 viruses were identified, no SPCSV related fragment was found. This may be the reason why so many virus fragments were detected from Xushu 18 but no symptom could be observed. Our results also indicated that most of the symptomless field-grown sweet potatoes were infected by several viruses. Usually, leaves are collected for virus detection in sweet potato [48,52]. However, based on the DGE (Figure 3) and RT-PCR (Figure 4) analyses in this study, we found that expression levels of most virus transcripts were unevenly distributed in different tissues. Most virus transcripts possess extremely low expression levels in young and mature leaves, but higher expression levels in fibrous roots and initial tuberous roots. This indicated that using leaves as a test sample may give false negative results, while fibrous root should be the optimal choice for virus detection in this crop.

Sweet potato is vegetatively propagated from tuberous roots or vines, and farmers usually take vines for propagation from the farm year after year. If the sweet potato is infected with viruses, they will be transmitted to the next generation and accumulate in this crop, resulting in significantly decreased yields. The virus expression analyses results described in this study indicated that high expression levels of most viruses in fibrous and initial tuberous roots may be the main reason for the germplasm decline and production decrease. For sweet potato, adventitious roots develop at the nodes of a vine cutting, and then some of these roots change their growth pattern and develop into tuberous roots [53]. Depending on the number of fibrous roots that will be induced to form tuberous roots, sweet potato plants will yield either a high root production or a low number of tuberous roots [53]. High virus expression levels in fibrous roots will adversely affect the development of the root system and then result in tuberous root initiation failure. A well-developed root system is a prerequisite for healthy plant growth [14,54–56] and is recognized as a key factor of high tuberous root yield [57]. The development failure of tuberous roots caused by virus infection will significantly decrease the total bio-mass production. Previous studies demonstrated that tuberous roots of virus-infected sweet potato form later and expand slower than virus-free ones [14,54]. Compared with that of healthy plants, virus-infected plants have a significantly higher respiration rate and lower photosynthetic rate [57,58], and are more easily infected by the fungal pathogens *Monilochaetes infuscans* and *Ceratocystis fimbriata*, and the nematode *Pratylenchus coffeae*. All these physiological characteristics will inevitably result in final yield loss.

Table 6. Viral species detected by deep sequencing of other 3 sweet potato variety.

Virus name	No. of read pairs		
	Xushu 18	Guangshu 87	Jingshu 6
SPFMV	6	66,464	127
SPVG	22	9,332	511
SwPLV	0	126	0
SPLCV	3	12	0
SPCFV	0	1,055	17
SPVB1	0	12	0
SPVB2	0	36	0
SPVB3	0	5	0
SPBV-A	0	2	0
SPBV-B	0	22	0
CymMV	20	1	0
CMV	0	0	3

Acknowledgments

We thank Saanya Sequeira at University of East Anglia for critical reading of the manuscript.

Author Contributions

Conceived and designed the experiments: YZZ HYW YHG. Performed the experiments: YHG XT XJL. Analyzed the data: YHG XT. Contributed reagents/materials/analysis tools: YHG XT XJL. Wrote the paper: YHG XT. Revised the manuscript: YHG XT YZZ HYW.

References

1. O'Brien PJ (1972) The Sweet Potato: Its Origin and Dispersal. American anthropologist 74: 342–365.
2. Srinivas T (2009) Economics of sweetpotato production and marketing. In: Loebenstein G, Thottappilly G, editors. The sweetpotato: Springer. 235–267.
3. Clark CA, Davis JA, Abad JA, Cuellar WJ, Fuentes S, et al. (2012) Sweetpotato viruses: 15 years of progress on understanding and managing complex diseases. Plant Dis 96: 168–185.
4. Zhang LM, Wang QM, Liu QC, Wang QC (2009) Sweetpotato in China. In: Loebenstein G, Thottappilly G, editors. The Sweetpotato: Springer Netherlands. 325–358.
5. De Vries CA, Ferwerda JD, Flach M (1967) Choice of food crops in relation to actual and potential production in the tropics. Neth J Agr Sci 15: 241–248.
6. Varshney RK, Glaszmann JC, Leung H, Ribaut JM (2010) More genomic resources for less-studied crops. Trends Biotechnol 28: 452–460.
7. Geddes AMW (1990) The relative importance of crop pests in sub-Saharan Africa.
8. Schaefers GA, Terry ER (1976) Insect transmission of sweet potato disease agents in Nigeria. Phytopathology 66: 642–645.
9. Clark CA, Smith TP, Ferrin DM, Villordon AQ (2010) Performance of sweetpotato foundation seed after incorporation into commercial operations in Louisiana. Horttechnology 20: 977–982.
10. Gao F, Gong YF, Zhang PB (2000) Production and deployment of virus-free sweetpotato in China. Crop Prot 19: 105–111.
11. Karyeija RF, Gibson RW, Valkonen JPT (1998) The significance of sweet potato feathery mottle virus in subsistence sweet potato production in Africa. Plant Dis 82: 4–15.
12. Stathers T, Namanda S, Mwanga ROM, Khisa G, Kapinga R (2005) Manual for Sweetpotato Integrated Production and Pest Management Farmer Field Schools in Sub-Saharan Africa. International Potato Center, Kampala, Uganda. 168.
13. Trenado HP, Orílio AF, Márquez-Martín B, Moriones E, Navas-Castillo J (2011) Sweepoviruses cause disease in sweet potato and related *Ipomoea spp.*: fulfilling koch's postulates for a divergent group in the genus *Begomovirus.* Plos One 6: e27329.
14. Wang QM, Zhang LM, Wang B, Yin ZF, Feng CH, et al. (2010) Sweetpotato viruses in China. Crop Prot 29: 110–114.
15. Zhang LM, Wang QM, Ma DF, Wang Y (2006) The effect of major viruses and virus-free planting materials on sweetpotato root yield in China. Acta Hortic 703: 71–78.
16. Qiao Q, Zhang ZC, Qin YH, Zhang DS, Tian YT, et al. (2010) First report of sweet potato chlorotic stunt virus infecting sweet potato in China. Plant Dis 95: 356–356.
17. Zhang LM, Wang QM, Ma DF, Wang Y (2005) Major viruses and effect of major virus diseases and virus-eliminating meristem culture on sweetpotato yield and quality in China. Acta Botanica Boreali-Occidentalia Sinica 25: 316–320 (in Chinese, English Abstarct).
18. Colinet D, Kummert J (1993) Identification of a sweet potato feathery mottle virus isolate from China (SPFMV-CH) by the polymerase chain reaction with degenerate primers. J Virol Methods 45: 149–159.
19. Colinet D, Kummert J, Lepoivre P (1997) Evidence for the assignment of two strains of SPLV to the genus Potyvirus based on coat protein and 3' non-coding region sequence data. Virus Res 49: 91–100.
20. Colinet D, Nguyen M, Kummert J, Lepoivre P, Xia FZ (1998) Differentiation among potyviruses infecting sweet potato based on genus- and virus-specific reverse transcription polymerase chain reaction. Plant Dis 82: 223–229.
21. Luan YS, Zhang J, An LJ (2006) First report of Sweet potato leaf curl virus in China. Plant Dis 90: 1111–1111.
22. Colinet D, Lepoivre P, Xia FZ, Kummert J (1996) Detection and identification of sweet potato viruses by the polymerase chain reaction. Agro Food Ind Hi Tec 7: 33–34.
23. Ateka EM, Barg E, Njeru RW, Thompson G, Vetten HJ (2007) Biological and molecular variability among geographically diverse isolates of sweet potato virus 2. Arch Virol 152: 479–488.
24. Ateka EM, Barg E, Njeru RW, Lesemann DE, Vetten HJ (2004) Further characterization of 'sweet potato virus 2': a distinct species of the genus Potyvirus. Arch Virol 149: 225–239.
25. Zhang YQ, Guo HC (2005) Research progress on the tip meristem culture of sweet potato. Chinese Agr Sci Bull 21: 74–76 (in Chinese, English abstract).
26. Li F, Zuo R, Abad J, Xu D, Bao G, et al. (2012) Simultaneous detection and differentiation of four closely related sweet potato potyviruses by a multiplex one-step RT-PCR. J Virol Methods 186: 161–166.
27. Mbanzibwa DR, Tairo F, Gwandu C, Kullaya A, Valkonen JPT (2012) First Report of Sweetpotato symptomless virus 1 and Sweetpotato virus A in sweetpotatoes in Tanzania. Plant Dis 96: 1430–1437.
28. Valverde RA, Clark CA, Valkonen JP (2007) Viruses and virus disease complexes of sweetpotato. Plant Viruses 1: 116–126.
29. Al Rwahnih M, Daubert S, Golino D, Rowhani A (2009) Deep sequencing analysis of RNAs from a grapevine showing Syrah decline symptoms reveals a multiple virus infection that includes a novel virus. Virology 387: 395–401.
30. Coetzee B, Freeborough MJ, Maree HJ, Celton JM, Rees DJG, et al. (2010) Deep sequencing analysis of viruses infecting grapevines: virome of a vineyard. Virology 400: 157–163.
31. Kreuze JF, Perez A, Untiveros M, Quispe D, Fuentes S, et al. (2009) Complete viral genome sequence and discovery of novel viruses by deep sequencing of small RNAs: a generic method for diagnosis, discovery and sequencing of viruses. Virology 388: 1–7.
32. Wang Z, Fang B, Chen J, Zhang X, Luo Z, et al. (2010) De novo assembly and characterization of root transcriptome using Illumina paired-end sequencing and development of cSSR markers in sweet potato (Ipomoea batatas). Bmc Genomics 11: 726.
33. Xie F, Burklew CE, Yang Y, Liu M, Xiao P, et al. (2012) De novo sequencing and a comprehensive analysis of purple sweet potato (Impomoea batatas L.) transcriptome. Planta 236: 101–113.
34. Tao X, Gu YH, Wang HY, Zheng W, Li X, et al. (2012) Digital gene expression analysis based on integrated de novo transcriptome assembly of sweet potato [Ipomoea batatas (L.) Lam]. Plos One 7: e36234.
35. Tao X, Gu YH, Jiang YS, Zhang YZ, Wang HY (2013) Transcriptome Analysis to Identify Putative Floral-Specific Genes and Flowering Regulatory-Related Genes of Sweet Potato. Biosci Biotechnol Biochem 77: 2169–2174.
36. Langmead B, Trapnell C, Pop M, Salzberg SL (2009) Ultrafast and memory-efficient alignment of short DNA sequences to the human genome. Genome Biol 10: R25.
37. Mortazavi A, Williams BA, McCue K, Schaeffer L, Wold B (2008) Mapping and quantifying mammalian transcriptomes by RNA-Seq. Nat Methods 5: 621–628.
38. Morrissy AS, Morin RD, Delaney A, Zeng T, McDonald H, et al. (2009) Next-generation tag sequencing for cancer gene expression profiling. Genome Res 19: 1825–1835.
39. Wang RL, Ding LW, Sun QY, Li J, Xu ZF, et al. (2008) Genome sequence and characterization of a new virus infecting Mikania micrantha H.B.K. Arch Virol 153: 1765–1770.
40. Kreuze JF, Karyeija RF, Gibson RW, Valkonen JPT (2000) Comparisons of coat protein gene sequences show that East African isolates of Sweet potato feathery mottle virus form a genetically distinct group. Arch Virol 145: 567–574.
41. Chiba S, Kondo H, Tani A, Saisho D, Sakamoto W, et al. (2011) Widespread endogenization of genome sequences of non-retroviral RNA viruses into plant genomes. PLoS pathog 7: e1002146.
42. AC't Hoen P, Ariyurek Y, Thygesen HH, Vreugdenhil E, Vossen RH, et al. (2008) Deep sequencing-based expression analysis shows major advances in robustness, resolution and inter-lab portability over five microarray platforms. Nucleic Acids Res 36: e141–e141.
43. Aritua V, Alicai T, Adipala E, Carey EE, Gibson RW (1998) Aspects of resistance to sweet potato virus disease in sweet potato. Ann Appl Biol 132: 387–398.
44. Yang C, Shang Y, Zhao J, Li C (1998) Produce techniques and practice of virus-free sweet potato. Acta Phytophylacica Sinica 25: 51–55.
45. Gibson RW, Mpembe I, Alicai T, Carey EE, Mwanga ROM, et al. (1998) Symptoms, aetiology and serological analysis of sweet potato virus disease in Uganda. Plant Pathol 47: 95–102.
46. Gibbs A, Mackenzie A (1997) A primer pair for amplifying part of the genome of all potyvirids by RT-PCR. J Virol Methods 63: 9–16.
47. Li F, Xu D, Abad J, Li R (2012) Phylogenetic relationships of closely related potyviruses infecting sweet potato determined by genomic characterization of Sweet potato virus G and Sweet potato virus 2. Virus Genes 45: 118–125.
48. Perez-Egusquiza Z, Ward LI, Clover GRG, Fletcher JD (2009) Detection of Sweet potato virus 2 in sweet potato in New Zealand. Plant Dis 93: 427–427.
49. Loebenstein G, Thottappilly G, Fuentes S, Cohen J (2009) Virus and phytoplasma diseases. The sweetpotato: Springer. 105–134.

50. Gibson RW, Mwanga ROM, Kasule S, Mpembe I, Carey EE (1997) Apparent absence of viruses in most symptomless field-grown sweet potato in Uganda. Ann Appl Biol 130: 481–490.

51. Ateka EM, Njeru RW, Kibaru AG, Kimenju JW, Barg E, et al. (2004) Identification and distribution of viruses infecting sweet potato in Kenya. Ann Appl Biol 144: 371–379.

52. Kokkinos CD, Clark CA (2006) Real-Time PCR sssays for detection and quantification of sweetpotato viruses. Plant Dis 90: 783–788.

53. Firon N, LaBonte D, Villordon A, McGregor C, Kfir Y, et al. (2009) Botany and physiology: storage root formation and development. In: G L, G T, editors. The sweetpotato Springer. 13–26.

54. Du XH, Zhan HJ, Xu QY, Wang QC, Niu YZ, et al. (1999) Effects of virus elimination on several physiological characteristics of sweetpotato. Plant Physiol Commun 35: 185–187 (in Chinese).

55. Khan MB, Rafiq R, Hussain M, Farooq M, Jabran K (2012) Ridge sowing improves root system, phosphorus uptake, growth and yield of Maize (Zea Mays L.) Hybrids. Measurements 22: 309–317.

56. Ma C, Naidu R, Liu F, Lin C, Ming H (2012) Influence of hybrid giant Napier grass on salt and nutrient distributions with depth in a saline soil. Biodegradation 23: 907–916.

57. Nie F, Xu YH, Qian J, Chen JX, Dong L (2000) Effects of virus-free techniques on development and growth of sweet potato. Chinese Agr Sci Bull 16: 13–15 (in Chinese, English abstract).

58. Chen XY, Chen FX, Yuan ZN, Zhuang BH (2001) Effect of virus-elimination on some physiological indices in sweet potato. J Fujian Agr Univ 30: 449–453 (in Chinese, English abstract).

17

Production of Destruxins from *Metarhizium* spp. Fungi in Artificial Medium and in Endophytically Colonized Cowpea Plants

Patrícia S. Golo[1,2]*, **Dale R. Gardner**[3], **Michelle M. Grilley**[2], **Jon Y. Takemoto**[2], **Stuart B. Krasnoff**[4], **Marcus S. Pires**[1], **Éverton K. K. Fernandes**[5], **Vânia R. E. P. Bittencourt**[1], **Donald W. Roberts**[2]

1 Departamento de Parasitologia Animal, Instituto de Veterinária, Universidade Federal Rural do Rio de Janeiro, Seropédica, RJ, Brazil, **2** Department of Biology, Utah State University, Logan, Utah, United States of America, **3** USDA, ARS, Poisonous Plants Research Laboratory, Logan, Utah, United States of America, **4** Biological Integrated Pest Management Research Unit, Robert W. Holley Center for Agriculture and Health, USDA-ARS, Ithaca, New York, United States of America, **5** Instituto de Patologia Tropical e Saúde Pública, Universidade Federal de Goiás, Goiânia, GO, Brazil

Abstract

Destruxins (DTXs) are cyclic depsipeptides produced by many *Metarhizium* isolates that have long been assumed to contribute to virulence of these entomopathogenic fungi. We evaluated the virulence of 20 *Metarhizium* isolates against insect larvae and measured the concentration of DTXs A, B, and E produced by these same isolates in submerged (shaken) cultures. Eight of the isolates (ARSEF 324, 724, 760, 1448, 1882, 1883, 3479, and 3918) did not produce DTXs A, B, or E during the five days of submerged culture. DTXs were first detected in culture medium at 2–3 days in submerged culture. *Galleria mellonella* and *Tenebrio molitor* showed considerable variation in their susceptibility to the *Metarhizium* isolates. The concentration of DTXs produced *in vitro* did not correlate with percent or speed of insect kill. We established endophytic associations of *M. robertsii* and *M. acridum* isolates in *Vigna unguiculata* (cowpeas) and *Cucumis sativus* (cucumber) plants. DTXs were detected in cowpeas colonized by *M. robertsii* ARSEF 2575 12 days after fungal inoculation, but DTXs were not detected in cucumber. This is the first instance of DTXs detected in plants endophytically colonized by *M. robertsii*. This finding has implications for new approaches to fungus-based biological control of pest arthropods.

Editor: Martin Heil, Centro de Investigación y de Estudios Avanzados, Mexico

Funding: This research was supported in part by grants from the United States Department of Agriculture (USDA)/Animal Plant Health and Inspection Service (APHIS); Cooperative agreement (CA) #13-8130-0114-CA. The financial support was received by DWR. The funders had no role in study design, data collection and analysis, decision to publish, or preparation of the manuscript.

Competing Interests: The authors have declared that no competing interests exist.

* Email: patriciagolo@gmail.com

Introduction

Despite concerns with negative impacts of chemical insecticides on human health, the use of these chemicals remains high. Consequently, the demand for alternatives is increased. Biological control of arthropod pests using entomopathogenic fungi is one promising alternative [1,2,3]. Entomopathogenic fungi from the genus *Metarhizium* are some of the most frequently studied biological control agents for use against insects and ticks [2,3,4].

Metarhizium spp. produce a wide array of small molecules including destruxins (DTXs), cyclic depsipeptides which are produced as well as by some other fungi, both insect (*Aschersonia*) and plant pathogens (*Alternaria*, *Trichothecium*) [5]. The effects of DTXs on insects include: tetanic paralysis [6,7], inhibition of DNA and RNA synthesis in insect cell lines [8], inhibition of Malpighian tubule fluid secretion [9], blocking H$^+$ ATPase activity [10], and suppression of insect defense responses [11,12,13,14,15]. DTXs also have antifeedant and repellent properties [16,17]. The insecticidal potential of these toxins has been confirmed in numerous reports of acute toxicity [5]. Despite demonstrated insecticidal activity of DTX, Donzelli et al. [18] showed that a

Metarhizium robertsii mutant with disrupted DTX synthetases was as virulent as the wild type strain when fungus conidia were topically applied to insect larvae. This supports the conclusions of a previous report that *Metarhizium* spp. isolates could be pathogenic for insects whether they had the ability to produce *in vitro* DTXs or not [19]. Although these compounds have been detected in moribund, infected hosts [20,21], DTXs reportedly have little or no impact on virulence as measured in whole-insect bioassays [18,19].

DTXs also have negative effects on insect behavior, for example inducing phagodepression and repellence [16,17]. *Metarhizium robertsii* (ARSEF 2575) is plant-rhizosphere competent and has endophytic capability [22,23,24,25]; accordingly, if DTXs produced inside *Metarhizium*-colonized plants induced antifeedant effects on arthropod pests of those plants, then the presence of DTXs *in planta* may afford enhanced levels of *Metarhizium*-associated biological control of these pest arthropods.

We report here a survey of virulence of 20 *Metarhizium* isolates against insect larvae, and the concentration of DTXs A, B, and E produced by these same isolates *in vitro* (submerged shake cultures). We then analyzed plants endophytically colonized by a high-DTX producing *M. robertsii* isolate and a low- or non-DTX producing *M. acridum* isolate [26,27] to search for DTXs in colonized plants.

Material and Methods

Fungal isolates

Twenty *Metarhizium* spp. isolates were used in the present study: 18 isolates from different regions of Brazil, one from the USA and one from Australia (Table 1). Fungal isolates were obtained from the Agriculture Research Service Collection of Entomopathogenic Fungal Cultures (ARSEF) (USDA-US Plant, Soil and Nutrition Laboratory, Ithaca, NY, USA). Stock cultures were grown on PDAY (potato dextrose agar plus 0.01% yeast extract) at 27°C for 14 days and then held at 4°C. Conidia for all experiments were produced on PDAY 60×60 mm Petri plates and incubated at 27°C for 14 days. Conidia were harvest by scraping using a bacterial loop and suspended in 0.01% Tween 80 in 15-mL centrifuge tubes (Modified polystyrene, Corning inc., Corning, NY, USA) and vigorously agitated (vortexed). Conidial viability was measured by placing a 50 μL drop of fungal suspension on a PDAY plate and germination was observed by compound microscope (400×) after 24 hours at 28°C.

In vitro production of DTXs and HPLC-UV analysis

For the analysis of *in vitro* DTXs production, fungal cultures were started with 1×10^6 conidia/100 mL CZAPEK-DOX Broth (BD Difco) with bactopeptone (0.5%) and incubated in 250-mL flasks at room temperature (~22°C) on a rotary shaker at 150 rpm for 1, 2, 3, 4, or 5 days. Control isolates were *M. robertsii* ARSEF

2575 (a high DTX producer) and *Metarhizium acridum* ARSEF 324 (a low or non DTX producer) [26,27]. Production of DTX in the culture supernatants was determined by quantitative HPLC-UV analysis of the major components (DTXs A, B and E). All solvents used in the current study were HPLC grade. Cultures were separated into fungus mycelium and supernatant by centrifugation at 1000 × g for 20 minutes. Mycelia were harvested, dried at 80°C for 48 hours, and weighed to obtain the amount of mycelial production for each isolate. Extraction of DTXs from culture supernatants was accomplished by loading 5 mL aliquots onto C18-SPE cartridges (100 mg; Agilent Bond Elut #12102001) that were previously conditioned with 10 column volumes of methanol followed by a similar volume of ultra-pure water. The loaded cartridges were rinsed with 10 mL ultra pure water and then eluted with 2 mL methanol [18].

Just prior to analysis the methanol extracts were diluted 1:1 with water and then 10-μL aliquots of extract were injected onto a reversed phase (RP) Betasil C18 column (100 mm×2.1 mm, Thermo Fisher) with a guard column of the identical phase. Elution was with a gradient of acetonitrile and water using a modular HPLC system (Shimadzu Corp., Kyoto, Japan). The linear gradient conditions using the solvents A (acetonitrile) and B (water) were: 0–10 min (25% A increased to 60% A); 10–13 min (isocratic 60% A); 13–15 min (60% A decreased to 25% A) at a flow rate of 0.3 mL min^{-1}. Detection was by UV absorbance at 220 nm. After the run was complete, the column re-equilibration time was 5 min. DTXs A, B, and E were measured using standard curves for each compound. DTXs A, B, and E standards were purified using methods based on those of Krasnoff et al. [28] and standard solutions prepared at 1 mg/mL in methanol. Calibration standards were prepared by dilution of 20 μL of each standard

Table 1. *Metarhizium* spp. isolates used in this study, including their hosts and origins (state and country).

Fungal Isolate	Host/Substrate	Origin	Species
ARSEF 324	*Austracris guttulosa* (Orthoptera: Acrididae)	QLD, Australia	*Metarhizium acridum*
ARSEF 552	Lepidoptera	MG, Brazil	*Metarhizium pingshaense*
ARSEF 724	*Cerotoma arcuata* (Coleoptera:Chrysomelidae)	GO, Brazil	*Metarhizium robertsii*
ARSEF 729	*Deois flavopicta* (Homoptera: Cercopidae)	GO, Brazil	*Metarhizium anisopliae sensu lato* (s.l.)
ARSEF 759	*Deois flavopicta* (Homoptera: Cercopidae)	GO, Brazil	*Metarhizium anisopliae* s.l.
ARSEF 760	*Cerotoma arcuata* (Coleoptera: Chrysomelidae)	GO, Brazil	*Metarhizium anisopliae* s.l.
ARSEF 782	*Deois flavopicta* (Homoptera: Cercopidae)	GO, Brazil	*Metarhizium anisopliae* s.l.
ARSEF 929	*Chalcodermus aeneus* (Coleoptera: Curculionidae)	GO, Brazil	*Metarhizium anisopliae* s.l.
ARSEF 1448	*Scaptores castanea* (Hemiptera: Cydnidae)	GO, Brazil	*Metarhizium pingshaense*
ARSEF 1449	*Deois flavopicta* (Homoptera: Cercopidae)	PA, Brazil	*Metarhizium anisopliae* s.l.
ARSEF 1882	*Tibraca limbativentris* (Hemiptera: Pentatomidae)	GO, Brazil	*Metarhizium anisopliae* s.l.
ARSEF 1883	*Tibraca limbativentris* (Hemiptera: Pentatomidae)	GO, Brazil	*Metarhizium anisopliae sensu stricto*
ARSEF 1885	*Diabrotica* sp. (Coleoptera: Chrysomelidae)	GO, Brazil	*Metarhizium anisopliae* s.l.
ARSEF 2211	Soil	SP, Brazil	*Metarhizium anisopliae* s.l.
ARSEF 2521	*Deois* sp. (Homoptera: Cercopidae)	PR, Brazil	*Metarhizium anisopliae* s.l.
ARSEF 2575	*Curculio caryae* (Coleoptera: Curculionidae)	SC, USA	*Metarhizium robertsii*
ARSEF 3479	(Coleoptera: Scarabaeidae)	DF, Brazil	*Metarhizium anisopliae* s.l.
ARSEF 3641	Soil	GO, Brazil	*Metarhizium anisopliae* s.l.
ARSEF 3643	Soil	GO, Brazil	*Metarhizium anisopliae* s.l.
ARSEF 3918	Soil	PR, Brazil	*Metarhizium anisopliae* s.l.

* USDA-ARS Collection of Entomopathogenic Fungal Cultures, Ithaca, NY.
Identifications were provided September 2012 by curator of ARSEF* Richard Humber.

stock solution into 0.940 mL of 50% methanol and then serial dilution to give standards at 20, 10, 5, 2.5, 1.25, 0.62 and 0.31 µg mL^{-1}. Limit of detection (LOD) was estimated to be 0.10 µg/mL based on a S/N ratio of 3 for UV detection at 220 nm.

Detection of DTXs in plants

(i) **Fungal inoculation of plants.** Seeds of cowpea (*V. unguiculata*) (organic seeds, Shangri-la Health Foods, Logan, UT, USA) and cucumber (*C. sativus*) ("Straight Eight" untreated organic seeds, Snow Seed, Salinas, CA, USA) were weighed individually and only those weighing between 0.2500 g and 0.2599 g for cowpeas, and 0.0240 g and 0.0249 g for cucumber were used. Seeds were surface sterilized by immersion in 95% ethanol for 2 minutes, rinsed in sterile deionized water followed by immersion in 30% hydrogen peroxide for 1 minute. Disinfected seeds were then rinsed 3 times in sterile deionized water [29]. These axenic seeds were kept overnight at 4°C to synchronize growth. After synchronization, seeds were immersed for 1 h in conidial suspensions (1×10^6 conidia mL^{-1} 0.01% Tween 80) of ARSEF 2575 or ARSEF 324. Seeds were then individually set on sterile, moist filter paper in Petri plates and kept at 25°C for 12 days with a photoperiod of 16:8 (L:D) (white fluorescent tubes [30]). Sterile water was added as needed to keep the filter paper moist. Uninoculated seeds (no-fungus control) were immersed in sterile deionized water containing 0.01% Tween 80 [23]. After 12 days, presence or absence of *M. robertsii* or *M. acridum* in plants was confirmed by incubating surface sterilized leaves, stems and roots on artificial medium. Surface sterilization was by immersion for 2 minutes in 0.5% sodium hypochlorite, 2 minutes in 70% ethanol, rinsed in sterile deionized water 3 times and dried using sterile filter papers. The outer edges of the leaves were dissected and discarded [31]. The remaining parts were cut into pieces and cultured on PDAY medium supplemented with 0.05% chloramphenicol in a 60 mm Petri plate. Three plates from each treatment (ARSEF 2575 exposed, ARSEF 324 exposed, or not-infected plants) were incubated with 2 or 3 pieces of leaf, stem or root per plate. The plates were examined daily for 7 days. Fungi growing from plant tissues were isolated and characterized morphologically according to Tulloch [32].

(ii) **Extraction and LC-MS/MS analysis.** After 12 days of growth, 10 plants of each treatment (ARSEF 2575 exposed; ARSEF 324 exposed; and not exposed) were frozen in liquid nitrogen and ground with mortar and pestle to a powder. To verify the accuracy of the DTX detection method, pure DTX standards (A, B, and E, 16.5 µg each) were mixed (before liquid nitrogen freezing and homogenization) into ten additional not-fungus-exposed 12-day-old plants. Methanol (5 mL) was added to each plant powder, followed by 15 mL ultra pure water. Plant suspensions were clarified by filtration (Whatman N° 1). Extractions of filtrates (20 mL) were carried out with C18 SPE cartridges (as described before).

The concentrations of DTXs A, B, and E, were measured by liquid chromatography-mass spectrometry (LC–MS). The LC-MS system consisted of a Betasil C18 RP HPLC column (100×2.1 mm, Thermo Fisher), coupled to a Surveyor MS Pump Plus, a Surveyor Auto Sampler Plus and a PDA UV–vis absorbance detector in-line with an LCQ Advantage Max mass spectrometer and electrospray (esi) ionization source (Thermo Electron Corp, San Jose, CA, USA). Sample injection size was 5-µL. The gradient-elution steps were the same as those used for LC-UV analysis (see section 2.2). Pseudomolecular ions [M+H]$^+$ of DTX A, B, and E were observed at m/z 578, 594 and 594 respectively with the following retention times: 7.05 min (DTX A), 9.12 min (DTX B) and 5.03 min (DTX E). DTXs A, B, and E

concentrations were measured using a standard curve for each compound prepared by serial dilutions as previously described (section 2.2); but with the lowest standard at 0.15 µg mL^{-1}. The limit of detection (LOD) with the LC-MS system was estimated to be 0.010 µg mL^{-1} based on a S/N ratio of 3.

Effect of DTXs on plant dry weight

After DTXs extraction, plant powders were held for 48 hours at 80°C and their dry weights' determined. Dry weights of fungus-treated plant groups and the non-treated groups were analyzed by analysis of variance (ANOVA) followed by the Tukey test with a significance level of 5% ($P \leq 0.05$) [33].

Insect virulence assays

Bioassays were performed with *G. mellonella* (waxworms) and *T. molitor* (mealworms). Conidia of each isolate (Table 1) were used to prepare fungal suspensions (section 2.1). Conidial concentrations were estimated by hemocytometer counts and adjusted to 1×10^7 and 1×10^5 conidia mL^{-1}.

Commercially produced *G. mellonella* larvae (last instar; 239 mg average weight) and *T. molitor* larvae (at least ninth instar; 95 mg average weight) (Fluker Farms, Port Allen, LA, USA) were treated either with 1×10^7 conidia mL^{-1} or 1×10^5 conidia mL^{-1}. Two groups of 8 last instar *G. mellonella* larvae and 2 groups of 10 *T. molitor* larvae were placed in 60×15 mm polystyrene Petri dishes lined with a 5.5 cm P4 filter paper (Fisherbrand, Porosity: Medium – Fine, Flow rate: Slow) moistened with 0.5 mL sterile distilled water. Each plate containing larvae was sprayed with 0.5 mL fungal suspension. Control plates were sprayed with 0.01% Tween 80 solution. Plates were incubated at 28°C and ≥80 RH. Insect mortality was assessed daily for 10 days. The bioassays were repeated 3 times.

Mean larval mortalities at 3 days with *G. mellonella* and 5 days with *T. molitor* were compared using the non-parametric Kruskal-Wallis test for statistical differences. Comparison between the mean mortalities was performed using Student-Newman–Keuls (SNK) test. Data analyses were conducted using BioEstat software, version 4.0. *P*-values less than 0.05 were considered to be significant [34].

Results

Conidial viability (percent germination) of all suspensions used in *in vitro* DTX production, plant-seed inoculations, and insect bioassays was at least 98%.

In vitro production of DTXs

Of the 20 *Metarhizium* spp. isolates examined in the current study, one (ARSEF 2575) was previously known to produce high levels of DTXs and one (ARSEF 324) to produce low levels of DTXs. In addition, in the present study, seven other isolates (ARSEF 724, 760, 1448, 1882, 1883, 3479, and 3918) did not produce DTXs *in vitro*. Among the DTXs producers (ARSEF 552, 729, 759, 782, 929, 1449, 1885, 2211, 2521, 3641, and 3643), production ranged from 0.31 mg DTX A/g dry weight (d.w.) of ARSEF 1885 mycelium to 32 mg DTX E/g d.w. of ARSEF 3643 mycelium, at 5 days after inoculation of conidial suspensions into liquid medium (Figure 1). Table S1 shows DTXs production *in vitro* represented by mg DTXs per L liquid media.

Generally, the earliest detection of DTXs in *in vitro* cultures was at day 3; the exception being ARSEF 759, which produced DTX E at day 2 (0.55 mg DTX E/g d.w. mycelium) (Figure 2). Two isolates (ARSEF 1885 and ARSEF 729) did not produce DTXs until 4 days in culture. The time course (from day 1 to day

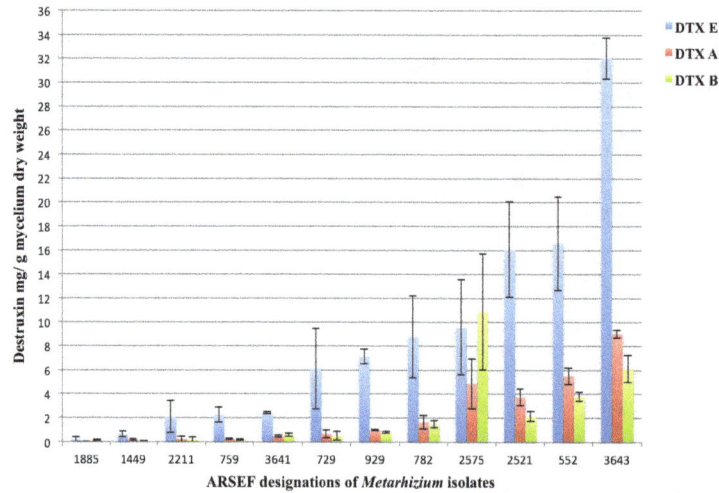

Figure 1. Destruxin (DTX) production by 12 *Metarhizium* spp. isolates *in vitro*. DTXs production is represented by mean values ± standard error after 5 days in submerged shaken cultures. Production of DTXs in supernatant of cultures was determined by quantitative HPLC analysis of the major components, viz., DTXs A, B and E. Cultures and assays were repeated 3 times.

5) of DTXs production by *M. anisopliae* s.l. (ARSEF 759) and *M. robertsii* (ARSEF 2575) (used as control isolate) is shown in Figures 2 and 3.

Detection of *Metarhizium*-produced DTXs in plants

Endophytic growth in 12-day-old cowpea (*V. unguiculata*) and cucumber (*C. sativus*) by *M. robertsii* and *M. acridum* was confirmed (Figure 4). In each case, the isolated fungus colonies presented the key morphological features consistent with *Metarhizium* isolates.

Detectable levels of DTXs A, B, and E were identified in combined roots, stems and leaves of cowpea plants cultured for 12 days after exposure of their seeds to *M. robertsii* conidia (Figure 5B). The concentrations of each compound followed by its respective standard error were: 5.73±0.29 µg DTX E/g d.w. cowpeas; 1.56±0.29 µg DTX A/g d.w. cowpeas; and 0.82±0.11 µg DTX B/g d.w. cowpeas. With cucumber, however, despite confirmation of *M. robertsii* endophytic colonization, no DTXs were detected in extracts of these plants. Also, no DTXs were detected in plants (cowpeas or cucumber) colonized by *M. acridum*, nor in control plants (not-infected plants). DTXs were detected in all positive controls (not-infected cowpea and cucumber plant tissues spiked with DTX standards) (Figure 5C).

Effect of DTXs on dry weights of plants

No differences in total dry weights were noted between endophytic *Metarhizium*-colonized and not-colonized plants (*P*≥ 0.05). Similarly, for both plant species (*V. unguiculata* and *C.*

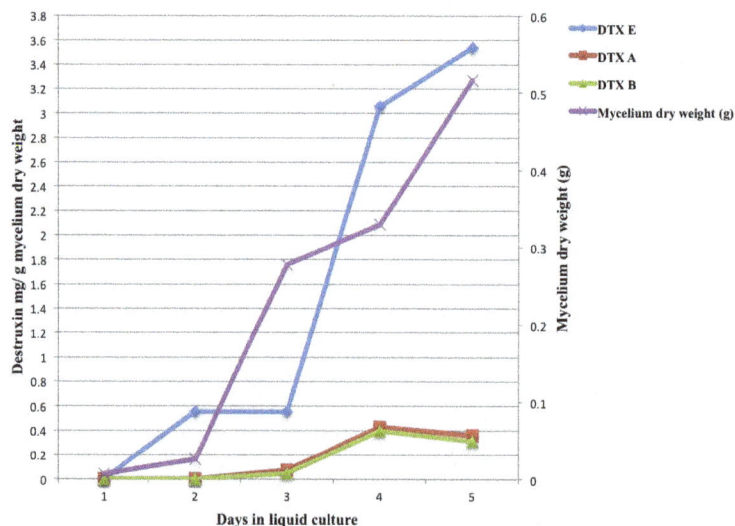

Figure 2. Time course of *in vitro* production of DTXs A, B, and E by *Metarhizium anisopliae* s.l. ARSEF 759. Destruxin concentrations in supernatants of submerged liquid cultures were determined by quantitative HPLC-UV analysis of the major components, viz., DTXs A, B and E. Values are expressed in mg DTXs per g dry weight mycelium.

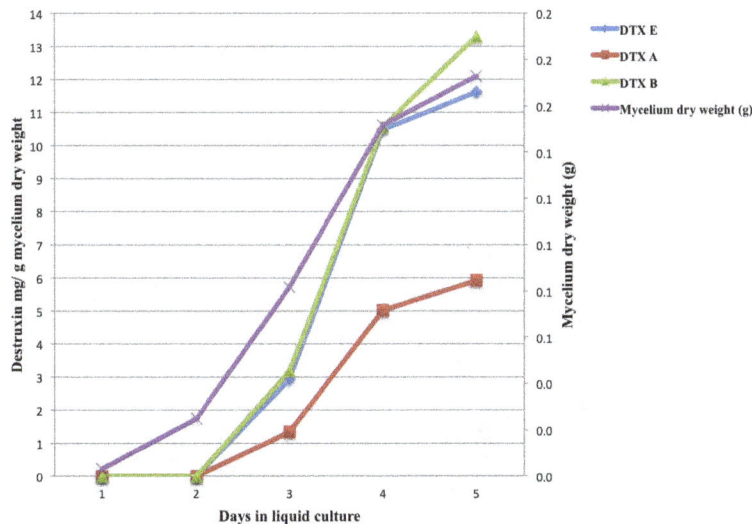

Figure 3. Time course of *in vitro* production of DTXs A, B, and E by *Metarhizium robertsii* ARSEF 2575. Destruxin concentrations in supernatant of submerged liquid cultures were determined by quantitative HPLC analysis of the major components, viz., DTXs A, B and E. Values are expressed in mg DTXs per g dry weight mycelium.

sativus) there were no statistical differences between dry weights of plants endophytically colonized by a fungus that produces DTXs (*M. robertsii*) and plants colonized by a non-DTXs producer (*M. acridum*).

Insect virulence assays

The virulence of 20 *Metarhizium* spp. isolates (Table 1) was surveyed using two different insect hosts: *G. mellonella* (waxworm) and *T. molitor* (mealworm). Natural mortality of untreated (control) *G. mellonella* larvae was always higher than with *T. molitor* larvae; e.g., waxworm control mortality reached 16.67% at day 5 while mealworm mortality was 1.6% at the same time. There were variations in the virulence of the isolates, and in the susceptibility of the different host species (Table 2). *T. molitor* larvae were less susceptible than *G. mellonella* larvae. For this reason, *T. molitor* bioassay data at day 5 after treatment were used for comparisons, and with *G. mellonella* day 3 data were used (Table 2). According to this evaluation system, several isolates (e.g., ARSEFs 724, 1448, 1885, 2575, 3641, and 3643) had similar levels of virulence for both species of insect.

With the highest concentration (10^7 conidia mL^{-1}), isolates ARSEF 724, 760, 1885, 2575, 3641, and 3643 caused 100% *G. mellonella* larval mortality at day 3 after treatment (Table 2). At day 5 after treatment with 10^7 conidia mL^{-1}, another 7 isolates (ARSEF 552, 729, 759, 782, 1448, 1449, and 2521) had already caused 100% *G. mellonella* mortality. In contrast, within the same 5 days, only 2 isolates (ARSEF 3643 and 1448) caused 100% mortality of *T. molitor*; however, 4 other isolates (ARSEF 724, 760, 1885, and 2575) were sufficiently virulent to cause more than 90% mortality of *T. molitor* larvae at day 5 after treatment (Table 2).

ARSEF 552, 724, 729, 759, 782, 1449, 1882, 1885, 2521, 2575, 3641, 3643, 3479, and 3918 caused $\geq$ 50% *Galleria* larval mortality with the low-concentration treatment at day 5 after treatment. The most virulent isolates were ARSEF 1449 (91.67% larval mortality ±8.33 standard error), ARSEF 3643 (90.63% larval mortality ±9.38 se), and ARSEF 3643 (84.38% larval mortality ±15.63 se).

With *Tenebrio* at day 5, only ARSEF 1885 caused $\geq$50% mortality (66.67% ±23.3 se) in the low-concentration treatment. ARSEF 3643 and ARSEF 724 were the second and third most virulent isolates with each causing 45% larval mortality (±23.6 se and 22.5 se, respectively) at day 5 after treatment with the low fungus concentration.

Discussion

The present study investigated 20 *Metarhizium* spp. isolates as to their virulence against two insect species and their levels of DTXs production in artificial liquid medium. A wide variation in DTXs production *in vitro* was observed among the 13 isolates in the present study (Figure 1). Two *Metarhizium* isolates were used to analyze fungus-colonized cowpea and cucumber plants for DTXs production. This is the first report of the presence of DTXs in cowpea plants colonized by the entomopathogenic fungus *M. robertsii*.

Not all fungal isolates tested here have been classified according to the Bischoff et al. [35] protocol, but based on the species names attributed to these isolates in the ARSEF catalog (Table 1), we note that the production of DTXs is not strictly correlated with *Metarhizium* species. For example: ARSEF 552 (*M. pingshaense*) and ARSEF 2575 (*M. robertsii*) are producers of DTXs *in vitro*, while isolates ARSEF 1448 (*M. pingshaense*) and ARSEF 724 (*M. robertsii*) are not. In the present study, *M. acridum* isolate ARSEF 324 did not produce detectable levels of DTXs after culture *in vitro* for 5 days; which is similar to the findings of Wang et al. [27] with this isolate *in vitro*. In contrast, Kershaw et al. [19] and Moon et al. [26] reported low levels of DTXs A and E production *in vitro* by isolate ARSEF 324 with longer incubation periods and higher temperatures.

Comparisons of DTXs production *in vitro* with virulence to insects of the 20 *Metarhizium* spp. isolates did not indicate a close association of the two traits. The most virulent isolates for *T. molitor* were ARSEF 3643, ARSEF 1448 (both isolates caused 100% mortality 5 days after treatment with 1×10^7 conidia mL^{-1}). Interestingly, ARSEF 3643 was the best DTX producer *in vitro*,

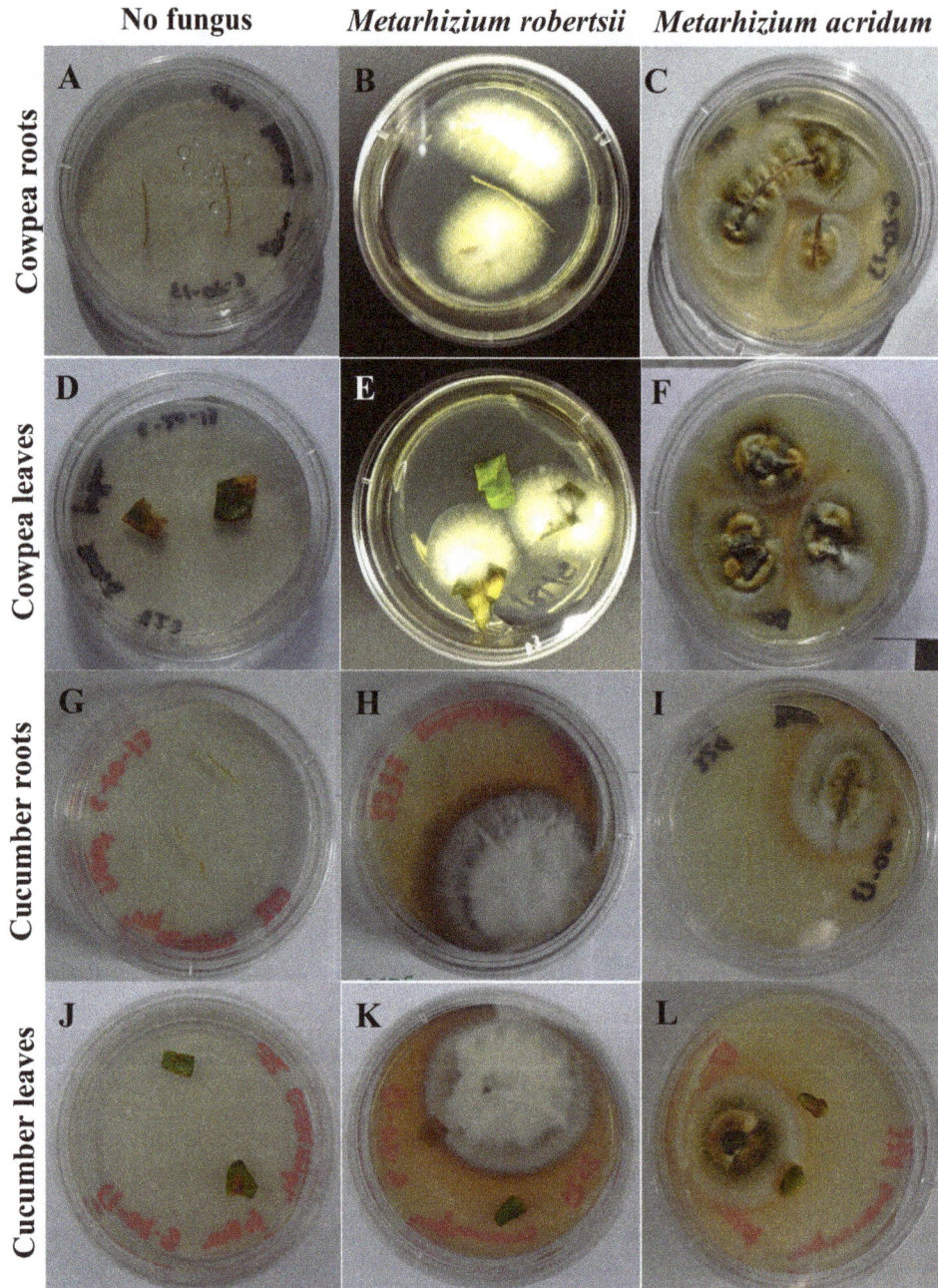

Figure 4. Re-isolation of *Metarhizium robertsii* **or** *M. acridum* **after their endophytic colonization of cowpeas (***Vigna unguiculata***) and cucumber (***Cucumis sativus***).** Control plants with no fungus inoculation (A, D, G, and J); *M. robertsii* growing from surface sterilized roots (B) and leaves (E) of cowpeas; *M. robertsii* growing from surface sterilized roots (H) and leaves (K) of cucumber. *M. acridum* growing from surface sterilized roots (C) and leaves (F) of cowpeas; and *M. acridum* growing from surface sterilized roots (I) and leaves (L) of cucumber. Note that the characteristic brownish-green conidia of *M. robertsii* were obscured by a layer of white mycelium, whereas the dark green conidia of *M. acridum* were more visible due to very little mycelial overlay.

while there were no detectable levels of DTXs produced by ARSEF 1448 in liquid culture. The same occurred with *G. mellonella*, i.e., of the 5 most virulent isolates (e.g., ARSEFs 724, 760, 1885, 2575, 3641, and 3643) two did not produce DTXs *in vitro* (ARSEF 724 and ARSEF 760), one was a very weak producer *in vitro* (ARSEF 1885), and three were good DTXs producers *in vitro* (ARSEF 2575, ARSEF 3641, and especially ARSEF 3643). Another trait that might relate to virulence is the first time (date) that DTXs were detectable in culture supernatants.

For most DTXs producing isolates, these compounds were detected in liquid culture on day 3 of fungal growth; however, ARSEF 759 had detectable levels of DTX E at day 2 in culture (Figure 2). ARSEF 759 did not demonstrate higher potency or a shorter lethal time in comparison to fungal isolates that only showed detectable levels at days 3 or 4 in culture (e.g., ARSEF 2575 and ARSEF 1885). These observations suggest that the presence or the absence of DTXs A, B, and E in *in vitro* culture

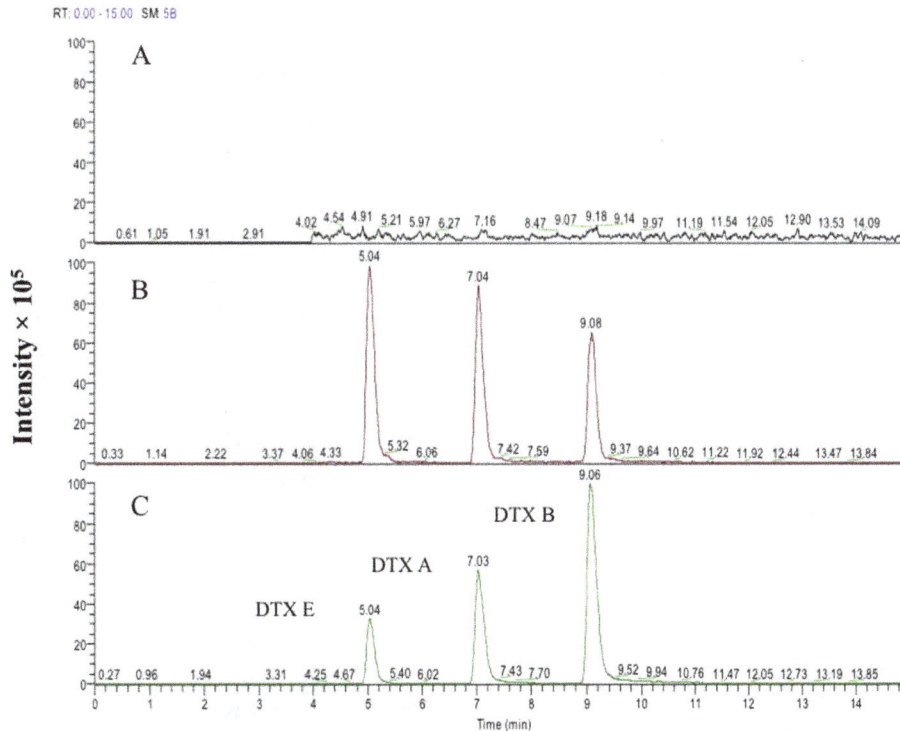

Figure 5. HPLC-MS analysis of cowpea extracts for destruxin (DTX) production. (A) Analysis of not colonized (free of fungus) plants (negative control); (B) plants endophytically colonized by *Metarhizium robertsii* ARSEF 2575; and (C) not-colonized plants spiked with DTX standards (positive control). The cowpea seeds, both fungus-inoculated and control (not colonized) were incubated on moist filter paper under optimal light (16L:8D) and temperature (25°C) conditions for 12 days at which time the germlings had developed roots, stems, cotyledons and two true leaves. DTXs were extracted from entire plants using methanol 100% and SPE-C18 cartridges.

supernatants had little or no correlation with percent mortality or speed of insect kill.

Arthropod pathogens such as *B. bassiana* [31,36,37]; *Lecanicillium lecanii* (= *Verticillium lecanii*) [37,38]; *Isaria farinosa* (= *Paecilomyces farinosus*) [39]; and *M. robertsii* [23] have been reported as endophytes. According to O'Brien [40], *M. acridum*, an acridid specialist, is not rhizosphere-competent; and Pava-Ripoll et al. [29] reported that germination of this fungal species in plant root exudates was significantly lower than with *M. robertsii* (= *M. anisopliae*). On the other hand, as reported in the current study, *M. acridum* colonized endophytically either cowpea or cucumber when surface sterilized seeds were inoculated with conidia in the laboratory. It currently is not known if spraying leaves of plants with this fungus will permit endophytic establishment in leaves, stems and roots.

The entomopathogenic fungus *B. bassiana* is an endophyte in naturally colonized plants [36], and also has been isolated after artificial inoculation in many important agricultural crops such as bananas, bean, coffee, corn, cotton, tomato and wheat [37]. Bing and Lewis [41] reported that tunneling in corn plants by *Ostrina nubilalis* larvae, the European corn borer, was reduced when plants were endophytically colonized by *B. bassiana*. Although the overwhelming majority of publications on the use of arthropod-pathogenic fungi against insects discuss the reduction of insect damage through insect death due to direct fungal infection by conidia, Vega et al. [36] suggested that this suppression of insect damage in response to *B. bassiana* plant colonization [41] may be the result of feeding deterrence or antibiosis. Such deterrence by some fungi is related to their production of metabolites. More

recently, Gurulingappa et al. [37] studied the effect of endophytes (*B. bassiana*, *L. lecanii* and *Aspergillus parasitucus*) on the reproduction and growth of *Aphis gossypii* and *Chortoicetes terminifera*. They reported that endophytes significantly reduced aphid reproduction and locust growth rate, but no direct mortality was observed. Amiri et al. [16] reported residual and antifeedant activities of DTXs A, B, and E when leaf discs of Chinese cabbage were immersed in these toxins and submitted to larvae of crucifer pests *P. xylostella* and *P. cochleariae*; as a result, leaf area ingested by these larvae was greater for untreated leaves than DTXs-treated leaves in doses higher than 3 µg/g [16]. According to the study [16] with crucifer pest larvae, the amount of DTXs detected in cowpeas in the current study should be slightly toxic. However, DTXs amounts in plants older than those that we studied probably would vary, and DTXs susceptibility of other insect species also are likely to vary. The mechanisms involved in feeding suppression of insects by contact and/or ingestion of DTXs remains unclear.

Metarhizium spp. have been indicated as mediators of interactions among plants, insects and soil: e.g., Behie et al. [22] showed that plants can receive significant amounts of nitrogen from *Metarhizium*-infected soil insects. Sasan and Bidochka [23] reported that endophytic establishment of *M. robertsii* in roots induced growth of plant roots and root hairs. The ability of some *Metarhizium* isolates to produce DTXs within plants, as reported here, suggests another potentially important benefit to plants from endophytic association with these fungi.

DTXs production by AP fungi in plants depends not only on the fungal isolate but also on the plant species. Our results showed that even when colonized with *M. robertsii* ARSEF 2575 (an isolate

Table 2. Mean mortality (%) ± standard error of *Tenebrio molitor* larvae 5 days after treatment, and *Galleria mellonella* 3 days after treatment.

Fungal Isolates	Tenebrio molitor		Galleria mellonella	
	Conidia concentration (mL^{-1})		Conidia concentration (mL^{-1})	
	1×10^7	1×10^5	1×10^7	1×10^5
ARSEF 3643	100.00±0.0 d	45.00±23.63 bce	100.00±0.0 f	14.58±8.33 a
ARSEF 1448	100.0±0.0 d	28.33±15.90 bcde	84.38±12.76 bdf	31.25±25.52 a
ARSEF 1885	96.67±1.7 cd	66.67±23.33 b	100.00±0.0 f	27.08±21.14 a
ARSEF 724	95.00±5.0 cd	45.00±22.55 bce	100.00±0.0 f	8.33±2.08 a
ARSEF 760	93.33±6.7 cd	13.33±7.26 ac	100.00±0.0 f	4.17±2.08 a
ARSEF 2575	93.33±6.7 cd	28.33±21.86 bc	100.00±0.0 f	20.83±20.83 a
ARSEF 1449	90.00±7.6 bcd	16.67±14.24 ac	97.92±2.08 df	8.33±5.51 a
ARSEF 782	88.33±9.3 bcd	16.67±7.26 bce	97.92±2.08 df	14.17±8.70 a
ARSEF 3641	86.67±8.8 bcde	33.33±20.28 bcde	100.00±0.0 f	53.13±38.27 a
ARSEF 2521	80.0±11.5 bcde	28.33±23.33 bce	68.75±31.25 bdefg	34.38±28.07 a
ARSEF 929	75.00±15.3 acd	16.67±14.24 ac	52.08±28.94 abd	20.83 12.67 a
ARSEF 759	66.67±20.3 abcd	3.33±3.33 ade	79.17±20.83 bcdf	16.67±9.08 a
ARSEF 552	63.33±11.7 abcd	0.00±0.0 a	89.58±10.42 bdf	0.00±0.0 a
ARSEF 2211	61.67±25.9 abd	5.00±2.89 ac	70.83±29.17 bdefg	25.00±18.75 a
ARSEF 729	46.67±14.8 abc	1.67±1.67 ade	64.58±26.60 bcdf	10.42±5.51 a
ARSEF 3918	40.00±30.6 abc	0.00±0.0 a	35.42±29.39 ab	53.13±38.27 a
ARSEF 1883	21.67±10.9 ab	0.00±0.0 a	33.33±18.52 ace	27.08±24.03 a
ARSEF 1882	16.67±14.2 ab	5.00±0.0 ac	35.42±18.52 ace	8.33±8.33 a
ARSEF 324	13.33±10.9 ae	0.00±0.0 a	31.25±15.73 acg	4.17±4.17 a
ARSEF 3479	5.00±5.0 a	1.67±1.67 ad	6.25±0.0 a	8.33±2.08 a
Control	1.67±1.7 a	1.67±1.7 ad	4.17±4.17 a	4.17±4.17 a

Bioassays were performed 3 times (using two replicates for each isolate) under controlled conditions (27°C), using new batches of larvae and conidia in each bioassay. Controls were treated with Tween 80 (0.01%) solution. Means followed by the same letter in a column do not differ statistically ($P \geq 0.05$) (Kruskal-Wallis test followed by Student-Newman-Keuls).

that produces DTXs *in vitro* and also in cowpeas), cucumber extracts did not have detectable levels of DTXs. A plant pathogen *Alternaria brassicae*, the causative agent of *Alternaria* blackspot, is known to produce DTX B that is used to facilitate plant colonization. DTX B is a selective toxin, in that only plant cultivars susceptible to the toxin are damaged by the fungus [5,42]. Resistant plants have enzymes that detoxify DTX B [42]. The current study did not investigate whether cucumber plants hydrolyzed DTX or if this host plant did not support DTX production.

Further studies on the effects of *per os* DTXs exposure in vertebrate organisms are needed to support the use of entomopathogenic fungi inoculated in crop seeds to control insect pests. In an instance where there is some hesitancy by regulating agencies about allowing DTXs in a food product, a non-DTXs producing isolate of *Metarhizium* could be selected for use in biological control on that crop to avoid such DTXs production, or plant cultivars that detoxify DTXs could be selected.

In planta production of secondary metabolites by endophytic *Metarhizium* may be an exploitable feature of this fungus in its use against agricultural arthropod pests. The production of DTXs in *M. robertsii*-colonized plants reported here clearly indicates that further investigation is warranted on the antifeedant or repellent properties of fungal metabolites expressed *in planta*.

Supporting Information

Table S1   Destruxin production (mg/L) by 12 *Metarhizium* spp. isolates *in vitro*. Destruxin production is represented by mean values ± standard error after 5 days in submerged shaken cultures.

Acknowledgments

We thank Coordenação de Aperfeiçoamento de Pessoal de Nível Superior (CAPES) and Conselho Nacional de Desenvolvimento Científico e Tecnológico (CNPq) from Brazil, and also the consistent interest of Larry E. Jech and R. Nelson Foster from USDA/APHIS (Phoenix, AZ) in the research reported here. We appreciate the advice of Daniel Cook from USDA/ARS (Logan, UT). Vania R.E.P. Bittencourt is a CNPq researcher.

Author Contributions

Conceived and designed the experiments: PSG DWR. Performed the experiments: PSG DRG MMG. Analyzed the data: PSG DRG MMG MSP. Contributed reagents/materials/analysis tools: JYT SBK EKKF VREPB DWR. Contributed to the writing of the manuscript: PSG DRG SBK DWR.

References

1. Bittencourt VREP, Massard CL, Lima AF (1992) Uso do fungo *Metarhizium anisopliae* (Metschnikoff, 1879) Sorokin, 1883, no controle do carrapato *Boophilus microplus* (Canestrini, 1887). Arquivo da Universidade Rural do Rio de Janeiro 15: 197–202.

2. Fernandes EKK, Bittencourt VREP (2008) Entomopathogenic fungi against South American tick species. Exp Appl Acarol 46: 71–93.

3. Roberts DW, St. Leger RJ (2004) *Metarhizium* spp., cosmopolitan insect pathogenic fungi: mycological aspects. Adv Appl Microbiol 54: 1–70.

4. Samish M, Ginsberg H, Glaser I (2004) Biological control of ticks. Parasitology 129: S389–S403.

5. Pedras MSC, Zaharia LI, Ward DE (2002) The destruxins: synthesis, biosynthesis, biotransformation, and biological activity. Phytochemistry 59: 579–596.

6. Samuels RI, Charnley AK, Reynolds SE (1988) The role of destruxins in the pathogenicity of 3 strains of *Metarhizium anisopliae* for the tobacco hornworm *Manduca sexta*. Mycopathologia 104: 51–58.

7. Samuels RI, Reynolds SE, Charnley AK (1988) Calcium channel activation of insect muscle by destruxins, insecticidal compounds produced by the entomopathogenic fungus *Metarhizium anisopliae*. Comp Biochem Physiol 90C: 403–412.

8. Quiot JM, Vey A, Vago C (1985) Effects of mycotoxins on invertebrate cells *in vitro*. Adv Cell Cult 4: 199–212.

9. James PJ, Kershaw MJ, Reynolds SE, Charnley AK (1993) Inhibition of desert locust (*Schistocera gregaria*) Malpighian tubule fluid secretion by destruxins, cyclic peptide toxins from the insect pathogenic fungus *Metarhizium anisopliae*. J Insect Physiol 39: 797–804.

10. Muroi M, Shiragami N, Takatsuki A (1994) Destruxin B, a specific and readily reversible inhibitor of vacuolar-type H+-translocating ATPase. Biochem Biophys Res Commun 205: 1358–1365.

11. Cerenius L, Thornqvist P, Vey A, Johansson MW, Soderhall K (1990) The effect of the fungal toxin destruxin E on isolated crayfish haemocytes. J Insect Physiol 36: 785–789.

12. Han F, Jin F, Dong X, Fan J, Qiu B, Ren S (2013) Transcript and protein analysis of the destruxin A-induced response in larvae of *Plutella xylostella*. Plos One 8: e60771–e60781.

13. Huxham IM, Lackie AM, McCorkindale NJ (1989) Inhibitory effects of cyclodepsipeptides, destruxins, from the fungus *Metarhizium anisopliae*, on cellular immunity in insects. J Insect Physiol 35: 97–105.

14. Vey A, Matha V, Dumas C (2002) Effects of the peptide mycotoxin destruxin E on insect haemocytes and on dynamics and efficiency of the multicellular immune reaction. J Invertbr Pathol 80: 177–187.

15. Vilcinskas A, Matha V, Götz P (1997) Inhibition of phagocytic activity of plasmatocytes isolated from *Galleria mellonella* by entomogenous fungi and their secondary metabolites. J Insect Physiol 43:475–483.

16. Amiri B, Ibrahim L, Butt TM (1999) Antifeedant properties of destruxins and their potential use with the entomogenous fungus *Metarhizium anisopliae* for improved control of crucifer pests. Biocontrol Sci Technol 9: 487–498.

17. Thomsen L, Eilenberg J, Esbjerg P (1996) Effects of destruxins on *Pieris brassicae* and *Agrotis segetum*. In: Smits PH, editor. Insect pathogens and insect parasitic nematodes. IOBC Bulletin 19: 190–195.

18. Donzelli BGG, Krasnoff SB, Sun-Moon Y, Churchill ACL, Gibson DM (2012) Genetic basis of destruxin production in the entomopathogen *Metarhizium robertsii*. Curr Genet 58: 105–116.

19. Kershaw MJ, Moorhouse ER, Bateman R, Reynolds SE, Charnley AK (1999) The role of destruxins in the pathogenicity of *Metarhizium anisopliae* for three species of insect. J Invertebr Pathol 74: 213–223.

20. Suzuki A, Kawakami K, Tamura S (1971) Detection of destruxins in silkworm larvae infected with *Metarhizium anisopliae*. Agr Biol Chem 35: 1641–1643.

21. Skrobek A, Shah FS, Butt TM (2008) Destruxin production by the entomogenous fungus *Metarhizium anisopliae* in insects and factors influencing their degradation. Biocontrol 53:361–373.

22. Behie SW, Zelisko PM, Bidochka MJ (2012) Endophytic insect-parasitic fungi translocate nitrogen directly from insects to plants. Science 336: 1576–1577.

23. Sasan RK, Bidochka M (2012) The insect-pathogenic fungus *Metarhizium robertsii* (Clavicipitaceae) is also an endophyte that stimulates plant root development. Am J Bot 99: 101–107.

24. St. Leger RJ (2008) Studies on adaptations of *Metarhizium anisopliae* to life in the soil. J Invertebr Pathol 98: 271–276.

25. Wyrebek M, Huber C, Sasan RK, Bidochka MJ (2011) Three sympatrically occurring species of *Metarhizium* show plant rhizosphere specificity. Microbiology 157: 2904–2911.

26. Moon Y-S, Donzelli BGG, Krasnoff SB, McLane H, Griggs MH, et al. (2008) *Agrobacterium*-mediated disruption of a nonribosomal peptide synthetase gene in the invertebrate pathogen *Metarhizium anisopliae* reveals a peptide spore factor. App Environ Microbiol 74: 4366–4380.

27. Wang B, Kang Q, Lu Y, Bai L, Wang C (2012) Unveiling the biosynthetic puzzle of destruxins in *Metarhizium* species. Proc Natl Acad Sci U S A 109: 1287–1292.

28. Krasnoff SB, Sommers CH, Moon Y-S, Donzelli BGG, Vandenberg JD, et al. (2006) Production of mutagenic metabolites by *Metarhizium anisopliae*. J Agric Food Chem 54: 7083–7088.

29. Pava-Ripoll M, Angelini C, Fang W, Wang S, Posada F, et al. (2011) The rhizosphere-competent entomopathogen *Metarhizium anisopliae* expresses a specific subset of genes in plant root exudates. Microbiology 157: 47–55.

30. Rangel DNE, Fernandes EKK, Braga GUL, Roberts DW (2011) Visible light during mycelial growth and conidiation of *Metarhizium anisopliae* produces conidia with increased stress tolerance. FEMS Microbiol Lett 315: 81–86.

31. Parsa S, Ortiz V, Vega FE (2013) Establishing fungal entomopathogens as endophytes: towards endophytic biological control. JoVE 74: e50360.

32. Tulloch M (1976) The genus *Metarhizium*. Trans Brit Mycol Soc 66: 407–411.

33. Ayres M, Ayres JR M, Ayres DL, Santos AAS (2007) BioEstat 5.0 - Aplicações Estatísticas nas Áreas das Ciências Biológicas e Médicas. Sociedade Civil Mamirauá, Tefé, Brazil, 380 p.

34. Sampaio IBM (2002) Estatística Aplicada à Experimentação Animal. Belo Horizonte, Brazil: FEPMVZ-Editora. 265 p.

35. Bischoff JF, Rehner SA, Humber RA (2009) A multilocus phylogeny of the *Metarhizium anisopliae* lineage. Mycologia 101: 512–530.

36. Vega FE, Posada F, Aime MC, Pava-Ripoll M, Infante F, et al. (2008) Entomopathogenic fungal endophytes. Bio Control 44: 72–82.

37. Gurulingappa P, Sword GA, Murdoch G, McGee PA (2010) Colonization of crop plants by fungal entomopathogens and their effect on two insect pests when *in planta*. Bio Control 55: 34–41.

38. Petrini O (1981). Endophytische pilze in *Epiphytischen araceae*, *Bromeliaceae* and *Orchidiaceae*. Sydowia 34: 135–148.

39. Bills GF, Polishook JD (1991) Microfungi from *Carpinus caroliniana*. Can J Bot 69: 1477–1482.

40. O'Brien TR (2008) *Metarhizium anisopliae*'s persistence as a saprophyte, genetic basis of adaptation and role as a plant symbiont. PhD Dissertation, University of Maryland. Available: http://drum.lib.umd.edu/handle/1903/8839. Accessed 2013 July 05.

41. Bing LA, Lewis LC (1991) Suppression of *Ostrinia nubilalis* (Hubner)(Lepdoptera: Pyralidae) by endophytic *Beauveria bassiana* (Balsamo) Vuillemin. Environ Entomol 20: 1207–1211.

42. Pedras MSC, Zaharia IL, Gai Y, Zhou Y, Ward DE (2001) *In planta* sequential hydroxylization and glycosylation of a fungal phytotoxin: avoiding cell death and overcoming the fungal invader. Proc Natl Acad Sci U S A 98:747–752.

18

Use of Two-Part Regression Calibration Model to Correct for Measurement Error in Episodically Consumed Foods in a Single-Replicate Study Design: EPIC Case Study

George O. Agogo[1,2]*, **Hilko van der Voet**[2], **Pieter van't Veer**[3], **Pietro Ferrari**[4], **Max Leenders**[5], **David C. Muller**[6], **Emilio Sánchez-Cantalejo**[7,8], **Christina Bamia**[9], **Tonje Braaten**[10], **Sven Knüppel**[11], **Ingegerd Johansson**[12], **Fred A. van Eeuwijk**[2], **Hendriek Boshuizen**[1,2,3]

1 National Institute for Public Health and the Environment, Bilthoven, The Netherlands, **2** Biometris, Wageningen University and Research Center, Wageningen, The Netherlands, **3** Department of Human Nutrition, Wageningen University and Research Center, Wageningen, The Netherlands, **4** Nutritional Epidemiology Group, International Agency for Research on Cancer, Lyon, France, **5** Department of Gastroenterology and Hepatology, University Medical Center Utrecht, Utrecht, The Netherlands, **6** Genetic Epidemiology Group, International Agency for Research on Cancer, 150 cours Albert Thomas, Lyon, 69008, France, **7** Andalusian School of Public Health, Granada, Spain, **8** CIBER de Epidemiología y Salud Pública (CIBERESP), Barcelona, Spain, **9** WHO Collaborating Center for Food and Nutrition Policies, Department of Hygiene, Epidemiology and Medical Statistics, University of Athens Medical School, Athens, Greece, **10** Department of Community Medicine, University of Tromsø, N-9037, Tromsø, Norway, **11** Department of Epidemiology, German Institute of Human Nutrition Potsdam-Rehbrücke, Potsdam, Germany, **12** Department of Odontology, Umeå University, Umeå, Sweden

Abstract

In epidemiologic studies, measurement error in dietary variables often attenuates association between dietary intake and disease occurrence. To adjust for the attenuation caused by error in dietary intake, regression calibration is commonly used. To apply regression calibration, unbiased reference measurements are required. Short-term reference measurements for foods that are not consumed daily contain excess zeroes that pose challenges in the calibration model. We adapted two-part regression calibration model, initially developed for multiple replicates of reference measurements per individual to a single-replicate setting. We showed how to handle excess zero reference measurements by two-step modeling approach, how to explore heteroscedasticity in the consumed amount with variance-mean graph, how to explore nonlinearity with the generalized additive modeling (GAM) and the empirical logit approaches, and how to select covariates in the calibration model. The performance of two-part calibration model was compared with the one-part counterpart. We used vegetable intake and mortality data from European Prospective Investigation on Cancer and Nutrition (EPIC) study. In the EPIC, reference measurements were taken with 24-hour recalls. For each of the three vegetable subgroups assessed separately, correcting for error with an appropriately specified two-part calibration model resulted in about three fold increase in the strength of association with all-cause mortality, as measured by the log hazard ratio. Further found is that the standard way of including covariates in the calibration model can lead to over fitting the two-part calibration model. Moreover, the extent of adjusting for error is influenced by the number and forms of covariates in the calibration model. For episodically consumed foods, we advise researchers to pay special attention to response distribution, nonlinearity, and covariate inclusion in specifying the calibration model.

Editor: Zhong-Ke Gao, Tianjin University, China

Funding: The authors did not receive specific funding for this study. However, the publication cost will be covered by Wageningen University and Research Centre, Biometris, P.O. Box 100, 6700 AC WAGENINGEN should the paper be accepted for a publication.

Competing Interests: The authors have declared that no competing interests exist.

* Email: george.agogo@wur.nl

Introduction

Dietary variables are often measured with error in nutritional epidemiology. In such studies, usual dietary intake is assessed with instruments such as, food frequency questionnaire and dietary questionnaire [1–3]. In these instruments, the queried period of intake ranges from several months to a year. As a result, these instruments are prone to error caused by difficulties to recall past intake of foods or food groups, the frequency of consumption, and the portion size. In general, the measurement error in usual dietary intake can either be systematic or random. Systematic error occurs when an individual systematically overestimates or underestimates dietary intake, whereas random error is due to random within-individual variation in reporting of dietary intake [1,4]. The random error attenuates the association between dietary intake and disease occurrence, whereas systematic error can either attenuate or inflate the association.

As a case study, we used the European Prospective Investigation on Cancer and Nutrition (EPIC) study. In EPIC, country-specific dietary questionnaires, hereafter DQ, were used to measure usual intake of various dietary variables or groups of dietary variables in different participating cohorts. With DQ measurements for usual intake, an association parameter estimate that relates usual intake to disease occurrence is often biased, mainly towards the null [4–6].

Regression calibration is the commonly used method to adjust for the bias in the association between usual intake and disease occurrence, due to measurement error in the DQ. Regression calibration involves finding the best prediction of true usual intake given DQ measurements and other error-free variables [7]. The prediction is further used as a proxy for true usual intake in the disease model that relates dietary intake to disease occurrence. Regression calibration requires a calibration sub-study, where unbiased measurements are taken. Some prospective studies therefore include a calibration sub-study that can either be internal or external. Internal calibration study consists of a random sample from the main study population, as was the case in the EPIC, whereas external calibration sub-study consists of subject not in the main-study but with similar design as the main-study [8]. In the calibration sub-study, unbiased reference measurements are collected by short-term reference instruments, such as food records or 24-hour dietary recalls. The reference measurements can be used as the response in the calibration model to predict true usual intake. In the EPIC study, regression calibration can also adjust for systematic error in DQ measurements due to the multicenter component of the EPIC study, as described in [9,10]. In the EPIC calibration sub-study, a 24-hour dietary recall, hereafter 24-HDR was used as the reference instrument. For each subject in the calibration sub-study in the EPIC, only one reference measurement was available [11]. For foods that are not consumed daily, 24-HDR measurements would contain many zeroes for many individuals. Handling these zeroes poses a challenge in the calibration model [12–15]. The excess zeroes can be handled with regression calibration in a two-step approach, where the consumption probability and the consumed amount on consumption days are modeled separately [13]. We refer to this model as two-part regression calibration.

The currently published studies on two-part regression calibration method require epidemiologic studies with at least two replicate reference measurements per subject [13–15]. Given the design of the EPIC study with a single measurement per individual, however, these calibration models cannot be applied directly. Moreover, the performance of the calibration models in a study design such as EPIC for episodically consumed foods has not been studied exhaustively. Further, the effect of variable selection on the performance of a two-part calibration model has not yet been studied fully. The standard theory of selecting covariates into the calibration model states that confounding variables in the disease model must be included in the calibration model together with the covariates that only predict dietary intake but not the risk of the disease [14,16].

To fill the aforementioned gaps, we developed a two-part regression calibration model to adjust for the bias in the diet-disease association, due to measurement error in self-reported episodically consumed foods, when each subject in the calibration sub-study has only a single reference measurement. The second goal was to assess the effect of reducing the number of variables in the two-part calibration model with the covariates selected based on the standard theory. As a working example, we studied the association between intakes of each of the three vegetable subgroups: leafy vegetables, fruiting vegetables, and root vegeta-

bles, on all-cause mortality as reported in the EPIC. We described how to handle the excess zeroes, the highly skewed-heteroscedastic non-zero reference measurements, non-linear relations in the calibration model, and how to select covariates into the calibration model. We showed that a suitably specified two-part calibration model adjusts for the bias in the diet-disease association caused by measurement error in self-reported intake in EPIC study. We further showed that the extent of adjusting for the bias is much influenced by how the calibration model is specified.

Materials and Methods

Ethics Statement

All participants who agreed to join the EPIC study signed an informed written consent. The study was approved by the Institutional Review Board of the International Agency for Research on Cancer and local institutional review boards of each participating center.

Study subjects

EPIC is an on-going multicenter prospective cohort study to investigate the relation between diet and the risk of cancer and other chronic diseases. The study consisted of 519,978 eligible men and women aged between 35 and 70 years and recruited in 23 centers in 10 Western European countries [11,17]. The 10 participating countries were: France, Italy, Spain, United Kingdom, Germany, The Netherlands, Greece, Sweden, Denmark, and Norway. The study populations comprised of heterogeneous groups. In most centers, study populations were based on general population while some consisted of participants in breast screening programs (Utrecht, The Netherlands; and Florence, Italy), teachers and school workers (France) or blood donors (certain Italian and Spanish centers). In Oxford, most of the cohort was recruited among subjects with interest in health or on vegetarian eating. Only women were recruited in France, Norway, Utrecht (The Netherlands) and Naples [18]. Information on usual dietary intake, lifestyle, environmental factors and anthropometry was collected from each individual at baseline. The dietary intake information was assessed with different dietary history questionnaires, food frequency questionnaires or a modified dietary history developed and validated separately in each participating country [17]. The questions asked in the questionnaires included the frequency of consumption over the past 12 months preceding the administration, categorized into the number of times per day, per week, per month or per year. A calibration sub-study was carried out within the entire EPIC cohort by taking a stratified random sample of 36,900 subjects. In the calibration sub-study, a 24-HDR was administered once per subject using a specifically developed software program (EPIC-SOFT) designed to harmonize the dietary measurements across study populations [19].

We used EPIC dietary intake data for leafy vegetables, fruiting vegetables and root vegetable sub-groups as a working example. We further assumed measurements on the 24-HDR (in g/day) as the unbiased reference measurements and those on the DQ as the biased main-study measurements. We excluded subjects with missing questionnaire data, missing dates of diagnosis or follow up, in the top and bottom 1% of the distribution of the ratio of reported total energy intake to energy requirement. We further excluded subjects with a history of cancer, myocardial infarction, stroke, angina, diabetes or a combination of these diseases at baseline. As a result, data for 430,215 subjects were eligible for the analyses. In the analysis, the data from the following centers were excluded: Umeå and Norway for leafy vegetables and Norway for fruiting vegetables. The decision to exclude these data was based

Table 1. Country-specific summary measures for the percentage of zero intake measurements reported on 24-HDR (% R = 0, non-consumers) and Pearson Correlation (ρ) for intake as measured by 24-HDR and DQ for leafy vegetables, fruiting vegetables and root vegetables.

Participating Countries	Leafy vegetables			Fruiting vegetables		Root vegetables	
	N	% R = 0	ρ	% R = 0	ρ	% R = 0	ρ
France	4735	42.8	0.17	44.4	0.10	71.6	0.06
Italy	3961	59.3	0.16	37.6	0.15	79.6	0.11
Spain	3220	48.9	0.34	31.7	0.22	76.1	0.12
UK	1313	68.2	0.16	40.8	0.19	59.3	0.23
Netherlands	4545	70.5	0.10	48.7	0.21	82.0	0.14
Greece	2930	67.9	0.10	29.5	0.13	83.2	0.03[ns]
Germany	4418	75.9	0.15	41.6	0.17	79.2	0.22
Sweden	[a]6132	70.5	0.19	34.9	0.24	67.2	0.17
Denmark	3918	77.4	0.09	41	0.21	61.8	0.40
Norway	[b]1798					58.5	0.12

EPIC Study, 1999–2000.
[a]N is 3132 instead of 6132 for leafy vegetables in Sweden because data from Umeå were excluded from analysis based of the inclusion criteria in EPIC;
[b]N refers to data for root vegetables only because data for Norway were excluded for leafy vegetable and fruit vegetable subgroups; [ns]means correlation is not statistically significant at α = 0.05, other correlation coefficients are highly significant with P<0.0001.

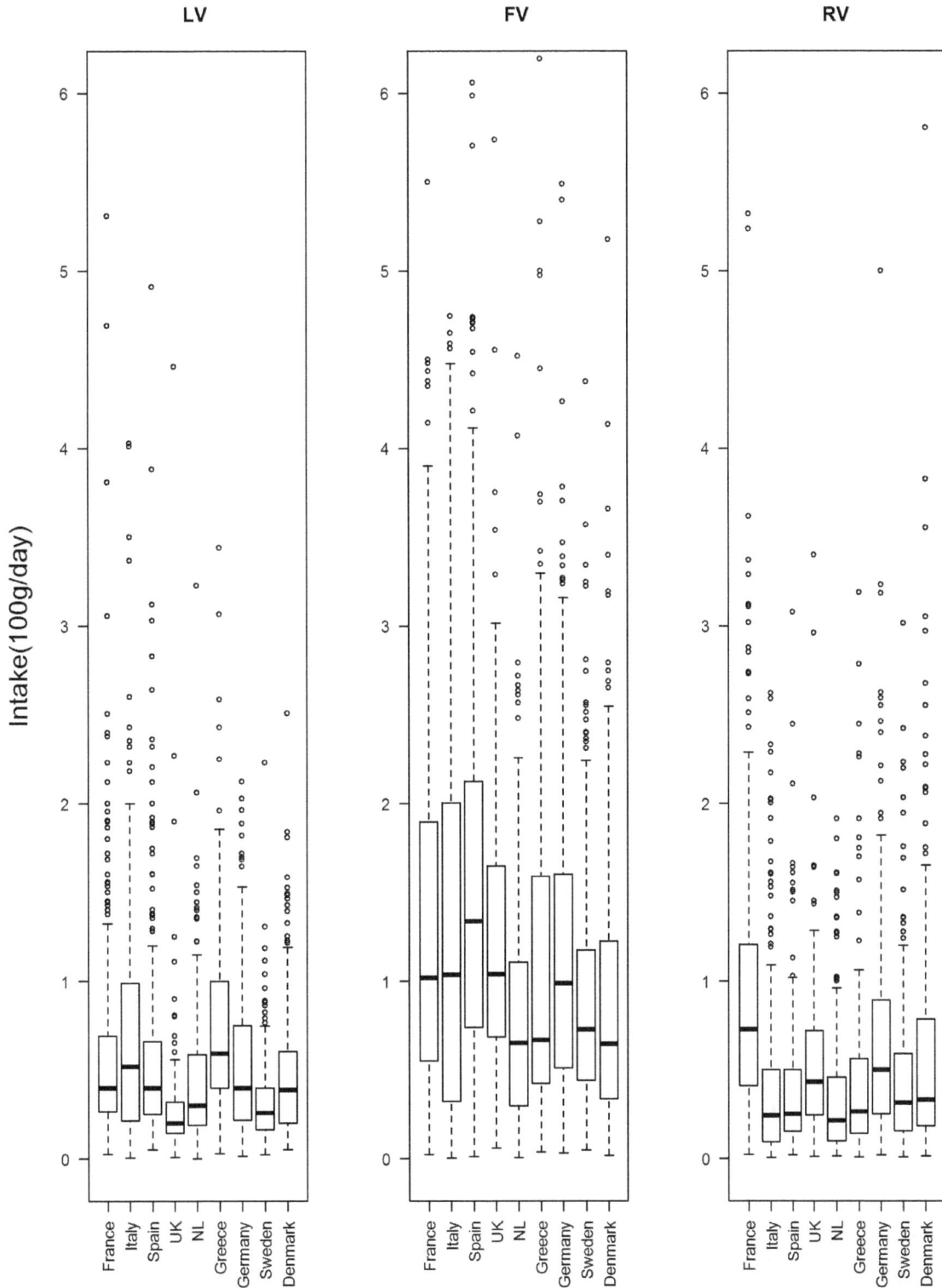

Figure 1. The boxplots for the distribution of intake of vegetable subgroups. The country-specific boxplots show the distribution of the consumed amount for those who reported consumption on the 24-HDR for leafy vegetables (LV), fruiting vegetables (FV) and root vegetable (RV) subgroups in the EPIC study, 1992–2000.

Disease model

In epidemiological studies, the interest is mainly in the association between an exposure and disease occurrence. In our working example, we were interested in the association between intake of vegetable subgroups and all-cause mortality. If the true usual intake of vegetable subgroups were known, then a generalized linear disease model would be:

$$\varphi\{E(Y|T,\mathbf{Z})\} = \beta_T T + \beta_{\mathbf{Z}}^T \mathbf{Z} \tag{1}$$

where Y is a disease outcome, here, an indicator for mortality, T is true usual dietary intake of a vegetable subgroup, $\mathbf{Z}$ is a vector of error-free confounding variables and φ is a function linking the conditional mean and the linear predictor. The coefficient β_T quantifies the association of interest and $\beta_{\mathbf{Z}}^T$ is a vector of coefficients for the confounding variables. If dietary intake is measured with error, then β_T would mostly be underestimated. Therefore, a researcher should adjust for the bias in estimating β_T due to measurement error in DQ.

Regression calibration model

Regression calibration is the most commonly used method to adjust for the bias in estimating β_T (i.e., diet-disease association) due to measurement error in the DQ. To describe regression calibration, we denote reference measurement from 24-HDR by R, main-study measurement from DQ by Q, and the covariates that only predict vegetable intake and not mortality by $\mathbf{C}$. Then, a set of all covariates that possibly relate to usual intake is given by $\mathbf{X} = \{\mathbf{Z},\mathbf{C}\}$. Regression calibration involves finding the best prediction of true usual intake given DQ measurement and other covariates [14]. The mean predictor from regression calibration is denoted by $E(T|Q,\mathbf{X})$. A major challenge in fitting the calibration model is that true usual intake is not only unobservable but also cannot be measured exactly. To circumvent this, a reference measurement is required in place of the latent true intake. The reference measurement should be unbiased for true intake, and should be measured with errors that are uncorrelated with the errors in the DQ measurements. We, therefore, made two strong assumptions. First, we assumed the short-term measurement from the 24-HDR to be an unbiased measurement for true usual intake. Second, we assumed the errors in the 24-HDR measurements to be uncorrelated with the errors in the DQ measurements. We denote the calibration model by:

$$E(T|Q,\mathbf{X}) = E(R|Q,\mathbf{X}). \tag{2}$$

We assumed in model (2) that measurement error in Q does not provide extra information about Y other than that provided by T. The measurement error in Q is, therefore, said to be non-differential. In model (2), R is modeled as a function of Q and $\mathbf{X}$ using standard regression methods, where a suitable distribution for the error terms and a suitable parametric form of each covariate in $\mathbf{X}$ is chosen.

In this work, we considered only the case of a single dietary intake variable measured with error. In our data, the correlation between the vegetable subgroups and the other confounders, as measured by the questionnaire, were low justifying their omission, as the contamination effect of the measurement error in these variables on the correction factor for our dietary intake of interest would be negligible.

Excess zeroes, heteroscedasticity and skewness in reference measurements

Vegetable subgroups considered in this study are not consumed daily. This results in many zero reference measurements reported on the 24-HDR. As a result, the reference measurements have a mixture of zeroes for non-consumers and positive intake for consumers. The excess zeroes pose challenge in regression calibration, with the reference measurements as the response. To handle these excess zeroes, we used a two-part approach to build a regression calibration model. In the first part, the consumption probability as reported in the 24-HDR is modelled. In the second part, the consumed amount on consumption occasion is modelled [13]. The first part involves discrete data and can be modeled either with logistic or probit regression, where the probability of consumption depends on a given set of covariates. In the second part, plausible family of densities for the consumed amount on consumption occasion can be assumed [20]. The GLM model for the consumption probability is parameterized as.

$\Pr(R > 0|Q,\mathbf{X}) = \phi^{-1}(\alpha_q Q + \alpha_{\mathbf{X}}^T \mathbf{X}) = \pi_{Q,\mathbf{X}}$. Similarly, the GLM model for the consumed amount is parameterized as $E(R|Q,\mathbf{X},R > 0) = g^{-1}(\beta_q Q + \beta_{\mathbf{X}}^T \mathbf{X}) = \mu_{Q,\mathbf{X}}$, where ϕ^{-1} can be either inverse-logit or inverse-probit function and g^{-1} can be an inverse of any plausible link function. Thus, the calibration model (2), adapted to two-part form to handle the excess zeroes in the response is parameterized as $E(R|Q,\mathbf{X}) = \phi^{-1}(\alpha_q Q + \alpha_{\mathbf{X}}^T \mathbf{X}) \times g^{-1}(\beta_q Q + \beta_{\mathbf{X}}^T \mathbf{X}) = \pi_{Q,\mathbf{X}}\mu_{Q,\mathbf{X}}$. The true usual intake can thus be predicted from this model. We denote the prediction from this two-part calibration model by

$$\hat{E}(R|Q,\mathbf{X}) = \hat{\pi}_{Q,\mathbf{X}}\hat{\mu}_{Q,\mathbf{X}}. \tag{3}$$

Another challenge is how to handle distribution for the consumed amount that is commonly right-skewed with heteroscedastic variance. To handle heteroscedasticity, we applied a generalized linear modeling (GLM) approach in a regression calibration context. In the GLM approach, the variance is linked to the mean as $\sigma^2(R|Q,\mathbf{X},R > 0) = \psi\{E(R|Q,\mathbf{X},R > 0)\}$, where ψ is a function that links the conditional variance with the conditional mean of reference measurement in the model for consumed amount, σ^2 denotes the conditional variance, and $E(.|.)$ denotes the conditional expectation [21]. The advantage of GLM approach is that the consumed amount can be predicted directly without transforming the data. To determine the optimal relation between the conditional variance and the conditional mean, the GLM model is parameterized using a class of power-proportional variance functions as follows: $\sigma^2(R|Q,\mathbf{X},R > 0) = \kappa\{E(R|Q,\mathbf{X},R > 0)\}^{\lambda}$, where κ denotes the coefficient of variation, λ is a finite non-negative constant. This power variance function can be rewritten in a linear log-form as follows:

$$\sigma(R|Q,\mathbf{X},R > 0) = a + b\log\{E(R|Q,\mathbf{X},R > 0)\} \tag{4}$$

where $a = (\log \kappa)/2$ and $b = \lambda/2$. In model (4), λ equals zero refers to a classical nonlinear regression with constant error variance, λ equals one refers to a Poisson regression with the variance that is proportional to the mean, where $k > 1$ indicates degree of over dispersion. Similarly, λ equals two with $k > 0$ refers to a gamma

LV

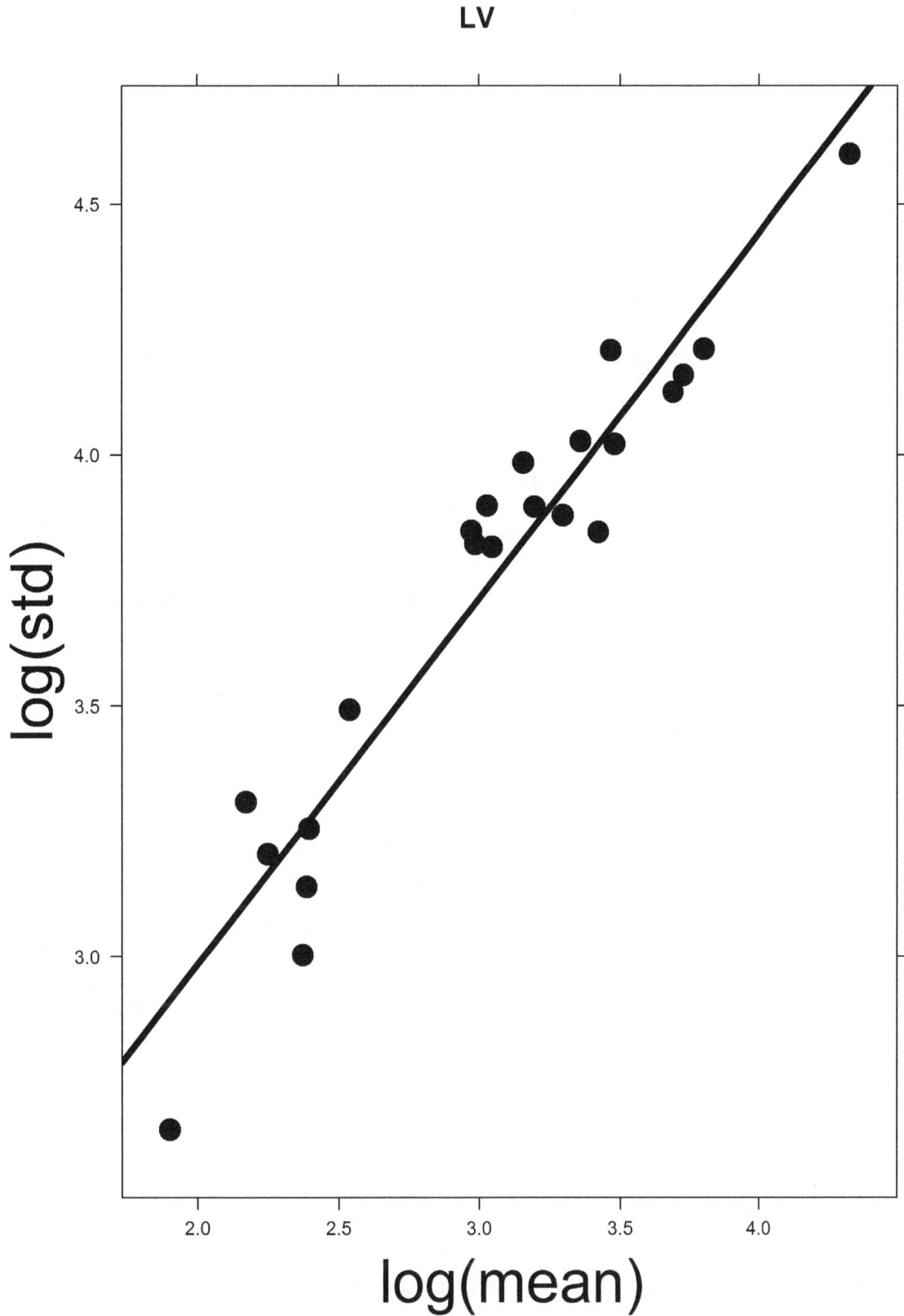

Figure 2. The variance-mean relation for Leafy vegetable intake. The graph shows a least squares regression line fitted to the scatterplots of the logarithm of center-specific standard deviation versus logarithm of center-specific mean of the consumed amount of leafy vegetables for those who reported consumption on the 24HDR in the EPIC Study, 1992–2000. The approximately linear regression line suggests a variance that increases with the mean.

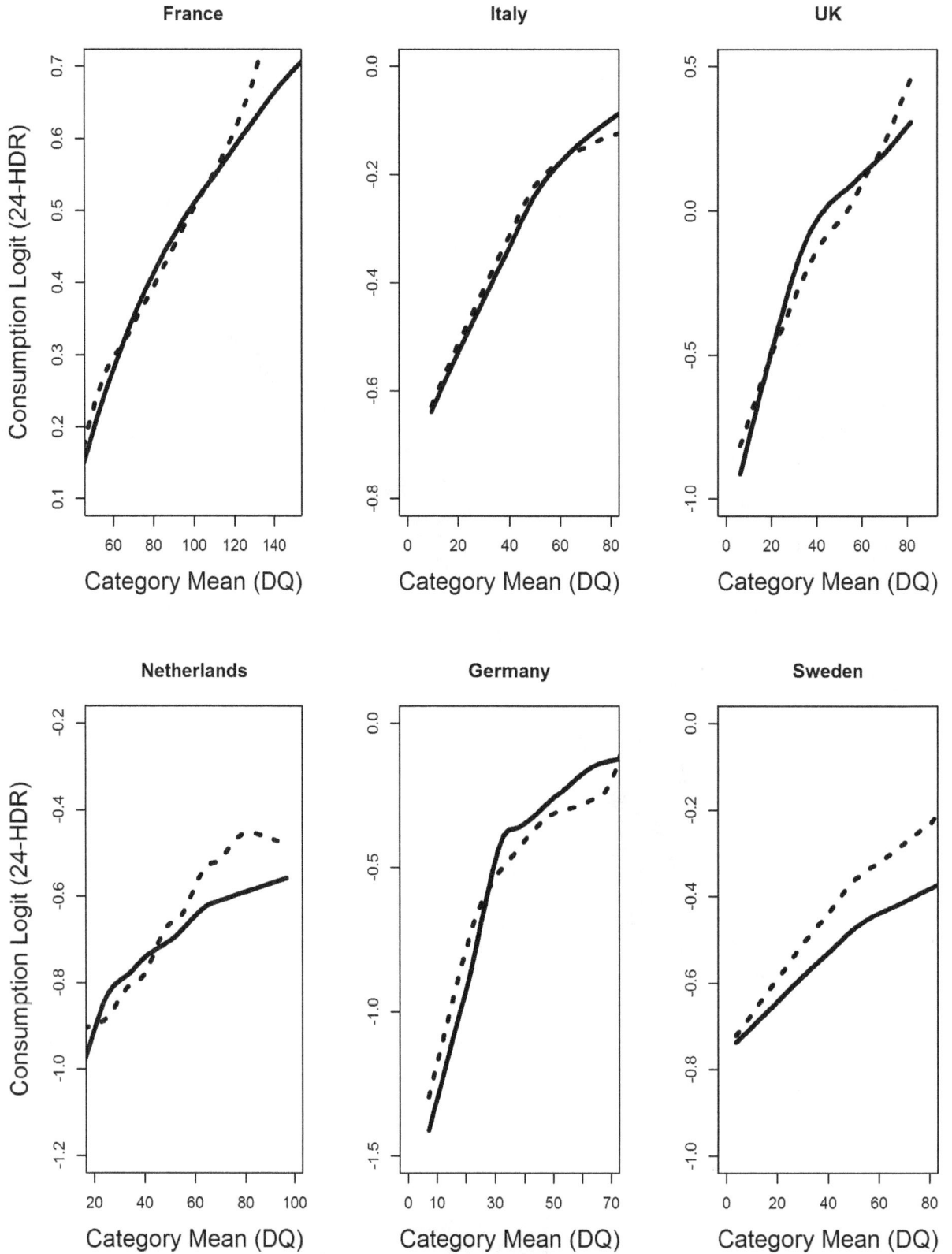

Figure 3. The empirical logit graph for Leafy vegetable intake. The graph shows loess curves fitted to 1) the scatterplots for the empirical logit (dotted line) and 2) the mean of the predicted logit from a logistic model with log-transformed DQ (thick line) against the DQ category-specific means for leafy vegetable intake in the EPIC Study, 1992–2000. The similarity in the two logit curves suggests that a log-transformed DQ is appropriate for the consumption probability part of the two-part calibration model.

Table 2. Significant covariates (marked ×) in the reduced two-part calibration models, after a backward elimination on each part of the standard two-part regression calibration model with transformed DQ and with other covariates selected using the standard way of variable inclusion.

Covariates	Leafy vegetables		Fruiting vegetables		Root vegetable	
	Part I	Part II	Part I	Part II	Part I	Part II
Main effects						
Q^t	×	×	×	×	×	×
BMI		×	×	×	×	×
Smoking status	×		×	×	×	×
Physical activity		×	×	×		×
Lifetime alcohol	×		×			
Education	×	×	×	×	×	
Age	×	×	×	×	×	×
Age^2				×		
Total energy			×	×		×
Weight			×	×		×
Center	×	×	×	×	×	×
Season		×	×	×	×	×
Sex	×	×		×	×	
Interaction terms						
Q^t * sex		×			×	
Q^t * age	×	×		×		×
Q^t * season			×		×	
Q^t * BMI				×	×	
Q^t * center	×	×	×	×	×	×

EPIC Study, 1992–2000.
Q^t is a transformed DQ; Part I, refers to consumption probability part of the two-part calibration model; Part II, refers to consumed amount part of the two-part calibration model;
*refers to an interaction term.

model with the standard deviation that is proportional to the mean [22]. To explore a suitable value for λ to identify the right GLM model, we plotted center-specific log-transformed standard deviation versus center-specific log-transformed mean, separately for each of the three vegetable subgroups as reported on 24-HDR in the EPIC study. Then λ is estimated as twice the slope of the fitted regression line. The GLM model considered here can accommodate family of densities with skewed (asymmetric) distributions. We chose to use graphical method to identify λ due to its simplicity as opposed to estimation methods such as the maximum likelihood (MLE).

Table 3. The area under the curve (AUC) from ROC curve for consumption probability (Part I), and root mean square error (RMSE) and mean bias for the consumed amount (Part II) of the standard and the reduced forms of two-part regression calibration models with transformed DQ.

Vegetable Subgroups	Part I		Part II	
	Models	AUC	RMSE[a]	Mean Bias[b]
Leafy	Standard	0.6846	66.841	0.0223
	Reduced	0.6843	64.578	0.0019
Fruiting	Standard	0.6305	118.823	0.0446
	Reduced	0.6304	110.415	−0.0334
Root	Standard	0.6413	68.626	0.0895
	Reduced	0.6408	66.524	0.0883

[a]$RMSE = \frac{1}{n}\sum_{i=1}^{n}\left(\hat{R}_i - R_i\right)^2$; [b]$mean\_bias = \frac{1}{n}\sum_{i=1}^{n}\left(\hat{R}_i - R_i\right)$

Nonlinearity and variable transformation

The relation between dietary intake variables is often nonlinear. To explore the form of relation between consumption probability as reported on 24-HDR and usual intake as reported on DQ, we applied two techniques: the empirical logit plot, and the nonparametric generalized additive model (GAM). With the empirical logit technique, we categorized DQ measurements, starting with the category of never-consumers followed by 10 g/day intake intervals. In each category, we computed the logit of consumption as reported on 24-HDR. The formula for the empirical logit transformation [20,23] of consumption used is given by

$$\log\left(\frac{y_i + 0.5}{n_i - y_i + 0.5}\right) \quad (5)$$

where y_i is the number of individuals who reported consumption on the 24-HDR and n_i is the number of individuals in the ith DQ-category. The addition of 0.5 to both the numerator and the denominator of the logit function serves to avoid indefinite empirical logit values when $y_i = n_i$ or $y_i = 0$, and this particular value minimizes the bias in estimating the log odds [20]. The estimated empirical logit is then plotted against the DQ category-specific means. We fitted a loess curve to the resulting scatterplots to have a visual inspection of the form of relation between the two variables [24]. We further made the empirical logit plots for each of the participating country in the EPIC study. With the nonparametric GAM technique, we obtained an optimal smoothing splines for the relation between the consumption probability, as reported on 24-HDR, and DQ and other continuous variables based on generalized cross validation criterion (GCV) [25]. We fitted the GAM model for consumption probability, assuming a binomial response and a logit link function using the mcgv package in R [26]. In the GAM model, we included confounding variables in the disease model ($\mathbf{Z}$). We used the partial prediction plot from the smoothed DQ component to identify plausible forms of parametric transformations for the DQ [27]. From the selected set of parametric transformations, Akaike Information Criterion (AIC) was used to identify the optimal transformation. Similar to the consumption probability part, we explored optimal form of DQ for the consumed amount part of the calibration model with the GAM approach.

Variables inclusion in the calibration model

The theory of regression calibration states that all confounding variables in the disease model must also be included in the calibration model in addition to the covariates that only predict dietary intake [14]. We used the same set of confounding variables in Agudo [3] that studied the relation between intake of vegetables and mortality in the Spanish cohort of EPIC. The eight confounding variables were: BMI (kg/m^2), smoking status (never, former, current smoker), physical activity index (inactive, moderately inactive, moderately active, active), lifetime alcohol consumption (g/day), level of education (none, primary, technical, secondary, university), age at recruitment (years), total energy (kcal), and sex (male/female).

The covariates that only predict intake as measured 24-HDR were selected based on their statistical significance in the calibration model (3). We included plausible two-way interaction terms of DQ measurements with the other covariates in the calibration model. We hereafter refer to each of the calibration model with covariates selected using the standard theory with the prefix "standard", here, standard two-part calibration model. The covariates are not only included once but twice in the two-part

calibration model (i.e., in each part of the two-part model), thus posing a threat to over fitting. Moreover, some disease confounding variables might not necessarily predict true usual intake conditional on DQ. We therefore conducted a backward elimination on the standard two-part calibration model based on a significance level α of 0.2. We chose 0.2 to ensure that no significant covariates are excluded from the model. We hereafter refer to each of the reduced version of the standard calibration model with the prefix "reduced", here, reduced two-part calibration model.

To assess the power of the probability part of the two-part calibration model to correctly discriminate consumers from non-consumers as reported by 24-HDR, we used the Area under the curve from the Receiver operating characteristic curve of the fitted logistic model [28]. For the consumed amount part, we assessed the predictive power of the model based on the root mean squared error and the mean bias [29]. In building the two-part calibration model, we conducted country-specific rather than center-specific regression calibration models to obtain stable estimates given the relatively smaller sample sizes in each center [10].

We also fitted other forms of regression calibration models to compare with the developed two-part calibration model. These forms of the calibration model include:

i.   A two-part calibration model similar to the developed one but with untransformed DQ. We hereafter refer to this model as "Two-part (untransformed DQ)". The aim of fitting this model was to assess the effect of nonlinearity on the performance of a two-part calibration model.

ii. A one-part calibration model with untransformed DQ and with the usual assumptions of a classical linear model. This is the calibration model commonly used by epidemiologists to adjust for the bias in the diet-disease association. In this model, two strong assumptions are made, namely, normality and linearity. The aim of fitting this calibration model was to quantify the inadequacy in adjusting for the bias in the diet-disease association when these assumptions are violated.

In each of the two forms of calibration models, we used the same set of covariates in each part of the standard two-part calibration but with different parametric forms of DQ as explained above. We conducted a backward elimination ($\alpha = 0.2$) on each of these forms of regression calibration models to obtain their reduced forms. Subsequently, we used a Cox proportional hazard model to study the association between usual intake of vegetable subgroups and all-cause mortality [30]. The Cox proportional hazards model was stratified by center and sex. To explore the form of relation between usual intake of each of the three vegetable subgroups and all-cause mortality in the Cox model, we plotted the log hazard ratio estimate against the DQ category-specific median intake [31].

We used bootstrap procedure to compute correct standard error for the log hazard ratio estimate. The bootstrap approach accounts for the uncertainty in the calibration process. We used center-stratified bootstrap procedure on the calibration sub-study. To each bootstrap sample, we added the main-study data and fitted regression calibration model to generate replicate versions of $E(R|Q,\mathbf{X})$ for each subject in the entire EPIC cohort [32]. To each replicate data, we fitted the Cox model yielding an estimate of log hazard ratio with a standard error. The within-calibration and between-calibration variances were combined using Rubin's formula to account for the uncertainty in the calibration process [33–35]. The Rubin's formula used to estimate the standard error for the log hazard ratio estimate is

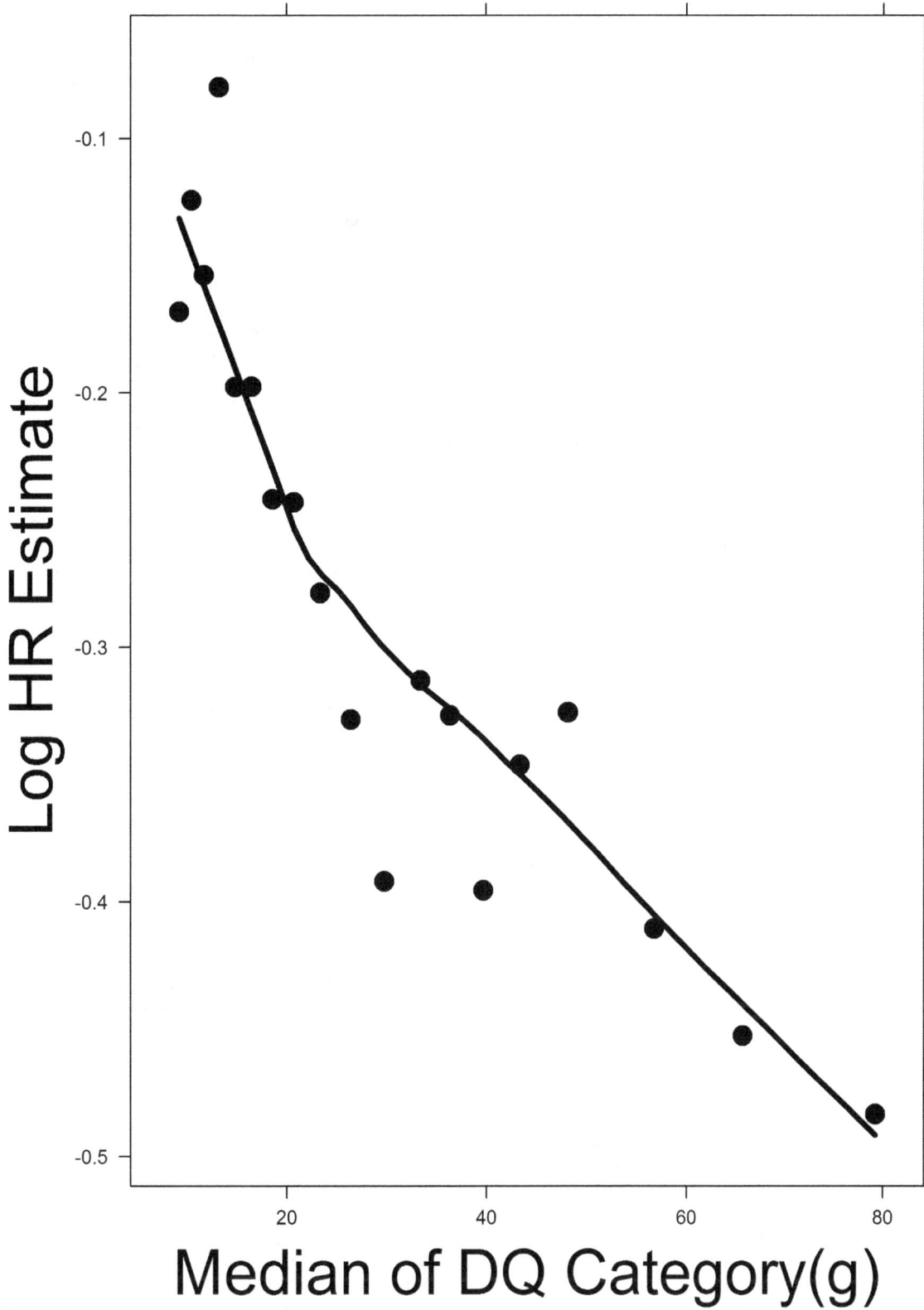

Figure 4. Linearity assessment in the Cox proportional hazards model for Leafy vegetables. The graph shows a smoothed curve fitted to the scatterplots of log hazard ratio estimate of leafy vegetable intake on all-cause mortality in each DQ category versus DQ category-specific median intake. The approximately linear downward trend suggests a possible linear relation and a beneficial effect of vegetable intake on the risk of all-cause mortality.

$$\sigma_T^2(\bar{\beta}) = \frac{1}{m}\sum_{i=1}^{m}\left(SE(\hat{\beta}_i)\right)^2 + \left(1+\frac{1}{m}\right)\left(\frac{1}{m-1}\right)\sum_{i=1}^{m}\left(\hat{\beta}_i-\bar{\beta}\right)^2 \quad (6)$$

where $\sigma_T^2(\bar{\beta})$ is the total variance of the mean of log hazard ratio estimate from m calibrated samples, $SE(\hat{\beta}_i)$ is the within-calibration standard error, and $\left(\frac{1}{m-1}\right)\sum_{i=1}^{m}\left(\hat{\beta}_i-\bar{\beta}\right)^2$ is the between-calibration variance.

We fitted a Cox proportional hazards model that ignores the measurement error in DQ. This method is hereafter referred to as the naïve method. In the naïve method, the DQ measurements were used to study the association between usual intake of a vegetable subgroup and all-cause mortality.

Results

Excess zeroes, heteroscedasticity and skewness in reference measurements

In Table 1, each of the three vegetable subgroups showed a high percentage of zero reference measurements as reported on the 24-HDR, especially for root vegetable subgroup in most of the participating countries. The rather high percentage of zero reference measurements suggests that these subgroups of vegetables are not consumed daily by everyone. The Pearson correlation coefficient for each of the three vegetable subgroups in each of the participating countries, as measured with 24-HDR and DQ, were rather low but mostly statistically significant. The boxplots for the distribution of the consumed amount on consumption events as reported on 24-HDR showed positive skewed distributions for these dietary variables (Figure 1). These exploratory findings suggested a need to properly handle the excess zeroes, to choose either a suitable distribution or a correct transformation for the consumed amount, as reported on 24-HDR in building a calibration model.

For each of the three vegetable subgroups, a linear trend was shown between the log of standard deviation and the log of the mean for the consumed amount (see Figure 2 for leafy vegetables). The linear trend is a clear evidence of a variance that increases with a mean (presence of heteroscedasticity). The slope (standard error) of least squares regression line fitted to the resulting scatterplots was estimated as 1.057 (0.085). For fruiting vegetables, the estimates were 0.994 (0.076) as shown in Figure S1. Likewise for root vegetables, the estimates were 1.021 (0.130) as shown in Figure S2. These slopes of the fitted lines were all close to the theoretical value of 1 for a GLM gamma model. Based on these exploratory findings, we chose a gamma GLM model for the consumed amount part of the two-part calibration model separately for each of the three vegetable subgroups. The correlation between each of the three vegetable subgroups ranged from 0.06 to 0.12 with total energy and from −0.07 to 0.05 with alcohol, as measured with DQ. These low correlations suggest minimal contamination effect of measurement error, hence justifying our choice not to adjust for the error in these variables.

Non-linearity and variable transformations

To explore the form of DQ in the part I model for the consumption probability, the loess curve fitted to the scatterplots of the empirical logit versus the DQ category-specific means presented in Figure 3 showed a nonlinear relation between the logit of consumption as reported on 24-HDR and the DQ reported intake (dotted lines). The partial prediction plots from the GAM approach showed similar behavior. From the plausible set of

parametric transformations for DQ, here, square-root and logarithmic, we chose log-transformed DQ based on the AIC criterion for each model fitted to country-specific data. As a result, we further fitted a logistic model with log-transformed DQ and computed mean of the predicted logit of consumption in each category of DQ. The loess curve fitted to the scatterplots of the mean of predicted logit against DQ category-specific means is shown in the same figure (continuous line). The similarity of the two loess curves suggested the aptness of log-transforming DQ in the part I model for consumption probability of leafy vegetables. The graphs for fruiting vegetables and root vegetables yielded similar results.

To explore the form of DQ in the part II model for the consumed amount part, we fitted a GAM model with gamma distributed error terms and a log link function (as suggested by exploratory results). Based on partial prediction plots for the smoothed DQ components and using the AIC criterion, we chose a square-root transformed DQ for both leafy vegetables and root vegetables subgroups, and a log-transformed DQ for fruiting vegetables.

Two-part regression model building

In addition to the confounding variables in the Cox model (under variables inclusion in the calibration model sub-section), season of DQ administration, center where the DQ was administered and the body weight of the participant were also included in the calibration model because they predicted intake of each of the three vegetable subgroups. Other covariates included in the standard two-part calibration model were the transformed DQ, and two-way interaction of transformed DQ with sex, age, season, BMI and center. We used the same set of covariates on each part of the standard two-part calibration model but with additional quadratic term for age at recruitment in the consumed amount part. In Table 2, we showed the remaining significant terms after a backward elimination on each part of the standard two-part calibration model separately for each of the three vegetable subgroups.

The areas under the curve from the ROC curve for the consumption probability part of the standard two-part calibration model and its reduced form were quite similar for each of the vegetable subgroups (Table 3). This suggest that some confounding variables and other two-way interaction terms of DQ with other covariates in the standard model do not necessarily predict the consumption probability and therefore should not be included in the calibration model.

A similar remark could be made for the consumed amount part of the model, based on the root mean squared error and the mean bias, which were quite similar.

The graphical exploration of the smoothed curve fitted to the scatterplots of the log hazard ratio estimate of dietary intake on all-cause mortality versus the DQ category-specific median intake showed approximately linear relations for each of the three vegetable subgroups as shown in Figure 4 for leafy vegetables, Figure S3 for fruiting vegetables, and Figure S4 for Root vegetables. We therefore assumed a linear term for DQ in the three fitted Cox proportional hazards models.

As expected, the log hazard ratio estimate for usual intake in the Cox model adjusted for measurement error in DQ were larger in absolute value than the naive estimate that ignores the measurement error. Similar remark was made for all the fitted forms of regression calibration models but the standard two-part calibration model with untransformed DQ (Table 4). The log hazard ratio estimates adjusted for the bias with the standard calibration models were smaller than those adjusted with the reduced

Table 4. Log hazard ratio estimate (standard error) per 100 g usual intake of each of the three vegetable subgroups, calibrated with each of the three forms of regression calibration models in their reduced and standard forms.

Vegetable Subgroups	Calibration methods	Reduced form		Standard form	
		$\hat{\beta}$ (s.e^a; s.e^b)	s.e ratio^c	$\hat{\beta}$ (s.e^a; s.e^b)	s.e ratio^c
Leafy	Naïve method	−0.144 (0.027)		−0.144 (0.027)	
	One-part linear calibration	−0.480 (0.090; 0.112)	1.24	−0.409 (0.083; 0.127)	1.53
	Two-part (untransformed DQ)	−0.395 (0.092; 0.183)	1.99	−0.174 (0.089; 0.278)	3.11
	Two-part (transformed DQ)	−0.509 (0.090; 0.292)	3.24	−0.461 (0.047; 0.160)	3.41
Fruiting	Naïve method	−0.094 (0.014)		−0.094 (0.014)	
	One-part linear calibration	−0.125 (0.031; 0.034)	1.11	−0.123 (0.031; 0.034)	1.11
	Two-part(untransformed DQ)	−0.161 (0.030; 0.034)	1.14	−0.109 (0.030; 0.073)	2.42
	Two-part (transformed DQ)	−0.255 (0.037; 0.108)	2.92	−0.228 (0.035; 0.131)	3.74
Root	Naïve method	−0.160 (0.026)		−0.16 (0.026)	
	One-part linear calibration	−0.342 (0.060; 0.082)	1.36	−0.305 (0.054; 0.077)	1.43
	Two-part(untransformed DQ)	−0.203 (0.088; 0.219)	2.49	−0.107 (0.060; 0.167)	2.78
	Two-part (transformed DQ)	−0.479 (0.070; 0.214)	3.06	−0.265 (0.056; 0.181)	3.23

s.e^a is the standard error ($\times 10^{-2}$) for $\hat{\beta}$ that does not account for the uncertainty in the calibration; s.e^b is the standard error ($\times 10^{-2}$) that accounts for the uncertainty in the calibration; s.e ratio^c is the ratio of s.e^b to s.e^a.

calibration models, e.g., -0.265 for the standard two-part (transformed DQ) and -0.479 for the reduced two-part (transformed DQ) calibration model per 100 g intake of root vegetables. The poor performance of the standard calibration models might be due to over fitting by covariates that did not significantly predict usual intake of vegetable sub-groups. The log hazard ratio estimate adjusted with the standard two-part calibration model was even smaller than the naïve estimate. This shows that a poorly specified calibration model can result in adjusted association estimates that are more biased than the unadjusted estimates. The standard error of the log hazard ratio estimate corrected for the uncertainty in the calibration was larger than the uncorrected one for each of the calibration models presented. This means that ignoring uncertainty in the calibration underestimates the standard error. The underestimation of standard error was more severe for the standard calibration models. Further, the log hazard ratio estimate calibrated with the reduced one-part linear calibration model was smaller than that obtained with the reduced two-part (transformed DQ) model. The seemingly poor performance of one-part linear calibration model suggests that a poorly specified calibration model does not adequately adjust for the bias in the diet-disease association. Further, the predicted intake values for some subjects not in the calibration sub-study, in some cases were rather unrealistic. The unrealistic predictions were mainly from the standard calibration model with untransformed DQ. The calibration models with untransformed DQ resulted in a much smaller log hazard ratio estimate than their counterparts with transformed DQ. This might be driven by extreme prediction from highly skewed DQ measurements in the calibration model. The effect of the extreme DQ values was further compounded by two factors: including the same covariate twice in the two-part calibration model and by the exponentiation effect due to the log link function used to fit the calibration model. As a result, we conducted a small sensitivity analysis where the unreasonably high predicted values were retained in the Cox model. Including these high predicted values resulted in massive change in the log hazard ratio estimate mainly with standard two-part calibration model with untransformed DQ. For leafy vegetables, for instance, including the unrealistic predictions from the standard two-part calibration model with untransformed DQ changed the estimate of log hazard ratio from -0.174 to -0.00518 per 100 g intake. In Table S1, we present the percentages of these unrealistic predictions, defined as extreme if it exceeded fivefold the ninety ninth percentile of the predicted usual intake. In the final analysis, we excluded these unrealistic values.

Discussion

In this work, we adapted a two-part regression calibration model initially developed for multiple 24-HDR measurements per individual for episodically consumed foods to a single replicate setting. We focused on dietary intake data that are skewed, heteroscedastic, and with substantial percentage of zeroes as reported on the 24-HDR. We further described how to explore and identify a suitable GLM model and a correct parametric form of a continuous covariate in the calibration model. As a result, we applied flexible GLM models that could simultaneously handle the skewness and heteroscedasticity in the consumed amount. Thus, we avoided complications resulting from data transformation. We chose the log link function to stabilize the variance and to ensure positive prediction for usual intake [36].

The standard way of including variables in the calibration model states that all confounding variables in the disease model and those that only predict dietary intake but not the disease occurrence must be included in the calibration model. Given the complexity of the two-part calibration model, some confounding variables in the disease model do not necessarily predict dietary intake. This could pose a threat to over fitting the calibration model. We further conducted a backward elimination on each part of the two-part calibration model separately. The reduced calibration model with only significant covariates outperformed its standard counterpart in adjusting for the association bias. Leaving out confounding variables from the calibration model is against the standard theory of regression calibration. Nevertheless, we argue that if the omitted covariates have no effect in the calibration model, they should be excluded and the calibration method should still be correct. We further found out that assuming linearity when it does not hold in a calibration model can pose a serious threat to the bias-adjustment of the association parameter. The association parameter estimate adjusted for the bias with a poorly specified calibration model can sometime be worse than the unadjusted estimates. Thoresen [37] also found, in a simulation study, that a less accurately specified calibration model can have a considerable impact on the degree of bias-adjustment. We observed that predicted values for some subjects not in the calibration sub-study were extremely large. The extreme predictions resulted mainly from standard calibration models with linear DQ as a covariate. In such a case, predictions are made outside the variable space on which the model is fitted. Due to the curse of dimensionality, the prediction space would extend more outside the variable space in the complex models.

The consumption probability and the consumed amount for episodically consumed foods may be correlated. In each of the fitted two-part calibration models, we accounted for this correlation partly by allowing covariates to overlap on both parts of the calibration model [14]. With only a single 24-hour recall measurement per subject, any further correlation cannot be estimated. In future studies, a sensitivity analysis can be performed to assess the effect of the unaccounted part of the correlation. This can be done by varying the magnitude of the assumed positive correlation between the consumption probability and the consumed amount.

A limitation of this study is that we made some strong assumptions. First, we assumed the 24-HDR to be unbiased measurement of true usual intake. Second, we assumed that the errors in the 24-HDR are uncorrelated with the errors in the DQ. However, previous studies have shown that these assumptions may not hold for dietary self-report instruments, and that, use of 24-HDR as a reference instrument for vegetable intake may be flawed [4,38–40]. The biomarker studies using doubly labelled water for energy intake and urinary nitrogen for protein intake suggest that self-reports on recalls or food records may be biased. This is because individuals may systematically differ in their reporting accuracy. Additionally, the errors in these short-term instruments are shown to be positively correlated with the errors in the DQ [41]. As a result, using 24-HDR as a reference instrument can seriously underestimate true attenuation [42]. Therefore, the results obtained with the 24-HDR as reference instrument should be interpreted with caution. Nevertheless, the bias in 24-HDR is reported to be substantially less severe than that in the DQ [38]. Thus, when there is no objective biomarker measurements for dietary intake, using 24-HDR may still provide the best possible estimation of true intake [14].

In summary, a correctly specified two-part regression calibration model, which fits the data better, can adequately adjust for the bias in the diet-disease association, when only a single reference measurement is available per individual. Further, the ability to adjust for the bias is influenced considerably by the form of the specified calibration model. We therefore advise researchers to pay special attention to calibration model specification, with respect to the response distribution and the form of the covariates.

Supporting Information

Figure S1 The variance-mean relation for Fruiting vegetables (FV). The graph shows a least squares regression line fitted to the scatterplots of the logarithm of center-specific standard deviation versus logarithm of center-specific mean of the consumed amount of fruiting vegetables for those who reported consumption on the 24HDR in the EPIC Study, 1992–2000. The approximately linear regression line suggests a variance that increases with the mean.

Figure S2 The variance-mean relation for Root vegetables (RV). The graph shows a least squares regression line fitted to the scatterplots of the logarithm of center-specific standard deviation versus logarithm of center-specific mean of the consumed amount of root vegetables for those who reported consumption on the 24HDR in the EPIC Study, 1992–2000. The approximately linear regression line suggests a variance that increases with the mean.

Figure S3 Linearity assessment in the Cox proportional hazards model for Fruiting vegetables. The graph shows a smoothed curve fitted to the scatterplots of log hazard ratio estimate of fruiting vegetable intake on all-cause mortality in each

DQ category versus DQ category-specific median intake. The approximately linear downward trend suggests a possible linear relation and a beneficial effect of fruiting vegetable intake on the risk of all-cause mortality.

Figure S4 Linearity assessment in the Cox proportional hazards model for Root vegetables. The graph shows a smoothed curve fitted to the scatterplots of log hazard ratio estimate of root vegetable intake on all-cause mortality in each DQ category versus DQ category-specific median intake. The approximately linear downward trend suggests a possible linear relation and a beneficial effect of root vegetable intake on the risk of all-cause mortality.

Table S1 Unrealistic predicted usual intake of vegetable subgroups. The table displays the maximum and the ninety-ninth percentile of predicted usual intake and percentage (number) of unrealistic predictions (i.e., unrealistic if greater than five times ninety-ninth percentile of predicted intake) using different forms of regression calibration models; each model in its standard form, that is, with the covariates selected using the standard theory, and also in the reduced form, that is, with covariates that significantly predict intake.

Acknowledgments

We are grateful to Dr. Paul Goedhart of Wageningen University and Research Center, Netherlands, for his advice on the statistical methodology used to answer the research question.

Author Contributions

Analyzed the data: GOA. Contributed to the writing of the manuscript: GOA HB HV FE PF ES CB SK DM ML PV TB IJ.

References

1. Kaaks R, Ferrari P, Ciampi A, Plummer M, Riboli E (2002) Uses and limitations of statistical accounting for random error correlations, in the validation of dietary questionnaire assessments. PUBLIC HEALTH NUTRITION 5: 969–676.
2. Willet W (1998) Nutritional Epidemiology. New York: Oxford University Press.
3. Agudo A (2004) Measuring intake of fruit and vegetables. Background paper for the Joint FAO/WHO Workshop on Fruits and Vegetables. Kobe, Japan: WHO.
4. Kipnis V, Subar AF, Midthune D, Freedman LS, Ballard-Barbash R, et al. (2003) Structure of dietary measurement error: results of the OPEN biomarker study. American Journal of Epidemiology 158: 14–21; discussion 22–16.
5. Kaaks R (1997) Biochemical markers as additional measurements in studies of the accuracy of dietary questionnaire measurements:conceptual issues. American Journal of Clinical Nutrition: 1232s–1239s.
6. Fraser GE, Stram DO (2001) Regression calibration in studies with correlated variables measured with error. American Journal of Epidemiology 154: 836–844.
7. Freedman LS, Midthune D, Carroll RJ, Kipnis V (2008) A comparison of regression calibration, moment reconstruction and imputation for adjusting for covariate measurement error in regression. Stat Med 27: 5195–5216.
8. Slimani N, Kaaks R, Ferrari P, Casagrande C, Clavel-Chapelon F, et al. (2002) European Prospective Investigation into Cancer and Nutrition (EPIC) calibration study: rationale, design and population characteristics. Public Health Nutrition 5: 1125–1145.
9. Ferrari P, Kaaks R, Fahey MT, Slimani N, Day NE, et al. (2004) Within- and between-cohort variation in measured macronutrient intakes, taking account of measurement errors, in the European Prospective Investigation into cancer and nutrition study. American Journal of Epidemiology 160: 814–822.
10. Ferrari P, Day NE, Boshuizen HC, Roddam A, Hoffmann K, et al. (2008) The evaluation of the diet/disease relation in the EPIC study: considerations for the calibration and the disease models. International Journal of Epidemiology 37: 368–378.
11. Slimani N, Kaaks R, Ferrari P, Casagrande C, Clavel-Chapelon F, et al. (2002) European Prospective Investigation into Cancer and Nutrition (EPIC) calibration study: rationale, design and population characteristics. Public Health Nutrition 5: 1125–1145.
12. Olsen M, Schafer J (2001) A Two-Part Random-Effects Model for Semicontinuous Longitudinal Data. Journal of the American Statistical Association 96: 730–745.
13. Tooze J, Midthune D, Dodd K, Freedman L, Krebs-Smith S, et al. (2006) A new method for estimating the usual intake of episodically consumed foods with application to their distribution. Journal of American Diet Association: 1575–1587.
14. Kipnis V, Midthune D, Buckman DW, Dodd KW, Guenther PM, et al. (2009) Modeling data with excess zeros and measurement error: application to evaluating relationships between episodically consumed foods and health outcomes. Biometrics 65: 1003–1010.
15. Zhang S, Midthune D, Guenther PM, Krebs-Smith SM, Kipnis V, et al. (2011) A New Multivariate Measurement Error Model with Zero-Inflated Dietary Data, and Its Application to Dietary Assessment. Ann Appl Stat 5: 1456–1487.
16. Carroll RJ, Ruppert D, Stefanski LA, Crainiceanu CM (2006) Measurement Error in Nonlinear Models. New York: Chapman & Hall/CRC.
17. Riboli E, Hunt KJ, Slimani N, Ferrari P, Norat T, et al. (2002) European prospective investigation into cancer and nutrition (EPIC): study populations and data collection. Public Health Nutrition 5: 1113–1124.
18. Riboli E, Hunt KJ, Slimani N, Ferrari P, Norat T, et al. (2002) European Prospective Investigation into Cancer and Nutrition (EPIC): study populations and data collection. Public Health Nutrition 5: 1113–1124.
19. Slimani N, Valsta L, Grp E (2002) Perspectives of using the EPIC-SOFT programme in the context of pan-European nutritional monitoring surveys: methodological and practical implications. European Journal of Clinical Nutrition 56: S63–S74.
20. McCullagh P, Nelder JA (1989) Generalized linear models. London; New York: Chapman and Hall. xix, 511 p.
21. Manning WG, Basu A, Mullahy J (2005) Generalized modeling approaches to risk adjustment of skewed outcomes data. Journal of Health Economics 24: 465–488.

22. Manning W, Mullahy J (2001) Estimating log models: to transform or not to transform. Journal of Health Economics 20: 461–494.

23. Cox DR (1970) The analysis of binary data. London,: Methuen. 142 p.

24. Weiss J (2006) Statistical Analysis. Ecology 145–Statistical Analysis. University of North Carolina, Chapel Hill: University of North Carolina.

25. Hastie T, Tibshirani R (1999) Generalized additive models. Boca Raton, Fla.: Chapman & Hall/CRC. xv, 335 p.

26. Wood SN (2012) Mixed GAM Computation Vehicle with GCV/AIC/REML smoothness estimation 1.7–22 ed.

27. Cai W (2008) Fitting Generalized Additive Models with the GAM Procedure in SAS 9.2. SAS Global Forum 2008. SAS Campus Drive, Cary, NC 27513: SAS Institute Inc.

28. Steyerberg EW (2009) Clinical prediction models: a practical approach to development, validation, and updating. New York, NY: Springer. xxviii, 497 p.

29. Hastie T, Tibshirani R, Friedman JH (2009) The elements of statistical learning: data mining, inference, and prediction. New York, NY: Springer. xxii, 745 p.

30. Cox DR (1972) Regression Models and Life-Tables. Journal of the Royal Statistical Society Series B-Statistical Methodology 34: 187-&.

31. Sainani KL (2009) Linearity assessment in the Cox model. Stanford University. pp. Linearity assessment in the Cox model.

32. Cassell DL (2007) Don't Be Loopy: Re-Sampling and Simulation the SAS® Way. SAS Global Forum: Statistics and Data Analysis. Corvallis, Orlando, Florida.

33. Rubin DB (2004) Multiple imputation for nonresponse in surveys. Hoboken, N.J.: Wiley-Interscience. xxix, 287 p.

34. Geert Molenberghs MGK (2007) Missing Data in Clinical Studies. West Sussex: John Wiley & Sons.

35. Boshuizen HC, Lanti M, Menotti A, Moschandreas J, Tolonen H, et al. (2007) Effects of past and recent blood pressure and cholesterol level on coronary heart disease and stroke mortality, accounting for measurement error. American Journal of Epidemiology 165: 398–409.

36. Raymond HM, Montgomery DC, Vining GG (2010) Generalized Linear Models With Applications in Engineering and the Sciences. Hoboken, New Jersey: John Wiley & Sons, Inc.

37. Thoresen M (2006) Correction for measurement error in multiple logistic regression: A simulation study. Journal of Statistical Computation and Simulation 76: 475–487.

38. Kipnis V, Midthune D, Freedman LS, Bingham S, Schatzkin A, et al. (2001) Empirical evidence of correlated biases in dietary assessment instruments and its implications. American Journal of Epidemiology 153: 394–403.

39. Natarajan L, Pu MY, Fan JJ, Levine RA, Patterson RE, et al. (2010) Measurement Error of Dietary Self-Report in Intervention Trials. American Journal of Epidemiology 172: 819–827.

40. Natarajan L, Flatt SW, Sun XY, Gamst AC, Major JM, et al. (2006) Validity and systematic error in measuring carotenoid consumption with dietary self-report instruments. American Journal of Epidemiology 163: 770–778.

41. Day NE, McKeown N, Wong MY, Welch A, Bingham S (2001) Epidemiological assessment of diet: a comparison of a 7-day diary with a food frequency questionnaire using urinary markers of nitrogen, potassium and sodium. International Journal of Epidemiology 30: 309–317.

42. Keogh RH, White IR, Rodwell SA (2013) Using surrogate biomarkers to improve measurement error models in nutritional epidemiology. Statistics in medicine 32: 3838–3861.

19

Seasonal Changes in *Thrips tabaci* Population Structure in Two Cultivated Hosts

Brian A. Nault*, Wendy C. Kain, Ping Wang

Department of Entomology, Cornell University, New York State Agricultural Experiment Station, Geneva, New York, United States of America

Abstract

Thrips tabaci is a major pest of high-value vegetable crops and understanding its population genetics will advance our knowledge about its ecology and management. Mitochondrial cytochrome oxidase subunit I (COI) gene sequence was used as a molecular marker to analyze *T. tabaci* populations from onion and cabbage fields in New York. Eight COI haplotypes were identified in 565 *T. tabaci* individuals collected from these fields. All *T. tabaci* were thelytokous and genetically similar to those originating from hosts representing seven plant families spanning five continents. The most dominant haplotype was NY-HT1, accounting for 92 and 88% of the total individuals collected from onion fields in mid-summer in 2005 and 2007, respectively, and 100 and 96% of the total in early fall in 2005 and 2007, respectively. In contrast, *T. tabaci* collected from cabbage fields showed a dynamic change in population structure from mid-summer to early fall. In mid-summer, haplotype NY-HT2 was highly abundant, accounting for 58 and 52% of the total in 2005 and 2007, respectively, but in early fall it decreased drastically to 15 and 7% of the total in 2005 and 2007, respectively. Haplotype NY-HT1 accounted for 12 and 46% of the total in cabbage fields in mid-summer of 2005 and 2007, respectively, but became the dominant haplotype in early fall accounting for 81 and 66% of the total in 2005 and 2007, respectively. Despite the relative proximity of onion and cabbage fields in the western New York landscape, *T. tabaci* populations differed seasonally within each cropping system. Differences may have been attributed to better establishment of certain genotypes on specific hosts or differing colonization patterns within these cropping systems. Future studies investigating temporal changes in *T. tabaci* populations on their major hosts in these ecosystems are needed to better understand host-plant utilization and implications for population management.

Editor: Cesar Rodriguez-Saona, Rutgers University, United States of America

Funding: The authors have no support or funding to report.

Competing Interests: The authors have declared that no competing interests exist.

* Email: ban6@cornell.edu

Introduction

Thrips tabaci Lindeman is a major insect pest of multiple crops worldwide including two high-value vegetable crops, onion, *Allium cepa* L., and cabbage, *Brassica oleracea capitata* (L.) [1]. *Thrips tabaci* is known to have multiple biotypes with different reproduction modes, host preferences and virus-transmission competencies [2–6]. Genetic analysis using molecular markers suggests that *T. tabaci* is a complex of cryptic species [7]. Genetic differences between *T. tabaci* populations from tobacco and populations from onion and leek have been documented [7], [8]. Understanding the genetic structure of *T. tabaci* populations in cropping systems will advance our knowledge about its population ecology, which can be important for developing effective management strategies [9].

In New York (USA), *T. tabaci* infestations in onion and cabbage fields are routinely managed with applications of insecticides. Insecticide resistance in *T. tabaci* populations in onion fields has been widespread [10], [11]. In contrast, control failures with similar insecticides have not been reported in New York cabbage fields, suggesting that populations in cabbage fields remain susceptible to these products. For example, neonicotinoid insecticides such as imidacloprid and acetamiprid continue to effectively control populations of *T. tabaci* in cabbage [12], but

have been ineffective against *T. tabaci* infestations in onion [13–15]. In addition, seasonal susceptibility shifts in *T. tabaci* populations in onion fields to insecticides have also been documented [10], [11]. Some populations were susceptible to insecticides in mid-summer, but tested resistant to them in early fall, and vice versa. These observations suggest that there may be host-associated differences among *T. tabaci* populations and that they may change during the season.

Knowledge about *T. tabaci* population genetics is important for developing management strategies, especially those that mitigate insecticide resistance. Mitochondrial DNA markers have been successfully used to identify genetic differentiation among *T. tabaci* populations [4–7], [16], [17]. In our study, molecular markers from the mitochondrial cytochrome oxidase subunit I (COI) gene were used to analyze *T. tabaci* populations collected from onion and cabbage fields in New York to examine 1) the differentiation between *T. tabaci* populations from these crops and how they relate to conspecifics from other reproductive modes, hosts and locations, and 2) seasonal changes in *T. tabaci* population genetic structure from onion and cabbage cropping systems. Our results revealed a commonality in the most abundant *T. tabaci* haplotypes from onion and cabbage fields and these haplotypes were genetically similar to those originating from an array of hosts and locations around the world. All haplotypes from onion and cabbage fields were in the same lineage as those that reproduce via thelytoky; none were in the arrhenotokous or "tobacco type" clades. The population genetic structure of *T. tabaci* populations in onion fields was similar between mid-summer and early fall, whereas the population structure in cabbage fields changed substantially between these periods.

Materials and Methods

Ethics statement

Vegetable growers gave us verbal permission to collect thrips on their privately owned land and no permits were required to collect this common insect. Sampling did not involve regulated, endangered or protected species.

Insect collection

Thrips tabaci adults were collected from commercial onion and cabbage fields in western New York in 2005 and 2007. Onion and cabbage fields were sampled from two of the largest onion and cabbage production regions in New York (Figure 1). Regions were separated from one another by at least 48 km and onion and cabbage fields within a region were separated by a minimum of 3.2 km. The landscape surrounding onion and cabbage fields differed. Onion fields were located on muck soil and ranged in size from 2 to 5 ha. These fields were bordered by several to many other onion fields in contiguous plantings that ranged from approximately 40 ha in Wayne Co., 240 ha in Yates Co, and 1,000 ha in Orleans/Genesee Counties. These onion monocultures were all bordered by woods. Cabbage fields were located on mineral soil and were often bordered by other crops such as corn, soybean, alfalfa and small grains and occasionally woods. Cabbage fields ranged from 12 to 20 ha.

Thrips tabaci populations were sampled from onion and cabbage fields during mid-summer and early fall. In mid-summer 2005 and 2007, sampling occurred between July 15–29 and July 17–26, respectively. In early fall 2005 and 2007, sampling occurred between Sept 8–19 and Sept 5–19, respectively. In 2005 and 2007, two of the same onion fields were sampled in mid-summer and early fall. The third onion field in each year was harvested early, so an adjacent onion field was sampled in early fall (Figure 1). One cabbage field was sampled both in mid-summer and early fall in 2005, whereas all others in 2005 and 2007 were harvested in August and cabbage fields in relative proximity to each other (e.g., within 3.2 km) were selected for sampling in early fall (Figure 1).

No more than one adult thrips was sampled per plant and plants were at least 1.5 m apart. Adults were removed using a fine-tipped paintbrush and then placed into vials containing 95% ethanol; adults from the same field and sampling period were placed into the same vial. The total number of *T. tabaci* adults sampled was 229 and 336 in 2005 and 2007, respectively. In onion, 129 and 165 adults were sampled in 2005 and 2007, respectively; in cabbage, 100 and 171 adults were sampled in 2005 and 2007, respectively. Voucher specimens are located at Cornell University's Department of Entomology, New York State Agricultural Experiment Station, Geneva, NY.

PCR amplification of a *COI* fragment from *Thrips tabaci*

Thrips preserved in ethanol were rinsed with de-ionized water to remove ethanol and then DNA from individual thrips was isolated with a rapid genomic DNA preparation method [18]. A single thrips was ground in 10 μl of lysis buffer containing 50 mM KCl, 2.5 mM $MgCl_2$, 0.45% Nonidet P-40, 0.45% Tween 20,

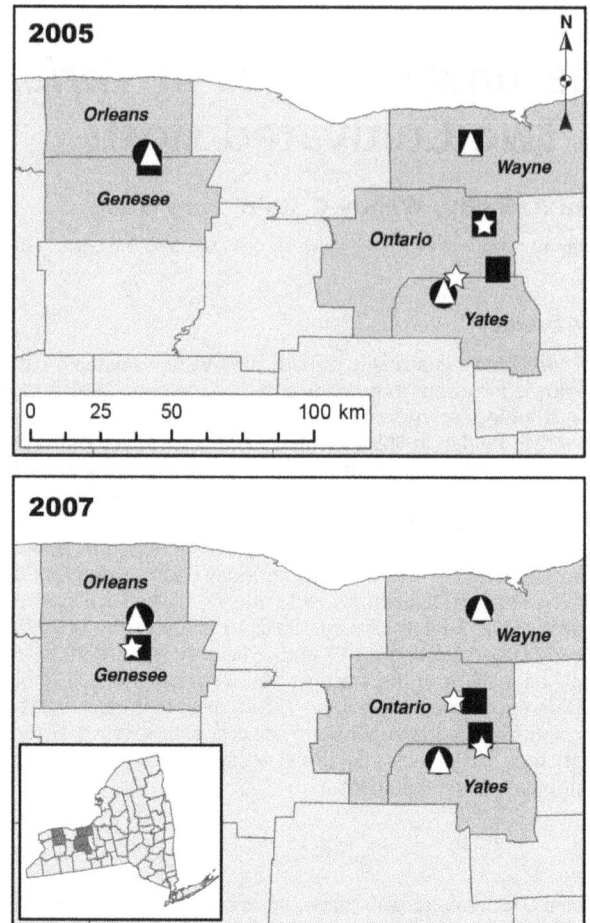

Figure 1. Map of onion and cabbage fields where *Thrips tabaci* adults were sampled. The following symbols indicate the season and crop sampled: ● = mid-season, onion; △ early fall, onion; ■ = mid-season, cabbage; * = early fall, cabbage.

0.01% gelatin and 60 μg/ml protease K in 10 mM Tris-HCl (pH 8.3), and incubated at 65°C for 30 min and then heated at 95°C for 15 min to inactivate the protease K activity in the lysate. The lysate was used as the DNA template for PCR amplification of a 706 bp fragment of mtCOI, using the universal COI primer pair for insects, LepF: ATTCAACCAATCATAAAGATATTGG and LepR: TAAACTTCTGGATGTCCAAAAAATCA [19]. Twenty five μl PCR reactions were prepared to contain 1 μl of thrips DNA prepared above, 2.5 U of Taq DNA polymerase (New England Biolabs, Beverly, MA), 0.4 mM of dNTPs, 0.4 μM of each primer and 2.5 μl of 10x Taq DNA polymerase buffer (New England Biolabs, Beverly, MA). The PCR reaction was performed by a denaturation incubation at 94°C for 1 min, followed by 40 cycles of 94°C for 30 sec, 52°C for 30 sec, and 72°C for 40 sec, and a final extension at 72°C for 10 min. The PCR products from each individual thrips were examined by 1% agarose gel electrophoresis to confirm the correct amplification of the COI fragment from the individuals analyzed.

DNA sequencing and sequence analysis

The PCR products were purified for DNA sequencing using a one-step enzymatic purification method [20]. Five μl of PCR product was mixed with 1 μl of enzyme solution containing 0.5 unit of shrimp alkaline phosphatase (USB, Cleveland, OH) and

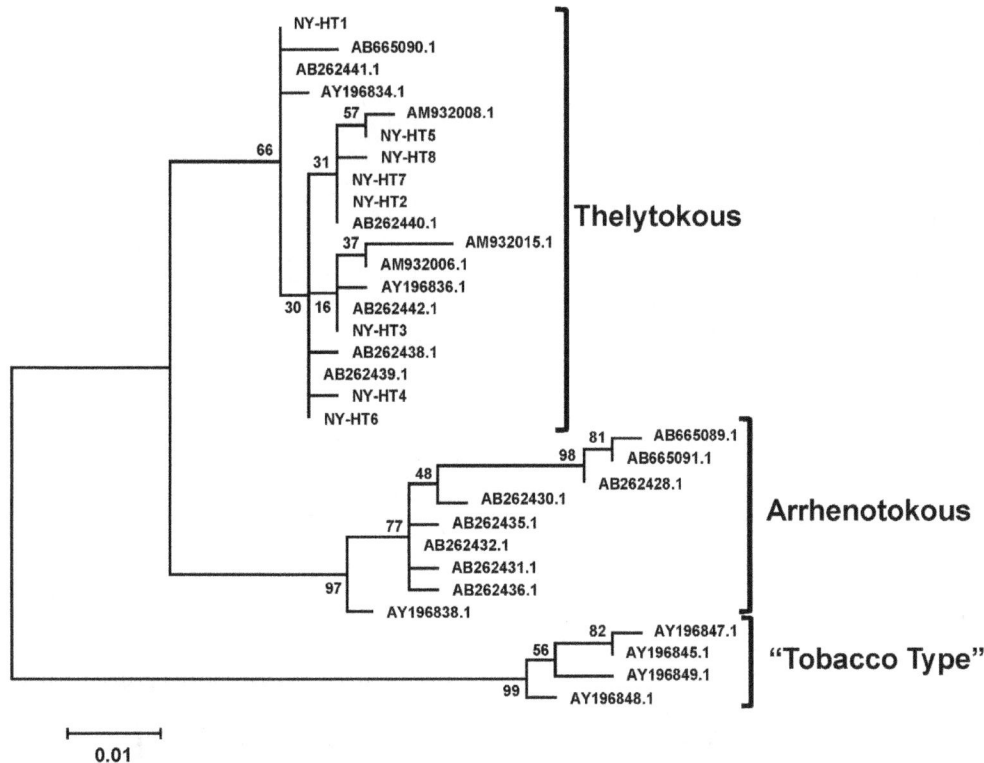

Figure 2. Phylogenetic tree of *Thrips tabaci* **COI sequences collected from onion and cabbage fields in New York (NY-HT1 to NY-HT8) relative to others with known reproductive mode.** The tree was generated by the maximum likelihood method based on the Jukes-Cantor model [23] using MEGA5 [24]. Sequence positions containing gaps and missing data were eliminated and bootstrap values (percentages of 1000 replicates) were shown next to the branches.

1.25 unit of exonuclease I (USB) and incubated at 37°C for 60 min, followed by incubation at 90°C for 10 min to inactivate the enzymes. The enzyme-treated PCR products were sequenced using the BigDye Terminator v3.1 Cycle Sequencing Kit (Applied Biosystems, Foster City, CA) following instructions provided by the manufacturer. The extension products from the sequencing reactions were purified by gel filtration with Sephadex G-50 (GE Healthcare, Pittsburgh, PA) using spin columns in a FiltrEX 96 well filter plate (Corning, Corning, NY) to remove unincorporated dye terminators. The final DNA sequence reading was performed on an Applied Biosystems 3730xl DNA Analyzer by the Genomics Facility in the Biotechnology Resource Center, Cornell University, Ithaca, NY. The COI fragments from the thrips individuals were sequenced in both strands.

DNA sequences from thrips individuals were assembled and aligned to identify single nucleotide polymorphisms (SNPs), using the DNASTAR Lasergene software suite (DNASTAR, Madison, WI). For analysis of the phylogenetic relationships of the *T. tabaci* COI haplotypes identified in this study and from public databases, a BLASTN search [21] against the NCBI GenBank [22] nucleotide database was performed to obtain COI sequences that had been deposited in the GenBank. For the identical sequences from different GenBank accessions, only one accession was selected and used for phylogenetic sequence analysis. Phylogenetic analyses were performed using the maximum likelihood method based on the Jukes-Cantor model [23] and molecular evolutionary genetics analysis software MEGA 5.2 [24]. For analysis of host plant association and seasonal changes in thrips population genetics, haplotype networks for the thrips populations were constructed based on statistical parsimony [25], using the software

TCS 1.21 [26]. For analysis of *T. tabaci* populations from cabbage and onion fields collected during the two seasons, the population fixation index F_{ST} was calculated using the software ARLEQUIN 3.5.1.2 [27] for all pairwise population comparisons at $P<0.01$.

Results

Thrips tabaci *COI* haplotypes

From 565 *T. tabaci* individuals sequenced, 10 SNP sites were identified within the 655 bp COI fragment (excluding the primer sequence regions) (Figure S1), yielding eight haplotypes, NY-HT1 to NY-HT8 (GenBank accession numbers: KF036290 to KF036297, respectively). Five haplotypes (NY-HT1, NY-HT2, NY-HT3, NY-HT5 and NY-HT8) were found in samples collected from onion and seven haplotypes (NY-HT1, NY-HT2, NY-HT3, NY-HT4, NY-HT5, NY-HT6 and NY-HT7) were found in samples collected from cabbage (Tables S1 and S2). The SNPs in the 655 bp fragment were all synonymous substitutions, except an A/G substitution at nucleotide position 98 (Figure S1), resulting in a change between two amino acid residues, isoleucine and valine.

Nucleotide polymorphisms at positions 355 and 515 (Figure S1) differentiate the two reproductive types of *T. tabaci*, arrhenotoky and thelytoky [4]. In all eight haplotypes, the nucleotides at 355 and at 515 were G and T, respectively, indicating that the *T. tabaci* individuals sampled in this study were all thelytokous (Figure S1). Further analysis of the eight haplotypes with the COI sequences from known reproductive types of *T. tabaci* showed that the eight haplotypes were in the same cluster with the thelytokous type (Figure 2).

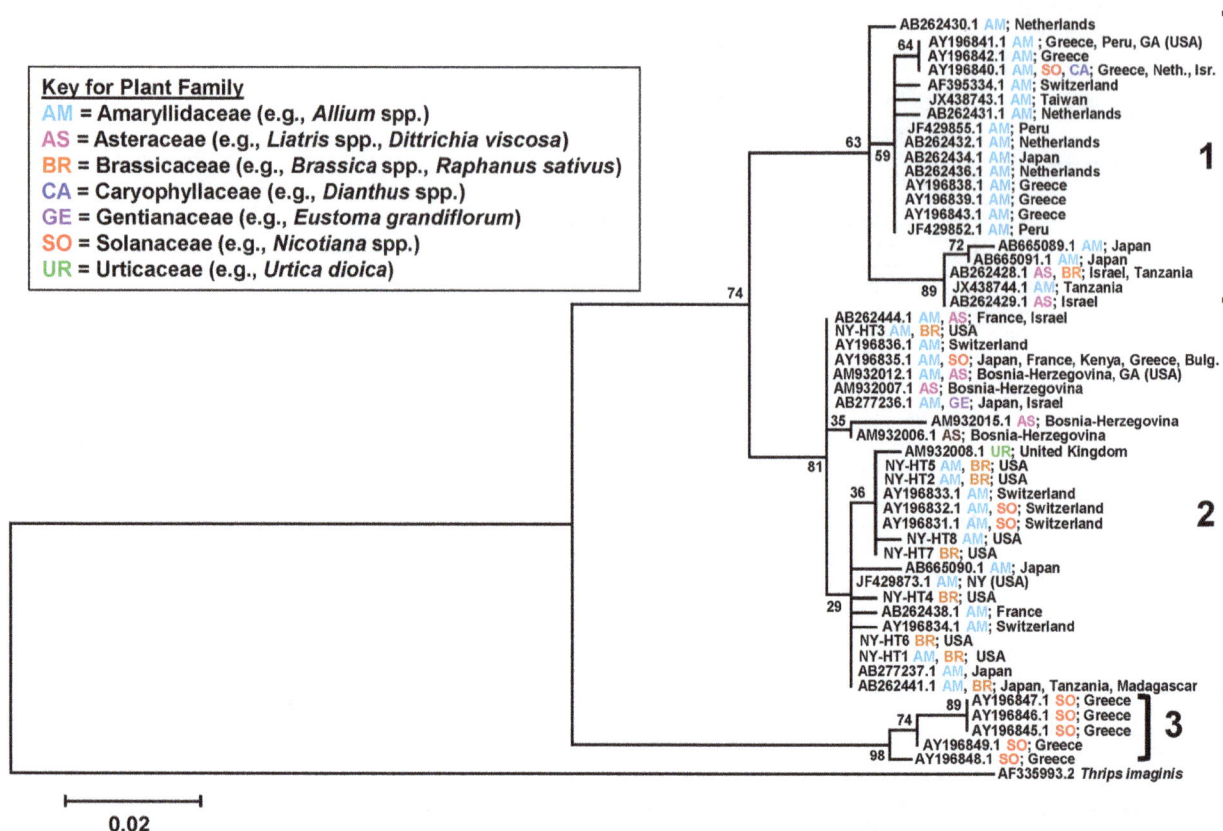

Figure 3. Phylogenetic tree of *Thrips tabaci* **COI sequences collected from onion and cabbage fields in New York (NY-HT1 to NY-HT8) relative to others collected from an array of host plants from around the world.** The Maximum Likelihood method was used for constructing the phylogenetic tree. Genbank accession number, host plant and country where the specimen originated are included for each entry in the tree. *Thrips imaginis* was included as an outgroup and the "tobacco type" as well. Bootstrap values from 1000 replicates are shown above branches.

Phylogenetic analysis of *Thrips tabaci* from different geographic regions and host plants

From GenBank, in total 133 COI sequences were found under *T. tabaci* by blast search as of Feb. 13, 2013. In these sequence accessions, 82 accessions were associated with geographic region and plant-host information of *T. tabaci* origins. Among the 82 sequences, 43 sequences remained after redundant (identical) sequences were combined and were chosen for this study. A phylogenetic tree of the eight *T. tabaci* COI haplotypes and the 43 sequences from GenBank was generated with the COI sequence from *T. imaginis* as an outgroup, by maximum likelihood analysis (Figure 3). The phylogenetic tree showed three distinct clades for the *T. tabaci* haplotypes. Haplotypes in Clade 1 and Clade 2 had a high degree of host plant diversity and geographical representation - seven plant families over five continents. Clade 1 includes haplotypes from the arrhenotokous strain and Clade 2 includes haplotypes from the thelytokous strain. Interestingly, Clade 1 shows a subdivision with bootstrapping support at 63%. The haplotypes in one branch are associated with three host plant families, but the haplotypes from the other branch were primarily found in host plants in the family Amaryllidaceae, except one haplotype found from two additional host plant families (Figure 3). Clade 3 was composed of haplotypes that have so far only been found in individuals from tobacco in Greece and is known as the "tobacco strain" [7].

Population structure of *Thrips tabaci* in onion fields and cabbage fields

Although eight haplotypes were identified in *T. tabaci* populations in onion and cabbage fields in New York, the haplotypes NY-HT1, NY-HT2 and NY-HT3 accounted for the vast majority of the individuals sampled, with NY-HT1 being the most abundant. *Thrips tabaci* COI haplotype networks indicated that the haplotype composition in mid-summer and early fall remained the same in years 2005 and 2007 (Figure 4). The majority of individuals collected from onion were NY-HT1, accounting for 92% and 88% of the total in mid-summer in 2005 and 2007, respectively, and 100% and 96% of the total in early fall in 2005 and 2007, respectively (Figure 4; Table S1). In contrast, haplotype networks of *T. tabaci* populations collected from cabbage fields showed a population structure changing dynamically from mid-summer to early fall (Figure 5). In *T. tabaci* populations from cabbage in mid-summer, the haplotype NY-HT2 was abundant, accounting for 58% and 52% of the total in 2005 and 2007, respectively, but in early fall this haplotype decreased to 15% and 7% of the total in 2005 and 2007, respectively (Figure 5; Table S2). A lower percentage of the individuals collected from cabbage in mid-summer were haplotype NY-HT1 (12% and 46% in 2005 and 2007, respectively), but a majority in early fall were NY-HT1 (81% and 66% in 2005 and 2007, respectively) (Figure 5; Table S2).

The population structural change of *T. tabaci* from mid-summer to early fall in cabbage fields also was clearly indicated by the F_{ST}

Onion

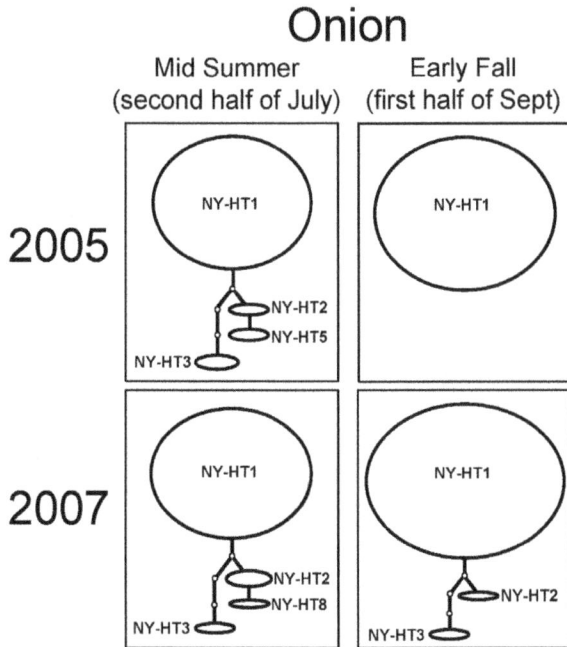

Figure 4. Haplotype networks for *Thrips tabaci* **COI sequences from populations collected from onion,** *Allium cepa,* **during mid-summer and early fall in New York in 2005 and 2007.** Each ellipse represents a certain haplotype (NY-HT1 to NY-HT8) and its size is proportional to the number of individuals from the population that it represents. Branches connecting the haplotypes represent the number of inferred mutations.

Cabbage

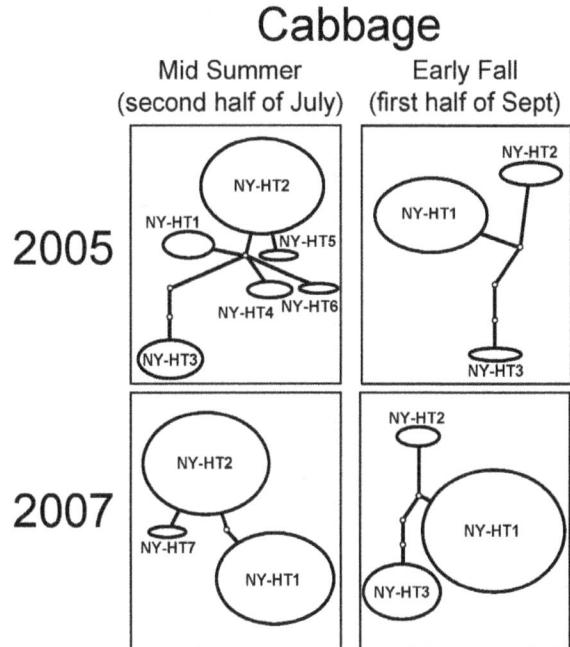

Figure 5. Haplotype networks for *Thrips tabaci* **COI sequences from populations collected from cabbage,** *Brassica oleracea,* **during mid-summer and early fall in New York in 2005 and 2007.** Each ellipse represents a certain haplotype (NY-HT1 to NY-HT8) and its size is proportional to the number of individuals from the population that it represents. Branches connecting the haplotypes represent the number of inferred mutations.

values (Table 1). The pairwise F_{ST} values between the mid-summer and early-fall populations in cabbage collected in 2005 and 2007 were 0.307 ($P<0.01$) and 0.244 ($P<0.01$), respectively. However, the F_{ST} values for the populations in onion were very low, 0.033 and 0.034 in 2005 and 2007, respectively, showing minimal population structural change between the two seasons. Pairwise comparisons of F_{ST} values between the *T. tabaci* populations indicated that *T. tabaci* populations in cabbage and onion had different genetic structures (Table 1). The greatest difference was between the mid-summer populations in cabbage and the early-fall populations in onion with F_{ST} values between 0.452 ($P<0.01$) and 0.551 ($P<0.01$).

Discussion

Since Brunner et al.'s [7] phylogenic analysis that identified three major genetic lineages for *T. tabaci* (i.e., Leek 1, Leek 2 and Tobacco strains), numerous similar analyses from around the world have followed [4], [6], [16], [17], [28]. All *T. tabaci* in our study were from a single leek clade that included individuals collected from an array of hosts representing seven plant families and multiple locations spanning five continents (clade number 2 in Figure 3). Like many other phylogenetic studies, *T. tabaci* from New York in our study differed substantially from those identified as the "tobacco strain" (clade number 3 in Figure 3). All haplotypes in our study were from the same lineage as those that reproduce via thelytoky. *Thrips tabaci* that reproduce via thelytoky differ genetically from those that reproduce via arrhenotoky [4], [16]. In regions of New York, North Carolina (USA) and Japan, both thelytokous and arrhenotokous populations of *T. tabaci* collected from onion exist [3–5]. In North Carolina, arrhenotokous populations were found only in the mountains in the western

region, while thelytokous ones were found at lower elevations in the central region of the state. In New York, arrhenotokous and thelytokous populations were identified in nearby onion fields within each of the major onion-producing regions. In Japan, thelytokous and arrhenotokous populations were documented from the same field. Subsequent studies on *T. tabaci* populations in Japan confirmed that populations of sexually and asexually reproducing populations of *T. tabaci* coexist within the same crop field [29].

Thrips tabaci populations changed considerably during the season in cabbage fields, but not in onion fields (Table 1; Figures 4 and 5). *Thrips tabaci* populations in onion fields were dominated by a single haplotype (NY-HT1) during mid-summer and early fall, whereas populations in cabbage fields included several common haplotypes that changed considerably from mid-summer to early fall. The haplotype NY-HT2 was noticeably abundant in cabbage fields in mid-summer, but became a minor type in early fall (Figure 5). Despite the proximity of onion and cabbage fields in the western New York landscape, *T. tabaci* populations appeared to differ seasonally within each cropping system. The *T. tabaci* population in cabbage fields underwent a genetic structural change from mid-summer to early fall, suggesting that some individuals such as those with COI haplotype NY-HT1 outcompete others and ultimately establish more successfully later in the season on cabbage plants or that dispersal and colonization patterns differ within cabbage and onion cropping systems, or perhaps both of these phenomena occur.

While onion and cabbage fields were grown in proximity (Figure 1), habitats immediately adjacent to these crops differed. In New York, onion is grown in muck cropping systems, which are essentially islands in the landscape bordered by woods or weedy areas known to harbor *T. tabaci* [30], [31]. Onion fields are rarely

Table 1. Pairwise F_{ST} values[1] for *Thrips tabaci* populations collected from cabbage and onion fields at two times during the season in New York (USA).

Year	*T. tabaci* population	Cabbage/Mid-summer	Cabbage/Early fall	Onion/Mid-summer
2005	Cabbage/Early fall	0.307*		
	Onion/Mid-summer	0.472*	0.040	
	Onion/Early fall	0.551*	0.198*	0.033
2007	Cabbage/Early fall	0.244*		
	Onion/Mid-summer	0.282*	0.148*	
	Onion/Early fall	0.452*	0.209*	0.034

[1]F_{ST} values were calculated using ARLEQUIN 3.5.1.2.
*P value<0.01, calculated by permutation test.

bordered by other cultivated hosts for *T. tabaci* such as alfalfa and small grains, both of which are grown adjacent to cabbage and are speculated to contribute to early-season infestations in cabbage fields in New York [32]. Although dispersal behavior of *T. tabaci* over long distances is poorly understood, if dispersal occurs over long distances (i.e., several km or more), it is possible that *T. tabaci* populations from onion may colonize cabbage and other crops late in the season following onion harvest in August and September. If so, this may explain the late-season occurrences of NY-HT1 in cabbage fields.

Although *T. tabaci* has a wide host range [33], host-associations involving onion/leek and tobacco have been documented for *T. tabaci* [7]. Our results provide additional evidence for the occurrence of host associations between different genetic lineages of *T. tabaci* and other cultivated hosts. The extent to which *T. tabaci* populations restrict themselves to specific crop hosts in space and time will likely influence the population dynamics related to the development and spread of traits such as insecticide resistance within and among different cropping systems in the same area. Future studies investigating population structures of *T. tabaci* on their major hosts in these ecosystems through time are needed to better understand host-plant utilization and its potential implications for population management.

Supporting Information

Figure S1 DNA sequence of the 655 bp *COI* fragment from *Thrips tabaci* collected in New York. Eight COI

haplotypes were identified from 565 *T. tabaci* individuals, based on the SNP site (R = A or G; Y = C or T). The nucleotides G and T boxed are specific for thelytokous *T. tabaci* [4].

Acknowledgments

The authors are grateful for G. Tetreau's assistance in statistical analyses of insect populations and for comments made by G. Kennedy and A. Jacobson on a previous draft of this manuscript.

Author Contributions

Conceived and designed the experiments: BAN PW. Performed the experiments: BAN WCK PW. Analyzed the data: PW. Contributed reagents/materials/analysis tools: WCK PW. Contributed to the writing of the manuscript: BAN WCK PW.

References

1. Parrella M, Lewis T (1997) Integrated pest management (IPM) in field crops. In: Thrips as crop pests. Lewis T, editor. New York: CAB International. 595–614.
2. Zawirska I (1976) Untersuchungen uber zwei biologische Typen von *Thrips tabaci* Lind. (Thysanoptera, Thripidae) in der VR Polen. Arch Phytopath Plant Prot 12: 411–422.
3. Nault BA, Shelton AM, Gangloff-Kaufmann JL, Clark ME, Werren JL, et al. (2006) Reproductive modes in onion thrips (Thysanoptera: Thripidae) populations from New York onion fields. Environ Entomol 35: 1264–1271.
4. Kobayashi K, Hasegawa E (2012) Discrimination of reproductive forms of *Thrips tabaci* (Thysanoptera: Thripidae) by PCR with sequence specific primers. J Econ Entomol 105: 555–559.
5. Jacobson A, Booth W, Vargo EL, Kennedy GG (2013) *Thrips tabaci* population genetic structure and polyploidy in relation to competency as a vector of Tomato spotted wilt virus. PLoS ONE 8(1): e54484. Doi:10.1371/journal.pone.0054484.
6. Westmore GC, Poke FS, Allen GR, Wilson CR (2013) Genetic and host-associated differentiation within *Thrips tabaci* Lindeman (Thysanoptera: Thripidae) and its links to *Tomato spotted wilt virus*-vector competence. Heredity 1–6. doi:10.1038/hdy.2013.39.
7. Brunner PC, Chatzivassiliou EK, Katis NI, Frey JE (2004) Host-associated genetic differentiation in *Thrips tabaci* (Insecta; Thysanoptera), as determined from mtDNA sequence data. Heredity 93: 364–370.

8. Jenser G, Szenasi A, Torjek O, Gyulai G, Kiss E, et al. (2001) Molecular polymorphism between population of *Thrips tabaci* Lindeman (Thysanoptera: Thripidae) propogating on tobacco and onion. Acta Phytopath et Entomol Hungarica 36: 365–368.
9. Kennedy GG, Storer N (2000) Life systems of polyphagous arthropod pests in temporally unstable cropping systems. Ann Rev Entomol 45: 467–493.
10. Shelton AM, Nault BA, Plate J, Zhao JZ (2003) Regional and temporal variation in susceptibility to lambda-cyhalothrin in onion thrips, *Thrips tabaci* (Thysanoptera: Thripidae), in onion fields in New York. J Econ Entomol 96: 1843–1848.
11. Shelton AM, Zhao JZ, Nault BA, Plate J, Musser FR, et al. (2006) Patterns of insecticide resistance in onion thrips (Thysanoptera: Thripidae) in onion fields in New York. J Econ Entomol 99(5): 1798–1804.
12. Shelton AM, Plate J, Chen M (2008) Advances in control of onion thrips (Thysanoptera: Thripidae) in cabbage. J Econ Entomol 101: 438–443.
13. Nault BA, Hessney ML (2006) Onion thrips control in onion, 2005. Arthropod Management Tests, 2006. 31: E39.
14. Nault BA, Hessney ML (2008) Onion thrips control in onion, 2006. Arthropod Management Tests, 2008. 33: E19.
15. Nault BA, Hessney ML (2010) Onion thrips control in onion, 2009. Arthropod Management Tests, 2009. 35: E13.

16. Toda S, Murai T (2007) Phylogenetic analysis based on mitochondrial COI gene sequences in *Thrips tabaci* (Thysanoptera: Thripidae) in relation to reproductive forms and geographic distribution. App Entomol Zool 42: 309–316.

17. Srinivasan R, Guo F, Riley D, Diffie S, Gitaitis R, et al. (2011) Assessment of variation among *Thrips tabaci* populations from Georgia and Peru based on polymorphisms in mitochondrial cytochrome oxidase I and ribosomal ITS2 sequences. J Entomol Sci 46: 191–203.

18. Tiewsiri K, Wang P (2011) Differential alteration of two aminopeptidases N associated with resistance to *Bacillus thuringiensis* toxin Cry1Ac in cabbage looper. Proc Natl Acad Sci USA 108: 14037–14042. doi:10.1073/pnas.1102555108.

19. Hebert PDN, Penton EH, Burns JM, Janzen DH, Hallwachs W (2004) Ten species in one: DNA barcoding reveals cryptic species in the neotropical skipper butterfly *Astraptes fulgerator*. Proc Natl Acad Sci USA 101: 14812–14817.

20. Dorit RL, Ohara O, Hwang CB-C, Kim JB, Blackshaw S (2001) Direct DNA sequencing of PCR products. Curr Protoc Mol Biol 56: 15.2.1–15.2.13.

21. Altschul SF, Gish W, Miller W, Myers EW, Lipman DJ (1990) Basic local alignment search tool. J Mol Biol 215: 403–410.

22. Benson DA, Cavanaugh M, Clark K, Karsch-Mizrachi I, Lipman DJ, Ostell J, Sayers EW (2013) Genbank. Nucleic Acids Res 41 (Database issue): D36–42.

23. Jukes TH, Cantor CR (1969). Evolution of protein molecules. In: Munro HN, ed. Mammalian Protein Metabolism., Academic Press, New York. 21–132.

24. Tamura K, Peterson D, Peterson N, Stecher G, Nei M, Kumar S (2011) MEGA5: Molecular evolutionary genetics analysis using maximum likelihood, evolutionary distance, and maximum parsimony methods. Mol Biol Evol 28: 2731–2739.

25. Templeton AR, Crandall KA, Sing CF (1992) A cladistics analysis of phenotypic associations with haplotypes inferred from restriction endonuclease mapping and DNA sequence data. III. Cladogram estimation. Genetics 132: 619–633.

26. Clement M, Posada D, Crandall K (2000) TCS: a computer program to estimate gene genealogies. Mol Ecol 9: 1657–1660.

27. Excoffier L, Lischer HE (2010) Arlequin suite ver 3.5: a new series of programs to perform population genetics analyses under Linux and Windows. Mol Ecol Resour 10: 564–567.

28. Kadirvel P, Srinivasan R, Hsu YC, Su FC, de la Pena R (2013) Application of cytochrome oxidase I sequences for phylogenetic analysis and identification of thrips species occurring on vegetable crops. J Econ Entomol 106: 408–418.

29. Kobayashi K, Yoshimura J, Hasegawa E (2013) Coexistence of sexual individuals and genetically isolated asexual counterparts in a thrips. Sci Rep 3: 3286; DOI:10.1038/srep03286.

30. Larentzaki E, Shelton AM, Musser FR, Nault BA, Plate J (2007) Overwintering locations and hosts for onion thrips (Thysanoptera: Thripidae) in the onion cropping ecosystem in New York. J Econ Entomol 100(4): 1194–1200.

31. Smith EA, DiTommaso A, Fuchs M, Shelton AM, Nault BA (2011) Weed hosts for onion thrips (Thysanoptera: Thripidae) and their potential role in the epidemiology of *Iris yellow spot virus* in an onion ecosystem. Environ Entomol 40: 194–203.

32. North RC, Shelton AM (1986) Ecology of Thysanoptera within cabbage fields. Environ Entomol 15: 520–526.

33. Morison GD (1957) A review of British glasshouse Thysanoptera. Trans R Entomol Soc Lond 109: 467–520.

20

Strain Differences in Fitness of *Escherichia coli* O157:H7 to Resist Protozoan Predation and Survival in Soil

Subbarao V. Ravva*, Chester Z. Sarreal, Robert E. Mandrell

Produce Safety and Microbiology Research Unit, U.S. Department of Agriculture, Agriculture Research Service, Western Regional Research Center, Albany, California, United States of America

Abstract

Escherichia coli O157:H7 (EcO157) associated with the 2006 spinach outbreak appears to have persisted as the organism was isolated, three months after the outbreak, from environmental samples in the produce production areas of the central coast of California. Survival in harsh environments may be linked to the inherent fitness characteristics of EcO157. This study evaluated the comparative fitness of outbreak-related clinical and environmental strains to resist protozoan predation and survive in soil from a spinach field in the general vicinity of isolation of strains genetically indistinguishable from the 2006 outbreak strains. Environmental strains from soil and feral pig feces survived longer (11 to 35 days for 90% decreases, D-value) with *Vorticella microstoma* and *Colpoda aspera*, isolated previously from dairy wastewater; these D-values correlated ($P<0.05$) negatively with protozoan growth. Similarly, strains from cow feces, feral pig feces, and bagged spinach survived significantly longer in soil compared to clinical isolates indistinguishable by 11-loci multi-locus variable-number tandem-repeat analysis. The curli-positive (C^+) phenotype, a fitness trait linked with attachment in ruminant and human gut, decreased after exposure to protozoa, and in soils only C^- cells remained after 7 days. The C^+ phenotype correlated negatively with D-values of EcO157 exposed to soil ($r_s = -0.683$; $P = 0.036$), *Vorticella* ($r_s = -0.465$; $P = 0.05$) or *Colpoda* ($r_s = -0.750$; $P = 0.0001$). In contrast, protozoan growth correlated positively with C^+ phenotype (*Vorticella*, $r_s = 0.730$, $P = 0.0004$; *Colpoda*, $r_s = 0.625$, $P = 0.006$) suggesting a preference for consumption of C^+ cells, although they grew on C^- strains also. We speculate that the C^- phenotype is a selective trait for survival and possibly transport of the pathogen in soil and water environments.

Editor: Dongsheng Zhou, State Key Laboratory of Pathogen and Biosecurity, Beijing Institute of Microbiology and Epidemiology, China

Funding: The work was funded by the U.S. Department of Agriculture (USDA), Agricultural Research Service CRIS project 5325-42000-046 and partly by the National Research Initiative Competitive Grant nos. 2006-55212-16927 and 2007-35212-18239 from the USDA National Institute of Food and Agriculture. The funders had no role in study design, data collection and analysis, decision to publish, or preparation of the manuscript.

Competing Interests: The authors have declared that no competing interests exist.

* Email: subbarao.ravva@ars.usda.gov

Introduction

Escherichia coli O157:H7 (EcO157) responsible for over 200 infections in a large multi-state outbreak [1] related to consumption of spinach was traced back to produce grown in central California coast [2]. Major multi-country outbreaks associated with produce indicate that pre-harvest contamination has occurred often, so it is critical to identify sources of pathogens and interventions for minimizing them [1]. Indeed, EcO157 isolates that are genetically indistinguishable from the 2006 spinach outbreak strain were isolated from feral swine, cattle, and water samples near spinach fields from the central coast of California [2]. These results indicate wide-spread occurrence of this pathogen, and minimizing pre-harvest contamination will require an understanding of the biological and environmental factors that regulate its proliferation and transport from animal reservoirs to watersheds and produce grown in proximity.

There are numerous habitats in the vicinity of produce production, each of which may affect survival of EcO157 differently. Although most EcO157 strains decrease rapidly in the soil and manure environments [3,4], and to lesser extent in water [5,6], a small proportion of cells remain viable for extended periods. Therefore, it is probable that most EcO157 cells within a population exposed to stressful environments outside the animal host fail to survive [7]. Nevertheless, survival of some cells in water and soil results in transport of pathogen by irrigation and wind, possibly leading to produce contamination.

A mechanism for increased survival of *E. coli* and other pathogenic bacteria is through proliferation within vacuoles [8,9] of protozoa in the environment. Passage through protozoa provides the pathogenic bacteria a survival advantage, possibly by aiding in their persistence in inhospitable aquatic environments such as chlorinated waters [8,10] and may also increase the virulence of human pathogens [11,12].

Direct evidence was reported for sequestration of EcO157 in vacuoles of protozoa isolated from store-bought spinach and lettuce [13]. Conversely, we reported that predation [14,15] was associated with rapid declines of EcO157 in dairy wastewater [16] and inhibition of predation resulted in enhanced growth of EcO157 [17]. We isolated *Colpoda aspera*, *Vorticella microstoma* and *Platyophrya* sp. from dairy wastewater that consumed EcO157 in preference to native bacteria [14]. Although it appears that some EcO157 strains escape predation selectively due to heightened natural anti-predatory defenses, it is unclear if such mechanisms are strain specific or sub-populations and/or phenotypes exist that are resistant to predation, thus extending survival.

Survival of specific strains of EcO157 in sufficient numbers to cause infection is associated with their intrinsic fitness traits and genetic makeup [18]. Often this involves the formation of biofilms, facilitated by the production of adhesins and polysaccharides [19,20]. One such adhesin, curli (C), along with the production of cellulose, has been shown to enhance bacterial adherence necessary for formation of biofilms [21,22]. C are thin aggregative fimbriae and act as a virulence factor by promoting attachment to eukaryotic cells [23,24]. C fimbriae, encoded by csgA, are expressed in response to low temperature, low oxygen, low osmolarity, and nutrient limitation [25].

Subpopulations of EcO157 have been reported to adapt to harsh environmental conditions [26]. Mutations that result in the C$^-$ phenotype may confer a selective advantage in surviving austere environments. C$^-$ strains survive up to 10,000 times better than C$^+$ strains under acidic conditions (pH 2.4) and this appears to occur by maintaining C$^-$ cells regardless of selection pressure [27]. Thus, it is possible that some environmental conditions cause selection of subpopulations with enhanced fitness to contaminate produce and, amplify enough to cause significant illness and an outbreak.

We described differences in predation of EcO157 by different protozoa isolated from dairy wastewater in a previous study [14]. In this study, we evaluate predation by V. microstoma and C. aspera of clinical and environmental strains of EcO157 that are highly related genotypically and associated with the 2006 spinach outbreak. In addition, we compared if specific phenotype subpopulations that evade protozoan predation are increased in fitness for survival in soil from produce field.

Materials and Methods

Ethics statement

Soil samples were provided by the Western Center for Food Safety and Security, University of California at Davis. No special permits were required as the soils were collected under cooperative agreements with produce growers and the samplings did not involve any endangered or protected species.

Strains of EcO157 used in this study

EcO157 strains (Table 1) were selected based on sample source and genetic similarities determined by multi-locus variable-number tandem-repeat analysis (MLVA) and reported in a previous study [28]. Clinical and environmental strains of four different MLVA types (Table S1) associated with the 2006 spinach outbreak were used in these comparisons. All but one of the strains was highly related by 11-loci MLVA; strain RM9834 differed from the others at 9 loci. All of them carry virulence genes, stx2 (Shiga-toxin), eae (intimin), and hly (hemolysin); and serotype specific genes, fliC (H7-antigen) and rfbE (O157-antigen) [28–30]. Purity of cultures was confirmed by plating on Rainbow agar (Biolog, Hayward, CA) containing novobiocin (20 μg/ml, Sigma-Aldrich) and tellurite (0.8 μg/ml, Invitrogen/Dynal) (Rainbow-NT).

Preparation of inoculums and enumeration of EcO57

Isolated colonies from Rainbow-NT agar were grown over-night in 10% Luria broth (LB, Fisher Scientific, PA) at 25°C and at 150 rpm on a gyratory shaker, cells were separated by centrifuging at 10,000×g for 5 min and washed twice in 0.01 M PBS (pH 7.0) and adjusted to OD$_{600}$ of 0.3 prior to inoculations to Sonneborn medium (Solution 1 of ATCC medium 802, http://www.atcc.org/Attachments/4018.pdf) or soil. Enumeration of EcO157 from soils or protozoa media was carried out by plating 100 μl volumes of 10-fold serial dilutions in 0.01 M PBS onto

Rainbow-NT agar and the bluish-grey colonies were counted after over-night incubation at 37°C. Some colonies were tested at random by real-time PCR to confirm the presence of O157-antigen specific gene rfbE, using a method described previously [28].

Enumeration of phenotypic variant subpopulations

The proportion of phenotypic variants of EcO157 strains, post-protozoan or soil exposure, expressing C was determined as described previously [27], with some modifications. Briefly, twenty isolated colonies from each of three replicates, confirmed as EcO157 on Rainbow-NT agar plates initially, were patched on LB agar without NaCl, but supplemented with 40 μg/ml of Congo red dye and 10 μg/ml of Coomassie brilliant blue G (Congo red agar). Thus, a total of 60 colonies for each strain of EcO157 were analyzed for determining the proportion of C variants. C$^+$ (red) and C$^-$ (white) colonies were tested at random to confirm the presence of rfbE [28]. The proportion of C subpopulations prior to soil or protozoan exposure were determined by plating serial dilutions of over-night growth from LB agar onto Congo red agar.

Consumption of EcO157 strains by protozoa

Consumption of EcO157 strains was determined using V. microstoma and C. aspera isolated previously [14] from dairy wastewater. Twenty-five milliliters of sterilized 10% Sonneborn medium in 0.01 M PBS (pH 7.0) supplemented with 5% 3-μm filtered and heat-killed wastewater was inoculated with 1×10^8 CFU/ml of EcO157 and 2×10^3 cells of protozoa per ml (50,000 bacteria/protozoa). A 7-day old growth of protozoa in Sonneborn medium was used as inoculum. Overnight growth of EcO157 strains in 10% LB broth, centrifuged and resuspended in 0.01 M PBS was used as the bacterial food source for protozoa. The populations of both EcO157 and protozoa during a 7-day incubation without agitation at 25°C were determined using methods described previously [14] except that 5-fold serial dilutions in 0.01 M PBS were used for counting protozoa by the MPN method. The comparisons were in triplicate and days for 90% decreases (D-value) of EcO157 as a result of consumption by protozoa were calculated. Stationary incubations aid in grazing of EcO157 by micro-vortexing and filter feeding by the ciliates. Two-way ANOVA coupled with Bonferroni post t-test (Prism 4.0; GraphPad Software, Inc., San Diego, CA) or Holm-Sidak pairwise multiple comparisons (Sigmaplot v11, Systat software, Inc., Chicago, IL) were used to compare differences in growth of protozoa and differential uptake of EcO157 by protozoa. Changes in C subpopulations as influenced by protozoa also were statistically analyzed.

Survival of EcO157 in soil from a produce field

Survival of nine EcO157 strains (Table 1) was monitored in the <45 μm fraction of fine soil (US standard sieve, 325 mesh, Hogentogler, Columbia, MD) collected from a produce field from lower Salinas Valley in Monterey County, CA (Farm R). Soil samples from this field were cultured by methods described previously for the presence of EcO157 [28]. One gram samples of soil in 4-ml screw capped glass vials (Wheaton Science Products, Millville, NJ) were inoculated with ~1×10^7 CFU of EcO157 cells in 160 μl and adjusted to moisture at 50% water holding capacity (26.1% moisture on dry weight basis) by adding 110 μl sterile distilled water. Vials containing the inoculated soils were capped loosely to allow aeration, incubated at 25°C and sampled at regular intervals for the enumeration of surviving EcO157 cells and for analyzing the proportion of C subpopulations as described above. The moisture loss at each sampling interval was monitored

Table 1. EcO157 strains associated with 2006 spinach outbreak.

Strain No.	MLVA type	Source	State	Details[a]
RM6441	176	Cow feces	CA	CDPH-FDLB, Paicines ranch
RM6103	163	Cow feces	CA	CDPH-MDL, Paicines ranch
RM6088	176	Cow feces	CA	CDPH-MDL, Paicines ranch
RM6096	163	Cow feces	CA	CDPH-MDL, Paicines ranch
RM6440	176	Cow feces	CA	CDPH-FDLB, Paicines ranch
RM6157	176	Feral pig feces	CA	Paicines ranch
RM6106	174	Feral pig feces	CA	Paicines ranch
RM6155[c]	163	Feral pig feces	CA	Paicines ranch
RM9834[c]	778	Soil	CA	CSREES Environmental Study, Ranch J[b]
RM9993	163	Spinach bag	PA	CDC
RM6067[c]	163	Spinach bag	PA	Pennsylvania Department of Health
RM6068	163	Spinach bag	PA	Pennsylvania Department of Health
RM9996	163	Spinach bag	PA	CDC
RM6331[c]	163	Clinical	OR	Oregon State Public Health Lab
RM6653[c]	163	Clinical	WI	CDC
RM6069[c]	163	Clinical	PA	Pennsylvania Department of Health
RM6654[c]	163	Clinical	NM	CDC
RM6657[c]	163	Clinical	UT	CDC

[a]CDPH-FDLB, California Department of Public Health – Food and Drug Laboratory Branch; CDPH-MDL, California Department of Public Health – Microbial Diseases Laboratory; CSREES, USDA Cooperative state Research, Education, and Extension Service; and CDC, Centers for Disease Control.
[b]Isolated repeatedly from pasture soil during a 45-day period during 2009. All other strains were isolated during the outbreak period.
[c]Strains used in the soil fitness study.

by weighing another set of vials containing un-inoculated soil. Soils were mixed thoroughly with sterile spatulas and ~100 mg soil samples by weight were removed and used to make 10-fold serial dilutions for enumeration of EcO157. D-values were calculated based on the decreases in EcO157 populations during a 7-day incubation period. All the comparisons were in triplicates and the data was analyzed statistically as described above. In addition, Spearman rank order correlations were used to compare the fitness of EcO157 in soil with resistance to predation; growth increases of protozoa and the proportion of C subpopulations.

Results

Strain differences in uptake and utilization of EcO157 for protozoan growth

Environmental and clinical strains of EcO157 associated with the 2006 spinach outbreak (Table 1) were fed to *V. microstoma* and *C. aspera* to determine if any strains resist predation or, alternatively, were consumed preferentially based on their genetic differences or source of isolation. Inter-strain differences in predation of EcO157 isolates were indicated by a wide range of D-values from 1 to 35 days in the presence of *Vorticella* and 4 to 26 days for *Colpoda* (Figure 1). Strain RM9834, isolated from pasture soil (Table 1), resisted predation by both *Colpoda* and *Vorticella*, as indicated by longer D-values (26 to 35 d) (Figure 1). All strains of EcO157 grew slightly (0.5 to 1 log increase) in the absence of predation during the 7-day incubation and thus D-values for EcO157 without protozoa are not shown.

Both protozoa grew by consuming EcO157; the log increases in protozoa correlated inversely with D-values (Figure 1, Table 2). Log increases of protozoan numbers were significantly higher ($P<0.001$; Table S2) with *Colpoda* compared to *Vorticella* (Figure 1).

Inherent strain differences and their sources of isolation influenced protozoan growth significantly ($P<0.001$) resulting from the consumption of EcO157 cells (Table S2). An isolate from soil, two from feral pig feces (RM6106, RM 6157) and two clinical isolates (RM6653, RM6069) resisted protozoan predation compared to strains isolated from cow feces and bagged spinach.

Tests with strains grouped based on MLVA typing (Table S1) indicated that genetic differences influenced significantly ($P<0.001$, Table S2) the consumption of EcO157 by protozoa (Figure 1). Single strains representing MLVA types 778 and 174 resisted predation and survived significantly longer ($P<0.001$) than the clinical outbreak strains of MLVA 163. These results indicate that strains that are highly related by MLVA can have functional fitness differences depending upon the environment to which it was exposed.

Protozoan exposure alters the proportion of C variant subpopulations of EcO157

Strains with significant differences ($P<0.001$, Table S2; Figure 2) in proportion of C variants were evaluated to check if protozoa consume subpopulations of EcO157 preferentially. The proportion of C^+ subpopulations decreased significantly ($P=0.006$) during the 7-day incubation with both protozoa and even more significantly with strains RM6441, RM9993 and RM6331 and RM6553 ($P<0.0001$). However, C^- strains (RM6103, RM6157, RM6155, RM6067; Figure 2) after the protozoan exposure remained C^- and were consumed with comparatively low growth increases (Figure 1). In addition, significant decreases in the proportion of C^+ variants occurred specifically with some strains in co-culture with *Vorticella* and with different strains with *Colpoda* (Figure 2, Table S2,). C^+ variants of RM6440 and RM6657 exposed to *Vorticella* decreased very significantly compared to

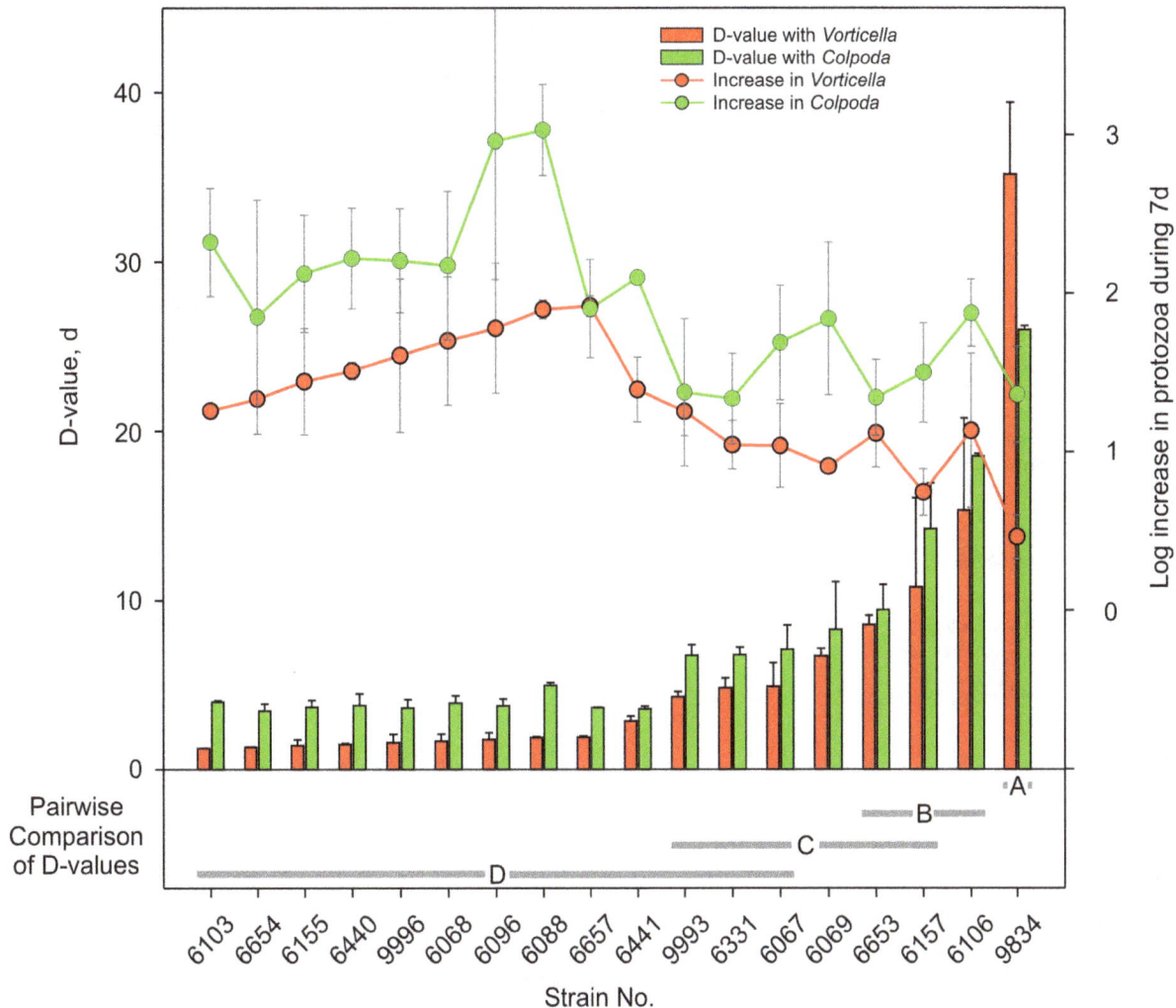

Figure 1. Inter-strain differences in preferential predation of EcO157 by protozoa from dairy wastewater. Survival of EcO157 populations measured as D-values (bar graph) and resultant increases in log numbers (line graph) of *V. microstoma* and *C. aspera* are shown.

Colpoda ($P<0.0001$). The proportion of C^+ variants in the presence of protozoa was correlated negatively with D-values ($P<0.05$; Table 3) of EcO157 and correlated positively with log increases of protozoa ($P<0.01$). Thus, log increases in protozoa correlated negatively with D-values of EcO157 strains ($P<0.05$; Table 3, Figure 1).

Strain differences in soil fitness and protozoan predation

The fate of EcO157 in soil was monitored to evaluate if strains that resist predation also survive longer in soil. Nine strains from fitness studies with protozoa (Table 1) were used in these comparisons; eight are genetically indistinguishable by MLVA typing from the major MLVA type of the spinach outbreak strain (MLVA 163). A pasture soil isolate, RM9834 (MLVA 778), was included to compare the soil fitness of strains highly related by MLVA, but from different sources, to an unrelated strain. Environmental strains survived longer in soil (high D-values) compared to clinical isolates (Table 3), but no correlation was apparent between soil persistence and resistance of EcO157 for predation by protozoa (Table 2). Strain differences were associated with significant variance in D-values ($P<0.0001$; Table 3). However, the soil isolate, RM9834, survived longer in soil and

also resisted predation significantly (Figure 1; Table 3). Similar to the predation resistance discussed above (Table 2 and S2), persistence of EcO157 in soil was also correlated negatively ($r_s = -0.683$; $P = 0.036$) with the proportion of C variants. Only C^- variants remained in soil after a 7-day exposure of 7 out of 9 strains. The proportions of C^+ and C^- variants of the other two strains (RM6657, RM6069) were not evaluated.

Discussion

EcO157 is prevalent in agricultural environments (such as dairies, feedlots) and has been detected frequently in environmental samples during and after the 2006 spinach outbreak [2,28]. The pathogen has been reported to persist in manure piles for nearly 2 years [31]. Conversely, some strains (including a strain linked to apple juice outbreak) disappeared from dairy wastewater in less than a day [16]. These results indicate that EcO157 strains respond differently to different environmental and biological factors. Indeed, one of the strains we used in this study, RM9834, was isolated repeatedly from a naturally contaminated dry pasture soil during a 45-day period indicating that at least some cells remained viable under harsh environmental stress. Similarly,

Table 2. Influence of C-variant proportions on the survival of EcO157 strains before or after exposure to protozoa and in soil from a spinach field.

Variables correlated	Correlation coefficient, r_s[a]	P-value	Significance[c]
C proportion prior to exposure and			
D-values with soil[b]	−0.683	0.036	*
D-values with *Colpoda*	−0.467	0.049	*
D-values with *Vorticella*	−0.129	0.603	NS
Growth increase of *Colpoda*	0.130	0.597	NS
Growth increase of *Vorticella*	0.450	0.060	NS
C proportion with *Colpoda* and			
D-values with *Colpoda*	−0.750	0.0001	****
Growth increase of *Colpoda*	0.625	0.006	**
C proportion with *Vorticella* and			
D-values with *Vorticella*	−0.465	0.05	*
Growth increase of *Vorticella*	0.730	0.0004	***
D-values of EcO157 with *Colpoda* and			
Growth increase of *Colpoda*	−0.554	0.017	*
D-values of EcO157 with *Vorticella* and			
Growth increase of *Vorticella*	−0.674	0.002	**
D-values of EcO157 in soil[b] and			
D-values with *Colpoda*	0.067	0.844	NS
D-values with *Vorticella*	−0.100	0.775	NS

[a]Spearman rank-order correlations.
[b]Fate of EcO157 in soil compared with 9 strains of different proportions of C (see Table 3 for proportion of C variants prior to exposure to soil).
[c]* = $P<0.05$, ** = $P<0.01$, *** = $P<0.001$, **** = $P<0.0001$ and NS = not significant.

bacterial pathogens that can avoid predation by protozoa and the capability of growing under low nutrient conditions and/or inhibitory chemicals in wastewater [15] have a higher probability of surviving under stressful conditions of produce production environments.

EcO157 strains from environmental sources such as soil and feral pig feces, in particular, resisted predation by both protozoa significantly (Figure 1, Table S2). These strains were co-isolated along with the outbreak strain from feral pigs and cow feces [2] during the 2006 spinach outbreak or, subsequently, from pasture soil. The soil isolate (RM9934) differed phylogenetically from the outbreak strains; tandem repeats at 9 out of 11 loci were different (Table S1). Environmental stresses can be associated with subtle changes in the MLVA tandem repeats resulting in increased phylogenetic diversity [32], but it is not known if environmental exposures may improve the chances of survival and resistance to predation of pathogenic EcO157.

Both protozoa we tested consumed strains from clinical samples, spinach and cow feces; *Vorticella* consumed EcO157 cells more rapidly (Figure 1, shorter D-values; $P=0.031$) compared to *Colpoda* (Table S2). Consumption of *E. coli* and EcO157 in preference to native bacteria, in soils [33] and dairy wastewater [14], respectively, has been reported, but inter-strain differences in consumption of EcO157 by environmental protozoa has not been described. In addition, we show that outbreak strains from clinical and environmental samples were consumed rapidly, compared to environmental strains from pig feces and soil, isolated from the same vicinity (Figure 1). These results indicate that protozoa may be a significant factor in eliminating many EcO157 soon after their release into dairy lagoons through feces.

Protozoa grew as they consumed EcO157 strains rapidly and, as expected, protozoan growth was correlated negatively with D-values of EcO157 (Figure 1, Table 3). However, *Colpoda* grew significantly (Figure 1, Table S2) compared to *Vorticella*, although *Vorticella* eliminated EcO157 more rapidly. To our knowledge, there have been no reports related to the growth of environmental protozoa after consumption of strains of EcO157. Significant increases in growth of both protozoa were observed for nearly half-of the strains (Figure 1) that are highly related genetically to the outbreak strain. In contrast, both protozoa grew significantly less ($P<0.001$) with isolates that resisted predation and were from feral pig feces and soil. Similarly, protozoa that can feed and grow on EcO157 when released into dairy ponds might explain why EcO157 were not isolated in a previous study from dairy wastewater [16].

In addition to genetic differences of EcO157 strains and their prior exposure to environmental stresses that influence their relative resistance to predation, intra-strain differences in C expression also influenced predation by protozoa. A high proportion of C^+ to C^- variants correlated negatively with D-values indicating significant consumption of EcO157 resulting in growth increases of both protozoa (Table 3). Thus, strains from cow feces, spinach bags and some clinical isolates with a high proportion of C^+ variants were rapidly consumed by both protozoa and decreased the proportion of C^+ variants very significantly ($P<0.0001$). In contrast, strains from feral pig feces and dry pasture soil were predominantly C^- (Figure 2) and resisted predation. Although no other data on predation resistance of subpopulations has been reported, a C-deficient EcO157 strain was reported to survive better on plant surfaces [34], presumably,

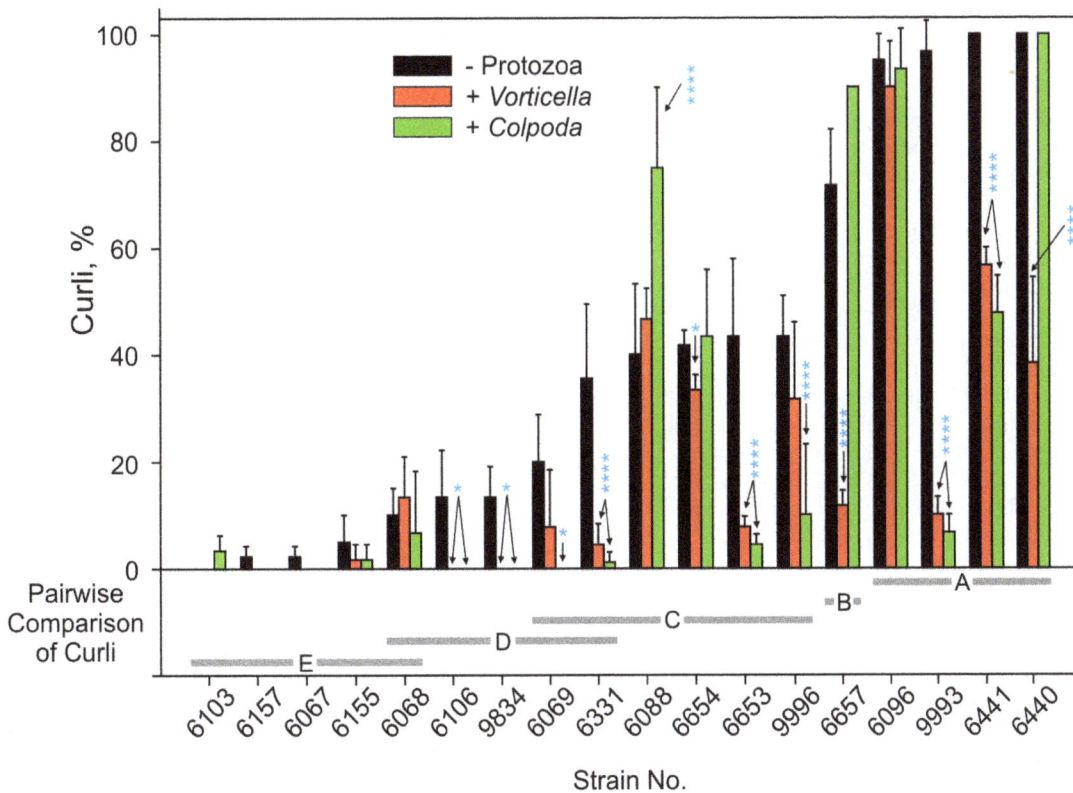

Figure 2. Effect of protozoan predation on curli-variant subpopulations of EcO157 strains. Proportion of C variants was monitored both in the presence or absence of protozoa. Values are averages of triplicates. Bonferroni post tests: * = $P<0.05$; ** = $P<0.01$; *** = $P<0.001$; **** = $P< 0.0001$.

because plant defense systems recognize C and influence the survival of subpopulations. Alternatively, some bacteria with hydrophobic cell surfaces resist predation [35] making it worth determining if similar mechanisms are responsible for predation resistance of C^- strains or subpopulations.

We hypothesize that EcO157 strains with increased percentage of C^- variants have a fitness mechanism to survive under hostile produce production soil environments with extremes in moisture and temperature conditions. Thus, we isolated RM9834 with 87% C^- subpopulations from dry pasture soil. Furthermore, a 7-day exposure of 3 clinical and 4 environmental strains (one of each

from cow feces, feral pig feces, soil and bagged spinach) in spinach field soil resulted in recovering only C^- subpopulations. This previous environmental exposure is probably associated with extended survival in soil when it is re-inoculated into soil, and may be due to C^- strains being isolated predominately from these environments; this is consistent with the proportion of C^+ variant subpopulations correlating negatively ($r_S = -0.683$, $P<0.05$) with soil D-values (Table 2). We speculate that naturally C^- environmental isolates can survive well in the moderately acidic soil used in this study (pH 5.4), and is consistent with significant acid stress-resistance reported for C^- variants of EcO157 strains

Table 3. EcO157 survival in soil and proportion of curli sub-populations of parental strains.

Strain No[a]	Source	D-value, days	Curli, %
RM6103	Cow feces	10.0±1.2	0±0
RM6155	Pig feces	12.3±0.6	5±5
RM6067	Spinach	10.3±2.5	2±2
RM9834	Soil	10.8±2.4	13±6
RM6069	Clinical	4.0±2.6	20±9
RM6331	Clinical	7.2±0.5	36±14
RM6653	Clinical	5.7±1.3	43±15
RM6654	Clinical	6.0±1.6	42±3
RM6657	Clinical	5.5±0.8	72±10

[a]Except for RM9834 (MLVA 778) all other strains are of MLVA 163.

[27]. Acid stress-resistance is critical for EcO157 surviving the acidic conditions in the stomach of ruminants and humans, and perhaps other environments, and it is intriguing that, generally, environmental strains survive acid stress much better than the isolated clinical or food strains associated with outbreaks [36]. Similarly, all clinical strains we analyzed decreased in numbers more rapidly from soil compared to environmental strains.

The fitness responses observed for survival of EcO157 under stressful conditions in soil and resisting predation by protozoa were quite similar. Both environments elicited a similar phenotypic response of C^- variants, predominately, surviving the exposure, a negative correlation of proportion of C^+ variants with D-values of EcO157 and a positive correlation with growth of both predators. Since 8 of the 9 strains used in these tests are genetically indistinguishable from the 2006 spinach outbreak strain by MLVA (163), soil fitness or predation resistance of these isolates cannot be attributed to genetic differences. However, the naturally persistent, but genetically different, soil isolate (MLVA778), which was 87% C^- variants, resisted predation by both protozoa and also survived longer in spinach field soil. Thus, it appears that most pathogenic EcO157 fail to survive in hostile soil and water environments, but selection and enrichment of C^- variants is one survival strategy that may relate to contamination of produce.

In summary, these results indicate that the environmental persistence of pathogenic EcO157 is associated with the fitness of clonal subpopulations, although some generalizations can be made on survival in soil and resistance to predation based on isolation source. Whereas populations of clinical strains associated with the spinach outbreak decreased rapidly in soils or were consumed rapidly by protozoa, environmental strains (from soil and feral pig feces) survived longer in soil and resisted predation. Our results suggest also that exposure of EcO157 strains to stress conditions (e.g. soil or protozoa) generally result in rapid die-off, but the subpopulations that do survive are usually C^-. Thus, the presence of C^- variants in a population of cells at some frequency appears to be a survival mechanism associated with EcO157 fitness in soil and on plant surfaces [34], resisting protozoan predation, and potentially, other environmental stresses or conditions to be discovered. The mechanisms of C variation (e.g. *rpoS*) and the role of C^+ and C^- variants in complex environments warrant further studies to determine whether selection of some variants are associated also with hypervirulent strains associated with outbreaks.

Supporting Information

Table S1 MLVA characteristics of EcO157 strains used. Number of tandem repeats at each of 11 loci are given.

Table S2 Two-way ANOVA comparisons of strain differences in fitness of EcO157 to protozoan predation, protozoan growth response to EcO157 consumption and influence of protozoan exposure on proportion of curli subpopulations.

Acknowledgments

We thank B. Hoar (UC Davis), E. Hyytiä-Trees (CDC), W. Keene (OR DPH), W. Chmielecki (PA DPH), J. Sabo (PA DPH), and S. Weagant (FDA) for providing strains; and Michael Cooley, Diana Carychao and Kimberley Nguyen for MLVA typing.

Author Contributions

Conceived and designed the experiments: SR. Performed the experiments: CS SR. Analyzed the data: SR CS. Contributed reagents/materials/analysis tools: SR RM. Wrote the paper: SR RM.

References

1. Mandrell RE (2009) Enteric human pathogens associated with fresh produce: Sources, transport and ecology. In: Fan X, Niemira BA, Doona CJ, Feeherry F, Gravani RB, Microbial Safety of Fresh Produce: Challenges, Perspectives and Strategies. Oxford, UK: IFT/Blackwell Publishing. pp. 3−42.

2. Jay MT, Cooley M, Carychao D, Wiscomb GW, Sweitzer RA, et al. (2007) *Escherichia coli* O157:H7 in feral swine near spinach fields and cattle, central California coast. Emerg Infect Dis 13: 1908−1911.

3. Berry ED, Miller DN (2005) Cattle feedlot soil moisture and manure content: II. Impact on *Escherichia coli* O157. J Environ Qual 34: 656−663.

4. Avery LM, Hill P, Killham K, Jones DL (2004) *Escherichia coli* O157 survival following the surface and sub-surface application of human pathogen contaminated organic waste to soil. Soil Biol Biochem 36: 2101−2103.

5. Scheuerman PR, Schmidt JP, Alexander M (1988) Factors affecting the survival and growth of bacteria introduced into lake water. Arch Microbiol 150: 320−325.

6. Wang G, Doyle MP (1998) Survival of enterohemorrhagic *Escherichia coli* O157:H7 in water. J Food Prot 61: 662−667.

7. Jones DL (1999) Potential health risks associated with the persistence of *Escherichia coli* O157 in agricultural environments. Soil Use Manag 15: 76−83.

8. King CH, Shotts EB Jr, Wooley RE, Porter KG (1988) Survival of coliforms and bacterial pathogens within protozoa during chlorination. Appl Environ Microbiol 54: 3023−3033.

9. Newsome AL, Scott TM, Benson RF, Fields BS (1998) Isolation of an amoeba naturally harboring a distinctive *Legionella* species. Appl Environ Microbiol 64: 1688−1693.

10. Jenkins MB, Fisher DS, Endale DM, Adams P (2011) Comparative die-off of *Escherichia coli* O157:H7 and fecal indicator bacteria in pond water. Environ Sci Technol 45: 1853−1858.

11. Rasmussen MA, Carlson SA, Franklin SK, McCuddin ZP, Wu MT, et al. (2005) Exposure to rumen protozoa leads to enhancement of pathogenicity of and invasion by multiple-antibiotic-resistant *Salmonella enterica* bearing SGI1. Infect Immun 73: 4668−4675.

12. Cirillo JD, Cirillo SL, Yan L, Bermudez LE, Falkow S, et al. (1999) Intracellular growth in *Acanthamoeba castellanii* affects monocyte entry mechanisms and enhances virulence of *Legionella pneumophila*. Infect Immun 67: 4427−4434.

13. Gourabathini P, Brandl MT, Redding KS, Gunderson JH, Berk SG (2008) Interactions between food-borne pathogens and protozoa isolated from lettuce and spinach. Appl Environ Microbiol 74: 2518−2525.

14. Ravva SV, Sarreal CZ, Mandrell RE (2010) Identification of protozoa in dairy lagoon wastewater that consume *Escherichia coli* O157:H7 preferentially. PLoS One 5: e15671.

15. Ravva SV, Korn A (2007) Extractable organic components and nutrients in wastewater from dairy lagoons influence the growth and survival of *Escherichia coli* O157:H7. Appl Environ Microbiol 73: 2191−2198.

16. Ravva SV, Sarreal CZ, Duffy B, Stanker LH (2006) Survival of *Escherichia coli* O157:H7 in wastewater from dairy lagoons. J Appl Microbiol 101: 891−902.

17. Ravva SV, Sarreal CZ, Mandrell RE (2013) Altered protozoan and bacterial communities and survival of *Escherichia coli* O157:H7 in monensin-treated wastewater from a dairy lagoon. PLoS One 8: e54782.

18. Topp E, Welsh M, Tien YC, Dang A, Lazarovits G, et al. (2003) Strain-dependent variability in growth and survival of *Escherichia coli* in agricultural soil. FEMS Microbiol Ecol 44: 303−308.

19. Cookson AL, Cooley WA, Woodward MJ (2002) The role of type 1 and curli fimbriae of Shiga toxin-producing *Escherichia coli* in adherence to abiotic surfaces. Int J Med Microbiol 292: 195−205.

20. Ryu JH, Beuchat LR (2005) Biofilm formation by *Escherichia coli* O157:H7 on stainless steel: Effect of exopolysaccharide and curli production on its resistance to chlorine. Appl Environ Microbiol 71: 247−254.

21. Saldaña Z, Xicohtencatl-Cortes J, Avelino F, Phillips AD, Kaper JB, et al. (2009) Synergistic role of curli and cellulose in cell adherence and biofilm formation of attaching and effacing *Escherichia coli* and identification of *Fis* as a negative regulator of curli. Environ Microbiol 11: 992−1006.

22. Biscola FT, Abe CM, Guth BE (2011) Determination of adhesin gene sequences in, and biofilm formation by, O157 and non-O157 Shiga toxin-producing *Escherichia coli* strains isolated from different sources. Appl Environ Microbiol 77: 2201−2208.

23. Kikuchi T, Mizunoe Y, Takade A, Naito S, Yoshida SI (2005) Curli fibers are required for development of biofilm architecture in *Escherichia coli* K-12 and enhance bacterial adherence to human uroepithelial cells. Microbiol Immunol 49: 875−884.

24. Uhlich GA, Keen JE, Elder RO (2002) Variations in the *csgD* promoter of *Escherichia coli* O157:H7 associated with increased virulence in mice and increased invasion of HEp-2 cells. Infect Immun 70: 395–399.

25. Barnhart MM, Chapman MR (2006) Curli biogenesis and function. Annu Rev Microbiol 60: 131–147.

26. Brzuszkiewicz E, Gottschalk G, Ron E, Hacker J, Dobrindt U (2009) Adaptation of pathogenic *E. coli* to various niches: Genome flexibility is the key. Genome Dyn 6: 110–125.

27. Carter MQ, Brandl MT, Louie JW, Kyle JL, Carychao DK, et al. (2011) Distinct acid resistance and survival fitness displayed by Curli variants of enterohemorrhagic *Escherichia coli* O157:H7. Appl Environ Microbiol 77: 3685–3695.

28. Cooley M, Carychao D, Crawford-Miksza L, Jay MT, Myers C, et al. (2007) Incidence and tracking of *Escherichia coli* O157:H7 in a major produce production region in California. PLoS One 2: e1159.

29. Fratamico PM, Bagi LK, Pepe T (2000) A multiplex polymerase chain reaction assay for rapid detection and identification of *Escherichia coli* O157:H7 in foods and bovine feces. J Food Prot 63: 1032–1037.

30. Desmarchelier PM, Bilge SS, Fegan N, Mills L, Vary JC Jr, et al. (1998) A PCR specific for *Escherichia coli* O157 based on the rfb locus encoding O157 lipopolysaccharide. J Clin Microbiol 36: 1801–1804.

31. Kudva IT, Blanch K, Hovde CJ (1998) Analysis of *Escherichia coli* O157:H7 survival in ovine or bovine manure and manure slurry. Appl Environ Microbiol 64: 3166–3174.

32. Cooley MB, Carychao D, Nguyen K, Whitehand L, Mandrell R (2010) Effects of environmental stress on stability of tandem repeats in *Escherichia coli* O157:H7. Appl Environ Microbiol 76: 3398–3400.

33. Casida LE (1989) Protozoan response to the addition of bacterial predators and other bacteria to soil. Appl Environ Microbiol 55: 1857–1859.

34. Seo S, Matthews KR (2012) Influence of the plant defense response to *Escherichia coli* O157:H7 cell surface structures on survival of that enteric pathogen on plant surfaces. Appl Environ Microbiol 78: 5882–5889.

35. Gurijala KR, Alexander M (1990) Effect of growth rate and hydrophobicity on bacteria surviving protozoan grazing. Appl Environ Microbiol 56: 1631–1635.

36. Oh DH, Pan YW, Berry E, Cooley M, Mandrell R, et al. (2009) *Escherichia coli* O157:H7 strains isolated from environmental sources differ significantly in acetic acid resistance compared with human outbreak strains. J Food Prot 72: 503–509.

21

Expression and Functional Characterization of the *Agrobacterium* VirB2 Amino Acid Substitution Variants in T-pilus Biogenesis, Virulence, and Transient Transformation Efficiency

Hung-Yi Wu[1,2], Chao-Ying Chen[2], Erh-Min Lai[1,2]*

1 Institute of Plant and Microbial Biology, Academia Sinica, Taipei, Taiwan, **2** Department of Plant Pathology and Microbiology, National Taiwan University, Taipei, Taiwan

Abstract

Agrobacterium tumefaciens is a phytopathogenic bacterium that causes crown gall disease by transferring transferred DNA (T-DNA) into the plant genome. The translocation process is mediated by the type IV secretion system (T4SS) consisting of the VirD4 coupling protein and 11 VirB proteins (VirB1 to VirB11). All VirB proteins are required for the production of T-pilus, which consists of processed VirB2 (T-pilin) and VirB5 as major and minor subunits, respectively. VirB2 is an essential component of T4SS, but the roles of VirB2 and the assembled T-pilus in *Agrobacterium* virulence and the T-DNA transfer process remain unknown. Here, we generated 34 VirB2 amino acid substitution variants to study the functions of VirB2 involved in VirB2 stability, extracellular VirB2/T-pilus production and virulence of *A. tumefaciens*. From the capacity for extracellular VirB2 production (ExB2$^+$ or ExB2$^-$) and tumorigenesis on tomato stems (Vir$^+$ or Vir$^-$), the mutants could be classified into three groups: ExB2$^-$/Vir$^-$, ExB2$^-$/Vir$^+$, and ExB2$^+$/Vir$^+$. We also confirmed by electron microscopy that five ExB2$^-$/Vir$^+$ mutants exhibited a wild-type level of virulence with their deficiency in T-pilus formation. Interestingly, although the five T-pilus$^-$/Vir$^+$ uncoupling mutants retained a wild-type level of tumorigenesis efficiency on tomato stems and/or potato tuber discs, their transient transformation efficiency in *Arabidopsis* seedlings was highly attenuated. In conclusion, we have provided evidence for a role of T-pilus in *Agrobacterium* transformation process and have identified the domains and amino acid residues critical for VirB2 stability, T-pilus biogenesis, tumorigenesis, and transient transformation efficiency.

Editor: Ching-Hong Yang, University of Wisconsin-Milwaukee, United States of America

Funding: This work was supported by a research grant from the National Science Council (NSC 101-2321-B-001 -033 -) and a grant from Academia Sinica to EML. The funders had no role in study design, data collection and analysis, decision to publish, or preparation of the manuscript.

Competing Interests: The authors have declared that no competing interests exist.

* Email: emlai@gate.sinica.edu.tw

Introduction

Agrobacterium tumefaciens is a Gram-negative plant pathogenic bacterium that causes crown gall disease in a wide range of plants [1]. *A. tumefaciens* can sense plant-released phenolic compounds (e.g., acetosyringone; AS) to activate the expression of virulence factors for infection. The VirA/VirG two-component system is responsible for the phenolics-induced virulence (*vir*) gene expression [2]. Transfer DNA (T-DNA) located on the tumor-inducing (Ti) plasmid is recognized and processed by the VirD1/VirD2 relaxosome-like protein complex and covalently linked with the VirD2 protein to form the T-strand. The T-DNA and effector protein substrates are transferred through the VirB/VirD4 assembled type IV secretion system (T4SS) into host plant cells [2–4].

The *A. tumefaciens* VirB/VirD4 T4SS consists of the envelope-spanning translocation channel and the extracellular T-pilus structure [5–8]. Accumulating biochemical and genetic data suggest a possible VirB/D4 T4SS assembly and T-DNA translocation pathway [9,10]. The T-DNA immunoprecipitation (TrIP) technique revealed that T-DNA was first recruited by the VirD4 coupling protein, then the substrate passed to the inner-membrane–associated ATPase VirB11. The VirD4/VirB4 ATPases provide energy for T-DNA transfer to the inner-membrane proteins VirB6/VirB8 followed by passage to VirB2/VirB9 presumably localized at the distal end of the T4SS transmembrane complex [10]. Recent cryo-electron microscopy (cryo-EM) and crystallographic structure studies of the *Escherichia coli* conjugative plasmid pKM101 revealed that the T4SS core complex consisted of 14 copies of each of the VirB7-like TraN, VirB9-like TraO, and VirB10-like TraF subunits forming two layers of a double-walled ring structure inserted in the inner and outer membranes [11,12]. Remarkably, eight T4SS proteins (VirB3–VirB11) encoded by the R388 conjugative plasmid can assemble into an approximately 3-MDa nanomachine spanning the double membranes, which was visualized and reconstructed by electron microscopy [13]. VirB10 may function dynamically to couple cytoplasmic-membrane ATPases with ATP energy to gate the outer-membrane translocation channel via a conformational switch. Interestingly, VirD4 coupling protein not only functions as a receptor for protein substrates [14]; a recent study revealed that DNA but not protein binding to VirD4 and VirB11 activates the VirB10 structural transition and enables DNA transfer [14]. This study also suggested that translocation of DNA and protein substrates through T4SS may be mechanistically distinct processes [14].

Agrobacterium T-pilus is composed of the major subunit VirB2 and the minor component VirB5 [7,15,16]. All VirB proteins (VirB1 to VirB11) but not the VirD4 coupling protein is essential for T-pilus biogenesis [6]. Pilin subunits typically undergo an additional post-translational modification reaction after removal of its N-terminal signal peptide, including acetylation of F-like pilin or cyclization of P-like pilin and T-pilin [17–20]. The 12.3-kDa VirB2 precursor is processed into a 7.2-kDa product by removal of a long N-terminal signal peptide in both *E. coli* and *A. tumefaciens*, but the cyclization occurs only in *A. tumefaciens* in a Ti-plasmid–independent manner [21]. Consistent with predicted topology [19], experimental evidence revealed that processed VirB2 consists of two hydrophobic trans-membrane domains linked by an intervening hydrophilic loop (residues 90–94) inside the cytoplasm and by a periplasmic loop formed by linkage between hydrophilic N- and C-termini at residues 48 and 121 [22].

In contrast to much better-defined functions for the T4SS translocation channel, the roles of T4SS extracellular pilus remain obscure. T4SS substrate translocation from bacteria into host cells likely requires close contact with target cells [23] and pili may play a role during this process. A recent study revealed that the *A. tumefaciens* VirB/D4 T4SS forms helically arranged foci around the bacterial cell that may help maximize effective contact and transfer of substrate to host cells [24–26]. Therefore, T-pili may help *A. tumefaciens* bind to plant cells for close contact [25]. Alternatively, pili may also provide a channel for substrate translocation through their hollow lumen [27,28]. The transfer of DNA via the T4SS pili could be observed by direct visualization of low-efficiency conjugal transfer events via *E. coli* F-pili when cells were separated up to 1.2 μm [29]. Furthermore, DNA could be detected in the F-pilus channels during conjugation [30] and the *Helicobacter pylori* T4SS substrate protein CagA could be detected at the tip of the pilus [31]. However, mutants that block biogenesis of the T-pilus but not substrate transfer could be isolated by amino acid substitution in several T4SS components such as VirB6, VirB9, VirB10 and VirB11 of *A. tumefaciens* [14,32–36]. Because VirB2 protein is a T4SS component required for substrate translocation, isolation of these "uncoupling" mutants suggested that intracellular VirB2 but not its assembled T-pilus is required for T4SS-mediated T-DNA/effector translocation. Thus, the role of T-pilus remains unknown.

In this study, we used site-directed mutagenesis to generate various VirB2 single amino acid substitution variants to identify the amino acid residues critical for VirB2 stability, extracellular VirB2/T-pilus production, and virulence. Notably, we isolated five T-pilus⁻/Vir⁺ uncoupling mutants that retain a wild-type level of tumorigenesis efficiency on tomato stems and/or potato tuber discs but are highly attenuated in transient transformation efficiency in *Arabidopsis* seedlings. These data suggest a role of T-pilus in the *Agrobacterium* transformation process.

Results

VirB2 family proteins comprise variable N-terminal signal peptides and conserved C-terminal processing products

To determine the amino acid residues critical for the function of VirB2 in *Agrobacterium* virulence and T-pilus production, we first compared the amino acid sequences of VirB2 homologs encoded by various agrobacteria and rhizobia (Figure 1). The N-terminal signal peptide of VirB2 is variable, but the mature processed T-pilin region (including periplasmic, trans-membrane and cytoplasmic domains) is highly conserved (Figure 1 and Table S1). A conserved motif PAxAQ at the processing site is critical for precise signal peptide removal and cyclization of processed T-pilin [17,21].

Identification of domains and amino acid residues critical for VirB2 stability, processing, and extracellular VirB2 production

For full complementation of *Agrobacterium* virulence in a *virB2* in-frame deletion mutant, *virB2* must be co-expressed with adjacent genes and driven by its native promoter [37]. Therefore, we cloned the DNA fragment containing the *virB* promoter and *virB1*, *virB2* and *virB3* genes (*virB*p-*B1-B2-B3*) into a broad host-range plasmid, pRL662, for expression of wild-type VirB2 and all VirB2 variants in the *virB2* in-frame deletion mutant (Δ*virB2*) derived from the *A. tumefaciens* wild-type C58 strain. The conserved or non-conserved amino acid residues near the processing site or different domains within the T-pilin region were randomly chosen for substitution with Alanine (A). *A. tumefaciens* cells induced for T-pilus production were scraped off the agar plate, resuspended in acidic phosphate buffer (pH 5.3), and centrifuged to obtain the S1 fraction. Cell pellets were resuspended again and subjected to shearing to obtain the S2 fraction enriched for T-pilus. Both intracellular and extracellular VirB2 levels were restored to wild-type levels in Δ*virB2* complemented with wild-type VirB2 (pVirB2), which was consistent with its full complementation by tumorigenesis analysis on tomato stems (Figure 2). Consistent with our previous study [21], we detected both pro-pilin (12.3-kDa unprocessed VirB2 precursor, named VirB2p) and T-pilin (7.2-kDa processed mature VirB2, named VirB2m) inside cells, but only processed T-pilin was detected extracellularly (Figure 2, also see Figure S1 for lower intensity of western blot signals). Extracellular VirB2 was more abundant in the S2 than S1 fraction, and all variants with no detectable VirB2 in the S2 fraction also did not produce any VirB2 signals in the S1 fraction (Figure 2 and S2). Because the detection of VirB2 in the sheared S2 fraction by western blot analysis agrees with the observation of T-pilus by electron microscopy and vice versa [6], we used western blot analysis of the S2 fraction as a first step to screen the mutant phenotype for T-pilus production.

For all five VirB2 variants with amino acid substitutions near the processing site (from P44 to G51), we detected both VirB2p and wild-type levels of VirB2m within the cells (Figure 2). Notably, extracellular VirB2 levels were reduced from the P44A and S49A variants and not detected from the A47V variant. Increased intracellular VirB2p from the P44A variant suggested a role of P44 for processing efficiency. The absence of extracellular VirB2 and slower migration of the intracellular VirB2m from the A47V variant relative to wild-type VirB2m implied that the A47V variant may undergo incorrect processing or cyclization, thus leading to defects in production of extracellular VirB2 or T-pilus. Strikingly, all variants except the R91A variant accumulated comparable intracellular VirB2m levels, in which only the unprocessed but not processed R91A variant could be detected (Figure 2). Because R91 is the sole positively charged residue within the cytoplasmic domain, substitution of Arginine 91 with Alanine may break the "positive inside rule" [38] and result in the instability of the R91A variant after processing.

Little or no extracellular VirB2 could be detected in all the mutants with amino acid substitutions in trans-membrane domain

1 (TM1) (Figure 3). In contrast, substitutions located in the N-terminal periplasmic domain (N-PP), trans-membrane domain 2 (TM2), and C-terminal periplasmic domain (C-PP) had differential effects on the accumulation of extracellular VirB2m. Therefore, these domains of processed VirB2 are all indispensable, but the integrity of the TM1 and its adjacent regions are most critical for production of extracellular VirB2m.

Identification of VirB2 variants uncoupling virulence and T-pilus biogenesis phenotypes

Tumorigenicity of each VirB2 variant was first determined by infection on tomato stems, and their virulence was evaluated by the occurrence or size of the tumors formed. From the capacity for extracellular VirB2 production (ExB2$^+$ or ExB2$^-$) and tumorigenesis on tomato stems (Vir$^+$ or Vir$^-$), the mutants were classified into three groups: ExB2$^-$/Vir$^-$, ExB2$^-$/Vir$^+$, and ExB2$^+$/Vir$^+$ (Figure 3). Although the combination of extracellular VirB2 production and virulence phenotypes theoretically can result in four groups of mutant phenotypes, we identified only three groups from our mutant pools. Group I mutants with the ExB2$^-$/Vir$^-$ phenotype led us to identify the amino acid residues crucial for both extracellular VirB2 production and tumorigenesis. These amino acid residues are mostly dispersed across all domains within the T-pilin region, but three (M88A, F89A, R91A) are located at the junction of TM1 and cytoplasmic domain (CP). Group II mutants with the ExB2$^-$/Vir$^+$ phenotype showed the amino acids residues critical for extracellular VirB2 production but dispensable for tumorigenesis. Strikingly, all mutants defective in virulence also lost extracellular VirB2 production (group I mutants) and all mutants capable of extracellular VirB2 production were virulent (group III mutants). These VirB2 amino acid residues required for *A. tumefaciens* to form a functional T4SS for substrate transfer may also be essential for production of extracellular VirB2.

Among the three groups of mutants, we were particularly interested in the five variants classified in the ExB2$^-$/Vir$^+$ group because these VirB2 variants (D55A, I85A, L94A, M107A, A110G) can incite the wild-type size of tumors on tomato stem but did not produce detectable extracellular VirB2. Although the loss of extracellular VirB2 production agrees with the loss of T-pilus biogenesis [6], we could not exclude that the failure to detect extracellular VirB2 may be due to the formation of short or structurally distinct T-pilus recalcitrant to isolation by shearing from the mutants. Thus, we negatively stained the *A. tumefaciens* cells expressing wild-type VirB2 and these five ExB2$^-$/Vir$^+$ variants by uranyl acetate and examined them by transmission electron microscopy (TEM) to observe the bacterial cells and associated surface structures. Similar to previous reports [6,7,17], we observed T-pilus as a rigid or semi-rigid long filament (500 nm to 2 μm) ~10-nm wide in the wild-type complemented strain Δ*virB2*(pVirB2) in an AS-induction–dependent manner (Figure 4A). In contrast, no T-pilus-like filament could be detected from these ExB2$^-$/Vir$^+$ mutants (Figure 4B–F). These results agree with the lack of extracellular VirB2 detected by western blot analysis and suggest that these five mutants represent an uncoupling phenotype of virulence and T-pilus biogenesis (Figure 2). Thus, we defined the *A. tumefaciens* strains expressing these five VirB2 variants (D55A, I85A, L94A, M107A, A110G) as T-pilus$^-$/Vir$^+$ uncoupling mutants.

T-pilus$^-$/Vir$^+$ uncoupling mutants show highly attenuated transient transformation efficiency in *Arabidopsis* seedlings

The evidence that the T-pilus$^-$/Vir$^+$ uncoupling mutant retains the ability to incite tumors on tomato stems suggested that the T-pilus may not be essential for virulence. To test whether these T-pilus$^-$/Vir$^+$ uncoupling mutants may quantitatively affect *Agrobacterium* virulence or transformation efficiency, we first used the quantitative tumor assay on potato tuber discs to determine their virulence. Two T-pilus$^-$/Vir$^+$ uncoupling mutants (L94A and A110G) and a randomly selected ExB2$^+$/Vir$^+$ mutant (G121A) all retained the wild-type tumorigenesis efficiency when infected with 10^6 or 10^4 cells (Figure S3). Because tumorigenesis is a complex and long process that may not be sensitive enough to detect quantitative differences in T-DNA transfer efficiency, we adapted a transient transformation assay recently developed in our laboratory [39] for further analysis. This system used the T-DNA–encoded β-glucuronidase *gusA* (*GUS*) gene as a reporter to quantitatively monitor transient transformation efficiency in *Arabidopsis* seedlings. The T-DNA vector pBISN1 harboring the *gusA-intron* [40] was transformed into an *A. tumefaciens* strain with virulence gene expression induced by acetosyringone (AS) before infection of 4-day-old seedlings. The seedlings were co-cultured with pre-induced *A. tumefaciens* in the presence of AS, and GUS activity was determined to monitor transient transformation efficiency at 3 days post-infection (dpi). Remarkably, all five T-pilus$^-$/Vir$^+$ uncoupling mutants showed highly reduced GUS stains in cotyledons with 3- to 4-fold lower GUS activity as compared to the wild type (Figure 5A and 5C). In contrast, five ExB2$^+$/Vir$^+$ mutants exhibited higher transient transformation efficiency than all T-pilus$^-$/Vir$^+$ uncoupling mutants, four showing efficiency comparable to that of the wild type (Figure 5B and C). These results suggest the importance of T-pilus in the *Agrobacterium* transformation process.

Transient transformation assays of *Arabidopsis* seedlings with wounded cotyledons

In contrast to tumor assays on tomato stems and potato tuber discs, with *A. tumefaciens* cells infected on wounded tissues, no intentional wounding was included during *A. tumefaciens* infection in *Arabidopsis* seedlings. Thus, we tested whether the T-pilus$^-$/Vir$^+$ uncoupling mutants could efficiently infect wounded tissue of *Arabidopsis* seedlings. Because cotyledons are highly transformed in the *Arabidopsis* seedling transient transformation assay, we wounded cotyledons of 7-day-old *Arabidopsis* seedlings by using a needle. Similar to the method used to infect wounded tomato stems and potato tuber discs, overnight-grown *A. tumefaciens* cells without AS induction were used to infect wounded cotyledons of *Arabidopsis* seedlings. As shown in Figure 6A, we detected large GUS stains extending from the wound site of cotyledons on co-culture with cells containing wild-type VirB2 or the ExB2$^+$/Vir$^+$ G121A variant, which also revealed wild-type tumorigenesis efficiency in potato tuber discs (Figure S3) and transient transformation efficiency in intact *Arabidopsis* seedlings (Figure 5A and C). In contrast, GUS signals were detected only at a focused wound site of cotyledons and not expanded to unwounded regions on co-culture with the two T-pilus$^-$/Vir$^+$ uncoupling mutants tested (L94A and A110G). As a control, 7-day-old seedlings were infected with pre-induced *A. tumefaciens* cells under unwounded conditions (Figure 6B). We detected 3- to 4-fold lower GUS activity in the uncoupling mutants as compared with *A. tumefaciens* cells producing wild-type VirB2 or the ExB2$^+$/Vir$^+$ G121A variant, which

```
                                        SP                    ↓              N-PP
                                     *    20    *    40    *    60    *
Agrobacterium tumefaciens C58 pTiC58     VirB2    : -MRCFERYRVHLNRLSLSNAVMRMVSGYAPVVGVMGWSIFSSGPAAAQSAGGGTDPATMVNNICTFILGPF : 71
Agrobacterium rhizogenes pRi1724         Riorf154 : -MAWLEGYRAPSKFRGLWHRAVRLIAPHVPSVTGAIGWSLFFCEPAAAQAAGG-TDPATMVNNICTFILGPF : 70
Agrobacterium tumefaciens pTiA6          VirB2    : -MRCFERYRLHLNRLSLSNAMMRVISSCAPSLGGAMAWSISSCGPAAAQAAGG-TDPATMVNNICTFILGPF : 71
Agrobacterium vitis S4 pTiS4             VirB2    : -MACFEKYPELSKIRDLRDRVVSLIVPHLPSVSGAIGWSLFFREPAAAQAAGG-TDPATMVNNICTFILGPF : 70
Rhizobium etli CFN42 p42a                VirB2a   : MMRCFERYRVHLNRLSLSNAVMRLVSGYAPVVGGMGWSMFSFGPAAAQSAGGGTDPATMVSNICTFILGPF : 72
Sinorhizobium medicae WSM419 pSMED02     VirB2    : -MASLNKHRARLSPVAFLKAMSRTAVDHAPAAGGAVWSIVCSGPATACVTGG-TDPATMVNNICAFILGPF : 70
Mesorhizobium cciceri bv. biserrulae chr. TrbC    : MIKPLKILRALKTRVPSPFSMLRMAAEYAPAAAAGVAWTVFSSGPAAAQVTGG-TDPAKMVQNICTFILGPF : 71
Agrobacterium radiobacter K84 chr.       AvhB2    : -----------------MMSRINIRAFGVIVAMITVLSIAMIEPAFAQSAG----IETVLQNIVTLLTCNV : 51
Agrobacterium tumefaciens C58 pAtC58     AvhB2    : -----------------MIISSRIRPVVASSVMAVAIIVTMVEPAFAQSAG----IETVLQNIVDMLTCNI : 50
Rhizobium etli CFN42 p42d                VirB2    : -----------------MISKAPIRPLAASTLMAAIVICLVEPAFAQAAG----IETVLQNIVDMLTCNI : 50
Sinorhizobium meliloti 1021 pSymA        VirB2    : -----------------MTFSSRIRPIAASTVMATAIMVTMVEPAFAQSAG----IETVLQNIVDMLTCNI : 50
                                                    a      6    PA AQ aG      t66 NI    6 G
```

```
                              TM-1        CP        TM-2          C-PP
                                   80        *        100       *        120
Agrobacterium tumefaciens C58 pTiC58     VirB2    : GQSLAVLGIVAIGISWMFGRASLGLVAGVVGGIVIMFGASFIGKILTGGG- : 121
Agrobacterium rhizogenes pRi1724         Riorf154 : GQSLAVLGIVAIGIVSWMFGRASLGLVAGVVGGIVIMFGASFIGQILTGGG : 121
Agrobacterium tumefaciens pTiA6          VirB2    : GQSLAVLGIVAIGISWMFGRASLGLVAGVVGGIVIMFGASFIGQILTGGS- : 121
Agrobacterium vitis S4 pTiS4             VirB2    : GQSLAVLGIVAIGVSWMFGRASLGLVAGVVGGIVIMFGASFIGQILTGGG : 121
Rhizobium etli CFN42 p42a                VirB2a   : GQSLAVLGIVAIGICLSWMFGRASLGLIAGVVGVVIMFGASFIGKILTGGG- : 122
Sinorhizobium medicae WSM419 pSMED02     VirB2    : GQSLAVLGIVAIGISWMFGRASLGLVAGVIGGIVIMFGASFIGKALIGAG- : 120
Mesorhizobium cciceri bv. biserrulae chr. TrbC    : GQSLAVLGLVAIGISWMFGRASLGLVAGVVGGIVIMFGASFIGKALIGAG- : 121
Agrobacterium radiobacter K84 chr.       AvhB2    : AKLLATIAVIIVGIAWMFGYLDLRKAAYVVLGIGILFGASQIVSTISGG-- : 100
Agrobacterium tumefaciens C58 pAtC58     AvhB2    : AKLLAVIAVIVICIAWMFGYMDLRRAGFWIIGIGGIFGATELVNTIVGA-- : 99
Rhizobium etli CFN42 p42d                VirB2    : ARLLAVIAVIIISIAWMFGYMDLRRAGFWIIGIGGIFGATELVNTIVGN-- : 99
Sinorhizobium meliloti 1021 pSymA        VirB2    : AKLLAVIAVIVICIAWMFGYMDLRRAGFWIIGIGGIFGATELVNTIVGS-- : 99
                                                    LAv6 66 6 6 WMFG   L    6 G6  6FGA3 6 t6 G
```

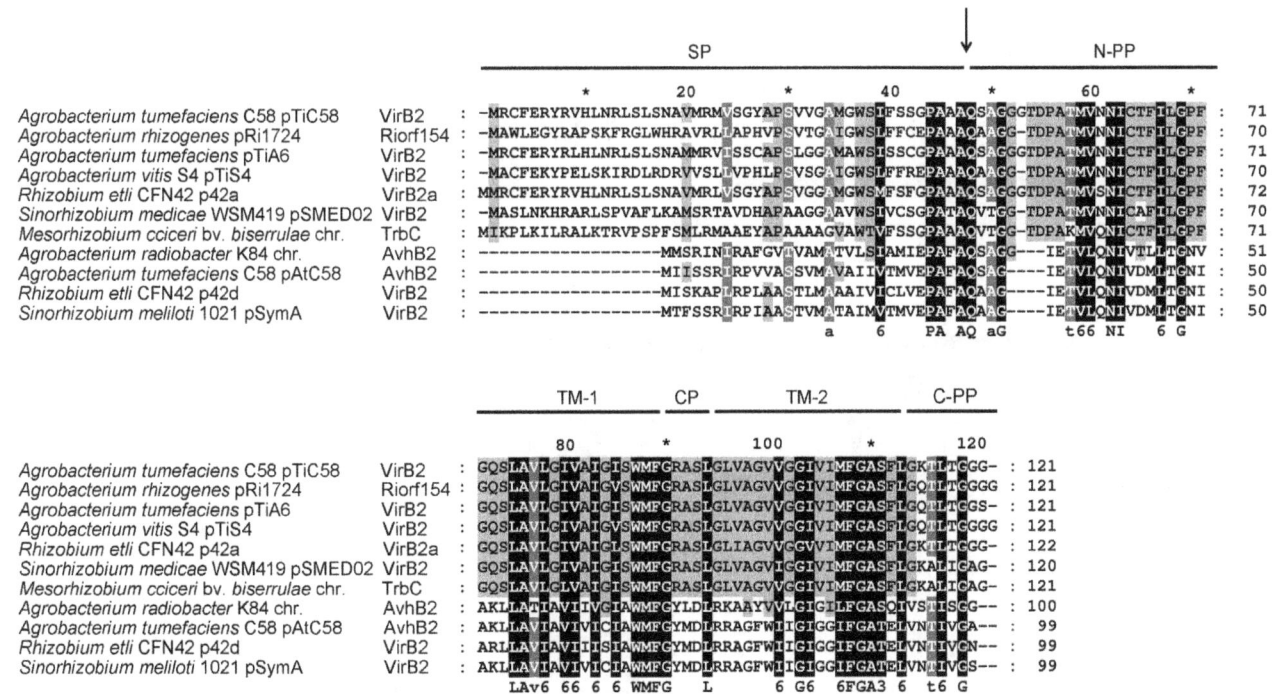

Figure 1. Amino acid sequence alignment of VirB2 family proteins. Multiple amino acid sequence alignment of VirB2 homologues with ClustalW2 [50]. The organism/plasmid name for each homolog is indicated on the left of the aligned sequence. UniProt accession numbers: *Agrobacterium tumefaciens* pTiC58, P17792; *Agrobacterium rhizogenes* pRi1724, Q9F5A1; *Agrobacterium tumefaciens* pTiA6, P05351; *Agrobacterium vitis* pTiS4, B9K417; *Rhizobium etli* p42a, Q2K2L1; *Sinorhizobium medicae* pSMED02, A6UMA7, *Mesorhizobium ciceri* chromosome (chr.), E8TGI0; *Agrobacterium radiobacter* K84 chromosome (chr.), B9JE70; *Agrobacterium tumefaciens* pAtC58, Q7D3S1; *Rhizobium etli* p42d, Q8KIM6; *Sinorhizobium meliloti* pSymA, Q92YZ4. The amino acid residues identical in all proteins are in black and those conserved in most but not all of the proteins are in gray. The arrow indicates the processing site of VirB2 encoded by pTiC58. Each region/domain of VirB2 is indicated: SP, signal peptide; TM-1, trans-membrane domain 1; CP, cytoplasmic domain; TM-2, trans-membrane domain 2; N-PP, N-terminal periplasmic domain; C-PP, C-terminal periplasmic domain.

efficiently infects both cotyledons and newly emerged true leaves not restricted to specific regions (Figure 6A and B). Because T-DNA transient transformation activity does not require the step(s) of T-DNA integration into the plant genome [40], T-pilus may play a role in the T-DNA transfer process at steps before T-DNA integration when infecting wounded and unwounded cotyledons of *Arabidopsis* seedlings.

Discussion

In this study, we identified the VirB2 key amino acid residues involved in VirB2 stability, extracellular VirB2/T-pilus production, and virulence of *A. tumefaciens* and discovered a role for the T-pilus in enhancing transient transformation efficiency. By screening 34 VirB2 variants for their ability to promote T-pilus production and tumorigenesis, we found that all mutants that are capable of producing extracellular VirB2 are virulent, and all mutants with loss of virulence (no tumor) also do not produce extracellular VirB2. Because no mutants with the ExB2$^+$/Vir$^-$ phenotype could be isolated in our screen, the VirB2 amino acid residue essential for forming a functional T4SS for substrate translocation may also be required for production of extracellular VirB2 and/or T-pilus.

Interestingly, the G119A variant generated in this study showed low amounts of extracellular VirB2 and full virulence phenotypes (Figure 2), whereas the substitution of Glycine 119 with Cysteine in pTiA6 VirB2 caused loss of virulence but retained function in

producing wild-type levels of extracellular VirB2 and T-pilus [22]. To clarify the phenotype discrepancy between pTiC58 G119A and pTiA6 G119C, we generated the pTiC58 VirB2 G119C amino acid substitution as for 34 other VirB2 variants and expressed it in C58 Δ*virB2*. By examining the levels of extracellular VirB2 in strains producing wild-type VirB2 and the G119A and G119C variants in parallel, all three independent G119C variants and the G119A variant produced low levels of extracellular VirB2 as compared with wild-type VirB2 (Figure S4). However, by examining all three independent pTiC58 G119C variants in three independent experiments, we did not observe any rigid and long T-pilus-like filament with ~10 nm diameter. In contrast, we detected the presence of T-pilus produced with the pTiC58 G119A variant, although the number of T-pilus observed was less frequent than with the wild-type VirB2 (Figure 4).

Transient transformation efficiency was comparable with the G119A variant and the wild type VirB2, but no transient GUS activity was detected in *Arabidopsis* seedlings infected with the G119C variant (Figure S5). Consistently, the G119A variant also retained the wild-type tumorigenicity in tomato stems and potato tuber discs, whereas the G119C variant showed highly attenuated induction of tumors on potato tuber discs (Figure S5A). Since G119A and G119C variants both caused reduced or abolished extracellular VirB2/T-pilus production, Glycine 119 may be critical for the efficient assembly of T-pilin into extracellular T-pilus. The nonpolar amino acid at position 119 seems critical for

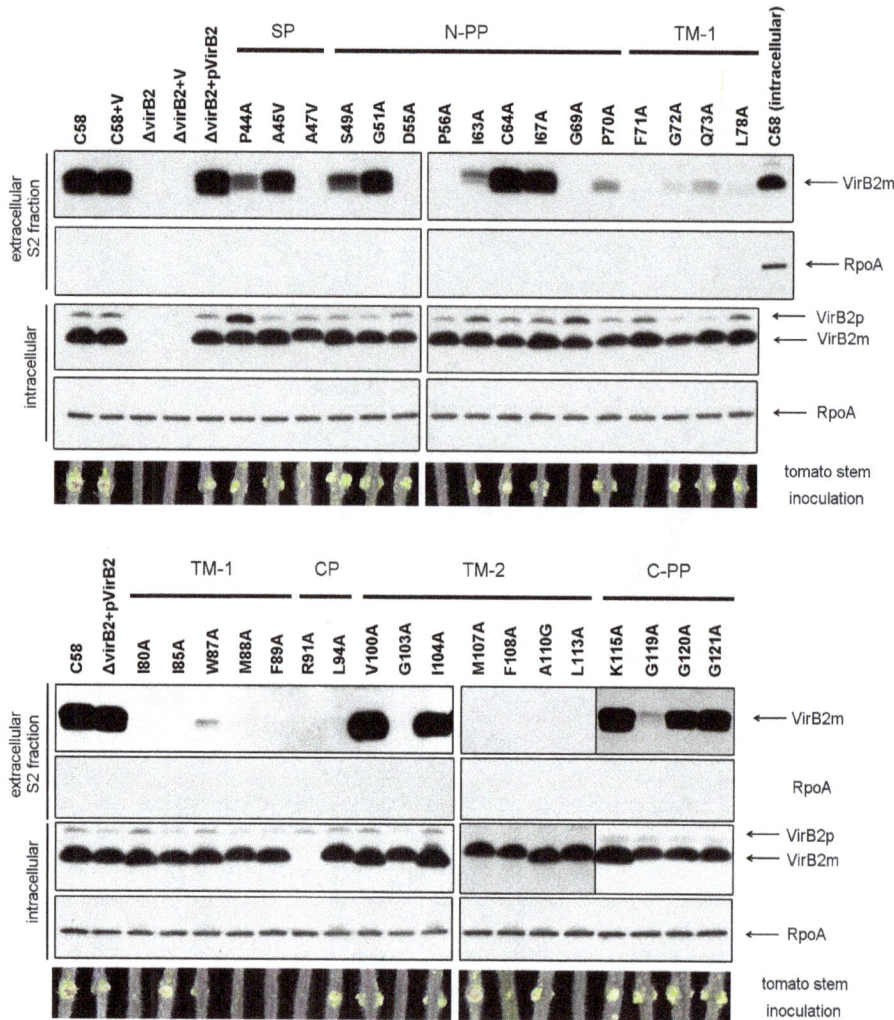

Figure 2. Western blot analysis of the intracellular and extracellular S2 fractions, and tumor assays on tomato stem of VirB2 variants. *A. tumefaciens* cells grown on acetosyringone (AS)-induced AB-MES (pH 5.5) agar at 19°C for 3 days [7] were collected to isolate intracellular proteins and extracellular S2 fractions. C58, *A. tumefaciens* wild type strain C58; V, empty vector pRL662; Δ*virB2*, *virB2* deletion mutant; Δ*virB2*(pVirB2), expression of wild type *virB*p-*B1-B2-B3* in Δ*virB2*. Western blot analysis with antisera against VirB2 B24 peptide or B23 peptide (for variants in C-PP) or RNA polymerase RpoA, as an internal control. Unprocessed VirB2 precursor is indicated as VirB2p and processed mature VirB2 as VirB2m. Shows representative tumor assay results on tomato stems. Similar results were obtained from at least three independent experiments (3–5 plants for each mutant in each independent experiment). Each region/domain of VirB2 is indicated as described in Figure 1.

efficient T-DNA and/or effector translocation because replacing nonpolar Glycine 119 with polar Cysteine but not nonpolar Alanine strongly suppressed the tumorigenesis/transformation efficiency. Although the loss of extracellular VirB2 in both S1 and S2 fractions is consistent with the lack of any observable long and rigid T-pilus-like filament in our five T-pilus⁻/Vir⁺ uncoupling mutants, these methods may have limitations to detect short T-pilus produced on bacterial cell surfaces. It is also possible that certain VirB2 variants may assemble T-pilus only during infection *in planta*. To this end, our data revealed the positive correlation of long and rigid T-pilus with the ability to induce high transient transformation efficiency in *Arabidopsis*, although we observed no dose-dependent correlation. This result highlighted a critical role of T-pilus involved in transient transformation efficiency in *Arabidopsis* seedlings, but T-pilus seems to be dispensable for inciting tumors in wounded plant tissues. Consistently, the T-pilus⁻/Vir⁺ un-

coupling mutants were isolated from amino acid substitution in other T4SS components including VirB6, VirB9, VirB10, and VirB11 [14,32–36].

The lack of T-pilus with the pTiC58 G119C variant is in contrast to the abundant T-pilus production from the pTiA6 G119C variant, which is also defective in substrate transfer based on the deficiency in IncQ plasmid transfer and tumorigeneisis on the wound site of *Kalanchoe daigremontiana* leaves [22]. However, all the Cysteine-substitution VirB2 variants created by Kerr and Christie [22] were generated in a VirB2 C64S (Cysteine 64 substitution by Serine) background to perform Cysteine labeling for mapping the VirB2 membrane topology. Although pTiA6 C64S variant exhibited near-wild-type substrate translocation frequency, whether the T-pilus⁺/Vir⁻ phenotype observed by the pTiA6 VirB2 C64S-G119C variant is solely caused by G119C substitution remains to be determined. Thus, the difference in T-pilus production phenotypes between pTiC58 and pTiA6 VirB2

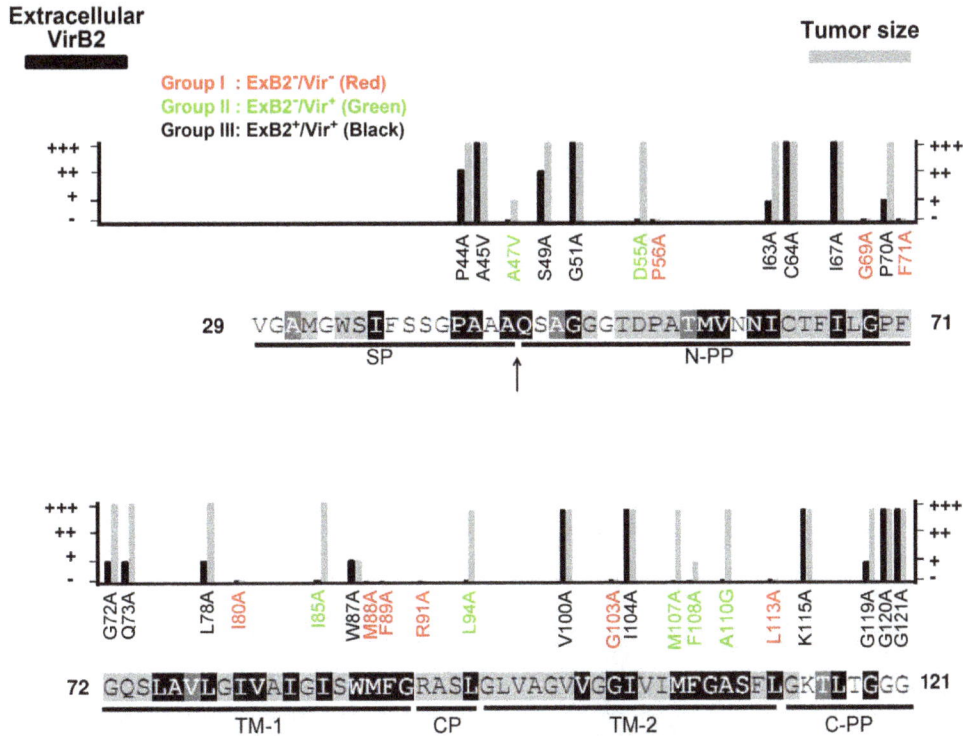

Figure 3. Phenotype summary of VirB2 variants. VirB2 amino acid substitutions are indicated with the levels of extracellular VirB2 (ExB2) production and occurrence or size of tumor formation on tomato stems (Vir) with wild type (+++), modest reduction (++), highly attenuation (+), or loss (−). VirB2 protein sequences with indicated conserved amino acid residues, regions/domains, and the processing site (indicated by an arrow) are presented as described in Figure 1. The 34 VirB2 variants are classified into three groups: ExB2⁻/Vir⁻, ExB2⁻/Vir⁺, and ExB2⁺/Vir⁺ shown in red, green and black, respectively.

G119C variants could be caused by this additional C64S mutation created in pTiA6. In addition, another four pTiC58 VirB2 Alanine substitution variants showed contrasting phenotypes in virulence or extracellular VirB2/T-pilus production as compared with pTiA6 VirB2 Cysteine substitution variants (Table S4). Thus, future work by creating the VirB2 single amino acid substitution with identical amino acid is required to confirm the phenotype discrepancy observed between two Ti plasmids. Replacing a specific amino acid residue by an amino acid with both similar and opposite biochemical features may be critical to unambiguously identify the role of each specific amino acid in contributing the observed phenotype.

Of note, pre-induction of *A. tumefaciens vir* gene expression before infection is critical for successful transient transformation in intact *Arabidopsis* seedlings without intentional wounding (data not shown). In contrast, this *vir* pre-induction is not required when infecting wounded cotyledons of *Arabidopsis* seedlings (Figure 6A). Since T-pilus⁻/Vir⁺ uncoupling mutants caused reduced transient transformation efficiency when infecting *Arabidopsis* seedlings with wounded or unwounded cotyledons, T-pilus may contribute to enhance *Agrobacterium* transient transformation efficiency at both wounded and unwounded infection conditions. However, in contrast to the more pronounced GUS activity detected further from wound sites of cotyledons infected with T-pilus⁺-producing strains, GUS signals were detected only at a focused wound site of cotyledons infected with the T-pilus⁻/Vir⁺ uncoupling mutants. Thus, T-pilus may be dispensable for infection at a wound site but critical for infecting unwounded tissues/cells. This possibility may also explain why these T-

pilus⁻/Vir⁺ uncoupling mutants retained their ability to incite tumor formation when infecting wounded tomato stems and/or potato tuber discs in this study and in wounded *K. daigremontiana* leaves in other studies [14,32–36].

In summary, this study provides compelling evidence for a role of T-pilus in the *Agrobacterium* transformation process, although the mechanisms involved remain unclear. Recent microscopy studies revealed the formation of T4SS helical array around the bacterial cell and observed a pilus-like structure connecting bacterial cells to the plant cell [24,25]. However, the Ti plasmid and T-pilus were found not required for *A. tumefaciens* attachment to the plant cell [41]. Thus, T-pilus may modulate plant responses to achieve optimal transformation efficiency. VirB5 was found localized in the T-pilus tip [42] and extracellular VirB5 can accelerate *Agrobacterium* transformation efficiency [43]. T-pilus may provide a vehicle to localize VirB5 or other substrates to the correct location and exert their functions on/in plants or the T-pilus/VirB2 may play a direct role. The exact role and molecular mechanisms underlying how T-pilus contributes to *Agrobacterium* transformation process await future investigation.

Materials and Methods

Bacterial strains, growth and T-pilus induction conditions

Bacterial strains and plasmids used are in Table S2. *A. tumefaciens* and *E. coli* were grown in 523 medium [44] at 28°C and LB at 37°C, respectively, with appropriate antibiotics. The plasmids were maintained by the addition of 50 µg/ml gentamycin (Gm)

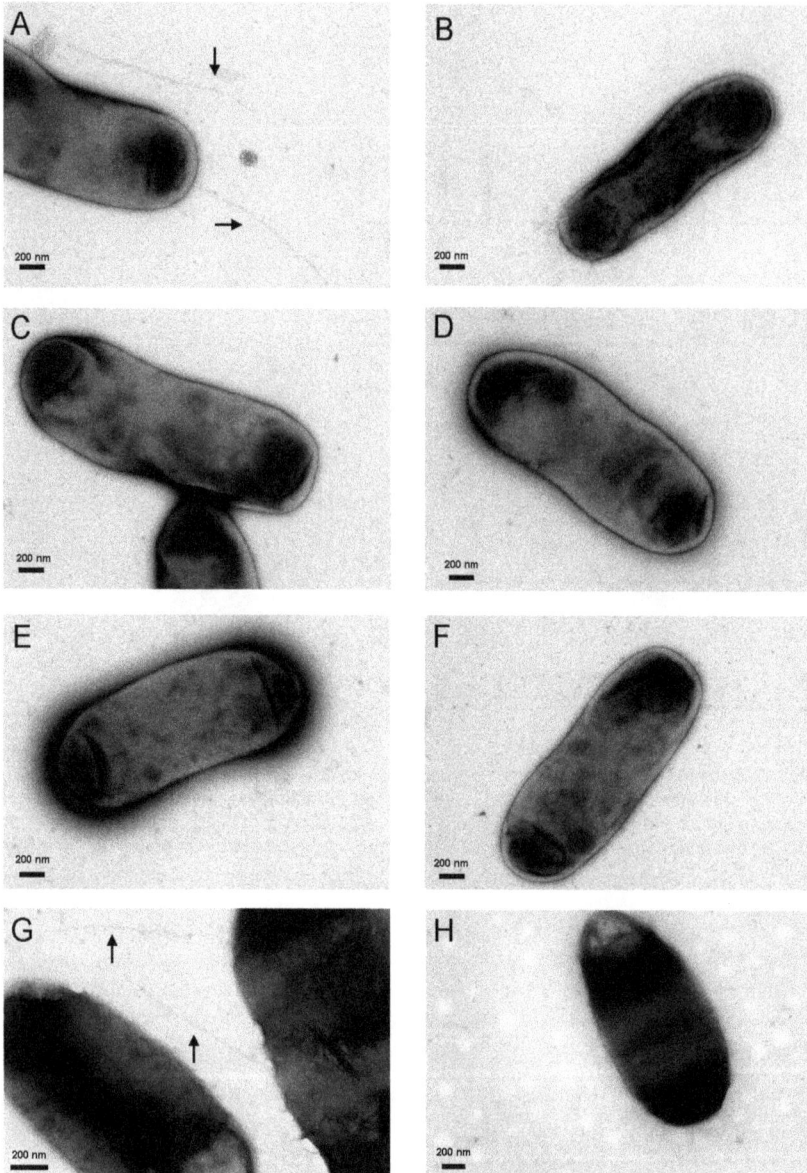

Figure 4. T-pilus observation by transmission electron microscopy (TEM). *A. tumefaciens* cells grown on T-pilus induction condition were collected and stained with 2% uranyl acetate to visualize T-pilus by TEM. Shows representative TEM image of *A. tumefaciens* strains producing wild-type VirB2 (A) or VirB2 variants D55A (B), I85A (C), L94A (D), M107A (E), A110G (F), G119A (G) and G119C (H). The rigid, long T-pilus is indicated by an arrow (A, G). Scale bar: 200 nm. All samples were examined for T-pilus formation by examining hundreds of bacterial cells from at least two independent experiments.

and 20 µg/ml kanamycin (Km) for *A. tumefaciens* and 20 µg/ml Km and 50 µg/ml Gm for *E. coli*. The T-pilus induction condition was as described [7]. Briefly, overnight culture of *A. tumefaciens* cells grown in 523 medium with appropriate antibiotics [44] were harvested and resuspended in liquid AB-MES minimal medium, pH 5.5 [7], with an adjusted OD_{600} of 0.1 for 4 hr without antibiotics. In total, 500 µl bacterial suspension was spreaded onto solid AB-MES medium with 200 µg/ml acetosyringone (AS) in a 150-mm Petri dish and incubated at 19°C for 3 days without antibiotics.

Construction of mutant strains and complementing plasmids

Gene replacement with the suicide plasmid pJQ200KS to generate the *virB2* in-frame deletion mutant (deletion of amino acid residues 4 to 113) followed a previous study [45]. For construction of pJQ-*virB2* for gene replacement, *virB2*-up (550-bp) and *virB2*-down (720-bp) DNA fragments were amplified by *pfu* polymerase and digested with SacI/SpeI and SpeI/XhoI separately and ligated into pJQ200KS at SacI/XhoI sites. For generating a plasmid to express *virB2* for complementation (pVirB2), a 2.1-kb fragment the *virB* promoter and *virB1*, *virB2* and *virB3* genes (*virB*p-*B1-B2-B3*) was PCR-amplified from the *A.*

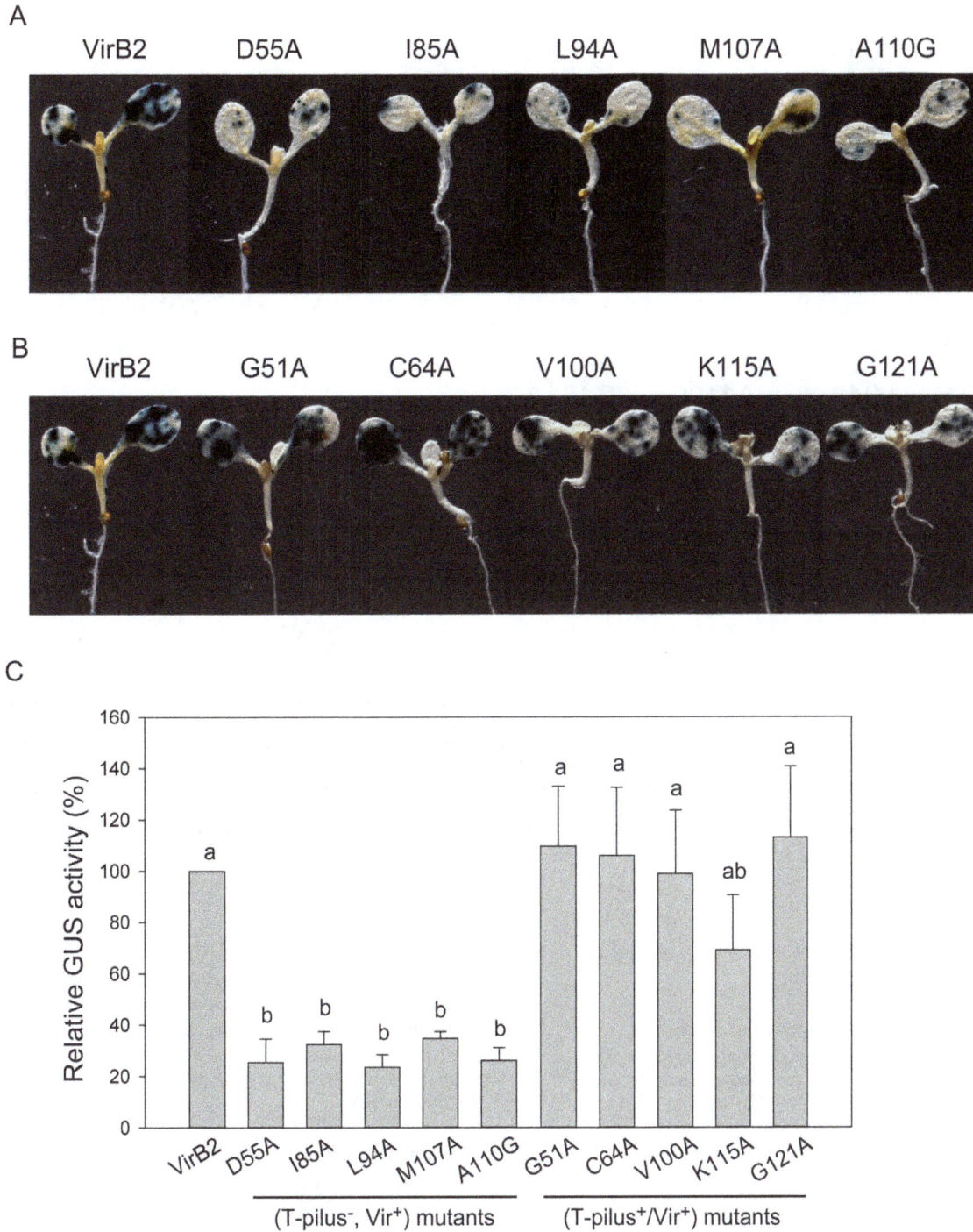

Figure 5. T-pilus⁻/Vir⁺ uncoupling mutants show highly attenuated transient transformation efficiency in *Arabidopsis* seedlings. *A. tumefaciens* strains expressing the wild type or variants of VirB2 harboring the T-DNA vector pBISN1 were used to infect 4-day-old *Arabidopsis* seedlings. GUS activity as a reporter of transient transformation efficiency was determined by GUS staining (A and B) or quantitative activity assay (C) at 3 dpi. (A) GUS staining of T-pilus⁻/Vir⁺ uncoupling mutants and (B) T-pilus⁺/Vir⁺ mutants. (C) Quantitative GUS activity of all mutants. Data are mean±SD of 4 biological repeats from 2 independent experiments (10 seedlings in each biological repeat). The data were analyzed by ANOVA for statistical classification, which revealed two groups (groups a and b) of strains differing in transient GUS activity.

tumefaciens C58 genome and ligated into pRL662 at SpeI/XhoI sites. For generating *virB2* mutants with single amino acid substitutions, corresponding primers in Table S3 were used to amplify mutated sequence templates, followed by further ampli-

fication by the universal primer pair VirBp-B1-SpeI-F and VirB3-XhoI-R to generate the 2.1-kb fragment for ligation into pRL662 at SpeI/XhoI sites. Sequences of the entire 2.1-kb fragment (*virB*p-

A

B

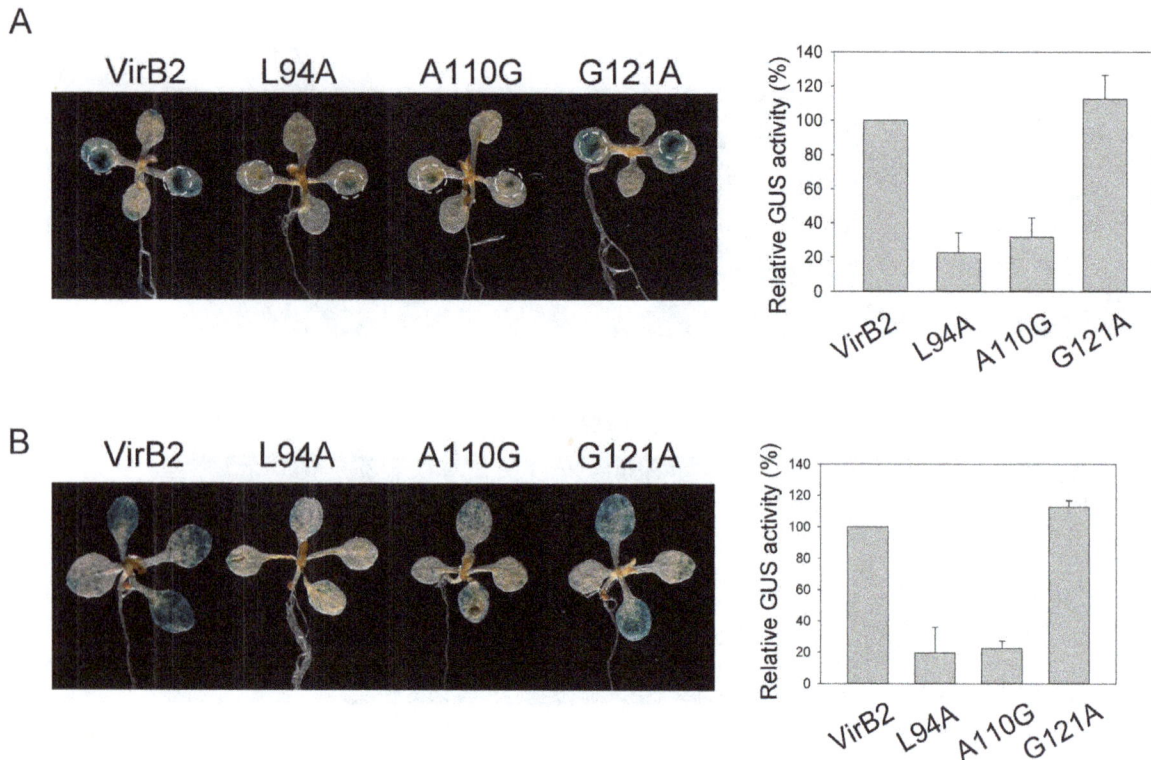

Figure 6. Transient transformation assay in *Arabidopsis* seedlings with or without wounded cotyledons. *A. tumefaciens* strains expressing the wild type or variants of VirB2, L94A (T-pilus⁻/Vir⁺), A110G (T-pilus⁻/Vir⁺) and G121A (T-pilus⁺/Vir⁺), harboring the T-DNA vector pBISN1 were used to infect 7-day-old *Arabidopsis* seedlings. GUS activity as a reporter of transient transformation efficiency was determined by GUS staining or quantitative activity assay at 3 dpi. (A) Infection of *Arabidopsis* seedlings with cotyledons wounded by a needle before infection. *A. tumefaciens* cells grown in 523 overnight culture without AS induction were used to infect the wounded *Arabidopsis* seedlings in the absence of AS. (B) Infection of *Arabidopsis* seedlings without intentional wounding, in which *A. tumefaciens* cells were pre-induced by AS for infection and co-cultured in the presence of AS. Data for quantitative GUS activity are mean±SD of four biological repeats from two independent experiments (10 seedlings in each biological repeat).

B1-B2-B3) were confirmed by DNA sequencing to ensure that no additional mutations occurred via PCR.

Isolation of intracellular and extracellular fractions

The procedure to isolate extracellular fractions was as described previously [7], with minor modifications. *A. tumefaciens* cells were scraped off a 150-mm AB-MES, pH 5.5, agar plate by adding 2 ml buffer A (10 mM phosphate buffer, pH 5.3); the resulting cell suspension was centrifuged (13000×g, 4°C, 10 min) to collect the supernatant, named the S1 fraction. The resulting pellet was resuspended again in buffer A to OD$_{600}$ 10 and divided into 1-ml aliquots. The cells were sheared through a 26-g needle syringe five times and harvested by centrifugation (13000×g, 4°C, 10 min) to collect the supernatant containing sheared T-pilus, named the S2 fraction. Both S1 and S2 fractions were collected and filtrated through a 0.22-μm low-protein-binding membrane (Minisart RC 15, Sartorius stedim biotech) to remove contaminating bacterial cells and precipitated by trichloroacetic acid (TCA) as described previously [45]. The sheared pellet was normalized to OD$_{600}$ 10 and designated as the intracellular fraction.

Tumor assay on tomato stems

Tomato cultivar FARMERS 301 from KNOWN-YOU SEED CO. (Kaohsiung, Taiwan) was grown at 23°C with a 16-/8-hr light/dark cycle. Two- to 3-week-old seedlings were wounded by use of a needle and inoculated with 5 μl *A. tumefaciens* cell suspension prepared as follows. *A. tumefaciens* cells were grown on a 523 agar plate at 28°C for 48 hr. Freshly grown colonies were re-suspended in 0.9% sodium chloride and adjusted to 10⁸ and 10⁶ CFU/ml for inoculation. Tumors were observed about 4 to 5 weeks after inoculation.

SDS-PAGE and western blot analysis

Proteins were separated by 12% or 16% tricine SDS-PAGE [46] followed by western blot analysis as described [45]. For each protein sample, an equivalent number of cells was mixed with an equal volume of 2x SDS-PAGE loading buffer (0.1 M Tris-Cl, pH 6.8, 4% SDS, 0.1% bromophenol blue, 20% glycerol, 200 mM dithiothreitol) and incubated at 100°C for 10 min before loading. Polyclonal antisera VirB2-B23 (against the N-terminal region of processed VirB2 T-pilin encoded by pTiC58), VirB2-B24 (against the C-terminal region of processed VirB2 T-pilin encoded by pTiC58) [47] and RNA polymerase α-subunit RpoA [48] were used as primary antibodies. Horseradish peroxidase-conjugated goat anti-rabbit immunoglobulin G (Chemichem) was the secondary antibody and chemiluminescence was detected by the Western Lightning system (Perkin Elmer, Boston, MA) with X-ray film (Amersham).

Electron microscopy

Procedures for negative staining were as described previously [6] with minor modifications. Briefly, *A. tumefaciens* cells grown under T-pilus induction conditions were collected, washed with pure water and re-suspend in 10 mM Tris buffer at pH 7.5. The bacterial suspension was deposited on a copper grid with carbon-Formvar film support for 1 min, rinsed with pure water for a few seconds, and stained with 2% uranyl acetate for 1 min. The samples were examined under a PHILIPS-CM100 transmission electron microscope (TEM) at 80 kV.

Transient transformation assay in *Arabidopsis* seedlings

The method for transient transformation assay in *Arabidopsis* seedlings was as described [39]. In brief, *Arabidopsis thaliana* mutant *efr-1* lacking the elongation factor Tu (EF-Tu) receptor (SALK_044334) was used. Seeds were sterilized in 50% bleach and 0.05% Trition X-100 for 10 min and washed with sterile water five times and incubated in a 4°C refrigerator for 3 days before germination. Seeds were germinated in 1-ml MS liquid medium (1/2 MS salts, 0.5% sucrose, pH 5.7) in a 6-well plate (10 seedlings in each well) at 22°C, 16-/8-hr light/dark cycle for 4 or 7 days (indicated as 4- or 7-day-old seedlings). For *A. tumefaciens vir* gene pre-induction and infection, overnight cultured *A. tumefaciens* cells were re-suspended to OD_{600} 0.2 in AB-MES, pH 5.5, with 200 μM AS for growth at 28°C for 14 to 16 hr. Cells were re-suspended in infection medium (1/2 MS salts, 0.5% sucrose, 50 μM AS, pH 5.7) and adjusted to OD_{600} 0.02 and co-cultured with *Arabidopsis* seedlings at 22°C for 3 days. After co-cultivation, seedlings were stained with X-Gluc staining solution for 6 to 12 hr at 37°C or GUS activity was measured according to the *Arabidopsis* protocol [49].

Tumor assay on potato tuber discs

Potato tumor assay was performed as described [45]. Briefly, overnight cultured *A. tumefaciens* cells were sub-cultured by 10X dilution and grown at 28°C to OD_{600} 1.0. Cells were washed with 0.9% sodium chloride and re-suspended in 0.9% sodium chloride at 10^8 and 10^6 CFU/ml. A total of 40 to 60 potato tuber disks were placed on water agar with each potato tuber disk infected with 10 μl bacterial culture and incubated at 22°C for 2 days. Disks were then placed on water agar supplemented with 100 μg/ml Timentin and tumors were scored with number of tumors/disc after incubation at 22°C for 3 to 4 weeks.

Supporting Information

Figure S1 Western blot analysis of the extracellular S2 fraction showing both high and low intensity. *A. tumefaciens* cells grown on AS-induced AB-MES (pH 5.5) agar at 19°C for 3 days [7] were collected to isolate the extracellular S2 fractions. C58, *A. tumefaciens* wild type strain; V, empty vector pRL662; Δ*virB2*, *virB2* deletion mutant; Δ*virB2*(pVirB2), expression of wild type *virB*p-*B1-B2-B3* in Δ*virB2*. Western blot analysis with antisera against VirB2 B24 peptide or B23 peptide (for variants in C-PP) or RNA polymerase RpoA, as an internal control. Unprocessed VirB2 precursor is indicated as VirB2p and processed mature VirB2 as VirB2m. Each region/domain of VirB2 is indicated as described in Figure 1. Western blot images with high (longer exposure time) and low intensity (shorter exposure time) are shown.

Figure S2 Western blot analysis of S1 fraction. *A. tumefaciens* cells grown on AS-induced AB-MES (pH 5.5) agar at 19°C for 3 days [7] were collected to isolate intracellular proteins and extracellular S1 fraction. C58, *A. tumefaciens* wild type strain; V, empty vector pRL662; Δ*virB2*, *virB2* deletion mutant; Δ*virB2*(pVirB2), expression of wild type *virB*p-*B1-B2-B3* in Δ*virB2*. Western blot analysis with antisera against VirB2 B24 peptide or RNA polymerase RpoA, as an internal control. Processed mature VirB2 as VirB2m.

Figure S3 Potato tumor assay of *A. tumefaciens* strains expressing wild-type VirB2 or variants of L94A (T-pilus⁻/Vir⁺), A110G (T-pilus⁻/Vir⁺) and G121A (T-pilus⁺/Vir⁺). *A. tumefaciens* cells at 10^8 and 10^6 CFU/ml were used for infection. The potato tuber disks were placed on water agar, infected with 10 μl of bacterial cultures, and incubated at 22°C for 2 days. Disks were placed on water agar supplemented with 100 μg/ml Timentin and incubated at 22°C. Tumors were scored after 3 weeks. Data are mean±SEM of number of tumors averaged from 40–60 disks. Similar results were obtained from at least two independent experiments.

Figure S4 Western blot analysis of the intracellular and extracellular S1 and S2 fractions of *A. tumefaciens* strains expressing wild-type VirB2, G119A, or G119C variants. *A. tumefaciens* cells grown on AS-induced AB-MES (pH 5.5) agar at 19°C for 3 days [7] were collected to isolate the intracellular and extracellular S1 and S2 fractions. *A. tumefaciens* strain producing wild-type VirB2, G119A variant, and three independent colonies of G119C variant (G119C-1,-2 or -3) were analyzed. Western blot analysis with antisera against VirB2 B23 peptide or RNA polymerase RpoA, as an internal control. Processed mature VirB2 is indicated as VirB2m.

Figure S5 Tumorigenesis and transient transformation assays of *A. tumefaciens* strains expressing wild-type VirB2, G119A, or G119C variants. (A) Potato tumor assay. *A. tumefaciens* cells at 10^8 and 10^6 CFU/ml were used for infection. The potato tuber disks were placed on water agar, infected with 10 μgl of bacterial cultures, and incubated at 22°C for 2 days. Disks were placed on water agar supplemented with 100 μg/ml Timentin and incubated at 22°C. Tumors were scored after 3 weeks. Data are mean±SEM of number of tumors averaged from 40–60 disks. Similar results were obtained from at least two independent experiments. (B) Transient transformation assay in *Arabidopsis* seedlings. *A. tumefaciens* strains expressing wild-type or variants of VirB2 harboring T-DNA vector pBISN1 were used to infect 4-day-old *Arabidopsis* seedlings. GUS activity as a reporter for transient transformation efficiency was determined by GUS staining or quantitative activity assay at 3 dpi. Data for quantitative GUS activity are mean±SD of four biological repeats from two independent experiments (10 seedlings in each biological repeat).

Acknowledgments

We thank the members of the Lai laboratory for stimulating discussion and Drs. Stanton Gelvin and Lan-Ying Lee for pBISN1. We also acknowledge the technical support of the Plant Cell Biology Core Laboratory and DNA Sequencing Laboratory at the Institute of Plant and Microbial Biology, Academia Sinica, for electron microscopy and DNA sequencing, respectively.

Author Contributions

Conceived and designed the experiments: HYW EML. Performed the experiments: HYW. Analyzed the data: HYW CYC EML. Contributed reagents/materials/analysis tools: HYW EML. Wrote the paper: HYW EML.

References

1. Smith EF, Townsend CO (1907) A plant-pumor of bacterial origin. Science 25: 671–673.
2. McCullen CA, Binns AN (2006) *Agrobacterium tumefaciens* and plant cell interactions and activities required for interkingdom macromolecular transfer. Annu Rev Cell Dev Biol 22: 101–127.
3. Alvarez-Martinez CE, Christie PJ (2009) Biological diversity of prokaryotic type IV secretion systems. Microbiol Mol Biol Rev 73: 775–808.
4. Gelvin SB (2010) Plant proteins involved in *Agrobacterium*-mediated genetic transformation. Annu Rev Phytopathol 48: 45–68.
5. Christie PJ (2001) Type IV secretion: intercellular transfer of macromolecules by systems ancestrally related to conjugation machines. Mol Microbiol 40: 294–305.
6. Lai EM, Chesnokova O, Banta LM, Kado CI (2000) Genetic and environmental factors affecting T-pilin export and T-pilus biogenesis in relation to flagellation of *Agrobacterium tumefaciens*. J Bacteriol 182: 3705–3716.
7. Lai EM, Kado CI (1998) Processed VirB2 is the major subunit of the promiscuous pilus of *Agrobacterium tumefaciens*. J Bacteriol 180: 2711–2717.
8. Thanassi DG, Bliska JB, Christie PJ (2012) Surface organelles assembled by secretion systems of Gram-negative bacteria: diversity in structure and function. FEMS Microbiol Rev 36: 1046–1082.
9. Baron C (2006) VirB8: a conserved type IV secretion system assembly factor and drug target. Biochem Cell Biol 84: 890–899.
10. Cascales E, Christie PJ (2004) Definition of a bacterial type IV secretion pathway for a DNA substrate. Science 304: 1170–1173.
11. Chandran V, Fronzes R, Duquerroy S, Cronin N, Navaza J, et al. (2009) Structure of the outer membrane complex of a type IV secretion system. Nature 462: 1011–1015.
12. Fronzes R, Schafer E, Wang L, Saibil HR, Orlova EV, et al. (2009) Structure of a type IV secretion system core complex. Science 323: 266–268.
13. Low HH, Gubellini F, Rivera-Calzada A, Braun N, Connery S, et al. (2014) Structure of a type IV secretion system. Nature 508: 550–553.
14. Cascales E, Atmakuri K, Sarkar MK, Christie PJ (2013) DNA substrate-induced activation of the *Agrobacterium* VirB/VirD4 type IV secretion system. J Bacteriol 195: 2691–2704.
15. Backert S, Fronzes R, Waksman G (2008) VirB2 and VirB5 proteins: specialized adhesins in bacterial type-IV secretion systems? Trends Microbiol 16: 409–413.
16. Schmidt-Eisenlohr H, Domke N, Angerer C, Wanner G, Zambryski PC, et al. (1999) Vir proteins stabilize VirB5 and mediate its association with the T pilus of *Agrobacterium tumefaciens*. J Bacteriol 181: 7485–7492.
17. Eisenbrandt R, Kalkum M, Lai EM, Lurz R, Kado CI, et al. (1999) Conjugative pili of IncP plasmids, and the Ti plasmid T pilus are composed of cyclic subunits. J Biol Chem 274: 22548–22555.
18. Kalkum M, Eisenbrandt R, Lanka E (2004) Protein circlets as sex pilus subunits. Curr Protein Pept Sci 5: 417–424.
19. Lai EM, Kado CI (2000) The T-pilus of *Agrobacterium tumefaciens*. Trends Microbiol 8: 361–369.
20. Lawley TD, Klimke WA, Gubbins MJ, Frost LS (2003) F factor conjugation is a true type IV secretion system. FEMS Microbiol Lett 224: 1–15.
21. Lai EM, Eisenbrandt R, Kalkum M, Lanka E, Kado CI (2002) Biogenesis of T pili in *Agrobacterium tumefaciens* requires precise VirB2 propilin cleavage and cyclization. J Bacteriol 184: 327–330.
22. Kerr JE, Christie PJ (2010) Evidence for VirB4-mediated dislocation of membrane-integrated VirB2 pilin during biogenesis of the *Agrobacterium* VirB/VirD4 type IV secretion system. J Bacteriol. 192: 4923–34.
23. Hayes CS, Aoki SK, Low DA (2010) Bacterial contact-dependent delivery systems. Annu Rev Genet 44: 71–90.
24. Aguilar J, Cameron TA, Zupan J, Zambryski P (2011) Membrane and core periplasmic *Agrobacterium tumefaciens* virulence Type IV secretion system components localize to multiple sites around the bacterial perimeter during lateral attachment to plant cells. MBio 2: e00218–00211.
25. Aguilar J, Zupan J, Cameron TA, Zambryski PC (2010) *Agrobacterium* type IV secretion system and its substrates form helical arrays around the circumference of virulence-induced cells. Proc Natl Acad Sci U S A 107: 3758–3763.
26. Cameron TA, Roper M, Zambryski PC (2012) Quantitative image analysis and modeling indicate the *Agrobacterium tumefaciens* type IV secretion system is organized in a periodic pattern of foci. PLoS One 7: e42219.

27. Lai EM, Kado CI (2002) The *Agrobacterium tumefaciens* T pilus composed of cyclic T pilin is highly resilient to extreme environments. FEMS Microbiol Lett 210: 111–114.
28. Wang YA, Yu X, Silverman PM, Harris RL, Egelman EH (2009) The structure of F-pili. J Mol Biol 385: 22–29.
29. Babic A, Lindner AB, Vulic M, Stewart EJ, Radman M (2008) Direct visualization of horizontal gene transfer. Science 319: 1533–1536.
30. Shu AC, Wu CC, Chen YY, Peng HL, Chang HY, et al. (2008) Evidence of DNA transfer through F-pilus channels during *Escherichia coli* conjugation. Langmuir 24: 6796–6802.
31. Kwok T, Zabler D, Urman S, Rohde M, Hartig R, et al. (2007) *Helicobacter* exploits integrin for type IV secretion and kinase activation. Nature 449: 862–866.
32. Banta LM, Kerr JE, Cascales E, Giuliano ME, Bailey ME, et al. (2011) An *Agrobacterium* VirB10 mutation conferring a type IV secretion system gating defect. J Bacteriol 193: 2566–2574.
33. Garza I, Christie PJ (2013) A putative transmembrane leucine zipper of *agrobacterium* VirB10 is essential for T-pilus biogenesis but not type IV secretion. J Bacteriol 195: 3022–3034.
34. Jakubowski SJ, Cascales E, Krishnamoorthy V, Christie PJ (2005) *Agrobacterium tumefaciens* VirB9, an outer-membrane-associated component of a type IV secretion system, regulates substrate selection and T-pilus biogenesis. J Bacteriol 187: 3486–3495.
35. Jakubowski SJ, Kerr JE, Garza I, Krishnamoorthy V, Bayliss R, et al. (2009) *Agrobacterium* VirB10 domain requirements for type IV secretion and T pilus biogenesis. Mol Microbiol 71: 779–794.
36. Jakubowski SJ, Krishnamoorthy V, Christie PJ (2003) *Agrobacterium tumefaciens* VirB6 protein participates in formation of VirB7 and VirB9 complexes required for type IV secretion. J Bacteriol 185: 2867–2878.
37. Berger BR, Christie PJ (1994) Genetic complementation analysis of the *Agrobacterium tumefaciens virB* operon: *virB2* through *virB11* are essential virulence genes. J Bacteriol 176: 3646–3660.
38. Bogdanov M, Xie J, Dowhan W (2009) Lipid-protein interactions drive membrane protein topogenesis in accordance with the positive inside rule. J Biol Chem 284: 9637–9641.
39. Wu HY, Liu KH, Wang YC, Wu CF, Chiu WL, et al. (2014) AGROBEST: an efficient *Agrobacterium*-mediated transient expression method for versatile gene function analyses in *Arabidopsis* seedlings. Plant Methods (In press).
40. Narasimhulu SB, Deng XB, Sarria R, Gelvin SB (1996) Early transcription of *Agrobacterium* T-DNA genes in tobacco and maize. Plant Cell 8: 873–886.
41. Li G, Brown PJ, Tang JX, Xu J, Quardokus EM, et al. (2012) Surface contact stimulates the just-in-time deployment of bacterial adhesins. Mol Microbiol 83: 41–51.
42. Aly KA, Baron C (2007) The VirB5 protein localizes to the T-pilus tips in *Agrobacterium tumefaciens*. Microbiology 153: 3766–3775.
43. Lacroix B, Citovsky V (2011) Extracellular VirB5 enhances T-DNA transfer from *Agrobacterium* to the host plant. PLoS One 6: e25578.
44. Kado CI, Heskett MG (1970) Selective media for isolation of *Agrobacterium, Corynebacterium, Erwinia, Pseudomonas*, and *Xanthomonas*. Trends Microbiol 60: 969–976.
45. Wu HY, Chung PC, Shih HW, Wen SR, Lai EM (2008) Secretome analysis uncovers an Hcp-family protein secreted via a type VI secretion system in *Agrobacterium tumefaciens*. J Bacteriol 190: 2841–2850.
46. Schagger H, von Jagow G (1987) Tricine-sodium dodecyl sulfate-polyacrylamide gel electrophoresis for the separation of proteins in the range from 1 to 100 kDa. Anal Biochem 166: 368–379.
47. Shirasu K, Kado CI (1993) Membrane location of the Ti plasmid VirB proteins involved in the biosynthesis of a pilin-like conjugative structure on *Agrobacterium tumefaciens*. FEMS Microbiol Lett 111: 287–294.
48. Lin JS, Ma LS, Lai EM (2013) Systematic Dissection of the *Agrobacterium* Type VI Secretion System Reveals Machinery and Secreted Components for Subcomplex Formation. PLoS One 8: e67647.
49. Julio Salinas JJS-S (2006) Arabidopsis Protocols, 2nd Edition (Methods in Molecular Biology) Humana Press.
50. Larkin MA, Blackshields G, Brown NP, Chenna R, McGettigan PA, et al. (2007) Clustal W and Clustal X version 2.0. Bioinformatics 23: 2947–2948.

Permissions

The contributors of this book come from diverse backgrounds, making this book a truly international effort. This book will bring forth new frontiers with its revolutionizing research information and detailed analysis of the nascent developments around the world.

We would like to thank all the contributing authors for lending their expertise to make the book truly unique. They have played a crucial role in the development of this book. Without their invaluable contributions this book wouldn't have been possible. They have made vital efforts to compile up to date information on the varied aspects of this subject to make this book a valuable addition to the collection of many professionals and students.

This book was conceptualized with the vision of imparting up-to-date information and advanced data in this field. To ensure the same, a matchless editorial board was set up. Every individual on the board went through rigorous rounds of assessment to prove their worth. After which they invested a large part of their time researching and compiling the most relevant data for our readers.

The editorial board has been involved in producing this book since its inception. They have spent rigorous hours researching and exploring the diverse topics which have resulted in the successful publishing of this book. They have passed on their knowledge of decades through this book. To expedite this challenging task, the publisher supported the team at every step. A small team of assistant editors was also appointed to further simplify the editing procedure and attain best results for the readers.

Apart from the editorial board, the designing team has also invested a significant amount of their time in understanding the subject and creating the most relevant covers. They scrutinized every image to scout for the most suitable representation of the subject and create an appropriate cover for the book.

The publishing team has been an ardent support to the editorial, designing and production team. Their endless efforts to recruit the best for this project, has resulted in the accomplishment of this book. They are a veteran in the field of academics and their pool of knowledge is as vast as their experience in printing. Their expertise and guidance has proved useful at every step. Their uncompromising quality standards have made this book an exceptional effort. Their encouragement from time to time has been an inspiration for everyone.

The publisher and the editorial board hope that this book will prove to be a valuable piece of knowledge for researchers, students, practitioners and scholars across the globe.

List of Contributors

Erica N. C. Renaud and Edith T. Lammerts van Bueren
Wageningen UR Plant Breeding, Plant Sciences Group, Wageningen University, Wageningen, The Netherlands

James R. Myers
Department of Horticulture, Oregon State University, Corvallis, Oregon, United States of America

Maria João Paulo and Fred A. van Eeuwijk
Biometris, Plant Sciences Group, Wageningen University, Wageningen, The Netherlands

Ning Zhu and John A. Juvik
Department of Crop Sciences, University of Illinois, Urbana, Illinois, United States of America

Karen R. Siegel and K. M. Venkat Narayan
Nutrition and Health Sciences, Laney Graduate School, Emory University, Atlanta, Georgia, United States of America
Hubert Department of Global Health, Emory University, Atlanta, Georgia, United States of America

Mohammed K. Ali
Hubert Department of Global Health, Emory University, Atlanta, Georgia, United States of America

Adithi Srinivasiah
Emory College, Emory University, Atlanta, Georgia, United States of America

Rachel A. Nugent
Department of Global Health, University of Washington, Seattle, Washington, United States of America

Meng-Xiao Lu, Xian-Jin Liu and Xiang-Yang Yu
Pesticide Biology and Ecology Research Center, Nanjing, Jiangsu, China
Key Laboratory of Food Safety Monitoring and Management of Ministry of Agriculture, Nanjing, Jiangsu, China

Wayne W. Jiang
Department of Entomology, Michigan State University, East Lansing, Michigan, United States of America

Jia-Lei Wang
Pesticide Biology and Ecology Research Center, Nanjing, Jiangsu, China

Qiu Jian
Institute for the Control of Agrochemicals, Ministry of Agriculture, Beijing, China

Yan Shen
Key Laboratory of Food Safety Monitoring and Management of Ministry of Agriculture, Nanjing, Jiangsu, China

Sudhakar V. S. Akella, William D. J. Kirk and James G. C. Hamilton
Centre for Applied Entomology and Parasitology, School of Life Sciences, Huxley Building, Keele University, Keele, Staffordshire, England, United Kingdom

Yao-bin Lu
Institute of Plant Protection and Microbiology, Zhejiang Academy of Agricultural Sciences, Hangzhou, Zhejiang, China

Tamotsu Murai
Laboratory of Applied Entomology, Faculty of Agriculture, Utsunomiya University, Utsunomiya, Tochigi, Japan

Keith F. A. Walters
Food and Environment Research Agency, Sand Hutton, York, North Yorkshire, England, United Kingdom

Jie Liu, Ying Xiao, Meiying Hu and Guohua Zhong
Laboratory of Insect Toxicology, and Key Laboratory of Pesticide and Chemical Biology, Ministry of Education, South China Agricultural University, Guangzhou, P.R. China

Yue He
Guangdong Zhuhai Supervision Testing Institute of Quality and Metrology, Zhuhai, P.R. China

Shaohua Chen
Guangdong Province Key Laboratory of Microbial Signals and Disease Control, South China Agricultural University, Guangzhou, P.R. China

Peiteng Shi, Jiang Zhang and Jingfei Luo
School of Systems Science, Beijing Normal University, Beijing, China

Bo Yang
Ministry of Commerce of the People's Republic of China, Beijing, China

Dref C. De Moura and Rickey Y. Yada
Biophysics Interdepartmental Group, University of Guelph, Guelph, Ontario, Canada
Department of Food Science, University of Guelph, Guelph, Ontario, Canada

Brian C. Bryksa
Department of Food Science, University of Guelph, Guelph, Ontario, Canada

Kang-Mo Ku and John A. Juvik
Department of Crop Sciences, University of Illinois at Urbana-Champaign, Urbana, Illinois, United States of America

Elizabeth H. Jeffery
Department of Food Science and Human Nutrition, University of Illinois at Urbana-Champaign, Urbana, Illinois, United States of America

Ana Lazar, Anna Coll, David Dobnik, Špela Baebler, Jana Žel and Kristina Gruden
Department of Biotechnology and Systems Biology, National Institute of Biology, Ljubljana, Slovenia

Apolonija Bedina-Zavec
Laboratory for Molecular Biology and Nano-biotechnology, National Institute of Chemistry, Ljubljana, Slovenia

Andrew W. Brown and David B. Allison
Office of Energetics, Nutrition Obesity Research Center, School of Public Health, University of Alabama at Birmingham, Birmingham, Alabama, United States of America

Shan-e-Ahmed Raza
Department of Computer Science, University of Warwick, Coventry, United Kingdom

Hazel K. Smith and Gail Taylor
Centre for Biological Sciences, Life Sciences, University of Southampton, Southampton, United Kingdom

Graham J. J. Clarkson
Vitacress Salads Ltd., Lower Link Farm, St Mary Bourne, Andover, United Kingdom

Andrew J. Thompson
Soil and Agri-Food Institute, School of Applied Sciences, Cranfield University, Bedford, United Kingdom

John Clarkson
School of Life Sciences, University of Warwick, Wellsbourne, United Kingdom

Nasir M. Rajpoot
Department of Computer Science and Engineering, Qatar University, Doha, Qatar

Ying Li
Key Laboratory of Optoelectronic Information and Sensing Technologies of Guangdong Higher Education Institutes, Jinan University, Guangzhou, Guangdong, China
Pre-university Department, Jinan University, Guangzhou, Guangdong, China

Xiao Ma and Minglu Zhao
Department of Optoelectronic Engineering, Jinan University, Guangzhou, Guangdong, China

Pan Qi
Department of Electronics Engineering, Guangdong Communication Polytechnic, Guangzhou, Guang

Jingang Zhong
Key Laboratory of Optoelectronic Information and Sensing Technologies of Guangdong Higher Education Institutes, Jinan University, Guangzhou, Guangdong, China
Department of Optoelectronic Engineering, Jinan University, Guangzhou, Guangdong, China

Mickaël Lecomte, Latifa Hamama, Linda Voisine, Cora Boedo, Claire Yovanopoulos, Melvina Gyomlai, Mathilde Briard, Philippe Simoneau, Pascal Poupard and Romain Berruyer
Agrocampus-Ouest, UMR 1345 IRHS, Angers, France
Université d'Angers, UMR 1345 IRHS, SFR QUASAV, Angers, France
INRA, UMR 1345 IRHS, Angers, France

Julia Gatto, Jean-Jacques Hélesbeux, Denis Séraphin and Pascal Richomme
Université d'Angers, UPRES EA921SONAS, SFR 4207 QUASAV, Angers, France

Luis M. Peña-Rodriguez
Unidad de Biotecnología, Centro de Investigación Científica de Yucatán, Mérida, Yucatán, Mexico

Frank Dunemann, Otto Schrader and Holger Budahn
Julius Kühn-Institut (JKI) - Federal Research Centre for Cultivated Plants, Institute for Breeding Research on Horticultural Crops, Quedlinburg, Germany

Andreas Houben
Leibniz-Institute of Plant Genetics and Crop Plant Research (IPK), Chromosome Structure and Function Laboratory, Gatersleben, Germany

Susanne Schreiter
Julius Kühn-Institut – Federal Research Centre for Cultivated Plants (JKI), Institute for Epidemiology and Pathogen Diagnostics, Braunschweig, Germany Leibniz Institute of Vegetable and Ornamental Crops Gro beeren/Erfurt e.V., Department Plant Health, Gro beeren, Germany

Martin Sandmann and Rita Grosch
Leibniz Institute of Vegetable and Ornamental Crops Gro beeren/Erfurt e.V., Department Plant Health, Gro beeren, Germany

Kornelia Smalla
Julius Kühn-Institut – Federal Research Centre for Cultivated Plants (JKI), Institute for Epidemiology and Pathogen Diagnostics, Braunschweig, Germany

Ying-Hong Gu, Xian-Jun Lai, Hai-Yan Wang and Yi-Zheng Zhang
College of Life Sciences, Sichuan University, Key Laboratory of Bio-resources and Eco-environment, Ministry of Education, Sichuan Key Laboratory of Molecular Biology and Biotechnology, Center for Functional Genomics and Bioinformatics, Chengdu, Sichuan, People's Republic of China

Xiang Tao
Chengdu Institute of Biology, Chinese Academy of Sciences, Chengdu, Sichuan, People's Republic of China

Patrícia S. Golo
Departamento de Parasitologia Animal, Instituto de Veterinária, Universidade Federal Rural do Rio de Janeiro, Seropédica, RJ, Brazil Department of Biology, Utah State University, Logan, Utah, United States of America

Dale R. Gardner
USDA, ARS, Poisonous Plants Research Laboratory, Logan, Utah, United States of America

Michelle M. Grilley and Jon Y. Takemoto
Department of Biology, Utah State University, Logan, Utah, United States of America

Stuart B. Krasnoff
Biological Integrated Pest Management Research Unit, Robert W. Holley Center for Agriculture and Health, USDA-ARS, Ithaca, New York, United States of America

Marcus S. Pires and Vânia R. E. P. Bittencourt
Departamento de Parasitologia Animal, Instituto de Veterinária, Universidade Federal Rural do Rio de Janeiro, Seropédica, RJ, Brazil

Éverton K. K. Fernandes
Instituto de Patologia Tropical e Saúde Pública, Universidade Federal de Goiás, Goiânia, GO, Brazil

Donald W. Roberts
Department of Biology, Utah State University, Logan, Utah, United States of America

George O. Agogo
National Institute for Public Health and the Environment, Bilthoven, The Netherlands Biometris, Wageningen University and Research Center, Wageningen, The Netherlands

Hilko van der Voet and Fred A. van Eeuwijk
Biometris, Wageningen University and Research Center, Wageningen, The Netherlands

Pieter van't Veer
Department of Human Nutrition, Wageningen University and Research Center, Wageningen, The Netherlands

Pietro Ferrari
Nutritional Epidemiology Group, International Agency for Research on Cancer, Lyon, France

Max Leenders
Department of Gastroenterology and Hepatology, University Medical Center Utrecht, Utrecht, The Netherlands

David C. Muller
Genetic Epidemiology Group, International Agency for Research on Cancer, 150 cours Albert Thomas, Lyon, 69008, France

Emilio Sánchez-Cantalejo
Andalusian School of Public Health, Granada, Spain, 8 CIBER de Epidemiología y Salud pública (CIBERESP), Barcelona, Spain

Christina Bamia
WHO Collaborating Center for Food and Nutrition Policies, Department of Hygiene, Epidemiology and Medical Statistics, University of Athens Medical School, Athens, Greece

Tonje Braaten
Department of Community Medicine, University of Tromsø, N- 9037, Tromsø, Norway

Sven Knüppel
Department of Epidemiology, German Institute of Human Nutrition Potsdam-Rehbrücke, Potsdam, Germany

Ingegerd Johansson
Department of Odontology, Umeå University, Umeå, Sweden

Hendriek Boshuizen
National Institute for Public Health and the Environment, Bilthoven, The Netherlands
Biometris, Wageningen University and Research Center, Wageningen, The Netherlands
Department of Human Nutrition, Wageningen University and Research Center, Wageningen, The Netherlands

Brian A. Nault, Wendy C. Kain and Ping Wang
Department of Entomology, Cornell University, New York State Agricultural Experiment Station, Geneva, New York, United States of America

Subbarao V. Ravva, Chester Z. Sarreal and Robert E. Mandrell
Produce Safety and Microbiology Research Unit, U.S. Department of Agriculture, Agriculture Research Service, Western Regional Research Center, Albany, California, United States of America

Hung-Yi Wu and Erh-Min Lai
Institute of Plant and Microbial Biology, Academia Sinica, Taipei, Taiwan
Department of Plant Pathology and Microbiology, National Taiwan University, Taipei, Taiwan

Chao-Ying Chen
Department of Plant Pathology and Microbiology, National Taiwan University, Taipei, Taiwan

Index